Directory of Atomic, Molecular, and Optical Scientists

Directory of Atomic, Molecular, and Optical Scientists

Committee on Atomic and Molecular Science
Board on Physics and Astronomy
Commission on Physical Sciences, Mathematics, and Resources
National Research Council

NATIONAL ACADEMY PRESS
Washington, D.C. 1986

NATIONAL ACADEMY PRESS   2101 Constitution Avenue, NW   Washington DC 20418

The National Research Council serves as an independent advisor to the federal government on scientific and technical questions of national importance. Established in 1916 under the congressional charter of the private, nonprofit National Academy of Sciences, the Research Council brings the resources of the entire scientific and technical community to bear on national problems through its volunteer advisory committees. Today the Research Council stands as the principal operating agency of both the National Academy of Sciences and the National Academy of Engineering and is administered jointly by the two academies and the Institute of Medicine. The National Academy of Engineering and the Institute of Medicine were established in 1964 and 1970, respectively, under the charter of the National Academy of Sciences.

LIBRARY OF CONGRESS CATALOG CARD NUMBER 86-62710

INTERNATIONAL STANDARD BOOK NUMBER 0-309-03696-8

Copyright © 1986 by the National Academy of Sciences

No part of this book may be reproduced by any mechanical, photographic, or electronic process, or in the form of a phonographic recording, nor may it be stored in a retrieval system, transmitted, or otherwise copied for public or private use, without written permission from the publisher, except for the purposes of official use by the United States Government.

Printed in the United States of America

COMMITTEE ON ATOMIC AND MOLECULAR SCIENCE

LLOYD ARMSTRONG, Johns Hopkins University, *Chairman*
RICHARD STEPHEN BERRY, University of Chicago
RICHARD BERSOHN, Columbia University
JOSEPH L. DEHMER, Argonne National Laboratory
GORDON H. DUNN, University of Colorado
RICHARD R. FREEMAN, AT&T Bell Laboratories
WILLIAM HAPPER, JR., Princeton University
ERIC J. HELLER, University of Washington
G. SAMUEL HURST, University of Tennessee
DANIEL KLEPPNER, Massachusetts Institute of Technology
JOSEPH MACEK, University of Nebraska
C. BRADLEY MOORE, University of California
FRANCIS M. PIPKIN, Harvard University
A. RAVI PRAKASH RAU, Louisiana State University
NORMAN H. TOLK, Vanderbilt University
R. CLAUDE WOODS, University of Wisconsin

DONALD C. SHAPERO, *Staff Director*
ARLENE P. MACLIN, *Staff Officer*

BOARD ON PHYSICS AND ASTRONOMY

HANS FRAUENFELDER, University of Illinois, *Chairman*
FELIX H. BOEHM, California Institute of Technology
RICHARD G. BREWER, IBM Corporation
DEAN E. EASTMAN, IBM Corporation
JAMES E. GUNN, Princeton University
LEO P. KADANOFF, University of Chicago
W. CARL LINEBERGER, University of Colorado
NORMAN RAMSEY, Harvard University
MORTON S. ROBERTS, National Radio Astronomy Observatory
MARSHALL N. ROSENBLUTH, University of Texas
WILLIAM P. SLICHTER, AT&T Bell Laboratories
SAM B. TREIMAN, Princeton University

DONALD C. SHAPERO, *Staff Director*
ROBERT L. RIEMER, *Staff Officer*
ARLENE P. MACLIN, *Staff Officer*
HELENE E. PATTERSON, *Staff Assistant*
SUSAN M. WYATT, *Staff Assistant*
BRENDA J. WILSON, *Senior Secretary*

COMMISSION ON PHYSICAL SCIENCES, MATHEMATICS, AND RESOURCES

HERBERT FRIEDMAN, National Research Council, *Chairman*
CLARENCE R. ALLEN, California Institute of Technology
THOMAS D. BARROW, Standard Oil Company, Ohio (retired)
ELKAN R. BLOUT, Harvard Medical School
BERNARD F. BURKE, Massachusetts Institute of Technology
GEORGE F. CARRIER, Harvard University
CHARLES L. DRAKE, Dartmouth College
MILDRED S. DRESSELHAUS, Massachusetts Institute of Technology
JOSEPH L. FISHER, George Mason University
WILLIAM A. FOWLER, California Institute of Technology
GERHART FRIEDLANDER, Brookhaven National Laboratory
EDWARD D. GOLDBERG, Scripps Institution of Oceanography
MARY L. GOOD, Allied Signal Corporation
J. ROSS MACDONALD, University of North Carolina, Chapel Hill
THOMAS F. MALONE, Saint Joseph College
CHARLES J. MANKIN, Oklahoma Geological Survey
PERRY L. MCCARTHY, Stanford University
WILLIAM D. PHILLIPS, Mallinckrodt, Inc.
ROBERT E. SIEVERS, University of Colorado
JOHN D. SPENGLER, Harvard School of Public Health
GEORGE W. WETHERILL, Carnegie Institution of Washington
IRVING WLADAWSKY-BERGER, IBM Corporation

RAPHAEL G. KASPER, *Executive Director*
LAWRENCE E. MCCRAY, *Associate Executive Director*

Preface

In 1984, the Committee on Atomic and Molecular Science decided to update the 1981 Directory of Atomic and Molecular Scientists in the United States. The Directory has been updated to include optical scientists and scientists from foreign countries; therefore, it has been retitled *Directory of Atomic, Molecular, and Optical Scientists*.

The purpose of this Directory is, first, to present an alphabetical listing of all respondents, with address, telephone number, experimental (E) and/or theoretical (T) emphasis, and stated research specialties and, second, to list all respondents who are in each of several broad specialty areas. The research specialties of each individual respondent is based upon an updated version of the 1981 specialties list.

A list of the Specialties Categories is given at the beginning of Part II of the Directory. These categories were selected in an attempt to devise a practical system of grouping people in areas of common interest. These specialties are not parallel characterizations of activity: they may stress the study of basic physical properties of atoms and/or molecules and their ions; they may stress development and/or applications of scientific techniques dependent on properties of atoms and/or molecules; they may stress use of A&M data or techniques in other areas of science and technology, such as plasmas or astrophysics; or they may stress the interactions of light with matter. Some specialties are generic or timeless; others represent more transient foci of current activity or emphasis.

This version of the Directory is the first updated version and it is the Committee's intent to continue to update the Directory periodically. Supplementary statements of specialty and/or status changes of listed individuals and names of scientists who have been inadvertently omitted or who have recently entered the field will be welcomed and will be used to improve the utility of the Directory. Inaccuracies should be brought to the attention of the Board on Physics and Astronomy/National Research Council; the tear sheet on the last page is provided for that purpose. In addition, any suggestions concerning the format or choice of specialty areas are welcome.

Funds for this project provided by the Department of Energy, Division of Chemical Sciences, Office of Basic Energy Sciences (Contract No. DE-FG05-85ER13326) and by the National Science Foundation (Grant No. PHY-8521040) through the National Research Council are gratefully acknowledged.

Contents

Part I

Alphabetical Listing of Atomic, Molecular, and Optical
  Scientists with Addresses and Specialties............................... 1

Part II

Listing of Atomic, Molecular, and Optical Scientists
  by Research Specialty 139

Preface

In 1984, the Committee on Atomic and Molecular Science decided to update the 1981 Directory of Atomic and Molecular Scientists in the United States. The Directory has been updated to include optical scientists and scientists from foreign countries; therefore, it has been retitled *Directory of Atomic, Molecular, and Optical Scientists*.

The purpose of this Directory is, first, to present an alphabetical listing of all respondents, with address, telephone number, experimental (E) and/or theoretical (T) emphasis, and stated research specialties and, second, to list all respondents who are in each of several broad specialty areas. The research specialties of each individual respondent is based upon an updated version of the 1981 specialties list.

A list of the Specialties Categories is given at the beginning of Part II of the Directory. These categories were selected in an attempt to devise a practical system of grouping people in areas of common interest. These specialties are not parallel characterizations of activity: they may stress the study of basic physical properties of atoms and/or molecules and their ions; they may stress development and/or applications of scientific techniques dependent on properties of atoms and/or molecules; they may stress use of A&M data or techniques in other areas of science and technology, such as plasmas or astrophysics; or they may stress the interactions of light with matter. Some specialties are generic or timeless; others represent more transient foci of current activity or emphasis.

This version of the Directory is the first updated version and it is the Committee's intent to continue to update the Directory periodically. Supplementary statements of specialty and/or status changes of listed individuals and names of scientists who have been inadvertently omitted or who have recently entered the field will be welcomed and will be used to improve the utility of the Directory. Inaccuracies should be brought to the attention of the Board on Physics and Astronomy/National Research Council; the tear sheet on the last page is provided for that purpose. In addition, any suggestions concerning the format or choice of specialty areas are welcome.

Funds for this project provided by the Department of Energy, Division of Chemical Sciences, Office of Basic Energy Sciences (Contract No. DE-FG05-85ER13326) and by the National Science Foundation (Grant No. PHY-8521040) through the National Research Council are gratefully acknowledged.

Contents

Part I

Alphabetical Listing of Atomic, Molecular, and Optical
Scientists with Addresses and Specialties............................... 1

Part II

Listing of Atomic, Molecular, and Optical Scientists
by Research Specialty 139

COMMISSION ON PHYSICAL SCIENCES, MATHEMATICS, AND RESOURCES

HERBERT FRIEDMAN, National Research Council, *Chairman*
CLARENCE R. ALLEN, California Institute of Technology
THOMAS D. BARROW, Standard Oil Company, Ohio (retired)
ELKAN R. BLOUT, Harvard Medical School
BERNARD F. BURKE, Massachusetts Institute of Technology
GEORGE F. CARRIER, Harvard University
CHARLES L. DRAKE, Dartmouth College
MILDRED S. DRESSELHAUS, Massachusetts Institute of Technology
JOSEPH L. FISHER, George Mason University
WILLIAM A. FOWLER, California Institute of Technology
GERHART FRIEDLANDER, Brookhaven National Laboratory
EDWARD D. GOLDBERG, Scripps Institution of Oceanography
MARY L. GOOD, Allied Signal Corporation
J. ROSS MACDONALD, University of North Carolina, Chapel Hill
THOMAS F. MALONE, Saint Joseph College
CHARLES J. MANKIN, Oklahoma Geological Survey
PERRY L. MCCARTHY, Stanford University
WILLIAM D. PHILLIPS, Mallinckrodt, Inc.
ROBERT E. SIEVERS, University of Colorado
JOHN D. SPENGLER, Harvard School of Public Health
GEORGE W. WETHERILL, Carnegie Institution of Washington
IRVING WLADAWSKY-BERGER, IBM Corporation

RAPHAEL G. KASPER, *Executive Director*
LAWRENCE E. MCCRAY, *Associate Executive Director*

COMMITTEE ON ATOMIC AND MOLECULAR SCIENCE

LLOYD ARMSTRONG, Johns Hopkins University, *Chairman*
RICHARD STEPHEN BERRY, University of Chicago
RICHARD BERSOHN, Columbia University
JOSEPH L. DEHMER, Argonne National Laboratory
GORDON H. DUNN, University of Colorado
RICHARD R. FREEMAN, AT&T Bell Laboratories
WILLIAM HAPPER, JR., Princeton University
ERIC J. HELLER, University of Washington
G. SAMUEL HURST, University of Tennessee
DANIEL KLEPPNER, Massachusetts Institute of Technology
JOSEPH MACEK, University of Nebraska
C. BRADLEY MOORE, University of California
FRANCIS M. PIPKIN, Harvard University
A. RAVI PRAKASH RAU, Louisiana State University
NORMAN H. TOLK, Vanderbilt University
R. CLAUDE WOODS, University of Wisconsin

DONALD C. SHAPERO, *Staff Director*
ARLENE P. MACLIN, *Staff Officer*

BOARD ON PHYSICS AND ASTRONOMY

HANS FRAUENFELDER, University of Illinois, *Chairman*
FELIX H. BOEHM, California Institute of Technology
RICHARD G. BREWER, IBM Corporation
DEAN E. EASTMAN, IBM Corporation
JAMES E. GUNN, Princeton University
LEO P. KADANOFF, University of Chicago
W. CARL LINEBERGER, University of Colorado
NORMAN RAMSEY, Harvard University
MORTON S. ROBERTS, National Radio Astronomy Observatory
MARSHALL N. ROSENBLUTH, University of Texas
WILLIAM P. SLICHTER, AT&T Bell Laboratories
SAM B. TREIMAN, Princeton University

DONALD C. SHAPERO, *Staff Director*
ROBERT L. RIEMER, *Staff Officer*
ARLENE P. MACLIN, *Staff Officer*
HELENE E. PATTERSON, *Staff Assistant*
SUSAN M. WYATT, *Staff Assistant*
BRENDA J. WILSON, *Senior Secretary*

Part *I*

Alphabetical Listing of
Atomic, Molecular, and Optical
Scientists with Addresses
and Specialties

AARTS, C. J.; Faculty of Science; Katholieke Universiteit; Toernooivnld I; Nijmegen 6525; Holland

ABELLA, Isaac D.; Ryerson Physics Laboratory; University of Chicago; 1100 E. 58th Street; Chicago IL 60637; USA; 312-753-8305/8307; E; 3.6; 5.4; Coherent transient spectroscopy; Atomic vapors, rare-gas plasmas, and molecular gases

ABI-GHANEM, Georges V.; Research Division; ARDI Corporation; PO Box 27113; Minneapolis MN 55427; USA; 612-377-2645; E/T; 5.2; 5.6; 5.8; 5.1; 2.1; State transition in liquids and solids; Structural effects of radiation in solid media (short-long term)

ABRAHAM, George; Research Department; Office of the Director; Naval Research Laboratory; Navy Department; Washington DC 20375; USA; 202-582-7210; 202-767-3521; E/T; 3.1; 3.10; 4.1; 2.1; 5.1; Solid state electronics/optics; Optical properties of semiconductors; Multivalued logic implementation

ABRAHAM, Neal B.; Department of Physics; Bryn Mawr College; Bryn Mawr PA 19010; USA; 215-645-5000; E/T; 4.1; Isotopic splitting in xenon; Amplified spontaneous emission; High-gain gas laser systems

ABRAHAMSON, Adolf A.; Department of Physics; City College of the CUNY; Convent Ave. at 138th St.; New York NY 10031; USA; 212-690-6827; 212-580-7953; T; 1.1; 1.3; Electron distribution in atoms and ions, including superheavies; Atom-atom, atom-ion and ion-ion interaction energies

ABRAMS, Richard L.; Optical Physics Division; Hughes Research Laboratories; Hughes Space & Communications Group; P.O. Box 92919; S50-X323; Los Angeles CA 90009; USA; 213-459-2865; E; 4.1; Atomic and molecular systems for use in lasers; Non-linear optical phenomena

ABREU, Raul A.; Centro de Fisica; IVIC; Apartado 1827; Caracas 1010A; Venezuela

ABUSALBI, Najib N.; Department of Chemistry; University of Houston; Central Campus; Houston TX 77004

ACKERHALT, Jay R.; MS-J569; Theoretical Division; Los Alamos National Laboratory; PO Box 1663; Los Alamos NM 87545; USA; 505-667-4615; 505-667-2097; T; 3.7; 3.10; 3.11; Chaos in quantum optics; Multiphoton dissociation in molecules; Propagation effects in nonlinear optics; Raman scattering and 4-wave mixing

ACKMAN, Paul J.; 3455 Emerald Street, Apartment 5; Torrance CA 90503; USA

ACQUISTA, Nicolo; Physics Division; National Bureau of Standards; Gaithersburg MD 20899; USA; 301-921-2011; E; 1.1; Highly ionized atoms--structure

ADAMS, William H.; Department of Chemistry; Rutgers University; New Brunswick NJ 08903; USA; 201-932-3758; T; 1.1; Atomic and molecular interactions--Quantum theory; Computational applications; Mathematical methods

ADASHKO, J. G.; 25 W. 81st Street; New York NY 10024; USA

ADELMAN, Saul J.; Department of Physics; The Citadel; Charleston SC 29409; USA; 803-792-6948; 803-792-5122; E/T; 1.1; 5.7; Atomic line identification lists--compilation of a bibliography; Astronomy applications

ADLERSTEIN, Joseph K.; 112 81st Avenue; Kew Gardens NY 11415; USA

ADRIAN, Frank J.; Applied Physics Laboratory; Johns Hopkins University; Johns Hopkins Road; Laurel MD 20810; USA; 301-953-7100 x2848; T; 3.8; 5.1; Radiation effects in membranes and thin films; Free radical structure and photochemistry; Surface enhanced Raman scattering

AFFATATO, Joseph F.; Fiber Optics; NSG America, Inc.; 1100 Clove Road, Suite L-0; Staten Island NY 10301; USA; 718-273-8681; 201-469-9650; 4.1; 3.3; 3.7; 3.11

AGASSI, Dan; Navy Surface Weapons Center, Code R-45; Silver Spring, MD 20903; USA

AGRAWAL, J.; Department of Natural Sciences and Mathematics; Atlanta Junior College; 1630 Stewart Street; Atlanta GA 30314; USA; 404-656-6366

AHEARNE JR., Daniel P.; 7721 Briarcliff Drive; Springfield VA 22153; USA

AHMAD, Shamshad; Department of Physics; State University of New York; Albany NY 12222; USA

AHN, Myong-Ku; Department of Chemistry; Indiana State University; Terre Haute IN 47809; USA; 812-237-2230; E; 5.2; 1.3; Simple molecules in liquids--transport properties; Zeolites--structure and properties

AISENBERG, Sol; 36 Bradford Road; Natick MA 10760; USA

AIZENMAN, Michael; Department of Mathematics; Rutgers University; New Brunswick NJ 08903; USA

AKERMAN, M. A.; Special Projects Department; Enrichment Technology Division; Union Carbide Nuclear Division; Oak Ridge National Lab.; PO Box P, MS322; Oak Ridge TN 37830; USA; 615-576-0139; E; 3.6; 4.1; Infrared and ultraviolet cross sections in spherical top molecules using lasers as the light source--measurements; Rare-gas halide lasers--impurities generated

AKINS, Daniel L.; Polaroid Corporation; 1265 Main Street W4-4A; Waltham MA 02154; USA; 617-890-7000; Ext. 4625/4647; E; 1.1; 5.1; Dye molecules--structure; Transient species on semiconductor substrates

AL-JOGHUL, Sami I.; Kufr Nema; Ramallah West Bank; Israel

ALBANY, H. J.; CEA (IVI); 33 Rue de la Federation; Paris 75015; France

ALBERS, Mark A.; Manville Corporation; PO Box 5108; Denver CO 80217; USA

ALBRECHT, A. C.; Department of Chemistry; Cornell University; Ithaca NY 14853; USA; 607-256-3990; E/T; 3.5; 5.2; Multiphoton molecular spectroscopy (absorption, ionization, Raman) in the condensed phase

ALBRECHT, Georg F.; Laboratory for Laser Energetics; University of Rochester; Rochester NY 14623; USA; 716-275-4754/5101; E; 3.6; 4.3; Picosecond/subpicosecond diagnostics of atomic/solid state events for development of fast devices (detectors, switches, pulse generators)

ALBRIDGE, Royal G.; Vanderbilt University; Box 1815 Station B; Nashville TN 37235; USA

ALBRITTON, Daniel L.; National Oceanic and Atmospheric Adm.; Aeronomy Laboratory; U.S. Department of Commerce; Radio 3522; 325 S. Broadway; Boulder CO 80302; USA; 303-497-5785; E; 2.3; 3.5; 5.6; Ion-molecule reactions; Molecular spectroscopy; Atmospheric chemistry

ALCARAZ, Ernest C.; Radiation Effects Group; JAYCOR; Suite 600; 205 S. Whiting Street; Alexandria VA 22304; USA; 703-823-1300; E; 4.1; 4.3; X-ray satellite test facility--development and implementation; Electro-optics--research and development

ALDER, Berni J.; Theoretical Physics Division; Lawrence Livermore National Lab.; PO Box 808; Livermore CA 94550; USA; 415-422-4384; T; 1.1; Quantum statistical mechanical calculation of properties of atoms, molecules and their interactions

ALDRIDGE III, Jack P.; Group X-7; Los Alamos National Laboratory; PO Box 1663 MSB257; Los Alamos NM 87545; USA; 505-667-8719; 505-672-9046; E; 2.1; 3.1; 3.2; 3.5; 4.2; High resolution laser molecular spectroscopy; Charge exchange and excitation studies; Photodetachment studies from high-velocity ions

ALEXANDER, James C.; 1006

San Benito; College Station TX 77840; USA

ALEXANDER, Millard H.; Department of Chemistry; University of Maryland; College Park MD 20742; USA; 301-454-2614; T; 2.1; Inelastic molecular collisions

ALEXEFF, Igor; Department of Electrical Engineering; University of Tennessee; Ferris Hall; Knoxville TN 37996; USA; 615-974-5467; E; 3.3; 5.3; Synthetic atoms in masers; Sub-millimeter microwave emission; Electrons in orbit around positive wires

ALGUARD, Mark J.; 1020 Waverley Street; Palo Alto CA 94301; USA

ALI, Abdul W.; Plasma Physics Division; U.S. Naval Research Laboratory; 4555 Overlook Avenue; Washington DC 20375; USA; 202-767-3762; T; 5.3; 5.6; 5.4; 2.3; 4.1; Physical phenomena in ionized gases --modeling; Radiation and charged particle interaction with air and other gases

ALI, Mahamed A.; Chemistry Department; Howard University; Washington DC 20059; USA; 202-636-6902; 301-649-3482; T; 1.1; 2.2; 3.1; 3.2; 5.4

ALLAN, Michael; Department of Engineering and Applied Science; Mason Laboratory; Yale University; New Haven CT 06520; USA; 203-436-3044; E; 2.1; Atomic negative ions with atoms and molecules in gas phase--low-energy, high-resolution collisions

ALLEN JR., Harry C.; Department of Chemistry; Clark University; 950 Main Street; Worcester MA 01610; USA; 617-793-7453; E; 1.1; 3.2; 3.3; 3.5; Molecular structure and bonding using EPR, UV, visible; IR spectroscopy and NMR

ALLEN JR., John E.; Code 691; NASA Goddard Space Flight Center; Greenbelt MD 20771; USA

ALLER, Lawrence H.; Department of Astronomy; University of California, LA; Los Angeles CA 90024; USA; 213-825-3515; E; 3.2; 5.7; Gaseous nebulae --intensities of forbidden lines in the spectra

ALLIS, William P.; Department of Physics; Research Labs. of Electronics; Massachusetts Institute of Technology; 77 Massachusetts Avenue; Cambridge MA 02139; USA; 617-253-2517; 617-876-7535; T; 5.3; 2.2; Electron velocity distributions and space-charge effects in plasmas

ALLISON, Paul; Group AT-2; Los Alamos National Laboratory; PO Box 1663; Los Alamos NM 87545; USA; 505-667-3509; E; 4.3; H- ion sources; Beam optics; Low energy accelerators

ALLISON JR., Robert W.; 6249 Shadelands Drive; San Jose CA 95123; USA

ALPER, Joseph S.; Department of Chemistry; University of Massachusetts, Boston; Dorchester MA 02125; USA; 617-287-1900x2411; T; 2.1; 2.3; Trajectory methods applied to energy transfer and kinetics in collisions of small molecules

ALTGILBERS, Larry L.; Advanced Technology; U.S. Army Missile Command; Redstone Arsenal AL 35810; USA; 205-876-4586; E/T; 4.2; 4.3; 4.1; 4.4; 3.3; Pulsed power for accelerators/RF; Accelerator for design; Dense plasma focus; Charged particle optics

ALTICK, Philip L.; Physics Department; University of Nevada; Reno NV 89557; USA; 702-784-6792; T; 2.2; 1.1

ALTMAN III, Joseph C.; Physics; Texas Christian University; PO Box 32915; T.C.U. Station; Fort Worth TX 76129; USA; 817-921-7375; E; 2.2; Single atomic-field Bremsstrahlung by electron impact; Double atomic-field Bremsstrahlung by electron impact; Inner-shell ionization by electron impact

ALTON, Gerald D.; Physics Division; Oak Ridge National Laboratory; PO Box X; Oak Ridge TN 37830; USA; 615-574-4782; E; 2.1; 3.1; 4.2; Electron loss and capture in heavy-particle, high-energy collisions; Electron photo-detachment experiments; Collisional detachment experiments; Metastable lifetime measurements

ALVAREZ JR., Raymond A.; Physics; Experimental Physics (E); Lawrence Livermore National Lab.; University of California; 808; Livermore CA 94550; USA; 415-422-9672; E; 5.3; 5.4

AMANO, Takayoshi; Herzberg Institute of Astrophysics; National Research Council; 100 Sussex; Ottawa Ontario K1A0R6; Canada; 613-990-0737; 613-990-0721; E; 3.3; 3.6; 5.7; 1.1; 3.5; Infrared and microwave spectra of ions and free radicals; Double resonance--MODR, IR-MW, IR-optical

AMBROSE, Rajkumar; Department of Physics; Texas Christian University; TCU Box 32915; Fort Worth TX 76129; USA

AMES, Donald P.; McDonnell Douglas Research Lab.; PO Box 516; St. Louis MO 63166; USA; 314-232-3254; E; 3.5; 3.6; 4.1; Chemical laser development; Polymers--molecular spectrometry; Laser beams scattering

AMEY, Stephen; 27 Cherry Orchard; Wotton Under Edge Glos GL12 7HT; England

AMIET, Peter F.; PO Box 6902; Lawrenceville NJ 08648; USA

AMME, Robert C.; Department of Physics; University of Denver; P.O. Box 10127; Denver CO 80210; USA; 303-871-3852; 303-871-2238; E; 1.3; 2.1; 4.3; 5.1; 5.6; Atomic and molecular beams applied to collision phenomena: Ionization, emission of visible, UV radiation; Atmospheric molecular ions (stratospheric clusters); Ultrasonic relaxation studies; Beam-surface phenomena

AMORUSO, Michael J.; 18 Crest Drive; Long Valley NJ 07853; USA

ANDERSEN, Nils O.; Institute of Physics; University of Aarhus; Aarhus DK8000; Denmark; 456-12-88-99; E/T; 2.1; 3.8; 4.2; 1.1; 2.2

ANDERSON, Alfred B.; Department of Chemistry; Case Western Reserve University; Cleveland OH 44106; USA; 216-368-3684; T; 1.1; 2.3; 3.5; 5.1; Molecular vibrational and electronic spectra, matrix effects; Molecular structures and reactions; Organometallic surfaces, catalysis, electrochemistry, corrosion

ANDERSON, Carl J.; IBM Watson Research Center; PO Box 218; Yorktown Heights NY 10598; USA

ANDERSON, Donald A.; 4125 SW 202; Aloha OR 97007; USA

ANDERSON, James B.; Department of Chemistry; Pennsylvania State University; 152 Davey Laboratory; University Park PA 16802; USA; 814-865-3933; E/T; 1.1; 2.3; 3.2; 3.8; 4.3; Monte Carlo methods in quantum mechanics; Potential energy; Surfaces for chemical reactions; Molecular dynamics of chemical reactions; Molecular beams from nozzle sources; Laser isotope separation; Resonance radiation imprisonment

ANDERSON, John M.; General Electric R&D Center; PO Box 8; Schenectady NY 12301; USA; 518-387-6357; E; 5.3; 5.4; Classical gaseous electronics

ANDERSON, Larry G.; Department of Chemistry; University of Colorado at Denver; 1100 Fourteenth Street; Denver CO 80202; USA; 303-556-2963; E/T; 2.3; 5.6; 3.2; 3.3; 3.6; Atmospheric reactions--reaction kinetics studies; Atmospheric processes--modeling; Atmospheric measurements

ANDERSON, Louis W.; Department of Physics; University of Wisconsin; Madison WI 53706; USA; 608-262-8962; E; 2.1; 5.3; Charge-changing collisions; Polarized ions for nuclear accelerators; Gas discharge kinetics; Optical pumping

ANDERSON, Richard J.; Division 8342; Sandia Labs.; PO Box 969; Livermore CA 94550; USA

ANDERSON, Richard J.; Department of Physics; University of Arkansas; Fayetteville AR 72701; USA; 501-575-2506; E; 2.2; 2.3; 3.6; Electron-impact excitation of atoms and molecules; Radiative lifetime measurements

ANDERSON, Robert W.; Webster Research Center; Xerox Corporation; Webster NY 14580; USA; 716-422-2487; E; 3.3; Picosecond spectroscopy

ANDERSON, Roger W.; Department of Chemistry NS2; Univ. of California, Santa Cruz; Santa Cruz CA 95064; USA; 408-429-2854; E/T; 2.1; 2.3; 4.3; 3.8; Electronic excitation and quenching in collisions; Chemiluminescent reactions; Quantum-coupled channel calculations; Crossed molecular beams

ANDERSON, Thomas G.; Department of Chemistry; Massachusetts Institute of Technology; 77 Massachusetts Avenue, Room 2-013; Cambridge MA 02139; USA; 617-253-4524; E/T; 2.1; 3.3; 3.7; 3.8; High-resolution IR spectroscopy; Molecular energy transfer; Laser-induced chemistry; Multiphoton dissociation

ANDERSON, Wayne; Department of Physics; U.S. Air Force Academy; USAF Academy CO 80840; USA; 303-472-2394; T; 5.1; Laser damage in optical coatings

ANDERSON, William R.; Ignition & Combustion Branch; Interior Ballistic Division; Ballistic Research Laboratory; U.S. Army; Aberdeen Prov'g Grnd MD 21005; USA; 301-278-6642; E; 5.5; 3.1; 3.2; 3.6; Spectroscopic techniques to investigate species concentration; Flames--temperatures and concentration; LIF techniques--uv/vis; Transition probabilities and quenching rates measured

ANDRA, Jurgen H.; Institut Fur Kernphysik; Domagkstr 71; Muenster 4400; West Germany

ANDRES, Ronald P.; Department of Chemical Engineering; Princeton University; Princeton NJ 08544; USA; 609-452-4591; E; 1.3; 2.1; Neutral atoms--low-energy collisions; Small clusters (dimer, trimer, etc.) of atoms and simple molecules--molecular properties

ANDRESEN, Bjarne B.; Physics Laboratory 2; University of Copenhagen; Universitetsparken 5; Copenhagen 2100; Denmark; +45-1-3531 33; T; 2.1; 1.1; 2.3; 5.1; Excitation mechanisms in beam-foil and sputtering; Differential cross sections inversion; Velocity dependent molecular orbitals

ANDREW, Kenneth L.; Department of Physics; Purdue University; West Lafayette IN 47907; USA; 317-494-5540; 317-494-5551; E; 3.2; Atomic emission spectroscopy; Spectra analysis; Fourier transform spectroscopy; Fabry-Perot interferometry; Theoretical calculations of spectra

ANDREWS, Hugh R.; Nuclear Physics Branch, ST 49; Physics Division; Chalk River Nuclear Laboratories; Atomic Energy of Canada; Chalk River Ontario K0J1J0; Canada; 613-584-3311 x2891; E; 5.1; 5.8; Energy loss phenomena in solids; Hyperfire interactions

ANDREWS, Lester S.; Department of Chemistry; University of Virginia; McCormick Road; Charlottesville VA 22901; USA; 804-924-3513; E; 5.2; Transient species and molecular ions in solid argon--spectroscopy

ANGELLO, Stephen J.; 112 Woodgate Road; Pittsburgh PA 15235; USA

ANHOLT, Robert E.; Department of Physics; Stanford University; Stanford CA 94305; USA

ANNIS, Brian K.; Chemistry Division; Oak Ridge National Laboratory; PO Box X; Oak Ridge TN 37830; USA; 615-574-5047; E; 2.1; Atomic and molecular collision dynamics--experimental investigation; Electron detachment collisions with negative ions in energy range of a few hundred volts

ANTAR, Ali A.; Institute of Materials Science; University of Connecticut; Storrs CT 06268; USA; 203-486-4915; E; Atomic and molecular physics--experimental; Ion implantation

ANTCLIFF, Richard R.; Research Applications Division; Systems Research Laboratories, Inc.; MS 168; NASA-Langley Research Center; Hampton VA 23665; USA; 804-865-2803; 804-865-3015; E; 3.10; 4.1; 5.5

ANTKIW, Stephen; 104 Blackman Road; Ridgefield CT 06877; USA

APPERSON, Gerald R.; Intermec, Inc.; PO Box N; 4405 Russell Road; Lynnwood WA 98036; USA

APPLETON, B. R.; Oak Ridge National Laboratory; Union Carbide Corporation; PO Box X; Oak Ridge TN 37830; USA; 615-574-6283; E; 4.2; 5.1; Atomic and molecular effects as modified in single-crystal; channels--exploitation; Ion-solid interactions

APT III, Jerome; Department of Earth and Space Sciences; Jet Propulsion Laboratory; California Institute of Technology; 4800 Oak Grove Drive; Pasadena CA 91103; USA; 213-792-2296; T; 5.6; Atomic and molecular data in planetary atmospheres research

ARCAND, Denis; 290 Cure Boivin; Boisbriand PQ J7G2A4; Canada

ARENTS, John S.; Department of Chemistry; City College of the CUNY; New York NY 10031; USA; 212-690-8404; T; 1.1; Quantum mechanical calculations on small molecules; Localized orbitals

ARMEN, George B.; Department of Physics; University of Oregon; Eugene OR 97403; USA

ARMENTROUT, Peter B.; Chemical Department; University of California; Berkeley CA 94720; USA; 415-642-4428; 415-642-4710; E; 1.1; 1.3; 2.3; 3.1; 4.4; Thermodynamics of transition metal clusters; Thermodynamics of transition metal compounds

ARMSTEAD, Robert L.; Department of Physics; USN Post Graduate School; Monterey CA 93940; USA

ARMSTRONG, J. A.; Physical Sciences Division; IBM Thomas J. Watson Research Center; PO Box 218; Yorktown Heights NY 10598; USA; 914-945-1270; T; 1.1; Multi-channel quantum-defect theory applied to two-electron atoms

ARMSTRONG JR., Lloyd; Department of Physics and Astronomy; Johns Hopkins University; Baltimore MD 21218; USA; 301-338-7362; 301-338-7375; T; 1.1; 1.4; 3.1; 3.7; 3.10; Laser-atom interactions; Relativistic and QED effects in atoms; Atomic structure calculations

ARNOLD, James O.; Physical Sciences Branch; NASA Ames Research Center; N230-3; Moffett Field CA 94035; USA; 415-965-6209; FTS-448-6209; T; 1.1; Small molecules by modern ab initio methods--properties

ARNOW, Clarence L.; 5041 Morse; Skokie IL 60076; USA

ARON, Paul R.; NASA Lewis Research Center 23-2; 21000 Brookpark Road; Cleveland OH 44135; USA

ARUNKUMAR, Koovappadi A.; Department of Electrical Engineering; University of Kentucky; Anderson Hall; Lexington KY 40506; USA

ASHKIN, Arthur; Physical Optics & Elec. Research Div.; Bell Telephone Laboratories; Crawfords Corner Road; Holmdel NJ 07733; USA; 201-949-2673; E; 4.1; Lasers; Nonlinear optics; Radiation pressure

ASHLEY, James C.; Health Physics Division; Oak Ridge National Laboratory; PO Box X; Oak Ridge TN 37830; USA; 615-574-6211; T; 5.2; 5.4; Electromagnetic theory applied to charged-particle energy-loss processes

ASMUS, John F.; Inst. Geophysics & Planetary Phys.; A-025; Univ. of California, San Diego; La Jolla CA 92093; USA; 619-452-2471; 619-453-0060; E; 5.1; 5.4; 4.1; 4.3; Artworks--laser cleaning; Rare-gas halide excimer lasers; Surface x-ray effects

ASSOUSA, George E.; Terrestrial Magnetism Division; Carnegie Institute of Washington; 5241 Broad Branch Road, NW; Washington DC

20015; USA; 202-966-0863; E

ATKINSON, George H.; Department of Chemistry; Syracuse University; Syracuse NY 13210; USA; 315-423-3238; E; 3.5; 3.6; 3.8; 4.1; Gas-phase polyatomic molecules--spectroscopy and photochemistry; Energy transfer; Tunable laser spectroscopy; Time-resolved resonance Raman spectroscopy

ATKINSON, J. B.; Department of Physics; University of Windsor; Windsor Ontario N9B3P4; Canada; 519-253-4232 x2652; E; 2.1; 3.6; 4.1

ATLAS, Susan R.; Chemistry; Chemical Physics; Harvard University; 336; 12 Oxford Street; Cambridge MA 02138; USA; 617-495-1990; 617-491-4841; T; 1.1; 3.6

AU, C. K.; Department of Physics and Astronomy; University of South Carolina; Columbia SC 29208; USA; 803-777-2488; 803-777-4121; T; 1.4; 2.3; 3.1; Perturbation theory; Photon-atom interactions; Quantum electrodynamics; Reactions in bound states

AUBREY, Bertrand B.; N.Y.C. Dept. of Environmental Protection; Bureau of Science & Technology; 382 Central Park West-Apt. 19N; New York, NY 10025; USA; 212-566-5906; 212-865-3938; E; 4.4; 3.1; 2.2; 2.1; 4.1

AUERBACH, Daniel J.; Department K33/281; IBM Research Laboratory; 5600 Cottle Road; San Jose CA 95193; USA; 408-256-1600; E; 2.1; 4.3; 5.1; Molecular beam scattering experiments

AUERBACH, Roy A.; Department of Chemistry; Tulane University; New Orleans LA 70118; USA; 504-865-4713; E/T; 2.3; 3.8; Inter- and intra-molecular interactions as studied through the use of subnanosecond light pulses

AUSLOOS, Pierre; Institute of Materials Research; National Bureau of Standards; Gaithersburg MD 20899; USA; 301-921-2783; E; 4.2; Ionizing radiation

AUYEUNG, John C.; Newport Corporation; PO Box 8020; Fountain Valley CA 92728; USA

AVCI, Recep; Dhahran Int. Airport; PO Box 144; UPM Box 2018; Dhahran; Saudi Arabia

AVERILL, Frank W.; Department of Science and Mathematics; Judson College; 1151 N. State Street; Elgin IL 60120; USA; 312-695-2500; T; 1.3; 1.1; 3.1; Electronic structure and total energy; Calculations on atom-cluster models; Interfaces

AVILES, Luis A.; PO Box 5985; College Station; Mayaguez 07009; Puerto Rico

AVOURIS, Phaedon; IBM Thomas J. Watson Research Center; PO Box 218; Yorktown Heights NY 10598; USA; 914-945-2722; E/T; 3.7; 3.8; 5.2; Multiphoton spectroscopy and chemistry; Luminescence and radiationless transitions in molecules

AVRETT, Eugene H.; Harvard-Smithsonian Astphy. Obsery; 60 Garden Street; Cambridge MA 02138; USA; 617-495-7423; FTS-830-7423; T; 5.7; Atomic and molecular data in radiative transfer studies in astrophysics

AYUB, S. M.; PO Box 3721; St. Paul MN 55165; USA

BABBITT, William R.; Gordon McKay Lab. Room 321; Harvard University; 9 Oxford Street; Cambridge MA 02138; USA; 617-495-2788; E; 3.10; 3.6; 4.1; Coherent transients studies of inhomogeneous and homogeneous broadening in atoms & molecules; Optical memories studies

BABCOCK, Lucia M.; Chemistry; Physical; Louisiana State University; Choppin Hall; Baton Rouge LA 70803; USA; 504-388-4694; 504-388-3361; E; 2.3; 4.4; 1.1; 2.1; 5.3; Gas phase ion-molecule chemistry; Flowing afterglow, SIFT; Temperature dependence of association reactions; Radiative stabilization

BACH, Richard L.; Chapelizod; 66 Belgrove Lawn; Dublin 20; Ireland

BACON, Frank M.; Applied Technology Division 2352; Sandia National Laboratories; Albuquerque NM 87185; USA; 505-844-3945; E; 4.3; 5.1; 5.4; Ion sources and beams; Ion interactions with molecules

BAE, Young K.; Molecular Physics Department; Chemical Physics Laboratory-PN0 73; SRI International; 333 Ravenswood Avenue; Menlo Park CA 94025; USA; 415-859-3814; E/T; 3.1; 2.1; 3.3; 3.6; 1.1

BAER, Thomas M.; Spectra Physics 2-15; 1250 W. Middlefield Road; Mountain View CA 94039; USA; 303-493-7387; E; 2.1; 3.6; Laser spectroscopy to study atomic and molecular collisions

BAER, Thomas; Department of Chemistry; University of North Carolina; Chapel Hill NC 27514; USA; 919-966-5433; E; 2.3; 3.1; 3.7; 1.2; 4.4; Photoionization of small molecules; Photoelectron--photoion coincidence spectroscopy; State-selected ions--reactions; Multiphoton ionization spectrometry

BAGLIN, Frank G.; Department of Chemistry; University of Nevada; Reno NV 89557; USA; 702-784-6041; 702-784-6651; E; 1.1; 2.1; 3.5; 5.2; High pressure inelastic light scattering of molecular fluids; Inelastic light scattering in molecular systems; Line width studies of molecular transitions

BAGUS, P. S.; Department K33/281; IBM Research Laboratory; 5600 Cottle Road; San Jose CA 95193; USA; 408-256-1600; T; 1.1; Ab initio molecular theory; Molecular calculations, accurate

BAILEY, Thomas L.; Department of Physics; University of Florida; Gainesville FL 32601; USA; 904-392-2142; E; 2.2; 2.3; Electron and ion impact studies

BAINUM, David E.; Academic Computer Center; Washburn University; 1700 College Street; Topeka KS 66621; USA; 913-295-6389; E; 4.2; 5.1; 5.8; Accelerated ion beams to study material properties, nuclear; excitations; production of radioisotopes; Photons, electrons, and neutrons--dosimetry of; Computer-based data acquisition development, and analysis systems lab use

BAIR, Edward J.; Department of Chemistry; Indiana University; Bloomington IN 47405; USA; 812-335-5437; E; 2.3; 3.1; 3.6; Time-resolved spectroscopy and photochemistry; Vibrational energy effects

BAIRD, James C.; Departments of Physics and Chemistry; Brown University; Providence RI 02912; USA; 401-863-3391 x2321; E/T; 2.1; Spin-aligned atomic hydrogen collision processes

BAKER, Howard C.; MSFA40; Rockwell International; Rocketdyne Division; 6633 Canoga Avenue; Canoga Park CA 91304; USA; T; 3.1; Atom--radiation phenomena--cooperative effects

BAKER, Oliver K.; Varian Physics Building; Stanford University; Stanford CA 94305; USA

BAKSHI, Pradip M.; Department of Physics; Boston College; Chestnut Hill MA 02167; USA; 617-552-3585; T; 3.1; 3.9; 3.7; 5.4; Hydrogenic atoms interaction with static and dynamic electric and magnetic fields; Plasma diagnostics applications; Non-perturbative techniques for matter-radiation interactions

BALDWIN, George C.; Physics Division; Los Alamos National Laboratory; PO Box 1663; Los Alamos NM 87545; USA; 505-667-7740; E/T; 5.8; 4.1; Development of gamma-ray lasers; Separation of nuclear isomers (this is not isotope separation!); Nuclear super-radiance in solids

BALIGA, Shankar B.; Department of Physics; Ohio State University; 174 W. 18th Avenue; Columbus OH 43210; USA; 614-422-8865; E/T; 5.2; 3.3; 3.10; Study of solid hydrogens

BALLARD, Stanley S.; Department of Physics; University of Florida; Gainesville FL 32611; USA; 904-392-0487; E; 3.3; Optical materials for IR region--properties

BALLING, Ludwig C.; Department of Physics; University of New Hampshire; Durham NH 03824; USA; 603-862-2829; E; 1.1; 3.6; Atomic structure and interatomic interactions--laser spectroscopy

BALOG, Katja; 31-APP 411; Morganbreede; Bielefeld 1 4800; West Germany

BAMIERE, Christine; Bureau Des Affairs; Scientifiques et Technology; 51 B De Latour Maubourg; Paris 75700; France

BAND, Hans E.; U.S. Air Force (Civilian); Hanscom AFB; AFCSC/OLMC; Bedford MA 01731; USA; 617-275-8575; 617-861-4569; E; 3.9; 3.6; 3.2; 3.3; 3.1; Far infrared laser spectroscopy; Stark effect tuning of infrared transitions; Atomic spectroscopy; Optical physics and electro-optics

BAND, Yehuda B.; Department of Chemistry; Natural Sciences; Ben-Gurion University; Beer-Sheva 84105; Israel; 972-59-690859; 972-57-664268; T; 1.1; 2.1; 3.1; 4.1; 5.3; Laser physics and chemistry; Thermodynamics and statistical mechanics; Nonlinear optics

BANNA, M. Salim; Department of Chemistry; Vanderbilt University; Nashville TN 37235; USA

BARASH, Yefim; 55 Charles Lindbergh Blvd.; Mitchell Field NY 11553; USA

BARAT, Michael; Batiment 351; Universite Paris SUD; Orsay 91405; France

BARDSLEY, J. N.; Department of Physics and Astronomy; University of Pittsburgh; Pittsburgh PA 15260; USA; 412-624-4359; T; 2.1; 2.2; 5.3; Low energy collisions involving electrons, ions, atoms, and molecules; Macroscopic properties of ionized gases--effect of such collisions

BARFIELD, Walter D.; 4647 Ridgeway Drive; Los Alamos NM 87544; USA

BARKER, Charles E.; Ginzton Lab.; Stanford University; PO Box 36; Stanford CA 94305; USA

BARKER, John R.; Atmospheric and Ocean Science; Space Physics; University of Michigan; Space Research Building; Ann Arbor MI 48109; USA; 313-763-6239; E/T; 2.3; 3.6; 3.8; 5.5; 5.6; Chemical kinetics; Energy transfer kinetics

BARNARD, A. Johannes; Department of Physics; University of British Columbia; Vancouver British Columbia V6T1W5; Canada

BARNES, Ramon M.; Department of Chemistry; Graduate Research Center Towers; University of Massachusetts; Amherst MA 01003; USA; 413-545-2294; E/T; 5.3; 5.4; 4.4; Spectrochemical analysis/electrical discharge sources (plasmas, spark, arc); Plasma discharge sources--fundamental studies

BARNES, Ronald; Barnes Development Company; PO Box 9081; Wichita KS 67277; USA; 316-722-4271; E/T; 5.4; High-energy arc light plasmas--optical radiation

BARNETT, Charles F.; Physics Division; Oak Ridge National Laboratory; PO Box X; Oak

Ridge TN 37830; USA; 615-574-4700; FTS-624-4700; E; 2.1; 4.2; Atomic collisions

BARNETT, Clarence F.; Consultant; P.O. Box 471; Phelps Rd. Rt. 4; Lenoir City TN 37771; USA; 615-986-3953; 615-574-4700; E; 2.1; 5.4; 5.5; 5.1; Atomic collisions; Plasma diagnostics; Plasma chemistry; Particle interactions with surfaces

BARR JR., Thomas A.; Directed Energy Directorate; U.S. Army Missile Command; DRSM-I-RHSS (Building 8978); Redstone Arsenal AL 35898; USA; 205-876-8181; E; 4.1; High-energy lasers; New lasers--kinetics analyses

BARRETO, Ernesto; Atmospheric Sciences Research Center; State University of NY, Albany; 1400 Washington Ave.; Albany NY 12222; USA; 518-442-3814; E; 1.1; 2.2; 5.5; 5.6; High pressure electrical discharges; Combustion of hydrocarbons; Aerosols, nucleation, charged; Fluid dynamics of electrons

BARRETT, John L.; Department of Physics; University of Missouri, St. Louis; 8001 Natural Bridge Road; St. Louis MO 63121; USA; 314-553-5280; E; 2.3; Metal vapors--charge-exchange scattering

BARRY, J. D.; Space & Communications Group; Hughes Aircraft Company; PO Box 92919; Building 373, MS A355; Los Angeles CA 90009; USA; 213-648-8647; T; 3.6; Laser physics

BARTELHEIMER, Darrel L.; Staff Syst Eng.; Dept 392 4E1; Link Flight Simulation Division; 2224 Bay Area Boulevard; Houston TX 77058; USA

BARTELL, Lawrence S.; Department of Chemistry; University of Michigan; Ann Arbor MI 48109; USA; 313-764-7375; E/T; 1.1; 2.2; 3.8; Molecular structure and vibrations by electron scattering; Laser-induced processes and supersonic jets; Electron holography (atomic and molecular images); Theoretical conformational analysis (molecular mechanics)

BARTIROMO, Rosario; Fusione; Fisica Fusione; Fisica Tokamak 3; Associazone Euratom-ENEA; P.O. Box 65; Enrico Fermi; Frascati Roma 100044; Italy; (06) 9400-5351; (06) 942-3396; E; 1.1; 2.2; 3.4; 4.1; 5.4; X-Ray spectroscopy of highly ionized atoms

BARTLETT, Rodney J.; Chemical Physics Division; Battelle Memorial Institute; 505 King Avenue; Columbus OH 43201; USA; 614-424-7275; T; 1.1; 2.3; Molecules and their interactions--quantum theory; Ab initio quantum chemistry; Many-body theory; Potential energy surfaces for molecules

BARTOLOTTI, Libero J.; Department of Chemistry; Physical Chemistry; University of Miami; Coral Gables FL 33124; USA; 305-284-6616; T; 1.1; Time-dependent variation-perturbation theory of non-linear susceptabilities of atoms and molecules--development and application; Density functional techniques in describing properties of atoms and molecules--development and application

BARTON JR., George W.; General Chemistry Division; Lawrence Livermore National Lab.; PO Box 808; Livermore CA 94550; USA; 415-422-6329; FTS-532-6329; E; 4.4; 5.1; Mass spectrometry; Surface science

BASBAS, George J.; Physical Review Letters; APS Editorial Office; American Physical Society; Box 1000; Ridge NY 11961; USA; 516-924-5533; T; 2.1; 3.4; 1.1; 5.2; 3.1

BASHKIN, Stanley; Department of Physics; University of Arizona; PAS 81; Tucson AZ 85721; USA; 602-621-2322; 602-624-1881; E; 1.1; 4.2; 5.1; Beam-foil spectroscopy; Atomic energy-level and diagrams--compilation

BASS, Arnold M.; Center for Analytical Chemistry; National Bureau of Standards; Room B-326; Gaithersburg MD 20899; USA; 301-921-2066; E; 3.1; 5.6; UV absorption parameters for molecules of atmospheric importance

BASSYOUNI, Ahmed H.; Department of Physics; Faculty of Science; University of Zagazig; Zagazig; Egypt

BATAY-CSORBA, Peter A.; 12 Dempsey Crescent; Willowdale Ontario M2L1Y5; Canada

BATES JR., Clayton W.; Department of Materials Science; Stanford University; Peterson 550 J; Stanford CA 94305; USA; 415-4974252

BATES, Harry E.; 6913-A Lachlan Circle; Baltimore MD 21239; USA

BATES JR., Richard D.; Department of Chemistry; Georgetown University; Washington DC 20057; USA; 202-625-3149; E; 1.1; 2.1; 3.6; 3.8; 5.2; Intermolecular interactions; Collision processes in the gas phase; Energy transfer phenomena; Transient complexes in liquids by magnetic resonance methods

BATTLESON, Kirk W.; Sandia National Laboratories; P. O. Box 969; East Avenue; Livermore CA 94550; USA; 415-422-2217; 415-462-1840; E; 4.3; 2.2; 3.2; 4.2; EBIS, Electron Beam Ion Search

BAUER, Ernest; 8109 Fenway Road; Bethesda MD 20817; USA

BAUER, Ernst; Physik Institut-Leibnizstr 4; Tech Univ. Clausthal; Clausthal-Zeller D-3392; West Germany

BAUER, Simon H.; Department of Chemistry; Baker Laboratories; Cornell University; Ithaca NY 14850; USA; 607-256-4028; 607-256-4032; E; 2.3; 3.2; 5.5; Structures of amorphous materials via synchrotron radiation; Gas-phase kinetics with emphasis on energy transfer rates & combustion; Shock tubes and lasers for chemical conversions

BAUGHCUM, Steven L.; Group CHM-4; Chemistry Division; Los Alamos National Laboratory; MS-J567; Los Alamos NM 87545; USA; 505-667-6838; 505-667-3758; E; 2.3; 3.6; 3.1; 3.2; 2.1; Small, gas-phase molecules--laser photodissociation studies; Atoms, radicals, and molecules --spectroscopy, kinetics and energy transfer; Laser based analytical techniques

BAUM, Guenter G.; Fakultaet Fuer Physik; Universitaet Bielefeld; 8640; Universitaetsstr; Bielefeld 48; West Germany; 521-106-5383; E; 2.2; 4.1

BAUMAN, Leslie E.; Physics and Chemistry, MHD Energy Center; Mississippi State University; PO Box 5167; Mississippi State MS 39762; USA; 601-325-2806; 601-325-2105; E; 5.5; 5.4; Temperature and electron density measurements of simulated coal-fired MHD; Test facilities by emission/absorption techniques.

BAUMAN, Robert P.; Department of Physics; University of Alabama, Birmingham, Birmingham AL 35294; USA; 205-934-3576/4736; E/T; 2.3; 5.3; Ion-molecule reactions; Electrical discharges

BAUR, James F.; GA Technologies, Inc.; P.O. Box 85608; San Diego CA 92138; USA; 619-455-3298; 619-455-9093; E; 5.4; 3.2; 3.6; Low-Z elements--spectroscopy of neutral and ionized species; Laser-particle beam-magnetic field interactions; Low-Z ion beams for plasma diagnostics, production of higher current

BAY, Zoltan L.; Department of Physics; The American University; Nebraska Avenue, N.W.; Washington DC 20016; USA; 301-986-9478; 202-885-2747; E/T; 1.1; 2.1; 3.6; 4.1; 5.7

BAYES, Kyle D.; Department of Chemistry; University of California, LA; Los Angeles CA 90024; USA; 213-825-2083; E; 2.3; 3.1; 5.5; 5.6; Gas-phase kinetics

BAYFIELD, James E.; Department of Physics and Astronomy; University of Pittsburgh; 100 Allen Hall; Pittsburgh PA 15260; USA; 412-624-5485; 412-624-1060; E; 1.2; 2.1; 3.6; 3.7; 3.10; Ion-atom charge exchange collisions; Microwave interactions with Rydberg atoms; Chaos in the driven bound electron; Deceleration of ion beams

BAYLIS, William E.; Department

of Physics; University of Windsor; Windsor Ontario N9B3P4; Canada; 519-253-4232x2653; T; 1.1; 2.1; 3.1

BEACH, Raymond J.; Columbia Radiation Lab.; 538 W. 120th Street; New York NY 10027; USA

BEAHN, Thomas J.; Lab. for Physical Sciences; 4928 College Avenue; College Park MD 20740; USA

BEARMAN, Gregory; Jet Propulsion Laboratory; 4800 Oak Grove Drive; Pasadena CA 91103; USA; 213-354-3776; E; 2.1; 4.3; Thermal metastable collisions studied with a beam apparatus

BEATY, Earl C.; (NBS); Mail Code 524.11; Boulder CO 80303; USA; E

BEAUCHAMP, Jesse L.; Noyes Laboratory; California Institute of Technology; Pasadena CA 91125; USA; 213-0795-6811; E; 2.3; 3.1; 3.6; 4.2; 5.1; Ion-molecule reactions; Photoionization mass spectrometry; Photoelectron spectra of free radicals; Reactions at gas surface interfaces; Ion-cyclotron resonance spectroscopy; Laser spectroscopy

BEAUDET, Robert A.; Department of Chemistry; University of Southern California; Los Angeles CA 90089; USA; 213-743-2297; E; 1.1; 3.3; 3.6; Microwave spectroscopy; Molecular structure; Laser fluorescence spectroscopy; Boranes and Carboranes

BEAUSOLEIL JR., R. G.; Department of Physics; Varian Laboratory of Physics; Stanford University; Stanford CA 94305; USA; 415-497-4357; 415-967-0419; E; 3.6; 3.10; 3.11; 4.1

BECHER, Jacob; Department of Physics; Old Dominion University; Hampton Boulevard; Norfolk VA 23508; USA; 804-440-4616; E/T; 3.2; 4.2; Ozone band studies; Atomic spectra

BECHIS, Kenneth P.; Surveillance and Space Technology; Signal and Sensor Sciences; The Analytic Sciences Corporation; 1 Jacob Way; Reading MA 01867; USA; 617-944-6850; T; 3.6; 4.3; Neutral particle beam propagation direction through fluorescence; measurements--application of laser pumping to measurements

BECHTEL, James H.; Physics Division; Genl Mtrs Research Laboratories; General Motors Technical Center; 12 Mile and Mound Roads; Warren MI 48090; USA; 313-575-7635; E; 2.3; 3.6; 5.5; Combustion reactions--chemistry and physics; Laser spectroscopy; Chemical kinetics; Nonlinear optics

BECK, Scott E.; 1925 East Cedar Street; Allentown PA 18103; USA

BECKEL, Charles L.; Department of Physics and Astronomy; University of New Mexico; Albuquerque NM 87131; USA; 505-277-2616/2449; T; 1.1; 1.3; Diatomic molecule vibrations; Biomolecules in stable conformations

BECKER, Kurt H.; Department of Physics; Lehigh University; Bethlehem PA 180154; USA; 215-861-4916; 215-861-3930; E; 2.2; 2.3; Electron collisions; Dynamics of dissociation processes

BECKER, Gordon E.; Surface Physics Research Division; Bell Telephone Laboratories; 600 Mountain Avenue; Murray Hill NJ 07974; USA; 201-582-2677; E; 5.1; Atomic and molecular beams at solid surfaces in vacuum--scattering and reaction

BECKER, Joseph F.; Department of Physics; San Jose University; San Jose CA 95119; USA

BECKER, Karl H.; Chemie/FB 9; Phys. Chemie; Univ. Wuppertal/FB 9; 100127; Gauss-STR 20; Wuppertal 1 NRW D-5600; West Germany; (0202)439-2666; E; 1.1; 2.3; 3.6; 4.4; 5.6; Atmospheric reactions, OH/HO2 kinetics, k-values, photo-oxidants; Radical reactions in flames, chemiluminescence, chemi-ionization; Photochemical processes, energy transfer, transition probabilities; Laser photolysis, laser induced fluorescence

BECKER, Michael F.; Department of Electrical Engineering; University of Texas; Austin TX 78712; USA; 512-471-3628; E/T; 3.7; 4.1; Nonlinear optical properties of molecules in the IR (harmonic generation, multiphoton absorption, nonlinear susceptibility)

BECKER, Richard L.; Physics Division; Oak Ridge National Laboratory; Building 6003; Oak Ridge TN 37830; USA; 615-574-4579; FTS-624-4579; T; 2.1; 1.2; Ion-atom collision theory

BEDARD, Fernand D.; 505 Hermleigh Road; Silver Spring MD 20902; USA

BEDELL, Louis R.; Department of Physics; Northeast Louisiana University; Monroe LA 71209; USA; 318-342-4133; 318-342-3120; E; 5.1; Surface physics; Low energy electron diffraction (LEED); Auger electron spectroscopy

BEDERSON, Benjamin; Department of Physics; New York University; 4 Washington Place; New York NY 10003; USA; 212-598-3682; 212-598-2803; E; 1.1; 2.2; 4.3; 3.6; 3.9; Atomic and molecular structure and interactions using beam techniques; Atom-photon interactions

BEERS, Brian L.; Beers Associates, Inc.; PO Box 2549; 11717 Bowman Green Drive; Reston VA 22090; USA; 703-437-0866; 703-435-5750; T; 2.2; 3.4; Applied X-ray photoemission physics; Applied electron-beam physics

BEHRING, William E.; Lab. for Astronomy and Solar Physics; Code 680.1; Goddard Space Flight Center; Greenbelt MD 20771; USA; 301-344-7410x6861; E; 3.2; 5.7; Solar spectroscopy with high spatial and spectral resolution; EUV spectra of highly ionized elements--production and analysis

BEHRINGER, Robert E.; Office of Naval Research; 1030 East Green Street; Pasadena CA 91106; USA; 213-795-5971; E; 2.1; 2.3; Quenching rates, branching ratios, cross sections of interest in atomic and molecular lasers

BEITING III, Edward J.; Department of Physics; Mississippi State University; PO Box 5167; Mississippi State MS 39762; USA; 601-325-2806; 601-325-2102; E; 3.10; 3.6; 4.1; 5.5; 1.2

BEKEFI, G.; Department of Physics; Massachusetts Institute/Technology; Room 36-213; Cambridge MA 02139; USA; 617-253-7330; E; 4.1; 4.2; 5.4; Relativistic electron beams; Free electron lasers; Magnetic fusion

BEL BRUNO, Joseph J.; Department of Chemistry; Dartmouth College; Steele Hall; Hanover NH 03755; USA; 603-646-2270; 603-646-2189; E/T; 2.3; 3.6; 3.7; 3.8; 4.4; Rotational relaxation in bulk gases via photoacoustic spectroscopy

BELFORD, R. L.; Noyes Chemistry Laboratory; University of Illinois; Urbana IL 61801; USA; 217-333-2553; E/T; 2.3; 5.3; Elementary reaction kinetics--shock tube studies

BELLINA JR., Joseph J.; 402 N. Esther; South Bend IN 46617; USA

BEN-REUVEN, Abraham; Tel-Aviv University; Tel Aviv 69978; Israel

BENARD, David J.; ARAP; U.S. Air Force Weapons Laboratory; Kirkland AFB; Albuquerque NM 87117; USA; 505-844-0196; E; Short wavelength chemical laser systems

BENDER, Charles F.; Chemistry and Material Science Div.; Lawrence Livermore National Lab.; PO Box 808; Livermore CA 94550; USA; 415-422-6340; E; Research management

BENDER, Peter L.; Quantum Physics Division (NBS); Joint Institute for Lab. Astrophysics; University of Colorado; Boulder CO 80309; USA; 303-492-6793; FTS-323-3846; E; 5.7; 3.6; 3.11; Geophysical applications of A&M physics; Precision measurements using A&M physics

BENDLER, John T.; General Electric Corporate R&D Center; PO Box 8; Room 4B7, Building

K1; Schenectady NY 12301; USA; 518-387-6632; 518-387-5874; T; 1.3; 2.3; Polymer dynamics; Main-chain structure and motion; Stochastic theory; Quantum modeling

BENEDICT, Lt. Col. Rettig P.; DEO; Defense Advanced Res Projects Agcy; 1400 Wilson Boulevard; Arlington VA 22209; USA; 202-694-3784; E/T; Program manager

BENESCH, William M.; Institute for Molecular Physics; University of Maryland; College Park MD 20742; USA; 301-454-3438; E; 3.2; 3.5; 5.6; 2.1; 5.3; Spectroscopy of small molecules and atoms, special emphasis on molecules of atmospheric interest and in the aurora

BENGSTON, Roger D.; Department of Physics; University of Texas, Austin; Austin TX 78712; USA; 512-471-3943; E; 5.4; 3.1; 5.3; 5.1; Transition probabilities and reaction rates in plasmas

BENNETT, Donald L.; Department of Physics; Royal Danish College of Pharmacy; Universitetsparken 2; Copenhagen 2100; Denmark; 01 370850; 01 421616; T; 1.4; Elementary particle physics; Quantum field theory; Gauge theories; Lattice gauge theories

BENNETT, Robert A.; 11042 Lakeview Drive; Whitehouse OH 43571; USA

BENNETT, Robert B.; Department of Physics; Central Washington University; Ellensburg WA 98926; USA; 509-963-2701; E; 1.1; 2.2; Ground bases spectroscopy of the upper atmosphere

BENNETT JR., William R.; Department of Physics; Yale University; New Haven CT 06520; USA; 203-436-8445; FTS-646-4709; E; 3.6; Gas lasers; Computer applications to applied science

BENSON, Richard C.; Applied Physics Laboratory; Johns Hopkins University; Johns Hopkins Road; Laurel MD 20707; USA; 301-953-6241; E; 3.6; 3.3; 4.4; 5.1; Raman spectroscopy; Outgassing measurements of microelectronic adhesives

BENSON, Sidney W.; Department of Chemistry; University of Southern California; University Park; Los Angeles CA 90007; USA; 213-743-2030; E/T; 2.3; 3.8; Kinetics; Photochemistry; Thermochemistry

BENTLEY, John; Radiation Laboratory; University of Notre Dame; Notre Dame IN 46556; USA; 219-239-5362; 219-239-6163; T; 1.1; 2.1; Calculation of molecular hyperfine coupling; Collisions of electronically excited atoms and molecules

BERG, Jacqueline O.; Chemical Physics Department; Applied Technology Division; TRW Inc., R1/1196; 1 Space Park; Redondo Beach CA 90278; USA; 213-536-1453; E/T; 4.1; 3.10; 3.6; 3.7; 3.2; Spectroscopic diagnostics; Chemical oxygen iodine laser development; Stimulated Brillouin scattering-phase conjugation

BERGE, Phillip O.; Department of Physics; Illinois State University; Normal IL 61761; USA

BERGEMAN, Thomas H.; Department of Physics; State Univ. of NY at Stony Brook; Stony Brook NY 11794; USA; 516-246-7962; T; 3.9; 1.3; 3.11; 3.7; Stark effect in simple atoms; Neutral atoms in magnetostatic traps-classical orbits and quantum levels; Atomic excitation in non-monochromatic laser fields; Microwave ionization of simple atoms

BERGER, Lev; 2115 Flame Tree Way; Hemet CA 92343; USA

BERGER, Martin J.; Center for Radiation Research; National Bureau of Standards; Gaithersburg MD 20899; USA; 301-921-1000; T; Radiation transport; Monte Carlo methods; Multiple scattering theory

BERGMANN, Otto; Department of Physics; George Washington University; Washington DC 20052; USA; 202-676-6485; T; 1.1; 5.2

BERGQUIST, James C.; Time and Frequency Standards Div.; National Bureau of Standards; 325 S. Broadway; Boulder CO 80303; USA; 303-497-5459/3276; E; 3.1; 3.6; 3.9; 4.2; Laser-cooled ions bound in an electromagnetic ion trap; A single, trapped electron--synchrotron radiation studies

BERI, Avinash C.; Department of Chemistry; University of Rochester; Rochester NY 14627; USA; 716-275-2511/0310; T; 5.1; Laser-stimulated surface processes, involving adsorption; desorption, migration and reaction of gas species with solid surfaces

BERK, Alexander; Atomic Physics Office; Astronomy & Solar Physics; NASA GSFC Laboratory; Sachs/Freeman Associates, Inc.; 1401 McCormick Drive; Landover MD 20785; USA; 301-344-8812; T; 1.1; 2.2; 3.9

BERKNER, Klaus H.; Accelerator and Fusion Research; Lawrence Berkeley Laboratory 50-149; 1 Cyclotron Road; Berkeley CA 94720; USA; 415-486-5501; E; 2.1; 4.2; 5.1; 5.4; Electron capture, loss, and impact ionization highly-stripped heavy ions; Negative hydrogen ion production at surfaces; Negative hydrogen ion stripping collisions in plasmas; Balmer emission cross sections

BERKOWITZ, Jeffery K.; GTE Sylvania; 100 Endicott Street; Danvers MA 01923; USA

BERKOWITZ, Joseph; Physics Division; Argonne National Laboratory; Building 203; 9700 S. Cass Avenue; Argonne IL 60439; USA; 312-972-4086; E; 3.1; Photoionization; Photoelectron spectroscopy

BERMAN, B. L.; Department of Physics; The George Washington University; Washington DC 20052; USA; 202-676-7192; 415-422-9674; E; 4.2; 5.1; 5.2; 5.7; 5.8; Channeling radiation; Transition radiation; Relativistic beam interactions with solids; Relativistic beam interactions with surfaces

BERMAN, Paul R.; Department of Physics; New York University; 4 Washington Place; New York NY 10003; USA; 212-598-3373; T; 3.6; 3.10; 3.8; Collisional processes occurring in the presence of radiation fields; Line shape studies

BERMUDEZ, Victor M.; Electronics Technology Division; Navel Research Laboratory; Code 6833; 4555 Overlook Avenue; Washington DC 20375; USA; 202-767-6728; 202-767-3896; E; 5.1; 4.1; Single-crystals in ultra-high vacuum--surface studies; Surfaces--electron and optical spectroscopy; Chemisorption phenomena; Surface electronic states and point-defects

BERNARD, Davy L.; Department of Physics; Nuclear Physics; Acadiana Research Laboratory; Univ. of Southwestern Louisiana; PO Box 44210; USL Station; Lafayette LA 70504; USA; 318-264-6692; E; 3.4; 4.1; 4.2; Particle-induced X-ray fluorescence analysis

BERNE, Bruce J.; Department of Chemistry; Columbia University; New York NY 10027; USA; T; 2.1; 2.3; Intramolecular energy transfer and relaxation; Molecular dynamics; Chemical dynamics

BERNEY, Charles V.; Department of Chemical Engineering; Massachusetts Institute of Technology; PO Box 66-571; 77 Massachusetts Avenue; Cambridge MA 02139; USA; 617-253-6526; 617-253-4569; E/T; 1.3; 4.2; 5.2; Submicroscopic structure in polymers by neutron and x-ray scattering; Molecular vibrations by neutron inelastic scattering

BERNHARDT, Anthony F.; Department of Physics; Laser Pantography; Lawrence Livermore National Lab.; PO Box 5508; Livermore CA 94550; USA; 415-423-7801; T/E; 3.6; 3.7; 3.8; 3.10; 4.1; Coherent atomic motion in a resonant standing wave

BERNHEIM, Robert A.; Department of Chemistry; Pennsylvania State University; 152 Davey Laboratory; University Park PA 16802; USA; 814-865-3642; E; 3.5; 3.6; 3.2; 3.3; 1.4; Molecular structure and spectroscopy of small molecules; Optical

--optical double resonance laser spectroscopy; Fourier transform spectroscopy; Spectroscopy of small molecular radicals, ions, and alkali dimers

BERNIUS, Mark T.; Nuclear Science; Baker Laboratory; Ward Lab.; Cornell University; 224; Ithaca NY 14853; USA; E/T; 4.4; 4.3; 4.2; 5.1; 2.1; Secondary ion mass spectrometry ion microscopy; ion-beam optics

BERNSTEIN, E. M.; Department of Physics; Western Michigan University; Kalamazoo MI 49008; USA; 616-383-1870; E; 2.1; 4.2; Accelerator-based atomic physics; High energy ion-atom collisions

BERNSTEIN, Elliot R.; Department of Chemistry; Colorado State University; Fort Collins CO 80523; USA; 303-491-6347 x5787; E/T; 1.1; 3.1; 3.2; 3.5; 3.6; Molecular jet optical/mass spectroscopy; Van der Waals molecular and ionic clusters; Intramolecular vibrational redistribution in VDW clusters; Vibrational predissociation in VDW clusters

BERNSTEIN, Richard B.; Department of Chemistry; University of California, LA; 405 Hilgard Avenue; Los Angeles CA 90024; USA; 213-206-0476; E; 2.1; 2.3; 3.6; 4.3; 3.8; Molecular reaction dynamics via molecular beam and laser techniques

BERREMAN, Dwight W.; Chemical Physics Division; Bell Telephone Laboratories; 600 Mountain Avenue; Murray Hill NJ 07974; USA; 201-582-6596; E/T; 3.4; X-ray optics

BERRY, H. G.; Physics Division; Argonne National Laboratory; 9700 S. Cass Avenue; Argonne IL 60439; USA; 312-972-4039; 312-972-4085; E; 4.3; 3.6; 4.2; 3.2; 2.1; Fast ion beam collisions and spectroscopy

BERRY, Herbert W.; 2649 E. Genesee Street; Syracuse NY 13224; USA

BERRY, Michael J.; Quantum Institute; Rice University; PO Box 1892; Houston TX 77001; USA; E/T

BERRY, R. Steven; Department of Chemistry; University of Chicago; 5735 S. Ellis Avenue; Chicago IL 60637; USA; 312-962-7021; T/E; 1.1; 1.3; 2.3; 3.7; 2.2; Few-body systems-structure and dynamics; Phase changes in finite systems; Resonant multiphoton ionization; Ion-ion neutralization and ion-pair formation

BERRY, Scott D.; Radiation Research; Physics; Murray Hill; AT&T Bell Laboratories; 600 Mountain Avenue; Murray Hill NJ 07974; USA; 201-582-5180; 201-582-3942; E; 1.3; 2.1; 1.2; Clusters-liquid metal ion sources; Electron transfer to the continuum-convoy electron production

BERSOHN, Richard; Department of Chemistry; Columbia University; 959 Havemeyer; New York NY 10027; USA; 212-280-2192; E; 3.1; 3.6; 4.2; 2.3; Photodissociation of molecules by lasers; Reaction of atoms with molecules

BEST, Philip E.; Department of Physics U-46; University of Connecticut; 2152 Hillside Road; Storrs CT 06268; USA; 203-486-2942; 203-486-4915; E; 1.1; 2.2; 3.4; 5.1; Scattering: (e, 2e)

BETTS, Jeanette A.; Chemical Physics; Applied Technology; Group Research Staff; TRW Space and Technology Group; MS R1/1196; 1 Space Park; Redondo Beach CA 90274; USA; 213-536-1453; 213-535-0547; E/T; 2.3; 3.2; 4.1; 5.3; Electronic transition lasers--development of new and different; Excimer lasers --kinetics and modeling; Pulsed HF&DF chemical lasers --kinetics and modeling

BETZ, Albert L.; Space Sciences Laboratory; University of California; Berkeley CA 94720; USA; 415-642-5604; E; 3.6; 5.7; Laser heterodyne spectroscopy for infrared astronomy

BEVELACQUA, Joseph J.; PO Box 166; Hummelstown PA 17036; USA

BEVERLY III, Robert E.; R.E. Beverly III and Associates; 1891 Fishinger Road; Columbus OH 43221; USA; 614-457-1242; E/T; 5.3; 5.4; 3.6; 4.1; Laser development and applications; Lasers --gas-discharge physics and optical pumping; Laser kinetic modeling; Computational physics

BEYER, Louis M.; Department of Physics; Murray State University; Murray KY 42071; USA; 502-762-2993; E; 3.4; 4.2; Proton-induced X-ray fluorescence

BEYERINCK, Herman C.; Physics Department; Plasma, Atomic & Molecular Physics; Eindhoven University of Technology; P.O. Box 513; Insulinde Laan; Eindhoven 5600MH; The Netherlands; 040-474094; 040-472550; E/T; 1.1; 2.1; 3.11; 4.1; 4.3; Excited atoms (t=2ons): intramultiplet mixing, Penning ionization, els.coll; Metastable atoms: ionization, dissociation, excitation transfer with molecule; Optical pumping: Rabi oscillations, adiabatic following; Beam sources: plasma sources, supersonic expansions

BHADRA, Kalidas; Department of Physics and Astronomy; Louisiana State University; Baton Rouge LA 70803; USA; 504-387-6970; T; 1.1; 2.2; Resonance structure determination in e-OVI collisions; Converged collisional results in terms of total differential parameters in e-He process

BHALLA, Chander P.; Department of Physics; Kansas State University; Cardwell Hall; Manhattan KS 66506; USA; 913-532-6786; T; 1.1; 2.1; 2.2; 3.4; Multiply-ionized atoms--de-excitation; Dielectronic recombination studies

BHALLA, Raj P.; Department of Physics; North Texas State University; Denton TX 76203; USA

BHARATHI, Sthanu M.; Department of Physics; Atomic Physics; Indian Institute of Technology; POWAI; Bombay Maharashtra 400076; India; (287) 586730; (287) 581421; E; 2.2

BHASAVANICH, Daun; Plasma and Nuclear Science; Applied Sciences; R&D Center; Westinghouse; 1310 Beulah Road; Pittsburgh PA 15235; USA; 412-256-1024; 412-372-6407; E; 5.4; 5.3; 5.5; 2.1; 4.4; Arc discharges: vacuum arcs, air, SF6 arcs; Circuit interruption; High current -high voltage; Arc plasma studies-plasma decay; Arced gap recovery, contact phenomena, high voltage switching; Surges, EMP, breakdowns, POW pressure noble gases breakdown process

BHASKAR, Natarajan D.; Department of Physics; Princeton University; Princeton NJ 08544; USA; 609-452-4400; E; Photochemistry; High-resolution spectroscopy; Optical pumping

BHATIA, Anand K.; NASA Goddard Space Flight Center; Code 680.1; Greenbelt MD 20771; USA; 301-344-8812; T; 1.1; 2.2; 5.7; Autoionization states--position and widths; Electrons and positrons scattering from atoms; Electron-impact excitation cross sections of ions; Astrophysics--applications

BHATTACHARYA, Ashok K.; Lighting Research & Technical Serv. Op.; Lighting Business Group; Lamp Phenom. Res. Lab.; G.E. Lighting Research Laboratory; Nela Park; Cleveland OH 44112; USA; 216-266-3404; 216-266-8920; E/T; 5.3; 5.4; 3.6; 2.3; 2.1; Atomic and molecular processes in high-pressure metal-halide discharges; Low-pressure discharges in rare-gas mixtures; Diagnostics of discharges

BHATTACHARYA, Samir K.; Department of Physics; Georgia State University; Atlanta GA 30303; USA; 404-658-2932; 404-658-2279; T; 1.2; 3.1; 3.9; 3.4; 5.7

BHAUMIK, Mani L.; Laser Technology Laboratory; Northrop Research and Tech. Center; 1 Research Park; Palo Verdes Peninsula CA 90274; USA; 213-377-4811; E; 3.6; 4.1; 5.3; Spectroscopy; Gas discharges; Superradiance

BICHSEL, Hans; Biomedical Research Division; Los Alamos National Laboratory; 1211 22nd Avenue, E; Seattle WA

98112; USA; 206-329-2792; FTS-843-3128; T; 1.1; 2.1; 2.2; 3.4; 5.8; Energy loss of fast charged particles, straggling; Stopping power data evaluation; Neutron dosimetry; Microdosimetry

BICKEL, William S.; Department of Physics; University of Arizona; Tucson AZ 85721; USA; 602-621-2524/2534; E; 3.2; 3.6; 4.2; 5.1; Polarization spectroscopy; Mean lives; Atomic spectra; Light scattering

BICKNELL, William E.; Group 51; Lincoln Laboratory; MIT; PO Box 73; Lexington MA 02173; USA; 617-862-5500; Ext-5589/7478; E; 3.6; 5.6; Laser physics; Atmospheric transmission in the IR

BIDELMAN, William P.; Warner and Swasey Observatory; Case Western Reserve University; Cleveland OH 44106; USA; 216-368-6699; E; 5.7; Line-identification in astronomical (principally stellar) spectra; Wavelength and energy-level data of atoms and molecules

BIEDENHARN, Lawrence C.; Department of Physics; Duke University; Durham NC 27706; USA; 919-684-8177; T; 1.1; Internal conversion processes in atoms; Angular correlations of such processes; Relativistic effects in atomic radiation

BIEN, Fritz; Aerodyne Research, Inc.; Crosby Drive; Bedford MA 01730; USA; 617-275-9400; E/T; 2.2; 4.1; 5.6; Laser media--development of; Upper atmosphere-electronic excitation; Ion-chemistry

BIENIEK, Ronald J.; Department of Physics; University of Missouri-Rolla; Rolla MO 65401; USA; 314-341-4781; T; 2.1; 3.1; 3.6; 3.8; 5.7; Atomic and molecular collisions involving electronic transitions; Collisionally induced ionization and electronic de-excitation; Spectroscopy of transitional collision complexes; Vibrational-rotational transitions in molecular collisions

BIENSTOCK, Sergio; Bell Telephone Lab.; 600 Mountain Avenue; Murray Hill NJ 07974; USA

BIERBAUM, Veronica M.; Department of Chemistry; University of Colorado; Campus Box 215; Boulder CO 80309; USA; 303-492-7081; 303-492-7396; E; 2.3; 4.4; 3.6; 5.6; Gas-phase ion-molecule reactions studies--the flowing afterglow technique; Rate constants, products, mechanisms; Infrared chemiluminescence, laser-induced fluorescence; Physical organic chemistry

BIGELOW, Nicholas P.; Department of Physics; Cornell University; Clark Hall; Ithaca NY 14853; USA

BIGIO, Irving J.; Group CHM-5; MS-J566; Los Alamos National Laboratory; PO Bx 1663 MS535; Los Alamos NM 87545; USA; 505-667-7748; E; 3.6; 1.3; 4.1; 5.3; Laser physics; Nonlinear optics

BILLARD, Laura A.; 80 St. George Street-Room 522; Toronto Ontario M5S1A1; Canada

BILLINGS, Bruce H.; International Technology Assoc. Inc; 7303 N. Marina Pacifica Drive; Long Beach CA 90803; USA

BINGHAM, Felton W.; 12608 Loyola Avenue, N.E.; Albuquerque NM 87112; USA

BINNS, Walter R.; Radiation Sciences Division; McDonnell Douglas Research Lab.; PO Box 516; St. Louis MO 63166; USA; 314-233-2546; E; 2.3; 4.1; Alkaline earths & possible laser applications--gas kinetics studies

BINSTOCK, Judith; MS F664 X4; Los Alamos National Lab.; PO Box 1663; Los Alamos NM 87545; USA

BIONDI, Manfred A.; Department of Physics and Astronomy; University of Pittsburgh; Pittsburgh PA 15260; USA; 412-624-4354; E; 2.1; 2.2; 2.3; 5.6; 5.7; Atomic collisions at low energy involving electrons, ions, excited; atoms molecules (especially electron-ion recombination, ion-molecule reaction collisions); Upper atmosphere atomic collision studies by ground-based optical technique

BIRELY, John H.; Chemistry, Earth and Life Sciences; Los Alamos National Laboratory; MSA102 Box 1663; Los Alamos NM 87545; USA; 505-667-5893; T; 2.3; 3.8; 3.6; 5.5; 5.6; Chemistry research; Metallurgy; Materials science

BIRNBAUM, George; Office of Nondestructive Evaluation; National Bureau of Standards; Gaithersburg MD 20899; USA; 301-921-3331; 2.1; 3.5; Collision induced absorption and light scattering; Spectral line and band shapes

BIRNBAUM, Milton; Lasers and Optics; Electronics Research Laboratory; The Aerospace Corporation; PO Box 92957; M2/246; Los Angeles CA 90009; USA; 213-648-6839; E/T; 4.1; Rare-earth spectroscopy in crystals and glasses--laser oriented

BISCHEL, William K.; Molecular Physics Department; Chemical Physics Laboratory; SRI International; 333 Ravenswood Avenue; Menlo Park CA 94025; USA; 415-326-5129; E; 3.6; 3.7; 3.10; 4.1; Optical and laser techniques; Stimulated Raman spectroscopy; Multiphoton laser Spectroscopy; Optical and laser techniques

BISSINGER, George A.; Department of Physics; East Carolina University; Greenville NC 27834; USA; 919-757-6320; 919-757-6739; E; 2.3; 3.4; 4.2; Accelerator based atomic collisions; Inner shell X-ray production; Charge transfer; Chemical effects in atomic collisions

BITTER, Manfred L.; Experimental; Plasma Physics Laboratory; Princeton University; 118 Dodds Lane; Princeton NJ 08540; USA; 609-924-9009; 609-683-2582; E; 5.4; Physics of highly ionized ions; X-ray diagnostics of tokamak plasmas

BJORKHOLM, John E.; Laser Science Research Department; Electronics Research Laboratory; AT&T Bell Laboratories; Room 4B-423; Crawfords Corner Road; Holmdel NJ 07733; USA; 201-949-3050; E/T; 3.6; 3.10; 3.7; 4.1; Resonance radiation pressure exerted on atoms--effects of and uses for

BJORKLUND, Gary C.; Department K32/80; IBM Almaden Research Center; 650 Harry Road; San Jose CA 95120; USA; 408-927-2424; E; 3.10; 3.6; 4.1; 3.7; 5.2; Laser absorption spectroscopy-development of novel techniques; Spectral hole burning in solids

BLACK, Graham; Chemical Physics Laboratory; SRI International; 333 Ravenswood Avenue; Menlo Park CA 94025; USA; 415-859-2677; 415-859-3122; E; 2.3; 1.3; 1,2; Electronically excited states of atoms and small molecules--production and properties; Ground-state atoms and radicals--reaction kinetics

BLACK, Truman D.; Department of Physics; University of Texas at Arlington; PO Box 19059; Arlington TX 76019; USA; 817-273-2266; 817-273-2450; E/T; 4.1; 1.3; 5.2; 3.3; Laser probing, acouto-optics, electro-optics; Paramagnetism of atoms and cluster of atoms in solids

BLACKMAN, Mark J.; 1642 Coloma Place; Wheaton IL 60187; USA

BLACKWELL, Richard J.; Department of Applied Physics; Cornell University; Clark Hall; Ithaca NY 14853; USA

BLADES, John D.; 18243 Verano Drive; San Diego CA 92128; USA; 619-451-4198; 619-487-0300; T; 5.2; 5.3; 5.4

BLAHA, Milan; Department of Physics and Astronomy; Lab. for Plasma & Fusion En. Study; University of Maryland; College Park MD 20742; USA; 301-454-7089; 301-454-3516; T; 1.1; 2.2; 5.4; Electron-atom scattering; Atomic transition probabilities and collision cross sections; Spectral line intensities

BLAIS, Normand C.; Chemistry; Los Alamos National Laboratory; PO Box 1663; Los Alamos NM 87545; USA; 505-667-4646; E/T; 2.1; 2.3;

5.5; 4.4; 3.6; Chemical reactions--dynamical features; Energy transfer; Molecular beam scattering; Ionization

BLAIS, Roger N.; The University of Tulsa; 600 South College Avenue; Tulsa OK 74104; USA

BLAKELY, John M.; Cornell University; Bard Hall; Ithaca NY 14850; USA

BLANKENSHIP, Dean M.; Department of Physics; University of Missouri-Rolla; Rolla MO 65401; USA

BLASS, William E.; Department of Physics and Astronomy; University of Tennessee; Knoxville TN 37996; USA; 615-974-3342; 615-691-4052; E/T; 3.1; 3.5; 3.6; 4.1; 5.6; Vibration-rotation resonance interactions; Doppler-limited and sub-doppler molecular spectry; Deconvolution of molecular spectra; Intensities and line shapes in molecular spectra

BLATT, Rainer; I. Inst. Fur Experimental Physik; University of Hamburg; Jungius-strasse 9; D-2000 Hamburg 36; F.R.G.; 040-4123 2406; E; 1.1; 3.10; 4.1; Ion trapping; Laser spectroscopy

BLATTNER, Richard J.; Charles Evans and Associates; 301 Chesapeake Drive; Redwood City CA 94063; USA; 415-369-4567; E; 4.2; 5.1; Ion-beam analysis techniques for materials characterization --sputtering; Problems related to the use of; Ion and electron interactions with solids

BLEWETT, John P.; Consultant; Brookhaven National Laboratory; 310 W. 106 Street; New York NY 10025; USA; 212-866-5767; 516-734-5514; E/T; 4.2; 4.3; 5.4; Electrical engineering

BLINT, Richard J.; Physical Chemistry; GM Research Laboratories; General Motors Technical Center; 12 Mile and Mound Roads; Warren MI 48090; USA; 313-575-7936; T; 5.5; Combustion processes

BLIVEN, Steven M.; Department of Engineering; Industrial Equipment Division; Research and Development; Veeco Instruments; 1611 Headway Circle, Bldg #1; Austin TX 78754; USA; 512-339-6020; 512-482-0496; E; 4.2; 4.3; 4.4; 5.3; State of the art accelerator development

BLOCH, Jeffrey J.; Department of Physics; Space Physics Laboratory; University of Wisconsin; 1150 University Avenue; Madison WI 53706; USA; 608-262-5916; E/T; 5.7; 3.4; 5.3; X-ray astronomical observations of the diffuse background; X-ray instrumentation; Multilayer X-ray optical elements

BLOEMBERGEN, Nicolaas; Department of Applied Sciences; Harvard University; Pierce Hall; Cambridge MA 02138; USA; 617-495-3336; E/T; 3.6; 4.1; Nonlinear spectroscopy; Nonlinear optics

BLOISE, Anthony; Physics Division; Block Engineering, Inc.; 19 Blackstone Street; Cambridge MA 02139; USA; 617-868-6050; E/T; 5.1; Thin films related to integrated optical circuits--surface studies; Near IR applications

BLOOM, Arnold L.; Components Division; Coherent, Inc.; Box 10321; Palo Alto CA 94303; USA; 415-493-2111; 415-858-2250; T; 3.6; 4.1; 5.1; Laser physics; Physical optics; Thin films

BLOOMFIELD, Louis A.; 1208 Knollwood Drive; Middletown NJ 07748; USA

BLUMBERG, Leroy N.; National Synchrotron Light Source; Brookhaven National Laboratory; 25 Brookhaven Avenue; Upton NY 11777; USA; 516-282-4600; 516-928-0466; E; 3.1; 4.2

BLUMBERG, William A.; Solid State Division; Lincoln Laboratory; MIT; PO Box 73; Lexington MA 02173; USA; 617-862-5500x7805; E; 3.3; Atoms and molecules in millimeter and submillimeter region--spectroscopy

BLUME, Martin; Department of Physics; Brookhaven National Lab.; Upton NY 11973; USA

BOBIN, Jean Louis; L.P.O.C.; Universite Paris 6; T12 E5; 4 Place Jussieu; Paris France 75230; France; 336-2525; E/T; 1.3; 2.1; 3.1; 3.6; 3.10; Dense plasmas; Laser interaction with gases and vapors; Atomic collisions

BOGAN, Denis J.; Chemistry Division; U.S. Naval Research Laboratory; Code 6180; 4555 Overlook Avenue, SW; Washington DC 20375; USA; 202-767-2766; E; 2.3; Hydrocarbon oxidation and oxygenated molecules--chemical kinetic; and spectroscopic studies

BOGGS, James E.; Department of Chemistry; University of Texas; Austin TX 78712; USA; 512-471-7525; T; 1.1; Molecular structure--quantum chemistry studies

BOGGY, Richard D.; 56 Hemlock Court; Milpitas CA 95035; USA

BOHLER, Christopher L.; Department of Physics; University of Missouri-Rolla; A27 Stately Mansions; Rolla MO 65401; USA

BOHN, Robert K.; Department of Chemistry; U-60, Rm. 151; University of Connecticut; 215 Glenbrook Rd.; Storrs CT 06268; USA; 203-486-3044; 203-486-2012; E; 1.1; 1.3; 3.3; 3.5; 3.6; Molecular structure and dynamics; Microwave spectroscopy; NMR Spectroscopy

BOKOR, Jeffrey; Bell Telephone Laboratories; Crawfords Corner Road; Holmdel NJ 07733; USA; 201-949-6834; E; 3.10; 3.7; 5.1; Nonlinear laser spectroscopy; Photochemistry

BOLLEN JR., Walter M.; Capitol II Office Building; 5503 Cherokee Avenue-Suite 201; Alexandria VA 22312; USA

BOLLINGER, John J.; Time and Frequency Division; National Bureau of Standards; MS 524-11; 325 Broadway; Boulder CO 80303; USA; 303-497-5861; E; 3.3; 3.6; 3.9; 1.4; Ion storage techniques

BOLORIZADEH, Mohammed A.; 722 G Street #4; Lincoln NE 68508; USA

BOLTON, John M.; Terenure; 63 Fortfield Road; Dublin 6; Ireland

BOLTON, Paul R.; 220 Oak Court; Menlo Park CA 94025; USA

BOMMANNAVAR, Arun S.; Physical Sciences Lab.; 3725 Schneider Drive; Stoughton WI 53589; USA

BONANNO, Regina E.; 225 Lakeside Drive, Apt. T-1; Greenbelt MD 20770; USA

BONCZYK, Paul A.; Energy Research Laboratory; United Technologies Research Center; Silver Lane; East Hartford CT 06108; USA; 203-727-7162; E; 3.6; 5.5; Atomic and molecular laser absorption and induced fluorescence; Spectroscopy with emphasis on high sensitivity detection; Trace quantities in laboratory scale flame media

BONESS, M. J.; Atomic and Molecular Physics Div.; Avco Everett Research Laboratory; 2385 Revere Beach Parkway; Everett MA 01867; USA; 617-389-3000/451; E; 3.6; 4.1; 5.3; Kinetic processes in gas discharges; Metastable atomic beams production; High power gas laser development

BONHAM, Russell A.; Department of Chemistry; Indiana University; Bloomington IN 47405; USA; 812-335-4843; E/T; 2.2; Electron impact spectroscopy

BONIN, Keith D.; Department of Physics; University of Maryland; College Park MD 20742; USA

BORER, William S.; Department of Chemistry; Harvard University; 12 Oxford Street; Cambridge MA 02138; USA; 617-497-1191; 617-495-4102; T; 2.1; 3.1; 5.4; 5.6; 5.7; Quantum calculation of inelastic transitions for large systems; Exact quantum calculation of reactive molecular collisions; Algorithm development for solution of differential equations

BORING, Arthur M.; Los Alamos Scientific Lab.; PO Box 1663; T11 MS 457; Los Alamos NM 87545; USA

BORING, John W.; Departments of Nuclear Engineering & Engineering Physics; University of Virginia; Thornton Hall; Charlottesville VA 22901; USA; 804-924-1059; 804-924-3213; E; 2.1; 5.1; Atomic collisions; Surface interactions

BORING, Michael; Group T-12; Theoretical Division; Los Alamos National Laboratory; PO Box 1663; Los Alamos NM 87545; USA; 505-667-6161; T; 1.1; Many-body effects in atoms; Relativistic electronic structure calculations for atoms and molecules

BORSELLA, Elisabetta; TIB; Applied Physics; Molecular Spectroscopy; ENEA; C.P. 65; Via Enrico Fermi; Frascati Roma 00044; Italy; 06/94005362; E; 1.1; 3.1; 3.6; 3.7; 3.8

BORST, Walter L.; Department of Physics and Astronomy; Southern Illinois University; Carbondale IL 62901; USA; 618-453-5166; E; 2.1; 2.2; Low-energy atomic collisions involving electrons, atoms, and upper atmospheric molecules; Metastable states-- excitation; Excitation transfer between atoms and molecules

BOSE, Subir K.; Department of Physics; Southern Illinois University; Carbondale IL 62901; USA; 618-536-2177; T; 1.3; 2.3; Polymer dynamics

BOSTANTZOGLOU, Stephanos; Filaretou 88/Kallithea; Athens; Greece

BOSWELL, Brian L.; Laboratory for Laser Energetics; University of Rochester; 250 East River Road; Rochester NY 14623; USA

BOTTGER, Gary L.; Research Laboratories; PMRD; Phototheory; Eastman Kodak Research Laboratories; 1999 Lake Avenue; Rochester NY 14650; USA; 716-477-5250; 716-477-4277; E; 1.3; 3.3; 3.5; 3.6; Photographic research; Vibrational spectroscopy; Chemical physics research management; Photophysics

BOTTRELL, Gerald J.; Department of Physics and Astronomy; University of Georgia; Athens GA 30602; USA; 404-542-8787; T; 2.1; 1.1; 5.7; 5.4; Charge transfer in low energy atom-ion collisions; Applications in charge balance in both astrophysical and fusion plasmas

BOUMAN, Thomas D.; Department of Chemistry; Southern Illinois University; Campus Box 1652; Edwardsville IL 62026; USA; 618-692-2042; 618-692-3155; T; 1.1; 3.1; 3.2; 3.3; Circular dichroism spectra of organic molecules; Localized-orbital analysis of computed properties; Electron correlation effects

BOWDEN, Charles M.; Research Directorate; U.S. Army Missile Laboratory; U.S. Army Missile Command; Redstone Arsenal AL 35809; USA; 205-876-3342; T; 3.6; 3.8; 4.1; Superradiance; Superfluorescence; Optical bistability; Laser-induced population dynamics in atoms and molecules

BOWERING, Norbert R.; Department of Physics; University of Texas; Austin TX 78712; USA

BOWERS, Michael T.; Department of Chemistry; University of CA, Santa Barbara; Santa Barbara CA 93106; USA; 805-961-2893; E/T; 1.3; 2.3; 3.6; 3.8; 4.4; Gas-phase ion chemistry; Energy disposal in ionic reactions; Cluster ions, formation, reactivity, photodissociation, structure; Interstellar chemistry, ultra cold reactions

BOWKER, Christopher; 555 E. El Camino Real; Sunnyvale CA 94087; USA

BOWLES, Edward L.; 15 Greylock Road; Wellesley Hills MA 02181; USA

BOWMAN, Charles D.; P-3; Physics Division; Los Alamos National Laboratory; D-449; Los Alamos NM 87545; USA; 505-667-3600; E/T; 4.2; 5.8; 5.8; Atomic and molecular physics in nuclear physics; Atomic and molecular excitation using neutrons

BOWMAN, Joel M.; Department of Chemistry; Illinois Institute of Technology; Chicago IL 60616; USA; 312-567-3229; T; 2.1; 2.3; 3.6; Molecular collision theory--reactive and non-reactive; Light interaction with large molecules

BOWMAN, Richard L.; Department of Physics and Earth Science; Elizabethtown College; Elizabethtown PA 17022; USA; 717-367-1151x320; E/T; 3.2; 3.5; Molecular circular dichroism spectra--calculation and interpretation; Synchrotron radiation used for UV circular dichroism spectra

BOYER, Keith; Office of the Director; Los Alamos National Laboratory; PO Box 1663 MS101; Los Alamos NM 87545; USA; 505-667-4753; E; 3.6; 5.4; Laser physics and technology; Plasma physics; Isotope separation

BOYER, Timothy H.; Department of Physics; City College of New York; New York NY 10031; USA; 212-690-6924; T; 1.1; Van der Waals forces between atoms and molecules --calculations; Zero-point energy in simple atomic models

BRADFORD JR., Robert S.; Chemical Physics Division; TRW, Inc.; MS R1-1184; 1 Space Park; Redondo Beach CA 90278; USA; 213-535-0547; E/T; 4.1; 4.2; Laser developement

BRADLEY III, Lee C.; 201 Somerset Street; Belmont MA 02178; USA

BRAITHWAITE, Wilfred J.; PO Box 49519; Austin TX 78765; USA

BRANDENBERGER, John R.; Department of Physics; Lawrence University; P.O. Box 599; College and Drew; Appleton WI 54912; USA; 414-735-6719; E; 3.6; 4.1; 2.1; 3.9; Quantum beat spectroscopy on noble gases using nitrogen-pumped dye lasers to measure g-values, disalignment cross sections, and lifetimes--excited states

BRANDT, Werner; Department of Physics; New York University; 4 Washington Place; New York NY 10003; USA; 212-598-3652; E/T; 2.2; 4.2; 5.1; Accelerator-based atomic physics; Positron solid state physics

BRANSCOMB, Lewis M.; Office, Chief Scientist; IBM Corporation; Old Orchard Road; Armonk NY 10504; USA; 914-765-6467; E; 3.1; 2.2; 1.1; Structure, spectra of negative ions; no longer active in field

BRAU, Charles D.; Applied Photochemistry Division; Los Alamos National Laboratory; PO Box 1663; Los Alamos NM 87545; USA; 505-667-7102; E/T; 3.6; 4.1; 4.2; Excimer lasers

BRAULT, James W.; National Solar Observatory; 950 N. Cherry Avenue; Tucson AZ 85726; USA; 602-325-9363; E; 3.2; 3.3; 1.1; 3.1; 5.7; High-resolution, high-accuracy spectra observation of atoms and molecules to provide a data base for astrophysical applications and standards

BRAVO, Jorge A.; Bustamante 189 Barranco; Lima 4; Peru

BRAY, Robert G.; Corporate Research Science Lab.; Exxon Research and Engineering Co.; PO Box 45; Linden NJ 07046; USA; 201-474-3319; E; 3.5; 3.8; Laser molecular spectroscopy; Laser photochemistry

BREAUX, Onezime P.; 19 McNay Court; Trotwood OH 45425; USA

BRECHA, Robert J.; 806 Park Boulevard; Austin TX 78751; USA

BRECKENRIDGE, W. H.; Department of Chemistry; University of Utah; Salt Lake City UT 84112; USA; 801-581-8024; E; 2.1; 3.8; Collisional deactivation of electronically excited atoms by simple molecules; Molecular photochemistry

BREIG, Edward L.; Center for Space Sciences; Physics Programs; University of Texas at Dallas; PO Box 688; Richardson TX 75080; USA; 214-690-2851; T; 5.6; Atomic and molecular science in the upper atmosphere and planetary atmospheres

BREINIG, Marianne; Department of Physics; University of Tennessee; Knoxville TN 37916; USA; 615-574-4794; 615-483-1953; E; 2.3; 4.2; Accelerator-related atomic physics investigating charge transfer and electron loss using heavy and highly ionized projectile

BRENN, Rudiger; Fak. f. Physik; Univ. Freiburg; Hermann-Herderstr 3; Freiburg 7800; West Germany; 0761-2-033664; E; 1.1; 2.1; 3.1; 3.4; 3.2

BRENNAN, James G.; Department of Physics; Catholic University of America; Washington DC 20064; USA; 202-635-5318; 301-394-2260; T; 1.1; 2.1; 4.2; 5.1; 5.2; Electronic stopping power theory; Inner-shell atomic ionization

BRENNER, Douglas M.; Chemistry Division; Brookhaven National Laboratory; Upton NY 11373; USA; 516-345-4374; E; 2.3; 3.8; Laser-induced chemical processes: dissociation, energy transfer, and photophysics

BREWER, Laurence R.; MIT; Room 26-237; Cambridge MA 02139; USA

BREWER, Leo; Department of Chemistry; University of California; Berkeley CA 94720; USA; 415-486-5946; E/T; 1.1; 3.2; 3.5; Atomic and molecular spectroscopy; Bonding theory

BREWER, Richard G.; K01-281; Research; IBM Research Laboratory; 5600 Cottle Road; San Jose CA 95193; USA; 408-256-2034; E/T; 3.6; 4.1; 3.10; 3.7; 5.2; Quantum optics; Laser spectroscopy; Coherent optical transients in atoms and molecules

BRICE, David K.; Ion Implantation Physics Div. 1112; Sandia National Laboratory; PO Box 5800; Albuquerque NM 87185; USA; 505-844-4708; T; 5.1; 5.2; Atomic collisions in solids; Ion implantation and radiation damage; Transport theory; Stopping powers

BRICKS, Bernard G.; Applied Sciences Laboratory; General Electric Space Division; PO Box 8555; Philadelphia PA 19101; USA; 215-962-2274; E; 4.1; 5.3; Fast pulse gas discharge phenomena in rare gas-metal vapor mixtures for metal vapor lasers

BRIDGES, William B.; Departments of Engineering & Applied Science; California Institute of Technology; Mail Code 12895; 1201 E. California; Pasadena CA 91125; USA; 818-356-4809; E/T; 3.6; 5.3; 3.1; 4.1; Spectroscopy; Mechanisms of ion lasers; Laser isotope separation in plasmas; Basic laser-gas discharge interactions

BRINK, Gilbert O.; Department of Physics and Astronomy; State University of NY, Buffalo; Amherst NY 14260; USA; 716-636-2005; E; 3.6; 1.1; 2.3; 5.4; 3.2; Dye-laser intracavity absorption

BRISTOW, Thomas C.; Optel Systems; 317 Main Street; E. Rochester NY 14445; USA; 716-385-6760; E; 4.1; 3.6; 5.1; High-Z ions spectroscopy in laser fusion research

BRITT, Edward I.; Rasor Associates Inc.; 253 Humboldt Ct.; Sunnyvale CA 94089; USA; 408-734-1622; 408-255-5178; E/T; 5.4; 3.1; 5.1; 5.5; 5.6; Thermionc energy conversion laser fusion research; Space power systems; Electric insulator performance; High temperature phenomena

BROADHURST, Martin G.; Polymers Division; Center Materials Science & Engineering; National Bureau of Standards; Gaithersburg MD 20899; USA; 301-921-3734; 301-977-8035; E/T; 1.3; 5.2; 1.1; Electrical, piezoelectric, pyroelectric, and ferroelectric properties of polymers--measurements and modeling; Equation of state of polymers, molecular-bulk relationships; Ionic conduction in solids

BRODIE, Ivor; Advanced Technology Division; Physical Electronics Laboratory; SRI International; 333 Ravenswood Avenue; Menlo Park CA 94025; USA; 415-859-4418; 415-859-3872; E/T; 2.2; 4.3; 4.4; 5.1; 5.5; Charge transfer in air as applied to electrophotography; Electron beam lithography; Field ionization sources for mass spectroscopy and ion beam lithography; Cathodoluminesence

BROMLEY, D. A.; Department of Physics; Wright Nuclear Structure Laboratory; Yale University; New Haven CT 06511; USA; 203-436-3026; E; 4.2; 5.1; 5.8; High-energy atomic physics; Atomic electron phenomena to probe nuclear interaction of ion beams with solids and surfaces

BROOKER, Murray H.; Chemistry; Phys-Chem; Memorial University of Newfoundland; St. John's NF A1B3X7; Canada; 709-737-8088; 709-737-8678; E/T; 1.1; 3.3; 3.5; 3.10

BROOKS, Charles L.; Chemistry; Carnegie-Mellon University; 4400 Fifth Avenue; Pittsburgh PA 15213; USA; 412-578-3176; T; 1.3; 2.3; 3.6; 5.2; Interfacing nonequilibrium statistical mechanics and molecular dynamics to study chemistry in biological systems; Computer simulations of condensed phase systems --classical and quantum; Quantum chemical dynamics in liquids and biopolymers

BROOKS, Neil H.; Building 13/456; Fusion Division; GA Technologies, Inc.; PO Box 81608; San Diego CA 92138; USA; 619-455-3979; E; 5.4; Spectroscopic techniques to deduce the impurity content of magnetically confined fusion plasmas

BROOKS, Philip R.; Department of Chemistry; Rice University; PO Box 1892; Houston TX 77251; USA; 713-527-8101x3266; 713-527-4844; E; 2.3; 3.8; 4.3; 2.1; Crossed molecular beams reactions; High laser intensities effects on the reaction process

BROOKS, Robert L.; Physics Division; Argonne National Laboratory; 9700 S. Cass Avenue; Argonne IL 60439; USA; 312-972-8878/4039; E; 3.2; 4.2; 5.1; Beam-foil spectroscopy

BROUDY, Robert M.; Honeywell E-O Division; 157 Robbins Drive; Carlisle MA 01741; USA; E; 3.1; 3.3

BROWER, Michael C.; Department of Physics; Lyman; Harvard University; Cambridge MA 02139; USA; 617-495-3386; E; 2.1

BROWN, Charles M.; E.O. Hulbert Center for Space Research; Naval Research Laboratory; 4555 Overlook Avenue, SW; Washington DC 20375; USA; 202-767-3578; E/T; 1.2; 3.2; 3.6; 4.2; 5.4; VUV spectroscopy; Laser heated plasma spectroscopy; Multichannel quantum defect theory; Solar and atmospheric spectra

BROWN, Ellen R.; Analysis and Evaluation; Strategic Warfare Directorate; Washington Analytical Services Co.; EG&G Inc.; 552; Dahlgren VA 22448; USA; 703-663-9492; T; 3.1; Photoabsorption calculations for open-shell atoms

BROWN, George S.; PO Box 4349; SSRP Bin 69; Stanford CA 94305; USA

BROWN JR., Howard H.; Department of Physics; New York University; 4 Washington Place; New York NY 10003; USA; 212-598-2739; E; 2.2; 2.3; 5.4; Electron-atom and atom-atom scattering; Plasma physics

BROWN, James C.; 4220 Oakland Avenue S.; Minneapolis MN 55407; USA

BROWN, Kenneth W.; The Aerospace Corporation; PO Box 92957; M1/109; Los Angeles CA 90009; USA

BROWN, Lorin W.; Deputy Director Research Programs; Office of Naval Research; 800 N. Quincy; Arlington VA 22217; USA; 202-696-4102; E; 3.6; 3.10; 5.6

BROWN, Lowell S.; Department of Physics, FM-15; University of Washington; Seattle WA 98195; USA; 206-543-8774; T; 1.4; 3.7; Physics of single stored ion or electron

BROWN, Matt D.; R-41; White Oak Laboratory; Naval Surface Weapons Center; Silver Spring MD 20910; USA; 202-394-2272; E; 3.4; 4.2; Inner-shell ionization by MeV light ions

BROWN, Nancy J.; Applied Science Division; Lawrence Berkeley Laboratory; University of California; Berkeley CA 94720; USA; 415-486-4241; FTS-451-4241; E/T; 1.3; 2,5; 6,8; 5,6; Theoretical and experimental chemical kinetics; Combustion chemistry

BROWN, Robert T.; WP/WJO; Los Alamos National Laboratory; PO Box 1663 MS634; Los Alamos NM 87545; USA; 505-667-2903; T; 5.4; Low-Z opacities

BROWNE, James C.; Department of Computer Science; University of Texas; 5.70-C Painter Hall; Austin TX 78712; USA; 512-471-5023; T; 1.1; 2.1; Scattering processes and structure--atomic and molecular

BROWNSTEIN, Kenneth R.; University of Maine; 123 Bennett Hall; Orono ME 04469; USA

BRUBAKER, Wilson M.; 1954 Highland Oak Drive; Arcadia CA 91006; USA; E; 4.4; 5.4; Private consultant; Isotope separation instrumentation (mass spectrometry)

BRUCE, Michael R.; Department of Physics; University of Texas, Austin; Robert Lee Moore Hall; Austin TX 78712; USA; 512-471-4151; E; 1; 1; 6; 1; 3; Multiphoton spectroscopy of rare gas halide excimers; Kinetics and laser spectroscopy of two photon excited xenon; Energy transfer in atomic molecular interactions; Kinetics of excited xenon and chloride following selective state laser excitation

BRUCH, Ludwig W.; Department of Physics; University of Wisconsin; 1150 University Avenue; Madison WI 53706; USA; 608-262-8968; T; 1.1; 5.1; Inert gas atoms interactions, free and in adsorption

BRUCH, Reinhard F.; Department of Physics; University of Nevada, Reno; Reno NV 89557; USA; 702-784-4920; E; 1,2; 1; 1,6; 1,2; 4,7; Auger effect autoionization; Charge exchange; Electron correlation; High resolution electron and EUV spectroscopy

BRUNNER, Timothy A.; Department of Physics; Massachusetts Institute of Technology; Room 26-251; Cambridge MA 02139; USA; 617-253-4169; E/T; 2.1; Rotationally inelastic collisions for diatomic sodium molecules with rare gas atoms--cross sections

BRYANT, Howard C.; Department of Physics and Astronomy; University of New Mexico; Albuquerque NM 87131; USA; 505-277-3044/2616; E; 3.1; 4.2; 3.2; 3.9; Negative hydrogen ion photodetachment

BRYNJOLFSSON, Ari; Radiation Laboratory; U.S. Army Natick R&D Command; 6 Kansas Street; Natick MA 01760; USA; 617-653-1000x2754; E/T; 4.2; 5.1; 5.2; Radiation effects on biological matter; Charged particles (electrons) and gamma rays--stopping of; Primary and secondary interactions--investigations

BU-ABBUD, George H.; W194 Nebraska Hall; Lincoln NE 68588; USA

BUCKSBAUM, Philip H.; Physics Division; 1D-466; AT&T Bell Laboratories; 600 Mountain Avenue; Murray Hill NJ 07974; USA; 201-582-3793; E; 1.4; 3.7; 3.10; 4.1; 5.1

BUDICK, Burton; Department of Physics; New York University; 4 Washington Place; New York NY 10003; USA; 212-598-2143; 212-598-7509; E; 3.4; 4.2; 5.8; 1.3; L subshells in intermediate Z elements--fluorescence yields

BUKOW, Hans; Experimentalphysik(III); Ruhr Universitat; 10 21 48; Bochum 4630; Germany (RFA); 0234 700-3575; E; 3.2; 3.1; 4.2; 5.1; 5.7; Beam foil spectroscopy; Ion-solid interaction

BULLOUGH, Robert K.; Department of Math.; Univ. Inst. of Science and Technology; Manchester M601QD; England

BUNCE, Daniel R.; Northrop Precision Production Div.; 100 Morse Street; Norwood MA 02062; USA

BURCH, David S.; Department of Physics; Oregon State University; Corvallis OR 97331; USA; 503-754-4631; E/T; 2.2; 5.3; Swarm analysis; Collisions; Optics

BURDGE, Geoffrey L.; Department of Defense; DIRNSA; 9800 Savage Road; Fort Meade MD 20755; USA; 301-796-6555; E/T; 4.1; Multiphoton mixing in alkali vapor cells

BURGDOERFER, Joachim E; Department of Physics; University of Tennessee; Knoxville TN 37996; USA; 615-974-7867; 615-576-8341; T; 2.1; 1.2; 5.2

BURKE, E. A.; 11 Indian Hill Road; Woburn MA 01801; USA

BURKE, Edward A.; 31 Belmont Parkway; Hempstead NY 11550; USA

BURKE, Philip G.; Department of Applied Mathematics; Queen's University of Belfast; Belfast BT71NN; N. Ireland; 0232 245133; T; 2.2; 3.1; 1.1; 5.3; 5.4

BURKHALTER, Philip G.; X-Ray Optics Branch; Code 6681; U.S. Naval Research Laboratory; 4555 Overlook Avenue, SW; Washington DC 20375; USA; 202-767-2993; E; 4.2; 5.4; Highly-ionized atoms from plasmas generated by pulsed-discharge devices such as exploded wire or gas-puff machines--spectroscopy

BURKHART, Craig W.; Department of Macromolecular Science; Olin Lab. for Materials; Case Western University; Cleveland OH 44106; USA

BURNHAM, Ralph; Code 6540; U.S. Naval Research Lab.; 4555 Overlook Avenue, SW; Washington DC 20375; 202-767-2175; E; 2.3; 3.6; 4.1; Laser physics; Chemical kinetics; Atomic and molecular processes such as radiation lifetimes, collisional kinetics, and laser development spectroscopy

BURNS, Donal J.; Department of Physics and Astronomy; University of Nebraska; Lincoln NE 68588; USA; 402-472-2773; E; 2.2; 4.2; 5.1; Electron scattering using coincidence techniques; Correlations in electron atom collisions

BURNS III, Jay; Department of Physics and Space Sciences; Florida Institute of Technology; 150 W. University Boulevard; Melbourne FL 32901; USA; 305-768-8098; E; 3.1; 2.2; 5.1; 3.2; 3.3

BURR, Alex F.; Department of Physics; New Mexico State University; Las Cruces NM 88003; USA; 505-646-3806; 505-646-3831; E/T; 1.1; 3.4; X-ray physics; Atomic energy level studies

BURRIS JR., John; Laboratory for Atmospheres; NASA/GSFC; Code 615; Greenbelt MD 20771; USA; E; 5.6; 4.1; 3.10

BURROW, Paul D.; Department of Physics and Astronomy; University of Nebraska; Lincoln NE 68588; USA; 402-472-2419; E; 2.2; 5.3; Resonances in low energy electron scattering

BUSHMAN, Gary; Engineering-Materials Division; Proteng; PO Box 5086; Mission Hills CA 91345; USA; E; 2.3; 5.1; Spectroscopy; Plasma-treating of substrates to desired chemical reaction mechanisms and resultant improved physical properties; Ion implants as a surface treatment

BUTERA, Robert J.; 5845 Marnell Avenue, Apt. 4; Mayfield Heights OH 44124; USA

BUTLER, J. K.; Department of Electrical Engineering; Southern Methodist University; Dallas TX 75275; USA; 214-692-3573; E/T; 4.1; Injection lasers; Optical waveguides

BUTLER, James E.; Chemical Division; U.S. Naval Research Lab.; 4555 Overlook Avenue, SW; Washington DC 20375; USA; 202-767-1115; E; 2.3; 3.6; 3.8; Reaction kinetics --laser studies; Laser chemistry; Radicals--laser detection

BUTLER, Scott E.; Atomic and Molecular Phys. Div.; Harvard-Smithsonian Center for Astrophys; 60 Garden Street; Cambridge MA 02138; USA; 617-495-4868; T; 1.1; 2.1; 2.3; 5.7; Low-energy charge transfer processes; Ab initio quantal calculations; Scattering theory; Astrophysical applications

BUTTRILL, Jr., Sidney E.; Charles Evans and Associates; 301 Chesapeake Drive; Redwood City CA 94063; USA; 415-369-4567; E; 4.4

BUTTS JR., Jesse J.; 19 Seaview Drive N.; Rolling Hills CA 90274; USA

BYER, Robert L.; Hansen Physics Laboratory; Stanford University; Stanford CA 94305; USA; 415-497-4028; E; 4.1; Nonlinear optics; Optical pumping spectroscopy

BYRON JR., Frederick W.; Department of Physics and Astronomy; Graduate Research Tower C; University of Massachusetts; Amherst MA 01003; USA; 413-545-2627; T; 2.2; Electron and positron-atom scattering; (elastic scattering, excitation, ionization, rearrangement)

BYSZEWSKI, Wojciech W.; GTE Lab., Inc.; 40 Sylvan Road; Waltham MA 02154; USA

CABLE, Peter G.; New London Lab.; Naval Underwater System Center; New London CT 06320; USA

CACAK, Robert K.; Department of Radiation Oncology; St. Paul Medical Center; 5909 Harry Hines Boulevard; Dallas TX 75235; USA; 214-879-2696; E; 3.1; 2.1; Interactions of high energy X-rays with matter

CADE, Paul E.; Department of Chemistry; University of Massachusetts; Amherst MA 01003; USA; 413-545-1350; T; 1.1; 1.3; 2.3; 5.4; Molecules--electronic structure; Electronic charge distribution--interpretation; Compton profiles; Positronium chemistry

CAI, Jialing; Department of Physics; New York University; Box 18; 4 Washington Place; New York NY 10003; USA

CAIRD, John A.; Process Technology Division; Bechtel National, Inc.; 50 Beale Street; San Francisco CA 94105; USA; 415-768-2730; T; 3.2; 3.8; High-temperature vapors--spectroscopic studies; Laser systems design and analysis for fusion, isotope separation and laser photochemistry

CALAPRICE, Frank P.; 201 Mandon Court; Princeton NJ 08540; USA

CALBICK, Chester J.; Department of Physics; Washington State University; Pullman WA 99163; USA

CALDWELL, C. Denise; Department of Physics; Sloane Laboratory; Yale University; New Haven CT 06520; USA

CALDWELL, Wallace C.; PO Box 28; Ankeny IA 50021; USA

CALEDONIA, George E.; Physical Sciences, Inc.; P.O. Box 3100; Research Park; Andover MA 01810; USA; 617-475-0930; E/T; 2.3; 4.1; 3.3; 5.3; 5.6; Examination and modeling of fluorescence from excited molecules; Modeling of gas laser systems, i.e., excitation, kinetics, etc.

CALLAN, Edwin J.; LOM, Ltd.; 4139 Windcross Lane; Orlando FL 32809; USA; 305-425-5257; T; 1.1; Atomic energy levels and transition rates; Nonlinear equation studies

CALLAWAY, David J.; Physics Department; Rockefeller University; 1230 York Avenue; New York NY 10021; USA; T; 5; Computational physics

CALLAWAY, Joseph; Department of Physics and Astronomy; Louisiana State University; Baton Rouge LA 70803; USA; 504-388-8400; T; 2.2; 1.3; Electron-atom and electron-ion scattering; (elastic, excitation, and ionization); Electronic structure of clusters

CAMPARA, James C.; 2501 Ruhland Avenue #D; Redondo Beach CA 90278; USA

CAMPBELL, David H.; AEDC Division; ARO, Inc.; Arnold AF Station TN 37389; USA; 615-455-2611x7200; E/T; 2.2; 3.6; Non-intrusive diagnostic techniques for application to aerospace testing--laser, electron beam

CAMPBELL, John L.; Department of Physics; University of Guelph; Guelph Ontario N1G2W1; Canada; 519-824-4120x2263; 519-824-4120x8397; E; 3.4; 5.8; 2.2; 4.2; PIXE: elemental analysis by proton-induced X-ray emission; EDX: energy-dispersive X-ray spectroscopy with Si(Li) and Ge detectors

CAMPION, Alan; Department of Chemistry; University of Texas; Austin TX 78712; USA; 512-471-3012; E; 3.6; 5.1; Surface chemistry; Laser spectroscopy

CANNON, Peter; Rockwell International; PO Box 1085; Thousand Oaks CA 91360; USA

CANTER, Karl; Department of Physics; Brandeis University; Waltham MA 02254; USA; 617-647-2886; E; 1.3; 2.2; 5.1; Positronium interactions with surfaces; Positron-helium collisions

CANTRELL, Cyrus D.; Group T-12; Los Alamos National Lab.; PO Box 1663 MS531; Los Alamos NM 87545; USA; 505-667-2897; E/T; 3.5; 3.7; 3.8; 4.1; Laser separation of isotopes; Laser induced chemical reactions; Molecular rotation-vibration spectroscopy; Multiphoton laser excitation of molecules and quantum optics

CANTRIL, Jerry M.; 6906 Storch Circle; Seabrook MD 20706; USA

CAPPELLER, Ulrich; Physics Institute University; Rentof 5; Marburg 355 0706; West Germany

CARBONE, Robert J.; NSP/IF; Los Alamos National Laboratory; PO Box 1663 MS530; Los Alamos NM 87545; USA; 505-667-6400; E; 4.1; Laser research and development

CARDINO, Mark J.; Surface Physics Division; Bell Telephone Lab.; 600 Mountain Avenue; Murray Hill NJ 07974; USA; 201-582-2418; E; 4.3; 5.1; Molecular beam-surface dynamics

CARDON, Bartley L.; KB-226; MIT Lincoln Laboratory; P.O. Box 73; Lexington MA 02173; USA; 617-863-5500 x2994; E/T; 1.1; 2.3; 3.1; 4.1; 5.6; Atomic oscillator strengths determination; Photoionization cross sections

CARLETON, Nathaniel P.; Harvard-Smithsonian Center for Astron.; 60 Garden Street; Cambridge MA 02138; USA; 617-495-7405; E; 5.7; 5.6

CARLISLE, Clinton B.; Department of Physics; University of Virginia; McCormick Road; Charlottesville VA 22901; USA

CARLSON, Edward H.; Department of Physics; Michigan State University; East Lansing MI 48823; USA

CARLSON, Lee R.; Lawrence Livermore National Lab.; PO Box 808 L467; Livermore CA 94550; USA; 415-422-3649; E; 3.1; 3.8; Laser photoionization, physics of atoms related to laser isotope separation processes

CARLSON, Nils W.; Department of Physics; Stanford University; Stanford CA 94305; USA; 4415-497-4357; E; 3.5; 3.6; Laser spectroscopy of excited states of diatomic molecules

CARLSON, Thomas A.; Chemistry Division; Oak Ridge National Lab.; Department of Energy; PO Box X; Oak Ridge TN 37831; USA; 615-574-4802; E; 3.1; 4.2; Angle-resolved photoelectron spectroscopy on atoms and molecules using synchrotron radiation

CARLSTEN, John L.; Physics Department; Montana State University; Bozeman MT 59717; USA; 406-994-6176; E; 3.10; 3.11; 3.7; Light scattering in gases with lasers

CARLSTON, Carl E.; 2466 W. Davis Avenue; Littleton OH 80120; USA

CARLTON, Terry S.; Department of Chemistry; Oberlin College; Oberlin OH 44074; USA; 216-775-8300; T; 1.1; Electron density in atoms; Repulsion between atoms

CARMICHAEL, Howard I.; Department of Physics; University of Arkansas; Fayetteville AK 72701; USA; 501-575-5929; 501-575-2506; T; 3.11; 3.10; 3.6; 1.4; Quantum statistics; Master equation methods, phase-space representations; Photon statistics: nonclassical effects, photon antibunching, squeezing; Nonlinear dynamics: bifurcation, instability and chaos in optical systems

CARMICHAEL, Ian; Radiation Laboratory; University of Notre Dame; Notre Dame IN 46556; USA; 219-283-4453; T; 1.1; 2.3; Ab initio calculations of spin densities and polarizabilities of free radicals and other photochemical and radiation chemical--transients

CARNAHAN, Byron; Department of Physics and Astronomy; University of Pittsburgh; Pittsburgh PA 15260; USA; E; 2.2; Electron impact induced dissociative excitation studies of various molecular gases resulting in the production of metastable fragments

CARO, Richard G.; Ginzton Lab.; Stanford CA 94305; USA

CARR, Herman Y.; Department of Physics; Rutgers University; PO Box 849; Piscataway NJ 08854; USA; 201-932-2538/2547; E; 1.1; Local magnetic field shifts from intramolecular and intermolecular; interactions--NMR studies

CARR, Percy H.; 424 Stanton Avenue; Ames IA 50010; USA

CARR JR., Robert W.; Department of Chemical Engr. and Materials Science; University of Minnesota; 151 Amundson Hall; Minneapolis MN 55455; USA; 612-376-4593; E; 2.3; 3.8; Excited state kinetics; Small molecule photochemistry and photophysics; Energy transfer; Free radical reactions--kinetics

CARRAGHER, Beverly A.; 11446 Lockwood Drive-Apt. 104; Silver Spring MD 20904; USA; 301-622-9401; T; 2.2

CARROLL, Clark E.; Department of Physics; St. John Fisher College; Rochester NY 14618; USA; 716-385-8192; T; 3.7; 3.10; Multiphoton effects; Nonlinear optics

CARROLL, James B.; 25 Welwyn Way; Rockville MD 20850; USA

CARTIER, Joan F.; 6612 Lexington Road; Austin TX 78731; USA

CARTWRIGHT, David C.; Fusion Research Applications; Defense Research Programs; Los Alamos National Lab.; Box 1663 MSE527; Los Alamos NM 87545; USA; 505-667-2097; T; 2.2; 3.1; 5.3; 5.6; 5.7; Electron scattering by atoms and molecules; Spectral properties of atoms and molecules; Applications of A&M physics to planetary atmospheres and astrophysics

CASHMAN, Robert J.; 830 Indian Road; Glenview IL 60025; USA

CASON, Charles M.; U.S. Army Missile Command; DRSMI-RHS; Building 8978; Redstone Arsenal AL 35809; USA; 205-876-8271; T; 3.6; 4.1; Electric lasers; Optics; Laser/matter interactions

CASTANO, Victor M.; Inst De Fisica Unam; AP 20-364; Deleg Alvaro Obregon; Mexico DF 01000; Mexico

CASTLEMAN JR., A. W.; Department of Chemistry; Pennsylvania State University; 152 Davey Laboratory; University Park PA; USA; 814-863-2549; E; 1.3; 2.3; 3.8; Ion and van der Waals clusters studies; Photochemistry/physics of clusters; Nucleation and surfaces; Solvation phenomena, Energy transfer; Kinetics of association reactions

CASTRO, George; K31/80D; Almaden Research Center; IBM Research Laboratory; 650 Harry Road; San Jose CA 95120; USA; 408-927-2400; E; 3.6; Research management

CATHEY, LeConte; Department of Physics; University of South Carolina; Columbia SC 29208; USA; 803-777-8104; E; 3.1; 3.6; Radiation and the interactions of radiation with matter measurements

CATTOLICA, Robert J.; Combustion Sciences Department; Combustion Physics Division; Sandia National Laboratory; Livermore CA 94550; USA; 415-422-2451; E; 5.5; 3.6; 4.1; 2.1; 1.1; Atomic and molecular science applications to the study of chemical kinetics; Laser-induced fluoresence spectroscopy for the measurement on species; Laser spectroscopic methods for simultaneous multidimensional observations; Atomic and molecular collision effects/interpretation of optical spectra

CATZ, Leonard A.; Department of Physics; University of Massachusetts, Boston; Harbor Campus; Boston MA 02125; USA; 617-287-1900x2357/2351; E; 3.4; Inner atomic transitions--spectroscopy; Angular correlations between characteristic atomic X-rays

CAUDANO, Roland; Department of Physics; FAC Universitaires NDP; 61 Rue Bruxelles; Namur B-5000; Belgium

CAVES, Thomas C.; Department of Chemistry; North Carolina State University; Raleigh NC 27650; USA; 919-737-2996; T; 1.1; 2.3; 5.2; Accurate effective potentials for valence electrons only; Chemical exchange from line-shapes of complex NMR spectra --extracting; rates and mechanisms

CECIL, Joseph N.; Engineering Department; Industrial Equipment Division; R&D Laboratory; VEECO Instruments; 1611 Headway Circle, Building #1; Austin TX 78760; USA; 512-339-6020; 512-442-4158; E; 4.2; 4.3; 4.4; 5.3; State-of-the-art accelerators development

CELOTTA, Robert J.; Radiation Physics Division; Electron Physics Group; National Bureau of Standards; Building 220 Rm B206; Gaithersburg MD 20899; USA; 301-921-2051; E; 2.2; 3.6; 4.1; 5.1.; Electron scattering from atoms and surfaces; Electron polarization; Electron, tunneling, and magnetic microscopy; Atoms on clean surfaces of metals

CENTER, R. E.; Mathematical Sciences NW, Inc.; 2755 Northup Way; Bellevue WA 98004; USA; 206-827-0460; E; 2.3; 3.6; 4.2; Kinetics relating to electrically excited visible lasers

CEPERLEY, David M.; Department of Physics; H Division; Lawrence Livermore National Lab.; University of California; 808; L 297; Livermore CA 94550; USA; 415-423-2825; T; 1.1; Intermolecular potentials calculations using Monte Carlo methods

CERCEO, J. Michael; I. S. C.; 8478B Tyco Road; Vienna VA 22180; USA

CERTAIN, Phillip R.; Theoretical Chemistry Institute; University of Wisconsin; 1101 University Avenue; Madison WI 53706; USA; 608-262-1511; T; 1.1; Theoretical chemistry; Quantum chemistry; Intermolecular forces; Electronic structure of molecules

CH'EN, Shang-Yi; Department of Physics; University of Oregon; Eugene OR 97403; USA; 503-686-4764; E; 3.8; 5.4; Spectral lines pressure broadening

CHACKERIAN JR., Charles; Astrophysical Experiments Branch; NASA Ames Research Center; Moffett Field CA 94035; USA; 415-694-6300; 415-694-5528; E/T; 3.5; 1.1; 2.1; 3.3; 5.6; Molecular spectroscopy; Intensities, shapes, positions; Molecular collisions; Planetary atmospheres

CHAKRABORTI, Parimal K.; Multidisciplinary Research Section; Modular Labs.; Bhabha Atomic Research Center; Bombay Maharastra 400085; India; 551-4910 x2685; 551-2114; E; 1.3; 2.1; 3.1; 3.6; 3.8; Nozzle beam characterization, condensation (E/T); Multiphoton dissociation and ionization (E)

CHAMBERLAIN, G. E.; Division 724.01; National Bureau of Standards; Boulder CO 80303; USA

CHAMBERLAIN, Joseph W.; Department of Space Physics and Astronomy; Rice University; PO Box 1892; Houston TX 77001; USA; 713-527-8101 x3641; T; 5.7; Basic atomic and molecular data applied to astrophysics

CHAMBERLIN, Edwin P.; Molecular Physics Division; Group MP-11; Los Alamos National Lab.; PO Box 1663 MS823; Los Alamos NM 87545; USA; 505-667-4593; E; 1.4; 2.1; 4.2; 4.3; Lamb-shift polarized ion sources

CHAMPAGNE, Louis F.; Laser Physics Branch; Optical Sciences Division; U.S. Naval Research Laboratory; Code 6540; 4555 Overlook Avenue, SW; Washington DC 20375; USA; 202-767-2512 x2507/2813; E/T; 5.3; 5.4; 4.1; 3.1; 2.2; Rate of the atomic and molecular species as absorbers of light in plasmas

CHAMPION, Kenneth S. W.; Atmospheric Sciences Division; Air Force Geophysics Laboratory; Hanscom AFB MA 01731; USA; 617-861-1033; E/T; 5.6; Research management

CHAMPION, R. L.; Department of Physics; College of William and Mary; Williamsburg VA 23185; USA; 804-253-4471; E; 2.1; Ions, atoms, and molecules--low-energy collisions (1-200 eV)

CHAN, Chia H.; Physics Division; University of Alabama; PO Box 1247; Huntsville AL 35807; USA

CHAN, F. T.; Department of Physics; University of Arkansas; Fayetteville AR 72701; USA; 501-575-2506; T; 2.1; 2.3; Electron-atom collisions; Charge-capture collisions

CHAN, I. Y.; Department of Chemistry; Brandeis University; Waltham MA 02254; USA; 617-647-2816; E; 3.3; 3.6; 5.2; Optically detected magnetic resonance, EPR, and ENDOR; Laser spectroscopy in solids and in supersonic jets

CHAN, Yau Wa; Science Center Physics; Chinese University of Hong Kong; Shatin New Terr.; Hong Kong

CHANCEY, Charles C.; Theoretical Physics; University of Oxford; 1 Keble Road; Oxford OX13NP; England; (0865) 53281; T; 1.1; 5.2; Jahn-Teller effect; Electron-phonon interactions

CHAND, Prakash; 13311 Broadmeade; Round Rock TX 78664; USA

CHANDER, Jagdish; Department of Physics and Astronomy; University of Wisconsin; Stevens Point WI 54481; USA; 715-346-3945; 715-341-1538; E; 1.1; 4.3; 2.1

CHANDLER, David; Department of Chemistry; University of Pennsylvania; Philadelphia PA 19104; USA; 215-898-3125; T; 2.3; Theoretical chemistry; Statistical mechanics

CHANEY, Robert; Surface Science Laboratories; 4151 Middlefield Road; Palo Alto CA 94303; USA; 415-493-0229; E; 3.6; 4.3; 4.4; 5.1; Surface science; Ion-gun development, laser physics; Electron and mass spectrometry

CHANG, Bo H.; Chung Ang University; Donjak Ku; 221 Heuk Suk Dong; Seoul 151 4303; Korea

CHANG, C. K.; Department of Chemistry; Christopher Newport College; Shoe Lane; Newport News VA 23606; USA; 804-599-7210; T; 3.3; 3.5; Microwave spectroscopy; Molecules--internal rotation

CHANG, Cheng-hui; Department of Physics; University of Arkansas; Fayetteville AR 72701; USA; 501-575-2506; E/T; 2.1; 3.6; Atomic collisions; Laser research

CHANG, David; 12118 Merewood Lane; Houston TX 77071; USA

CHANG, Edward S.; Department of Physics and Astronomy; Hasbrouck Laboratory; University of Massachusetts; Amherst MA 01003; USA; 413-545-0969; T; 2.2; 3.1; 1.2; 1.1; 3.6; Low energy electron-molecule scattering; Atoms and molecules--photoionization; Molecular spectroscopy; Rydberg states of high angular momentum

CHANG, Sin-Tarng; Department of Physics and Astronomy; University of Oklahoma; Norman OK 73019; USA

CHANG, Tai Y.; Department of Chemistry; Ford Motor Company Scientific Lab.; PO Box 2053; Room E31983; Dearborn MI 48121; USA; 313-322-7105; T; 3.8; 5.6; Chemical reactions in atmospheric photochemistry

CHANG, Tu-nan; Department of Physics; University of Southern California; University Park; Los Angeles CA 90007; USA; 213-743-6707; T/E; 1.1; 2.2; 3.1; Photon-atom and electron-atom interactions; Atomic structure; Many-body problems; Photon-atom interactions in metal vapors

CHANG, William S. C.; Electrical Engineering and Computer Science; C-014; University of CA, San Diego; La Jolla CA 92093; USA; 619-452-2737; E; 4.1; Opto-electronic devices

CHANIN, Lorne M.; Department of Electrical Engineering; University of Minnesota; 123 Church Street SE; Minneapolis MN 55455; USA; 612-373-5525; E; 5.3; 5.4; Glow discharges, flowing afterglows; Plasma chemistry; Discharges--mass and radiation analysis

CHANTRY, Peter J.; Nucleonic, Gas Physics & High-Voltage Tech. Division; Westinghouse R&D Center; 1310 Beulah Road; Pittsburgh PA 15235; USA; 412-256-3675; E/T; 2.2; 5.3; Electron-molecule interactions relevant to electrical insulation, gas discharges and arcs

CHAPMAN, Sally; Department of Chemistry; Barnard College; 606 W. 120th Street; New York NY 10027; USA; 212-280-3377/2098; T; 2.1; 2.3; Molecular reaction dynamics-classical modeling of the collision dynamics between small molecules in the gas phase to understand the relationship between the interaction and the dynamics

CHAPPELL, Richard F.; Computer Science Department; University of South Carolina; 1561 Brockwall Drive; Columbia SC 29206; USA; 803-782-4392; 803-777-2628; T; 1.1; Computer modeling of ESR hyperfine interactions

CHARATIS, George; Fusion Experiments; Fusion Experiments Division; KMS Fusion, Inc.; 3621 S. State Road; Ann Arbor MI 48106; USA; 313-769-8500x343; E; 2.3; 3.1; 3.4; 3.6; 5.1; Laser fusion: Laser-atomic, electron-atomic and atomic-atomic interactions; X-ray spectroscopy

CHARBONNIER, Francis M.; McMinnville Division; Hewlett-Packard Co.; 1700 South Baker Street; McMinnville OR 97218; USA

CHASE, Walter E.; 239 Auburndale Avenue; Auburndale MA 02166; USA

CHATURVEDI, Ram P.; Department of Physics; State University of NYC, Cortland; P.O. Box 2000; Cortland NY 13045; USA; 607-753-2011x2821; E; 2.1

CHEN, Albert S.; PO Box 9800; Berkeley CA 94709; USA

CHEN, Augustine C.; Department of Physics; St. John's University; Jamaica NY 11439; USA; 212-969-8000; T; 1.1; 3.9; Theoretical hydrogen atom studies--using group theoretical methods; Stark effect

CHEN, C. H.; Health & Safety Research Division; Photophysics; Oak Ridge National Laboratory; PO Box X; ORNL; Oak Ridge TN 37830; USA; 615-574-5895; E; 1.2; 1.3; 1.6; 1.4; 3.6; Laser spectroscopy; Molecular beams; Chemical kinetics; Accelerator physics

CHEN, C. L.; Gas Physics Division; Westinghouse R&D Center; 1310 Beulah Road; Pittsburgh PA 15235; USA; 412-256-7739; E; 2.2; 4.3; 5.1; Electron attachment cross sections; Photon--enhanced dissociative electron attachment; Radiation detectors

CHEN, Che-Jen; Jet Propulsion Laboratory; California Institute of Technology; 4800 Oak Grove Drive; Pasadena CA 91103; USA; 213-354-4857; E; 2.1; 4.1; 5.3; Collision phenomena in gas laser media

CHEN, Chin-Lin; School of Electrical Engineering; Purdue University; West Lafayette IN 47907; USA; 317-494-3525; E/T; 3.6; 4.1; 1.1; Physical optics

CHEN, Hao-lin; Laser Isotope Separation Project; Lawrence Livermore National Lab.; PO Box 808; Livermore CA 94550; USA; 415-422-6198; E; 2.1; 2.2; 3.1; 3.8; Photoionization physics; Atomic and molecular energy transfer; Laser-induced chemistry; Electron impact --physics

CHEN, Kuo-In; Room 36-213; Plasma Fusion Center; Massachusetts Institute of Technology; Cambridge MA 02139; USA; 617-253-8141; 617-253-2521; E; 5.4; 3.2

CHEN, Mang; Department of Physics; University of Pittsburgh; 100 Allen Hall; Pittsburgh PA 15260; USA

CHEN, Mau-Hsiung; Department of Physics; University of Oregon; Eugene OR 97403; USA; 503-343-5313; T; 3.4; Atomic inner-shell processes; Auger and X-ray emission rates; Multiply-ionized atoms--energy and transition rates

CHEN, Tian-Jie; Department of Physics; Beijing University; Beijing; Peoples Rep Of China

CHEN, Ying C.; GTE Lab.; 40 Sylvan Road; Waltham MA 02254; USA

CHEN, Yuan-Liang; Department of Physics; Baylor University; Waco TX 76798; USA

CHENG, Julian; 67 Candace Lane; Chatham NJ 07928; USA

CHENG, Kwok-Tsang; A-Division; Lawrence Livermore National Lab.; PO Box 808; Livermore CA 60439; USA; 415-423-8659; T; 1.1; 3.1; Relativistic atomic structure theory; Interaction of radiation with atoms

CHENG, Stephen S.; 5909 Post Road; Bronx NY 10471; USA

CHENG, Vianney K.; Institute of Nuclear Energy Research; Lung Tan; Taiwan 325 0471; Republic of China

CHERRINGTON, Blake E.; Department of Electrical Engineering; University of Florida; Gainesville FL 32611; USA; 904-392-0913; T; 5.1; 5.3; 5.4; Plasma chemistry for etching and deposition

CHERRY, William H.; 24 Dempsey Avenue; Princeton NJ 08540; USA

CHEUNG, Wan Y.; Department of Chemistry; Cornell University; Ithaca NY 14850; USA

CHIANG, Joseph F.; Department of Chemistry; State University of NYC, Oneonta; Oneonta NY 13820; USA; 607-431-3181; E/T; 1.1; 3.3; 3.6; Molecular structure by electron diffraction; Microwave spectroscopy; Molecular structure --molecular beam study; Laser spectroscopy

CHIEN, K. R.; Applied Physics Division; TRW, Inc.; 1 Space Park; Redondo Beach CA 90278; USA; E; 4.1; 5.3; Laser optics, small signal gain; Electron temperature and electron density in high pressure discharge

CHILDS, Wendell A.; Department of Physics; U.S. Military Academy; West Point NY 10996;
USA; 914-938-3901; 914-446-5464; E; 3.4; 4.2; 5.8; X-ray studies with small accelerator; Radioactive emissions, direct and indirect

CHILDS, William J.; Physics Division; Argonne National Laboratory; 9700 S. Cass Avenue; Argonne IL 60439; USA; 312-972-4042; E; 1.1; 3.6; Laser fluorescence and laser-rf double-resonance studies of atomic and molecular beams; Atomic and molecular structure

CHILUKURI, Santaram; Department of Physics; Union College; Barbourville KY 40906; USA; 606-546-4151; E; 2.3; 3.2; 3.5; Atomic and molecular spectroscopy; Kinetics; Lasers

CHIMENTI, Robert J.; Corporate Research Laboratories; Exxon Research and Engineering Co.; Clinton Township-Rte 22 East; Annandale NJ 08801; USA; 201-730-2830; E; 3.6; 5.3; 4.1; 3.2; Atomic and molecular physics in low energy plasmas; Laser physics; Plasmas with surfaces--interaction

CHIN-BING, Stanley A.; 3619 Bauvais Street; Metairie LA 70001; USA

CHING, Wai-Yim; Department of Physics; Univ. of Missouri, Kansas City; Kansas City MO 64110; USA; 816-276-1604; T; 1.1; Atomic wave functions in theoretical condensed-matter physics and molecular orbital calculations

CHIU, Lue-Yung C.; Department of Chemistry; Howard University; Washington DC 20059; USA; 202-636-6882; T; 2.1; 2.3; 3.9; Radiative interactions of atoms and molecules; Collisional energy transfer; Magnetic properties of atoms and molecules

CHIU, Ying-Nan; Department of Chemistry; Catholic University of America; Washington DC 20064; USA

CHIVIAN, Jay S.; Vought Corporation; PO Box 225907; MS 220-12; Dallas TX 75265; USA

CHMARA, Frank; Peabody Scientific; PO Box 2009; Peabody
MA 01960; USA

CHO, Chung Won; Physics; Memorial University of Newfoundland; Elizabeth Avenue; St. John's Newfoundland AIB3X7; Canada; 709737-8850; E; 3.10; 3.6; 3.7; 3.3; 1.1; Stimulated laser light scattering in molecular liquids and compressed gases; Infrared absorption spectroscopy arising from collision-induced effects; Lidar measurements on atmospheric constituents; Determination of quadropole moments and polarizabilities of simple molecule

CHO, Hwa S.; Department of Physics; Ulsan Institute of Technology; PO Box 18; Ulsan; Republic of China

CHO, Hyuck; Physics Department; University of Arizona; Tucson AZ 85721; USA; 602-621-4272; E; 1.1; 2.2; 4.4

CHOE, Hong-Soo; Department of Physics; Teachers College; Gyeongsang National University; Jinju Gyeongnam 620 5721; Korea

CHOI, Byung-Ho; Department of Physics; University of California, Riverside; Riverside CA 92521; USA; 714-7875339; T; Heavy ion-atom scattering; Atom-molecule scatter; Electron-molecule scattering; Atom-surface scattering; molecular -structure; Photon-molecule interactions

CHOI, Chan-Kyoo; School of Nuclear Energy; Purdue University; West Lafayette IN 47907; USA

CHOI, Sang-Il; Department of Physics and Astronomy; University of North Carolina; Chapel Hill NC 27514; USA; 919-933-1384; T; 5.2; Atoms and molecules in solid or on solid surface; Electronic structure

CHONG, Yee-Ping; Department of Physics; University of Maryland; College Park MD 20742; USA; 301-454-3536; E; 3.6; 5.4; Plasma spectroscopy; Thomson scattering

CHOPRA, Dev R.; Department of Physics; East Texas State University; Commerce
TX 75428; USA; 214-886-5484; 214-886-5488; E; 2.3; 3.4; 5.1; Solid surfaces--soft X-ray spectroscopy; Surface science; Electronic band structure determination; Thin films

CHOU, Mau-Song; Corporate Research Laboratories; Exxon Research and Engineering Co.; PO Box 45; Linden NJ 07036; USA; 201-233-7013; E; 5.5; Free radicals in flames via laser-induced fluorescence and Raman scattering-- laser diagnostics; Laser-modified flame chemistry

CHOU, Yu; Room 13-3145; Massachusetts Institute of Technology; Cambridge MA 02139; USA

CHOW, Peter P.; 4920 Colonial Drive; Golden Valley MN 55416; USA

CHOYKE, Wolfgang J.; Solid State Science; Westinghouse R&D Center; 1310 Beulah Road; Pittsburgh PA 15235; USA; 412-256-1428; 412-624-2532; E; 4.3; 5.1; Energetic ion beams and surface interactions; CVD and plasma deposition phenomena studied with UHV diagnostics

CHRISTIAN, Wolfgang; Department of Physics; Davidson College; Davidson NC 28036; USA

CHRISTIANSEN, Walter H.; Department of Aeronautics and Astronautics; University of Washington; Seattle WA 98195; USA; 206-543-6224; E; 3.8; 4.1; Laser development and applications; Isotope separation

CHRISTOPHOROU, Loucas G.; Health and Safety Research Div.; Oak Ridge National Laboratory; PO Box X; Oak Ridge TN 37830; USA; 615-574-6199; FTS-624-6199; E; 2.2; 2.3; 3.1; 3.8; 5.3; Atomic, molecular, and high voltage physics; Chemical physics; Radiation physics and chemistry; Electron and ion physics; Photophysics

CHU, Ben; Department of Chemistry; State Univ. of New York; Stony Brook NY 11794; USA; 516-246-7792; 516-246-5065; E; 1.3; 3.10;

4.1; Chemical and biophysics by spectroscopy; Polymers

CHU, Chen-Ning; PO Box 1-3-329; Lungtan Taiwan 325; Republic of China

CHU, Danny D.; PO Box 688; N B 1-1 U T Dallas; Richardson TX 75080; USA

CHU, Shih-I; Department of Chemistry; University of Kansas; Lawrence KS 66045; USA; 913-864-4094; T; 1.1; 2.1; 3.7; 5.7; Molecular astrophysics; Atomic and molecular collisions in fields; Multiphoton excitation, ionization and dissociation; Intramolecular dynamics in van der Waals molecules; Many-body resonances via complex-coordinate approaches

CHU, Steven; Chemical Physics Division; Bell Telephone Laboratories; 600 Mountain Avenue; Murray Hill NJ 07964; USA; 201-582-2092; E; 1.3; Positronium spectroscopy

CHU, Wei-Kan; IBM Corporation; Hopewell Junction NY 12533; USA; 914-897-8698; E; 4.2; 4.3; 5.1; 5.2; Ion stopping power in solids and gases; Ion beam analysis; backscattering; Channeling; Ion implantation

CHUANG, Shu-Yuan; 1905 San Jacinto Drive; Arlington TX 76012; USA

CHUDNOVSKY, David V.; Department of Math; Columbia University; New York NY 10027; USA

CHUNG, Kwong T.; Department of Physics; North Carolina State University; PO Box 5367; Raleigh NC 27650; USA

CHUNG, Sunggi; Department of Physics; University of Wisconsin; 1150 University Avenue; Madison WI 53706; USA; 608-262-7476; T; 1.1; 1.2; 2.2; 3.1; 5.6; Electron-impact excitation of molecules to electronically excited states

CHUPKA, William A.; Sterling Chemistry Laboratory; Yale University; 225 Prospect Street; New Haven CT 06520; USA; 203-436-8132; E; 2.2; 2.3; 3.1; 3.2; 4.4; Photoionization mass spectrometry; Photoelectron and electron--impact spectroscopy; UV, optical and IR spectroscopy, Ion-molecule reactions; Chemiionization phenomena

CHURCH, David A.; Department of Physics; Texas A&M University; College Station TX 77843; USA; 409-845-2841; 409-845-0317; E; 2.3; 3.1; 3.6; 4.2; 5.1; Near-thermal multi-charged-ion charge transfer to atoms and molecules; Ion collision excitation --coherence and anisotropy; Ion-storage experiments; Ion spectroscopy

CHUTJIAN, Ara; Molecular Chemistry and Physics Section; Jet Propulsion Laboratory; California Inst. of Technology; 4800 Oak Grove Drive; Pasadena CA 91103; USA; 213-354-7012; E; 2.1; 2.2; 3.1; Atomic ion-atomic neutral scattering; Photoionization; Electron-neutral scattering; Electron-ion scattering

CIPOLLA, Sam J.; Department of Physics; Creighton University; Omaha NE 68178; USA; 402-280-2133; 402-280-2835; E; 2.1; 3.4; 5.1; Heavy ion collisions with atoms in thick targets; Low-energy x-ray spectroscopy

CLARK, Arnold F.; L-Division 379; Lawrence Livermore National Lab.; PO Box 808; Livermore CA 94550; USA; 415-422-1049; E; 3.4; 4.2; Soft X-ray detection and measurements with bent crystal spectrometers

CLARK, Brian O.; 252 Farmingville Road; Ridgefield CT 06877; USA

CLARK, Charles W.; Far Ultraviolet Group; Radiation Physics Division; Center for Radiation Research; National Bureau of Standards; Gaithersburg MD 20899; USA; 301-921-2071; T; 1.1; 1.2; 2.2; 3.1; 3.6; Rydberg states in external fields; Electron correlation effects in atomic structure and collisions; Core- and multiply-excited states

CLARK, David A.; P-9; LANL; PO Box 1663; MS-H805; Los Alamos NM 87544; USA

CLARK, John H.; Department of Chemistry; University of California; Berkeley CA 94720; USA

CLARK, Kenneth C.; Department of Physics FM-15; University of Washington; Seattle WA 98195; USA; 206-543-5868; E; 5.6; 3.1; 2.3; 2.3; Optical aeronomic auroral and airglow processes; Spectrometric detection and imaging techniques

CLARK, Robert E.; Applied Theoretical Physics; X-6 MS B226; Los Alamos Scientific Lab.; PO Box 1663; Los Alamos NM 87545; USA; 505-667-7667; T; 2.2; Theoretical cross sections or collision strengths calculations for excitation of highly charged ions by electron impact

CLARK JR., William M.; Optical Physics Division; Hughes Research Laboratories; 3011 Malibu Canyon Road; Malibu CA 90265; USA; 213-456-6411x452; E; 4.3; Hydrogen maser clocks--design and characterization; Diode concepts research for magnetometer devices

CLEGG, Thomas B.; Department of Physics and Astronomy; Triangle Universities Nuclear Lab.; University of North Carolina; 176 Phillips Hall 039A; Chapel Hill NC 27514; USA; 919-962-3016; 919-684-8158; E; 2.2; 4.2; Polarized ion source development using atomic physics measurements

CLEMENTI, Enrico; Department D55; Building 996-2; IBM Corporation; PO Box 390; Poughkeepsie NY 12602; USA

CLENDENIN, James E.; Accelerator Physics Division; Stanford Linear Accelerator Center; P.O. Box 4349; Bin 12; Stanford CA 94305; USA; 415-854-3300x2962; E; 2.1; 4.3; Atomic beams; Polarized electron beams; Charged particle beam dynamics

CLIFFORD, H. James; Department of Physics; University of Puget Sound; Tacoma WA 98416; USA

CLOUGH, Shepard A.; Optical Physics Division; U.S. Air Force Geophysics Lab.; Hanscom AFB MA 01731; USA; 617-861-3654; T; 1.1; Small molecules; Energy level transition strengths, widths and shapes

COBB, Donald D.; Earth & Space Science Division; Los Alamos National Laboratory; University of California; D 446; Los Alamos NM 87545; USA; 505-667-6722; T; 3.1; 3.2; 5.7

COBBLE, James A.; Fusion Energy Division; Oak Ridge National Laboratory; PO Box Y; Oak Ridge TN 37830; USA; 615-574-0978; E; 4.3; 5.4; Barium-ion beam diagnostic research; fusion plasma research via atomic spectroscopy

COBURN, John W.; Department K33/801; IBM Almaden Research Center; 650 Harry Road; San Jose CA 95120; USA; 408-256-7322; E; 5.1; 5.3; Discharge-surface interactions; Physics and chemistry of low pressure glow discharges

COCKE, C. L.; Department of Physics; J. R. Macdonald Lab.; Kansas State University; Manhattan KS 66506; USA; 913-532-6779; 913-532-6786; E; 2.1; 4.2; Atomic collisions with high energy or highly charged ionic projectiles

CODE, R. F.; Department of Physics; Erindale College; McLennan Physical Laboratories; University of Toronto; 60 St. George Street; Toronto Ontario M5S1A7; Canada; 416-978-5220; 416-828-5353; E; 5.2; 4.1; 3.10; 3.6; 3.2; Electro-optical properties of diamond UV photoswitches; Nonlinear optics of liquid crystals; NMR (Nuclear Magnetic Resonance) in molecular solids; Biomedical applications of NMR

CODY, Regina J.; Astrochemistry Branch; Lab. for Extraterrestrial Physics; NASA/Goddard Space Flight Center; Code 691; Greenbelt MD 20771; USA; 301-344-8242; E; 3.1; 3.7; 3.6; 5.6

COENSGEN, F. H.; M-Division; Lawrence Livermore National Lab.; PO Box 808; Livermore CA 94550; USA; 415-422-1166; E; 5.4; Energetic plasmas confinement of interest in controlled fusion

COFFEY, Brian J.; 300 W. 109 Apt. #6M; New York NY 10025; USA

COFFMAN, Robert E.; Department of Chemistry; University of Iowa; Iowa City IA 52242; USA; 319-353-5997; E/T; 1.1; EPR, Magnetic interactions; SCF calculations with STO's; Spin-spin and spin-orbit coupling related integrals singlet-triplet splittings and open shell calculations

COGGIOLA, Michael J.; Molecular Physics Laboratory; SRI International; 333 Ravenswood Avenue; Menlo Park CA 94025; USA; 415-326-6200; E; 2.1; Low energy atom-atom collision studies including charge-exchange; reactions, inelastic energy transfer and ion pair formation

COHEN, E. R.; Rockwell International Science Center; 1049 Camino Dos Rios; Thousand Oaks CA 91360; USA; 805-498-4545; T; 1.4; 3.8; Consistency of fundamental constants --analysis; Collision-induced absorption in simple molecular species

COHEN, James S.; Theoretical Division; Group T-12; Los Alamos National Laboratory; Box 1663 MSJ569; Los Alamos NM 87545; USA; 505-667-5982; T; 1.1; 1.3; 2.1; Scattering: atom-atom and atom-molecules; Negative muon-normal atom interactions; Molecular structure

COHEN, Leslie; Code 6650; Radiation Technology Division; U.S. Naval Research Laboratory; 4555 Overlook Avenue, S.W.; Washington DC 20375; USA; 202-767-2572; E; 3.8; 4.2; Up conversion of radiation by the interaction of electron of electron beams and electromagnetic fields in dielectric media

COHEN, Martin J.; PCP, Inc.; 2155 Indian Road; West Palm Beach FL 33409; USA

COHEN, Ronald B.; Analytical Science-Chemistry and Physical Lab.; The Aerospace Corporation; PO Box 92957; Los Angeles CA 90009; USA; 213-648-5946; E; 2.3; 3.6; Atom-molecule and molecule-molecule reactions--chemical kinetics measurements; Laser analytical detection

COHEN, Stanley; Physics and Atmosphere Science; Drexel University; 32nd & Chestnut Sts.; Philadelphia PA 19104; USA; 215-895-2730; 215-895-1876; E; 3.9; 3.1; 4.2; Response of H-resources to high DC E-fields; Longer/Atomic beam; photo detriment of H; Use of LAMPF as a source of relatistic H-beam; Doppler tuning of photon source

COHN, Gerald E.; Department 93F (Building AP-9); Diagnostics (ADD) Division; Abbott Laboratories; Routes 43 & 137; Abbott Park IL 60064; USA; 312-937-4117; E; 4.1; 5.1; 3.2; 3.11; Thin film optics; Optical waveguides and coatings

COK, David R.; 10 Sibley Place-Apt. 3; Rochester NY 14607; USA

COLDWELL, Robert L.; Physics and Astronomy Department; Space Astronomy Laboratory; University of Florida; Gainesville FL 32611; USA; 904-392-0793; 904-392-5450; T; 1.1; 3.1; 5.1; Variational Monte-Carlo methods for finding fully correlated wavefunctions; Simulation of Gamma-ray detector systems; Curve fit codes to find ratios of satellite line components in fusion data

COLE, Gulnar B.; Pupin Lab.; Columbia University; PO Box 86; New York NY 10027; USA

COLE, Stephen K.; Po Box 112; Route 3; Chapel Hill NC 27514; USA

COLEMAN, Paul D.; Department of Electrical Engineering; Electro-Physics Laboratory; University of Illinois; 1406 West Green Street; Urbana IL 61801; USA; 217-333-2765; E/T; 3.10; 4.1; 3.8; 3.6; 3.3; Optical pumping of molecular lasers; Molecular chemical lasers; Nonlinear optics, Raman effect; Laser spectroscopy

COLEMAN, Paul G.; School of Math. and Physics; University of East Anglia; Norwich Norfolk NR152E; England; 0603-56161x2592; E; 2.2; 5.1; Positron surface studies

COLLINS, C. B.; Center for Quantum Electronics; Univ. of Texas, Dallas; PO Box 688; Mail Station NB1.1; Richardson TX 75080; USA; 214-690-2864; E; 2.3; 3.1; Three-body ion-molecule and Penning reactions occurring at atmospheric pressures --measurement and characterization; State-selective photolysis of simple molecules --study

COLLINS, Frank G.; Space Institute; University of Tennessee; Tullahoma TN 37388; USA

COLLINS, George J.; Department of Electrical Engineering; Colorado State University; Fort Collins CO 80523; USA; 303-491-8513; E/T; 3.8; 4.1; 5.3; 5.1; 4.4; Laser chemistry; Plasma chemistry; Microelectronics fabrication; Materials processing

COLLINS, Lee A.; Group T-4 MSB212; Theoretical Division; Los Alamos National Laboratory; PO Box 1633 MS212; Los Alamos NM 87545; USA; 505-667-2100; 505-667-5751; T; 2.1; 2.2; 3.1; 5.4; 5.7; Ab Initio methods for scattering calculations; Dense plasmas and line broadening

COLLINS, Robert J.; Department of Electrical Engineering; University of Minnesota; 123 Church Street, SE; Minneapolis MN 55455; USA; 612-373-5238; E; 3.6; 4.1; Solid and gas lasers and their applications

COMAS, James; Ion Implantation Tech. Section; U.S. Naval Research Laboratory; Code 6812; 4555 Overlook Ave., SW; Washington DC 20375; USA; 202-767-2146; E; 4.2; 5.1; 5.2; Ion implantation doping of semiconductors; Radiation damage; Sputtering; Energy loss mechanisms

COMASKEY, Brian; LLNL; P.O. Box 808; Livermore CA 95205; USA; 415-422-6092; E; 3.6; 3.1; 3.9; 4.1; 3.7

COMLY JR., Jack C.; 235 Kimberly Lane; Los Alamos NM 87544; USA

COMMINS, E. D.; Department of Physics; University of California; Berkeley CA 94720; USA; 415-642-2321; E; 1.4; Parity violation in atoms due to neutral weak currents

COMPAAN, Alvin; Department of Physics; College of Arts and Sciences; Kansas State University; Cardwell Hall; Manhattan KS 66506; USA; 913-532-6786; E; 4.1; 5.2; 3.10; Wave mixing (coherent Raman scattering); Laser spectroscopy-Raman scattering; Laser sputtering in semiconductors

COMPANION, Audrey L.; Department of Chemistry; University of Kentucky; Lexington KY 40506; USA; 606-257-1779; 606-278-5640; T; 1.3; 5.1; Metal hydride clusters--molecular orbital studies; Vibrations of small molecules on surfaces; Storage of hydrogen by metals

COMPTON, Robert N.; Chemical Physics Division; Oak Ridge National Laboratory; Oak Ridge TN 37830; USA; 615-574-6233; E; 3.1; 3.7; 4.2; Laser multiphoton ionization and photodetachment; Nozzle beam expansion; Negative ion physics

CONDIT, W.; Magnetic Fusion Energy Division; Lawrence Livermore National Lab.; PO Box 808; Livermore CA 94550; USA; 415-422-9810; E; 5.4; Atomic physics data in CTR spectroscopy

CONE, Rufus L.; Department of Physics; Montana State University; Bozeman MT 59717; USA; 406-994-3614; E; 4.1; Laser design; Gases and condensed matter--coherent Raman spectroscopy

CONTI, Ralph S.; Department of Physics; Randall Laboratory; University of Michigan; 400 E. University; Ann Arbor MI 48109; USA; 313-763-3464; E; 1.4; Electron and positron--G-2 measurement

CONWAY, John G.; Materials and Molecular Research Div.; Lawrence Berkeley Laboratory; University of California; 1 Cyclotron Road; Berkeley CA 94720; USA; 415-486-5141; E; 1.1; 3.2; 5.3; Actinides and rare earth spectroscopy; Solid and solution spectroscopy

COOK, Donald L.; Pulsed Power Systems Div. 4253; Sandia National Laboratories; PO Box 5800; Albuquerque NM 87185; USA; 505-844-5826; T; 4.2; 4.3; Light ion beam inertial fusion systems--conceptual reactor design

COOKE, Chatham M.; Department of Electrical Engineering and Computer Science; Massachusetts Institute of Technology; 155 Massachusetts Ave.; Cambridge MA 01239; USA; 617-253-2591; E/T; 5.3; Ionization development in gases by avalanche processes; Avalanche instabilities and eventual breakdown

COOKE, William E.; Department of Physics; University of Southern California; University Park; Los Angeles CA 90007; USA; 213-743-6211; E/T; 1.2; 3.2; 4.1; Autoionizing atomic states spectroscopy; Atomic Rydberg states--properties; Nonlinear optical phenomena in atomic vapors

COONEY, John; Department of Physics and Atmospheric Science; Drexel University; Philadelphia PA 19104; USA; 215-895-2707; E; 3.6; Quantum optics; Spectroscopy

COONEY, Patrick J.; Department of Physics; Millersville University; Millersville PA 17551; USA; 717-872-3770; 717-872-3411; E; 4.2; 2.1; Molecular ion dissociation induced by collisions; Ion-solid interactions; Molecular ion structure

COOPE, Dan; Sensors Division; NL Drilling Technology; PO Box 60087; Houston TX 77025; USA; 713-987-9561; E/T; 4.3; Radiation and radiation detection sensors

COOPER, C. Dewey; Department of Physics; University of Georgia; Department of Physics; Athens GA 30602; USA; 404-542-2485; E; 3.7; 1.1; 3.10; Photoelectrons-angular distribution; Molecular electron affinity measurements; Molecules-multiphoton ionization

COOPER, Charles B.; Department of Physics; University of Delaware; Newark DE 19711; USA; 302-451-2661; E; 5.1; 4.4; Low energy sputtering and particle-solid interactions

COOPER, Gary W.; Department of Chemical and Nuclear Engineering; University of New Mexico; Farris Engineering Center 209; Albuquerque NM 87131; USA; 505-277-5431; T; 4.1; 5.3; Gas laser media--basic processes; Laser processes to guide discharges--study

COOPER, John W.; National Bureau of Standards; Gaithersburg MD 20899; USA; 301-921-2001; E/T; 3.1; 3.6; 3.9; 5.4; Photoabsorption in atomic systems; Electric field effects on photoabsorption

COOPER, John; Department of Physics; Joint Inst. for Laboratory Astrophysics; University of Colorado; Campus Box 440; Boulder CO 80309; USA; 303-492-7813; FTS 320-3869; E/T; 5.4; 3.6; 3.7; 3.8; 3.10; Collisional redistribution of laser radiation; Pressure broadening; Collision induced radiation; Radiative transfer

COOPER, Walter; Phototheory Division; Eastman Kodak Research Labs.; B-81; Rochester NY 14650; USA; 716-722-0223; E; 3.8; 5.1; Spectral sensitizing dyes--molecular basis of luminescence, chemiluminescence and aggregation; Pigments in resin-coated papers--molecular photochemistry

COOPER, William S.; Magnetic Fusion Energy Group; Lawrence Berkeley Laboratory; Building 4; 1 Cyclotron Road, Room 210; Berkeley CA 94720; USA; 415-486-5011; E; 2.2; 4.3; 5.1; 5.4; Hydrogen atoms, molecules and ions--interactions with themselves and with wells for modeling of ion source plasmas; Electron capture, loss and excitation used in diagnosing beam, properties of neutral atoms by analysis of radiation emitted

COPELAND, Gary E.; Department of Physics; Old Dominion University; Norfolk VA 23508; USA; 804-440-4614; E/T; 1.1; 3.3; High resolution IR spectroscopy; Lifetimes of excited states; Molecular theory

COPLAN, Michael A.; Institute for Physical Science and Technology; MPSE; University of Maryland; CSS Building; College Park MD 20742; USA; 301-454-5352; E/T; 1.1; 2.3; 4.4; 5.7; Electron impact ionization; Charge exchange collisions; Mass spectrometry, electron spectroscopy

CORBIN, Robert J.; 2424 Kingston; Ponca City OK 74601; USA

CORDARO, Richard B.; Department of Physics; University of Arizona; Tucson AZ 85721; USA; 602-621-6820; E; 2.1; 4.2

CORDERMAN, Reed R.; Energy and Environment Division; Brookhaven National Laboratory; Building 480; Upton NY 11973; USA; 516-345-7731; E; 2.3; 3.2; 4.2; 5.3; Ion-molecule reactions; Glow discharge plasma species--characterization; Free radical species spectroscopy

CORLISS, Charles H.; Forest Hills Laboratory; 2955 Albemarle Street, NW; Washington DC 20008; USA; 202-362-6085; E; 1.1; Oscillator strengths determination of spectral lines of atoms and ions

CORLISS, Edith L.; 2955 Albemarle Street, NW; Washington DC 20008; USA

CORNWELL, C. D.; Department of Chemistry; University of Wisconsin; 1101 University Avenue; Madison WI 53706; USA; 608-262-7534; E; 3.3; 3.5; NMR spectroscopy; Nuclear spin relaxation in gases

CORONADO, Manuel J.; Ohio University; Mill Street, Apartment B-8; Athens OH 45701; USA

CORONGIU, Giorgina; 46P/MS 258; Data Systems Division; IBM Corporation; 390; Neighborhood Road; Kingston NY 12401; USA; 914-385-0424; T; 1.1; 5.2

CORSON, Bayard R.; 3897 Park Drive; Carlsbad CA 92008; USA

CORSON, Dale R.; Cornell University; 615 Clark Hall; Ithaca NY 14853; USA

COSBY, Philip C.; Molecular Physics; Physical Sciences; Chemical Physics; SRI International; 333 Ravenswood Avenue; Menlo Park CA 94025; USA; 415-859-5128; E; 3.1; 3.6; 2.2; 2.3; 1.1; Molecular ions--structure and dynamics

COTT, Donald W.; CDIF/Project and Test Engineering; Analysis Section; Montana Energy and MHD R&D Inst.; PO Box 3809; Butte MT 59701; USA; 406-494-6382; T/E; 5.5; MHD duct flows modeling and analysis and MHD-type combustion processes

COUILLAUD, Bernard J.; Department of Physics; Stanford University; Varian Building; Stanford CA 94306; USA; 415-497-4357; 415-493-2111; E/T; 1.4; 3.6; 4.1; Doppler free spectrometry; QED checks; Nonlinear optics

COULTER, Claude A.; Q-4 Department; MS-E541 Division; Los Alamos National Laboratory; Los Alamos NM 97545; USA; 505-667-4964; T; 3.1; 3.6; 3.10; 3.11; Atom-field interactions; Superradiance

COULTER, Philip W.; Department of Physics and Astronomy; University of Alabama; PO Box 1921; University AL 35486; USA; 205-348-5050; T; 2.2; 3.1; 3.7

COVENEY, Peter V.; Theoretical Chemistry Department; University of Oxford; 1 South Parks Road; Oxford Oxon OX1 3TG; England; 53303x440; 59201x40; T; 1.4; 2.1; Theory of microscopic irreversibility

COVEY, Joel F.; 232 S. Jenkins Street; Alexandria VA 22304; USA

COWAN, Robert D.; Theoretical Division; Group T-4; Los Alamos National Laboratory; PO Box 1663 MS212; Los Alamos NM 87545; USA; 505-667-5139; 505-662-5588; T; 1.1; Atomic structure and spectra

COWGILL, Donald F.; Neutron Devices and Technology; Division 2352; Sandia National Laboratories; Albuquerque NM 87185; USA; 505-844-7480;

E/T; 4.2; 5.1; 5.2; Intense ion irradiation--effects on solubility and migration of hydrogen in metals; Gas-surface interactions

COWLEY, Charles R.; Department of Astronomy; Dennison Bldg.; University of Michigan; Ann Arbor MI 48109; USA; 313-764-3437; 313-663-3534; E/T; 1.1; 5.7; Atomic lines identification in stellar spectra; Atomic oscillator strengths; Spectral line shapes in stars; Stellar abundances

COX, Donald P.; Department of Physics; University of Wisconsin; 1150 University Avenue; Madison WI 53706; USA; 608-262-5916; T; 5.7; Astrophysical plasmas--cooling and spectra

COX JR., Hollace L.; Electrical Engineering Department; Speed Scientific School; University of Louisville; Louisville KY 40292; USA; 502-588-6289; 502-426-6779; E/T; 4.1; 5.1; 3.6; 3.10; 3.11; Nonlinear optical applications in integrated optical devices; Ultra fast laser spectroscopy-solid state materials research; Materials studies for optical phase conjugation techniques; Development of optical communications systems

CRAIG, Norman C.; Department of Chemistry; Oberlin College; Oberlin OH 44074; USA; 216-775-8664; 216-774-7574; E; 1.1; 3.3; Unusual molecules and organic ions--IR and Raman spectroscopy; Normal coordinate analysis

CRAMER, W. H.; Division of Chemistry; National Science Foundation; Washington DC 20550; USA

CRAMPTON, Stuart B.; Department of Physics and Astronomy; Williams College; Williamstown MA 01267; USA; 413-597-2247 x2482; E/T; 1.1; 5.1; 2.1; 4.3; Hydrogen atoms with surfaces and each other at liquid helium temperatures --magnetic resonance studies

CRANDALL, David H.; Experimental Plasma Research Branch; Applied Plasma Physics; Office of Fusion Energy; U.S. Department of Energy; Building 6003; ER 542 GTN; Washington DC 20545; USA; 301-353-3421; FTS-233-3421; E; 2.1; 2.2; 4.2; 5.4; Low energy collisions between multiple charged ions and electrons and between ions and atoms; Atomic physics and diagnostics in high temperature plasma

CRANE, John K.; Laser Isotope Separation Program; Electronic Engineering Division; Lawrence Livermore National Lab.; PO Box 808; Livermore CA 94550; USA; 415-422-0420; 415-423-8556; E; 3.8; 3.6; 1.2; 3.9; 2.1; Laser isotope separation of elements excepting activities; Processes to ionize Rydberg states; Lifetime measurements of Rydberg states

CRASEMANN, Bernd; Department of Physics; University of Oregon; Eugene OR 97403; USA; 503-686-4754; T/E; 3.1; 3.4; 4.2; Atomic innershell processes, especially Auger and radiative transitions; Relativistic and many-body effects; Synchrotron radiation used in atomic physics research

CRAWFORD JR., Bryce; Department of Chemistry; University of Minnesota; 207 Pleasant Street, S.E.; Minneapolis MN 55455; USA; 612-373-9947; E/T; 3.3; 1.1; 3.5; 3.1; 4.1; Vibrational IR and Raman spectroscopy with focus on intensities, bands shapes and relaxation

CRAWFORD, Edward A.; Research Division; Spectra Technology Inc.; 2755 Northup Way; Bellevue WA 98004; USA; 206-827-0460; E; 5.4; 5.1; 5.3; Spectroscopic and atomic physics processes of interest to fusion; Diagnostics and radiative loss processes

CRAWFORD, Oakley H.; Health and Safety Research Division; Oak Ridge National Laboratory; P.O. Box X; Oak Ridge TN 37830; USA; 615-574-5048; T; 2.1; 2.2; 5.2; 5.1; 4.2; Interaction of energetic particles with solids; Optical radiation generated by bombardment of solids with atoms; Atomic physics of channeled ions; Electron transport in dielectrics under irradiation or at high pressure

CREIGHTON, John R.; Chemistry and Material Science; Chemical and Math'l Science Division; Lawrence Livermore National Lab.; University of California; PO Box 808 MS L-35; Livermore CA 94550; USA; 415-423-7949; FTS 543-7949; T; 2.3; 5.5; Combustion systems--numerical integration of chemical kinetics mechanisms with fluid mechanics to understand behavior

CREMERS, C. J.; Department of Mechanical Engineering; University of Kentucky; Lexington KY 40506; USA; 606-257-2661; E; 5.4; Arc-discharges spectroscopy

CRIM, F. F.; Department of Chemistry; University of Wisconsin; 1101 University Avenue; Madison WI 53706; USA; 608-263-7364; E; 2.1; 2.3; Molecular energy transfer and chemical reactions in highly vibrationally excited molecules

CRITTENDEN JR., Eugene C.; Department of Physics; Code 61 CT; Naval Postgraduate School; Monterey CA 93940; USA

CROMER, Christopher L.; Atomic and Plasma Radiation Div.; National Bureau of Standards; A167, Physics Building; Gaithersburg MD 20899; USA; 301-921-2011; E/T; 1.1; 3.1; 3.4; 3.6; 3.10; Spectroscopy of highly excited atomic states; Intense laser interactions with matter

CROOKS, Geoffrey B.; PO Box 5347; Lincoln NE 68505; USA

CROSLEY, David R.; Molecular Physics Lab.; SRI International; 333 Ravenswood Avenue; Menlo Park CA 94025; USA; 415-859-2395; E; 2.3; 3.6; 5.5; Small radicals --laser spectroscopy; Energy transfer and chemical kinetics; Laser diagnostic development; Combustion problems

CROSS, Jon B.; CHM-2; Chemistry Division; Los Alamos National Laboratory; Box 1663 MSG738; Los Alamos NM 87545; USA; 505-667-4646; E; 2.1; 2.3; 5.1; 4.3; 3.8; Atom-surface interaction studies (reactive and nonreactive); Oxygen and nitrogen atom high kinetic energy (5-10 ev) beam source dev.; Optical damage studies produced by atom interactions with surfaces; Atom interactions with spacecraft surfaces

CROSS JR., R. J.; Department of Chemistry; Yale University; PO Box 6666; New Haven CT 06511; USA; 203-436-2446; E/T; 2.1; 2.3; Inelastic scattering--vibrationally and rotationally; Molecular-beam studies of chemical reactions

CROSSWHITE, Henry M.; Chemistry Division K121; Argonne National Laboratory; 9700 S. Cass Avenue; Argonne IL 60439; USA; 312-972-3637; E/T; 3.2; 3.6; 5.2; Optical spectroscopy, emphasizing actinide atomic and crystal energy level analysis, using conventional spectrographs supplemented by laser and digital electronic technology

CROST, Munsey E.; Apt. 53; Gables Apartments-Walnut Street; Neptune NJ 07753; USA

CRUME JR., E. C.; Fusion Energy Division; Oak Ridge National Laboratory; PO Box Y; Oak Ridge TN 37830; USA; 615-574-1348; T; 5.4; Ionization, recombination, excitation, and charge transfer rate coefficients used in numerical simulations of plasma transport

CSAVINSZKY, Peter; Department of Physics; University of Maine; Orono ME 04469; USA; 207-581-1031; T; 1.1; Pseudopotential approach to the highly-excited states of alkali atoms; Density--functional calculations involving atoms

CUE, Nelson; Department of Physics; State University of NY, Albany; Albany NY 12222; USA; 518-457-8450; E/T; 2.2; 3.1; 4.2; 4.3; 5.1; MeV-molecular beams with thin solid foils--interactions; Ion beams (including electron and positron) interactions with solids; Photon and electron interactions with molecules

CULVER, William H.; Optelecom, Inc.; 15940 Luanne Drive; Gaithersburg MD 20760; USA;

301-948-4232; E/T; 4.1; Atomic and molecular science used in laser development; Optical fiber communication technology

CUMMINGS, Frank E.; Department of Chemistry; Atlanta University; Atlanta GA 30314; USA; 404-577-7594; T; 1.1; Interatomic forces and energies; Interactions near dissociation in diatomics

CUMMINS, Sally E.; NASA Goddard Institute for Space Studies; 2880 Broadway; New York NY 10025; USA; 212-678-5608; T/E; 3.3; 3.5; 5.7; Molecular rotation spectra of molecular clouds in the galaxy--analysis of data obtained with radio telescopes

CUNNINGHAM, Augustine J.; Department of Physics; University of Texas, Dallas; PO Box 688; Richardson TX 75080; USA; 214-690-2885; E; 1.1; 2.2; 2.3; 5.3; Electron-ion recombination; Vacuum ultraviolet oscillator strengths and shock tube kinetics; Absorption and fluorescence spectroscopy; High-pressure reaction kinetics

CUNNINGHAM, David L.; Laser Enrichment Division; Exxon Nuclear Co., Inc.; 2955 George Washington Way; Richland WA 99352; USA; 509-375-7253; E; 2.1; 3.1; 3.6; 5.3; Metal vapor flow diagnostics using laser techniques; Thermal collision processes; Photoionization

CUOMO, Jerome J.; IBM Research; PO Box 218; Yorktown Heights NY 10598; USA

CURELARU, Irina M.; Department of Math., Science and Engineering; University of Utah; Salt Lake City UT 84112; USA

CURL, Robert F.; Department of Chemistry; Rice University; PO Box 1892; Houston TX 77001; USA; 713-527-4816; 713-527-8101x3479; E; 3.6; 3.8; Reacting molecular beams--laser excitation

CURNUTTE JR., Basil; Department of Physics; Macdonald Laboratory; Kansas State University; Cardwell Hall; Manhattan KS 66506; USA; 913-532-6786; 913-539-5634; E; 1.3; 1.4; 3.2; 3.4; 4.2; Highly ionized atoms spectroscopy, usually of metastable states; Ion beam studies; Forbidden decay processes

CURRENT, David H.; Department of Physics; Central Michigan University; Brooks Hall; Mt. Pleasant MI 48859; USA; 517-774-3219; E; 2.1; Energy transfer mechanisms in molecular vapors

CURRY, Bill P.; VKF/Space; Calspan Field Services, Inc.; Arnold AF Station TN 37324; USA; 615-454-7200; T/E; 3.6; 5.1; 5.3; 2.1; Light scattering theory applied to particle and flow diagnostics; Electromagnetics

CURTIS, Earl C.; Chemical Laser Technology Department; Advanced Programs Division; Rocketdyne; 6633 Canoga Avenue; Canoga Park CA 91304; USA; 213-700-4893; E/T; 4.1; 5.5; 3.3; High energy lasers; Rocket engine performance; Infrared spectra

CURTIS, Lorenzo J.; Department of Physics and Astronomy; University of Toledo; 2801 W. Bancroft; Toledo OH 43606; 419-537-2341; E/T; 1.1; 3.2; 4.2; Beam-foil spectroscopy; Term value regularities --semi-empirical studies; Term values and transition probabilities--calculation

CURTIS, Paul M.; Dept. of Mech'l and Aerospace Engineering; Princeton University; Engineering Quadrangle; Princeton NJ 08544; USA; 609-452-5233; E; 2.3; 5.5; Gas-phase and heterogeneous kinetic processes in a catalytically combusting flow --investigations

CURTISS, Charles F.; Department of Chemistry; University of Wisconsin; 1101 University Avenue; Madison WI 53706; USA; 608-262-1511; T; 2.1; 0.0; Statistical mechanics; Transport phenomena in gases; Molecular scattering

CZARNIK, John W.; Strategic Planning; Texaco, Incorporation; 2000 Westchester Avenue; White Plains NY 10650; USA

CZECHANSKI, James; 1310 North Oak Street #505; Arlington VA 22209; USA

CZUCHLEWSKI, Stephen J.; Laser Division; Group L-9; Los Alamos National Lab.; PO Box 1663 MS535; Los Alamos NM 87545; USA; 505-667-6088 x5385; E/T; 3.7; 4.1; Pulse propagation in carbon dioxide laser gain media; Polyatomic molecules--multiphoton absorption

CZYSZ III, Michael F.; Department of Physics and Astronomy; University of Georgia; Athens GA 30602; USA

DABBOUSI, Osama B.; Department of Physics; University of Petroleum and Minerals 95; Box 144; Dhahran Intl Airport; Dhahran 31932; Saudi Arabia

DAGATA, John A.; Department of Chemistry; Chemistry Division; Naval Research Laboratory; Washington DC 20375; USA; E/T; 1.2; 3.1; 3.8

DAGDIGIAN, Paul J.; Department of Chemistry; Johns Hopkins University; 34th and Charles Streets; Baltimore MD 21218; USA; 301-338-7438; E; 2.1; 2.3; 3.6; Diatomic electronic spectroscopy; Gas-phase reactions under single collision conditions--laser fluorescence and chemiluminescence studies; State-to-state cross-sections in rotationally and vibrationally inelastic collisions

DAGENAIS, Mario; Gordon McKay Laboratory; Harvard University; 9 Oxford Street; Cambridge MA 02138; USA; 617-495-4466; E; 3.6; 5.1; Non-linear vapors and solids spectroscopy

DAHLER, John S.; Department of Chemistry and Chemical Engineering; University of Minnesota; Minneapolis MN 55455; 612-373-5483; 612-373-2305; T; 2.1; 2.2; 2.3; Electron-atom inelastic collisions; Penning and associative ionization; High-energy (.5-200 keV) atom-atom and ion-atom collisions

DALEY, Howard L.; 2100 Via Estrada; Carrollton TX 75006; USA

DALGARNO, Alexander; Harvard-Smithsonian Center for Astrophysics; 60 Garden Street; Cambridge MA 02138; USA; 617-495-4403; FTS-830-7238; T; 5.6; 5.7; 5.4; 2.1; 2.3; Atomic and molecular theory; Astrophysical and atmospheric applications

DANESE, John B.; 425 Center Street N.; Vienna VA 22180; USA

DANG, Richard K.; Joint Institute for Lab. Astrophysics; University of Colorado; Boulder CO 80309; USA; 303-492-7518; E; 2.2; Electron impact excitation on atoms

DANIELE, Joseph J.; Reprographic Technology Group; Scanned Image Sector; Xerox Corporation; Rochester NY 14644; USA; 716-422-2053; E; 4.3; Opto-electronic sensors and sources

DANIELEWICZ JR., Edward J.; Electronics Research Laboratory; The Aerospace Corporation; PO Box 92957; Los Angeles CA 90009; USA; 213-648-6986; E; 3.3; 3.5; polyatomic molecules optically pumped with CO_2 laser--IR/far IR double; resonance spectroscopy

DANIELS, Robert L.; Avionics Division; McDonnell Douglas Corp.; 19121 Mesa Drive; Villa Park CA 92667; USA; 714-637-2903; E/T; 3.5; Molecular spectroscopy

DARACK, Sheldon; Bell Labs.; 600 Mountain Ave., Room 1E 420; Murray Hill NJ 07974; USA

DARDIS, John G.; 1332 Pinetree Road; McLean VA 22101; USA

DAREWYCH, J. W.; Department of Physics; York University; Downsview Ontario M3J1P3; Canada; 416-667-3843; T; 2.2; 2.1

DAS, Mukunda P.; Department of Physics; Sambalpur University; Jyoti Vihar; Sambalpur 768017; India

DAS, Pankaj K.; Department of Electrical Systems Engineering; Rensselaer Polytech Institute; Troy NY 12181; USA

DAS, Tara P.; Department of Physics; Colorado State University; Fort Collins CO 80523; USA; 303-491-7272/6206; T; 1.1; Atomic properties with emphasis on hyperfine interaction--relativistic many-body theory

DASCH, Cameron J.; Physics Department; Gen'l Motors Research Laboratories; General Motors Technical Center; 12 Mile and Mound Roads; Warren MI 48090; USA; 313-575-2901; E; 5.5; 3.6; 4.1; 2.1; Atomic Raman spectroscopy; Molecular energy transfer; Combustion diagnostics

DASSEN, H. W.; Department of Physics and Computing; Herzberg Inst. of Astrophysics; Wilfrid Lawrier University; Room 1057; Waterloo Ontario N2L3C5; Canada; 519-884-1970x2436; E/T; 2.2; Electron impact phenomena-energy loss spectroscopy; Electron-photon coincidence measurements, excitation functions; Theoretical scattering-calculations, computational aspects; Non-linear partial differential equations and (atomic & molecular) physics

DATLA, Raju U.; Atomic and Plasma Radiation; National Bureau of Standards; Gaithersburg MD 20899; USA; 301-921-2356; 301-454-3536; E; 2.2; 5.4; Rate coefficients measurements of excitation, ionization, and recombination of high Z ions due to electron collisions; Diagnostics of high temperature plasmas using plasma spectroscopy

DATZ, Sheldon; Chemistry and Physics Division; Oak Ridge National Laboratory; PO Box X; Oak Ridge TN 37830; USA; 615-574-4984; E; 2.3; 4.2; 5.2; Collisional electron detachment; Chemically reactive collisions; Ion-solid interactions; Ion-channeling phenomena; Electron and positron channeling radiation; Merged electron-ion beam experiments

DAUDEY, Jean-Pierre L.; Lab. De Physique Quant; Univ. P Sabatier; 118 Route De Narbonne; Toulouse 31077; France

DAUGHERTY, Jack D.; Avco Everett Research Laboratory; 2385 Revere Beach Parkway; Everett MA 02149; USA; 617-389-3000; E/T

DAVENPORT, John E.; Physical Organic Chemistry; Chemistry Laboratory; Physical Sciences; SRI International; 333 Ravenswood Avenue; Menlo Park CA 94025; USA; 415-859-3279; E; 2.3; 3.1; 3.2; 3.5; 5.6; Absorption spectrum and quantum yields for ozone photolysis--temperature dependence; Spectroscopy, chemistry and dynamics of excited atoms and molecules

DAVIDOVICH, Luiz; Dept. de Fisica; Pontif Univ. Catolica; Rue M Sao Vicente 225; Rio De Janeiro 22451; Brazil

DAVIDOVITS, Paul; Department of Chemistry; Boston College; Chestnut Hill MA 02167; USA; 617-552-3617; E; 2.3; 5.6; Reaction rate and molecular dynamics studies of atomic reactions; Processes involved in acid rain formation

DAVIDSON, Ernest R.; Department of Chemistry; Indiana University; Bloomington IN 47405; USA; 812-335-6013; T; 1.1; 1.2; 3.9; Molecular structure and wave functions

DAVIDSON, Gilbert; Photo Metrics, Inc.; 4 Arrow Drive; Woburn MA 01801; USA

DAVIDSON, Lewis A.; 116 Partridge Drive, NE; Concord NC 28025; USA

DAVIDSON, Steven A.; Physics; JILA; University of Colorado; Boulder CO 80309; USA; 303-492-6839; 303-4421388; E; 3.1; 4.1; Collisional excitation transfer: Independance of products of Na(3PJ) + Na(3PJ1); Optical pumping at high vapor pressure of Rb

DAVIES, D. Kenneth; Westinghouse Research Lab.; Beulah Road; Pittsburgh PA 15235; USA

DAVIS, Christopher C.; Department of Electrical Engineering; University of Maryland; College Park MD 20742; USA; 301-454-6847; E; 2.1; 3.6; 4.1; Molecular energy transfer and relaxation in the gas phase using both optical heterodyne methods and laser-induced fluorescence

DAVIS, David S.; Steward Observatory; Lunar and Planetary Laboratory; University of Arizona; Tucson AZ 85721; USA; 602-621-6960; E; 5.7; 3.3; 3.2; 1.1; 5.6; Planetary atmospheres and galactic emission sources --astronomical spectroscopy

DAVIS, H. Ted; Department of Chemical Engr. and Materials Science; University of Minnesota; Minneapolis MN 55455; USA; 612-373-2299;

E/T; 2.2; 2.3; 5.3; Electron scattering theory; Kinetic and interfacial theory; Molecular dynamics of transport and fluid structure; Electrons in gases--drift velocity

DAVIS, Jack; Plasma Radiation Division; U.S. Naval Research Lab.; Code 4720; 4555 Overlook Avenue, SW; Washington DC 20375; USA; 202-767-3278; 202-767-2921; T; 2.2; Electron impact excitation and ionization of complex atoms and ions; Autoionization and dielectronic recombination

DAVIS, James I.; Laser Isotope Separation Program; Lawrence Livermore National Lab.; PO Box 808; Livermore CA 94550; USA; 415-422-6182; E/T; 3.8; Laser isotope separation; Program director

DAVIS, Jay C.; Physics Department; E-Division; LLNL; University of California/DOE; P.O. Box 808; Livermore CA 94550; USA; 415-422-4513; E; 4.2; 5.1; 5.4; Accelerator-based nuclear analytical techniques; Depth profiling of hydrogen isotopes

DAVIS, Lawrence W.; Department of Physics; University of Idaho; Moscow ID 83843; USA; 208-885-6745; E/T; 3.6; 4.1; 3.10; Atomic and molecular systems using absorption and light scattering techniques--laser spectroscopy

DAVIS, Lloyd C.; 3212 Katherine; Dearborn MI 48124; USA

DAVIS, Richard F.; Polaroid Corporation; 1265 Main Street W4-2A; Waltham MA 02254; USA

DAVIS, Richard W.; Department of Physics; US Air Force Academy; USAF Academy CO 80840; USA; 303-472-2394; T; 3.6; Atomic vapor laser isotope separation; Rare gas halide lasers

DAVIS, Robert H.; Department of Physics; Florida State University; Tallahassee FL 32306; USA

DAVIS, Robert W.; Department of Chemistry; Memorial University of Newfoundland; St. John's Newfoundland A1B3X7; Canada; 709-737-8904; 709-737-8772; E; 1.1; 3.3; 3.5; 3.9

DAVIS, Sumner P.; Department of Physics; University of California; Berkeley CA 94720; USA; 415-642-4857/7166; E; 3.2; 3.5; 5.7; Diatomic molecular spectroscopy of astrophysical interest; Atomic spectroscopy of heavy ions

DAVIS, William A.; Fusion Energy Division; Oak Ridge National Laboratory; PO Box Y MS 2; Building 9201-2; Oak Ridge TN 37830; USA; 615-576-3737; E; 5.4; Plasmas; Neutral beam diagnostics

DAVISON, Sydney G.; Department of Applied Mathematics; Waterloo University; Waterloo Ontario N2L3G1; Canada

DAWSON, Horace R.; Department of Physics; Angelo State University; San Angelo TX 76909; USA; 915-942-2242; E; 2.1; Cross sections measurement for charge transfer in ion-atom collisions

DAY, Frank H.; 450 W. 31st Street; Riviera Beach FL 33404; USA

DAY, Michael H.; 4M 312; Bell Labs.; Crawford Corners Road; Holmdel NJ 07733; USA

DE BRUYN, John R.; Department of Physics; University of British Columbia; 6224 Agriculture Road; Vancouver British Columbia V6T2A6; Canada

DE LA VEGA, Jose R.; Department of Chemistry; Villanova University; Villanova PA 19085; USA; 215-645-4879; T; 1.1; 2.3; 5.2; Ab initio molecular orbital calculation of intermediate size molecules; Proton exchange and proton tunnelling; NMR line analysis of systems with proton exchange; Potential energy surfaces in molecular re-arrangements--calculation

DE LUCIA, Frank C.; Department of Physics; Duke University; Durham NC 27706; USA; 919-684-8232; E; 3.3; 4.1; Millimeter and submillimeter spectroscopy; Molecular lasers; Quantum electronics

DE MARTINI, Francesco; Instituto Fisica G Marconi; Piazza Aldo Moro 2; Roma 00185; Italy

DE RIJK, Waldemar G.; Fixed Restorative Dentistry; Dental Materials; University of Maryland at Baltimore; 666 West Baltimore St.; Baltimore MD 20211; USA; 301-528-3507; 301-921-3336; E/T; 1.3; Dentistry, materials research; Dentistry, clinical procedures; Dentistry diagnostic techniques, biological; Interaction with radiation

DEAKYNE, Carol A.; Ionospheric Physics; Air Force Geophysics Laboratory; Hanscom AFB MA 01731; USA; 617-861-5031; T; 1.1; 1.3

DEBEER, David P.; Physics Department; Columbia Radiation Laboratory; Columbia University; P.O. Box 31; 538 W. 120th Street; New York NY 10027; USA; 212-280-3273; E; 3.10; 3.6; 2.1; 3.7; 1.2

DECARLO JR., J. L.; 213 Glenwood Lane; Port Jefferson NY 11777; USA

DECIUS, John C.; Department of Chemistry; Oregon State University; Corvallis OR 97331; USA; 503-754-2371; E/T; 1.1; 2.1; Molecular structure and dynamics; Intermolecular forces and vibrational spectroscopy; Vibrational energy transfer

DEFOTIS, Gary C.; Department of Chemistry; The College of William and Mary; Williamsburg VA 23185; USA; 804-253-4671; E; 1.1; Molecular properties such as g-factors and zero-field splitting; parameters--use and determination; Intramolecular and intermolecular exchange interactions

DEGENKOLB, Eugene; Advanced Integrated Circuit Technology; Components Research Laboratory; GTE Laboratories; 40 Sylvan Road; Waltham MA 02254; USA; 617-466-2407; 617-466-2503; E; 5.4; 5.1; 5.3; 3.2; 4.4; Plasma discharges used for etching of microstructures for; electrical circuit applications; Optical emission and mass spectometry used for process monitoring; Surfaces contamination from plasma-wall interactions

DEHMELT, Hans G.; Department of Physics; University of Washington; Seattle WA 98195; USA; 206-543-2779; E; 3.2; 3.3; High resolution many-ion spectroscopy

DEHMER, Joseph L.; Fund. Molecular Phys. and Chem. Section; Environmental Research Division; Argonne National Laboratory; Building 203; 9700 S. Cass Avenue; Argonne IL 60439; USA; 312-972-4194; E/T; 3.1; 3.6; 1.1; 1.2; 3.8; Molecular photoionization in VUV; Multiphoton ionization of molecules; Angle-resolved photoelectron studies; Instrument development

DEHMER, Patricia M.; Radiological and Environmental Research Division; Argonne National Laboratory; Box 203-C153; 9700 S. Cass Avenue; Argonne IL 60439; USA; 312-972-4187; T; 3.1; 3.5; 3.6; 3.7; 3.8; Molecular spectroscopy using photoionization mass spectrometry; Photoelectron spectrometry; Multiphoton ionization

DEHOPE, William J.; 7026 Schroll; Lakewood CA 90713; USA

DEKOVEN, Benjamin M.; Chemistry Division; U.S. Naval Research Lab.; Code 6110; 4555 Overlook Ave, SW; Washington DC 20375; USA; 202-767-3455; E; 2.3; 3.6; 3.8; Ion molecule reactions; Jet-cooled molecules--laser spectroscopy; Collision phenomena

DEL BENE, Janet E.; Department of Chemistry; Youngstown State University; Youngstown OH 44555; USA; 216-742-3466; T; 1.1; Hydrogen bonding and ion-molecule associations --molecular orbital; Correlation energy studies on associated species

DELFINO, Michelangelo; Research and Development Laboratory; Fairchild Corporation; 4001 Miranda Avenue; Palo Alto CA 94304; USA; 415-4937250; E; 5.1; 5.2; High power laser applications to semiconductor VLSI processing; Ion-implementation damage analysis by optical methods

DELGADO-BARRIO, Gerardo; Instituto De Estructura De La Material; Theoretical Atomic and Molecular; CSIC; Serrano 123; Madrid 28006; Spain; 4112962; 2619800; T; 1.1; 2.1; 3.1

DELOS, John B.; Department of Physics; The College of William and Mary; Williamsburg VA 23185; USA; 804-253-4471; T; 2.1; 1.2; 3.9; Atomic and molecular collision theory

DELSANTO, Pierpaolo; Code 5834; Naval Research Lab.; Washington DC 20375; USA

DELVAILLE, John P.; High Energy Astrophysics Division; Harvard-Smithsonian Center Astrophys; 60 Garden Street; Cambridge MA 02138; USA; 617-495-7144; E; 3.4; X-ray transmission diffraction gratings for applications in monochromators and spectrometers

DELY, Alex; Department of Physics; University of Arizona; Tucson AZ 85721; USA; 602-621-6819; 602-294-2944; E; 4.2; 4.4; 3.6; 3.3; 4.3; Strategic Defense Initiative laser work; Secondary ion mass spectroscopy

DEMIR, Oktay; Department of Physics; Suffolk University; Boston MA 02146; USA; 617-723-5700X603; 617-739-9092; E/T; 5.1; 5.3; 4.1; 3.8; 1.3; Energy transfer between molecules; Polar molecules; Ferro electric molecules; Fiber-optics

DEMSKE, David L.; 5977 Jacobs Ladder; Columbia MD 21045; USA

DENDRAMIS, Achille L.; Analytical/Physical Measurements; Corporate Research Division; Miami Valley Laboratories; Procter and Gamble Co.; PO Box 39175; Cincinnati OH 45247; USA; 513-245-2775; 513-245-2830; E; 4.1; 4.3; 5.1; 3.6; Interfaces--laser-induced fluorescence/Raman spectroscopy

DENES, Louis J.; Gas Lasers Division; Westinghouse R&D Center; 1310 Beulah Road; Pittsburgh PA 15235; USA; 412-256-3678; E/T; 4.1; 5.3; Excimer lasers; Attaching gas mixtures--sheath properties; Discharges stability analysis

DENKER, John S.; Cornell University; G-3 Clark Hall; Ithaca NY 14853; USA

DENNE, G. Boel; Plasma Physics Lab.; Princeton University; PO Box 451; James Forrestal Campus; Princeton NJ 08544; USA

DEPRISTO, Andrew E.; Department of Chemistry; Ames Laboratory; Iowa State University; Ames IA 50011; USA; 515-294-9924; T; 2.3; 2.1; 5.1; Chemical dynamics; Scattering theory; Gas-solid surface interactions

DERKITS, Conchita R.; 16 Columbus Avenue; New Providence NJ 07974; USA

DERR, Vernon E.; Department of Commerce; Environmental Research Laboratories; Nat'l. Oceanographic and Atmospheric Ad; P.O. Box R/E; 325 S. Broadway; Boulder CO 80303; USA; 303-497-6000; E/T; 3.1; 5.6; Atmospheric radiation balance, climate effects; Gaseous and particulate absorptions; Atmospheric and environmental applications of A&M

DESCLAUX, Jean-Paul; Cen/Grenoble DRF/LIH; 85X; Grenoble; Cedex 38041; France

DESERIO, Robert; Department of Physics; Oak Ridge National Laboratories; University of Tennessee; PO Box X; Building 5500; Oak Ridge TN 37830; USA; 615-574-4793; 615-974-7823; E; 2.1; 1.4; 3.9; 1.3; 3.2

DESHMUKH, Pranawa C.; Department of Physics; Indian Institute of Technology; Madras 600036; India

DESLATTES, Richard D.; Quantum Metrology Group; Center for Basic Standards; National Bureau of Standards; Physics Bldg-Rm A141; Gaithersburg MD 20899; USA; 301-921-2061; E; 3.4; 4.2; 1.4; 3.1; 5.4; Precision measurement methods development; Gamma-ray spectroscopy; X-Ray optics

DESPLAT, Jean-Louis; Rasor Associates Inc.; 253 Humbolt Court; Sunnyvale CA 94089; USA; 408-734-1622; E/T; 5.1; Surface physics; Electron emission; Cesium vapor thermionic converter

DETEMPLE, Thomas A.; Department of Electrical Engineering; 200 EERL; University of Illinois; 1406 W. Green; Urbana IL 61801; USA; 217-333-3094; 217-333-2287; E; 3.6; 3.7; 3.9; 3.10; 4.1; Non-linear optics; Semiconductor optoelectronics; Superradiance; Excited state kinetics of atoms; Laser modeling

DEUTCH, John M.; Department of Chemistry; Massachusetts Institute of Technology; 77 Massachusetts Avenue; Cambridge MA 02139; USA; 617-253-1479; T; 1.3; 3.6; Light scattering; Polymer chemistry

DEUTSCH, Lynne K.; 93 Medford Street; Arlington MA 02174; USA

DEUTSCH, Peter W.; Graduate Ctr. for Cloud Physics Research; University of Missouri, Rolla; 109F Norwood Hall; Rolla MO 65401; USA; 314-341-4362; T; 1.3; Water and sulfur molecular clusters--computational studies

DEVOE, Ralph G.; Department K01-281; IBM Research Laboratory; 5600 Cottle Road; San Jose CA 95193; USA; 408-256-6445; E; 1.4; 3.6; 4.1; High resolution spectroscopy of trapped atoms; Absolute optical frequency measurements; Frequency stabilized lasers and standards

DEVOR, Donald P.; Optical Physics Division; Hughes Research Laboratories; 3011 Malibu Canyon Road; Malibu CA 90265; USA; 213-456-6411; E; 5.2; Atomic and chemical physics of solids; Radiative and non-radiative transitions of ions in various hosts; Atomic species on bulk properties--effects

DEVOTO, R. Stephen; Magnetic Fusion Energy Division; Lawrence Livermore National Lab.; PO Box 808; Livermore CA 94550; USA; 415-422-9825; T; 5.4; Collision cross section data used for fusion plasma reactions

DEVRIES, Paul L.; Department of Physics; Miami University; Oxford OH 45056; USA; 513-529-2234; T; 2.1; 2.3; 3.6; 3.7; 3.8; Chemical physics; Laser induced collision processes; Effects of very short laser pulses on collision systems; Time-dependent quantum phenomena

DEWITT, Robert N.; Code F-12; Naval Surface Weapons Center/DL; Dahlgren VA 22448; USA

DEXTER, Richard N.; Department of Physics; University of Wisconsin; 1150 University Avenue; Madison WI 53711; USA; 608-263-5546; E; 5.1; 5.2; 5.4; Plasmas--spectroscopic diagnostics; Ion-induced desorption of atoms; Neutrals from proton and other ion collisions with solids--energy spectrum

DEYOUNG, Russell J.; Space Systems Division; NASA Langley Research Center; 160; Hampton VA 23665; USA; 804-865-3781; E; 1.1; 2.1; 3.1; 4.1; 5.3; Laser research on solar pumped lasers; Interaction of blackbody radiation with molecules

DEZAFRA, Robert L.; Department of Physics; State University of New York; Stony Brook NY 11794; USA; 516-246-7954; 516-246-6580; E; 5.6; 3.3; 5.5; Remote sensing via mm-wave spectroscopy of stratospheric trace gases; Environmental and atmospheric changes via anthropogenic pollutants

DHARAMSI, Amin N.; Department of Electrical Engineering; Old Dominion University; Hampton Boulevard; Norfolk VA 23508; USA; 804-440-4467; 804-440-4499; E/T; 3.6; 2.3; 4.1; 3.8; 3.1; Laser development

DIANA, Leonard M.; Department of Physics; University of Texas, Arlington; 720 Briarwood Boulevard; Arlington TX 76013; USA; 817-273-2455; 817-275-5763; E; 2.2; 4.3; 5.2; 1.3

DICK, Charles E.; X-Ray Physics Department; Ionizing Radiation Division; Center for Radiation Research; National Bureau of Standards;

P.O. Box C-216; Building 245; Gaithersburg MD 20899; USA; 301-921-2201; E; 2.2; 3.4; 4.2; 5.8

DICK, Warren J.; 2685 W County Rd. H2 Apt. 208; Mounds View MN 55112; USA

DIELS, Jean-Claude; Center for Laser Studies; University of Southern California; University Park; Los Angeles CA 90007; USA; 213-743-5360; E/T; 3.6; 3.7; Ultrashort light pulses/vapors interactions; Coherence in multiphoton excitation of molecules

DIESTLER, D. J.; Department of Chemistry; Purdue University; West Lafayette IN 47907; USA; 317-494-8897; T; 2.1; 2.3; 3.6; 3.8; Collision-induced dissociation; Charge transfer, chemical reactions in condensed phases; Laser-induced desorption of molecules from solid surfaces; Energy-transfer in atom-surface collisions

DIETRICH, Daniel D.; Physics E-Division; Lawrence Livermore National Lab.; PO Bx 808 L-401; Livermore CA 94550; USA; 415-422-7868; E; 3.6; 4.2; 5.4; Plasma spectroscopy, specializing in line shifts and widths; Precision measurements in few-electron highly-stripped heavy ions

DILL, Dan; Department of Chemistry; Boston University; 675 Commonwealth Avenue; Boston MA 02215; USA; 617-353-4277; FTS-223-5259; T; 2.2; 3.1; Electron-molecule scattering calculations; Molecular photoionization/ fragmentation

DILLINGHAM, Thomas R.; Department of Physics; Kansas State University; Manhattan KS 66506; USA

DILLON, Michael A.; 1110 Grove Street; Downers Grove IL 60515; USA

DILSNER, Paul A.; 901 Armand Street; N. Bellmore NY 11710; USA

DIMAURO, Louis F.; Department of Physics and Astronomy; Louisiana State University; Baton Rouge LA 70803; USA; 504-388-6852; E/T; 1.1; 1.2; 1.3; 3.6; 3.7; Atomic and molecular physics-Quantum beats

DINES, Eugene L.; Department of Physics; University of California; Davis CA 95616; USA

DISTEFANO, Thomas H.; IBM Watson Research Center; PO Box 218 12-208; Yorktown Heights NY 10598; USA

DITTNER, Peter F.; Physics Division; Oak Ridge National Lab.; Building 5500; PO Box X; Oak Ridge TN 37830; USA; 615-574-4789; E; 2.2; 4.2; Merged ion-electron beams used for measurement of ion-electron recombination and/or ionization cross-sections

DIXIT, Shamasundar N.; Arthur Noyes Lab. for Chemical Phys.; Cal Tech; Pasadena CA 91125; USA; 818-356-6523; 818-356-6537; T; 3.7; 3.6; 3.10; 3.11; 1.2; Atomic and molecular multiphoton ionization

DIXON, David A.; Central Research and Development Department; E.I. DuPont de Nemours and Co.; Experimental Station E328; Wilmington DE 19898; USA; 302-772-2619; T; 1.1; 3.3; 4.4; Ab initio molecular orbital theory to study the effect of electronic structure on chemical properties; Applications to fluorocarbons, polymer modeling, proton affinities, vibrational force fields, molecular structure

DIXON, Dwight R.; Department of Physics and Astronomy; Brigham Young University; Provo UT 84602; USA; 801-378-2341; E; 4.2; 5.1; Protons, deuterons, etc. in passing through thick foils--multiple; scattering

DIXON, Raymond D.; LANL; PO Box 1663; MS-G770; Los Alamos NM 87545; USA

DIXON, Robert H.; Optical Sciences Division; U.S. Naval Research Lab.; Code 6505; 4555 Overlook Ave., SW; Washington DC 20375; USA; 202-767-2826; E; 5.4; Laser produced plasmas--atomic spectroscopy

DOBBS, Gregory M.; Chemical Physics Group; Power and Industrial Systems Tech.; United Technologies Research Center; MS 90; Silver Lane; East Hartford CT 06108; USA; 203-727-7145; 203-727-7391; E; 2.3; 4.1; 5.3; Laser spectroscopy; Laser diagnostics of combustion and flames; Coherent antistatic Raman spectroscopy (CARS)

DOBRIN, Richard; Medical Mobile Service; 1160 Midland Avenue; Bronxville NY 10708; USA

DODD, Jack G.; Department of Physics and Astronomy; Colgate University; Hamilton NY 13346; USA; 315-824-1000; E/T; 3.6; 4.1; Coherent radiation; Quantum electronics; Experimental laser spectroscopy; Microscopic techniques

DODHY, Adila; Chemical Physics Section; Oak Ridge National Lab.; Building 45005-Rm 5116; Oak Ridge TN 37830; USA

DOERING, J. P.; Department of Chemistry; Johns Hopkins University; Baltimore MD 21218; USA; 301-338-7445; E; 2.2; Electron scattering; Electronic and ionic collision phenomena

DOGLIANI, Harald O.; 13612 Verbena Place; Albuquerque NM 87112; USA

DOLLINGER, Richard E.; Department of Electrical Engineering; State University of NY, Buffalo; 4232 Ridge Lea; Amherst NY 14226; USA; 716-831-3166; E; 5.4; Vacuum arc plasmas--temperatures, densities and percent ionization

DONAHUE, D. J.; Department of Physics; University of Arizona; Tuscon AZ 85721; USA; 602-621-2480; E; 4.2; 4.4; Radioisotope analysis with accelerators

DONAHUE, Joey B.; Group MP-7; Meson Physics Division; Los Alamos National Laboratory; University of California; 1663; Los Alamos NM 87545; USA; 505-667-2856; FTS-843-2856; E; 3.1; 4.2; 3.9; 1.1; 1.4; Relativistic hydrogen and beams; Laser photodetachment; Special relativity tests

DONNALLY, Bailey; Department of Physics; Lake Forest College; Lake Forest IL 60045; USA; 312-234-3100-X231; E; 1.1; 2.1; 2.2; Atomic collisions; Atomic structure; Polarized electron and proton sources; Particle-surface interactions involving atoms, electrons and positrons

DONNELLY, Keith E.; 18 Gladys Road; West Hill Ontario M1C1C6; Canada

DOODLESACK, Gary A.; 34 Ashland Street; Medford MA 02155; USA

DOOLEN, Gary D.; LANL; MS F669; Los Alamos NM 87544; USA; 505-667-3784; E; 1.1; 2.2; 3.1; 3.10; 5.4

DORAIN, Paul B.; Department of Chemistry; Brandeis University; Waltham MA 02254; USA; 617-647-2819; E/T; 4.1; Raman scattering and laser development

DORFMAN, Leon M.; Department of Chemistry; Ohio State University; Columbus OH 43210; USA; 614-422-5792; E; 2.3; 3.8; 4.2; Chemical kinetics; Fast reaction studies by pulse radiolysis; Radiation chemistry

DORMAN, Charles F.; Univ. Science and Arts of Oklahoma; PO Box 3247; Chickasha OK 73018; USA

DOSCHEK, George A.; Solar-Terrestrial Relationships Branch; Space Science Division; U.S. Naval Research Lab.; Code 4170; 4555 Overlook Ave, SW; Washington DC 20375; USA; 202-767-3527; E/T; 5.7; 5.4; X-ray and XUV spectroscopic research in: highly ionized astrophysical, plasmas, solar plasmas, V and laboratory plasmas

DOUGAL, Arwin A.; Department of Electrical Engineering; University of Texas, Austin; ENS 112; Austin TX 78712; USA; 512-471-3068; 512-471-1851; E/T; 3.6; 4.1; Laser-induced ionization in super--pressure gases; Laser instrumentation

DOUGHTY, Ben M.; Department

of Physics; East Texas State University; Commerce TX 75428; USA; 214-886-5485; E; 2.1; 4.2; 10-150 keV protons on rare gases--atomic collisions; Charge transfer into excited states

DOUGLAS-HAMILTON, D. H.; AVCO Everett Research Laboratory; 2385 Revere Beach Parkway; Everett MA 02149; USA; 617-389-3000; E/T; 5.3; Discharge physics

DOUTHAT, Daryl; Department of Physical Sciences; Kennedy-King College; Chicago IL 60621; USA; 312-962-3382; E/T

DOVERSPIKE, Lynn D.; Department of Physics; The College of William and Mary; Williamsburg VA 23185; USA; 804-253-4471; E; 2.1; Low energy atomic collisions

DOW, John D.; Department of Physics; University of Notre Dame; Notre Dame IN 46555; USA; T; 1.1; 1.3; 3.1; 3.2; 3.4

DOW, William G.; 915 Heatherway; Ann Arbor MI 48104; USA

DOWE JR., R. Michael; P.O. Box 12139; 7210 Country Club Drive; La Jolla CA 92037; USA; 614-453-9500; 614-454-0270; E; 4.1; 4.2; 3.3; 5.6; 4.1; Accelerator-based A&M physics for CPB/FEL research; Beam technology; IR, RF, microwave technology; Atmospheric science--high altitude chemical release phenomenology

DOWELL, Jerry T.; Physical Sciences Division; IRT Corporation; PO Box 80817; San Diego CA 92138; USA; 714-565-7171-X365; E; 2.2; 3.3; Electron attachment in polyatomic molecules; Free radicals-microwave spectroscopy

DOWLING, Jerome M.; Chemistry and Physics Division; The Aerospace Corporation, M2/251; PO Box 92957; Los Angeles CA 90009; USA; 213-648-7558; T; 5.6; 3.3; 4.1; Optics, remote sensing; Atmospheric physics and chemistry

DOWS, David A.; Department of Chemistry; University of Southern California; University Park; Los Angeles CA 90089; USA; 213-743-2794; E/T; 1.1; 3.5; 3.8; 1.3; Molecular spectroscopy; Molecular crystal properties; Photochemistry

DOYLE, Holly T.; Harvard-Smithsonian Center for Astrophysics; 60 Garden Street; Cambridge MA 02138; USA; 617-495-7000; E/T

DOZIER, Charles M.; Condensed Matter and Radiation Sciences; Condensed Matter Physics Branch; Code 6680; U.S. Naval Research Lab.; 4555 Overlook Avenue, SW; Washington DC 20375; USA; 202-767-2154; 202-767-2549; E; 5.1; Radiation deposition in materials

DRACHMAN, Richard J.; Atomic Physics Office; Lab. for Astronomy and Solar Physics; NASA Goddard Space Flight Center; Code 680.1; Greenbelt MD 20771; USA; 301-344-5273; T; 1.3; 2.2; 5.7; 1.2; Positron and atom or atomic ion interactions, especially for exotic systems; like positronium and positronium hydride--astrophysical applications; High Rydberg states of helium

DRAKE, Charles W.; Department of Physics; Oregon State University; Corvallis OR 97331; USA; 503-754-4569; E; 1.2; 3.7; Rydberg atoms; Electromagnetic radiation--statistical properties

DRAKE, Gordon W.; Department of Physics; University of Windsor; Windsor Ontario N9B3P4; Canada; 519-253-4232; T; 1.1; 1.3; 1.4; 3.1; 3.2

DRAKE, J. M.; Laser Enrichment Division; Exxon Nuclear Co., Inc.; 2955 George Washington Way; Richland WA 99352; USA; 509-375-7325; E; 2.3; 3.5; 4.1; Dye molecules with appplication to dye lasers--preparation, spectroscopic and photophysical characterization

DREIZLER, Reiner M.; Inst Fur Theoret Physik; Universitaet; Robert Mayer Str. 8-10; Frankfurt 01 6000; West Germany; 069-7982550; T; 1.1; 2.1; 5.4; Density functional methods

DRENTJE, A. G.; Kernfysisch Versneller Inst.; Ryksuniversiteit Groningen; Zernikelaan 25; Groningen 9747AA; Holland

DRESS, William B.; Route 2; Lenoir City TN 37771; USA

DRESSER, Miles J.; Department of Physics (2814); College of Science and Arts; Washington State University; 1245 Physical Sciences; Pullman WA 99164; USA; 509-335-4663; 509-335-9531; E; 5.1; 2.2; Atomic and molecular interactions with solid surfaces; Electron and photon induced effects in surface layers

DRESSLER, Kurt; Physical Chemistry Laboratory; ETH-Zentrum; Zurich CH8092; Switzerland; (1) 256 4441; E/T; 1.1; 1.2; 3.1; 3.2; 3.5; Diatomic molecules: electronic spectroscopy, nonadiabatic effects; Ab initio calculations: excited states of molecular hydrogen; Molecular solids: luminescence and energy transport

DREYFUS, Russell W.; Physical Science Division; IBM Thomas J. Watson Research Center; PO Box 218; Yorktown Heights NY 10598; USA; 914-945-1807; E; 4.1; Cold cesium cation beams and lasers

DRISCOLL JR., Timothy J.; Avondale Research Center; U.S. Bureau of Mines; 4900 LaSalle Road; Avondale MD 20782; USA; 301-436-7528; E; 5.2; Particle interactions in solid oxides due to ion bombardment

DROBNY, Vladimir F.; BPD; ICO; Tektronix Corporation; PO Box 500; Beaverton OR 97077; USA; 503-627-6261; 503-644-3799; E/T; 1.1; 5.1; 5.2; 2.3; 4.4; Bipolar process development; Vacuum deposition; Plasma processing; Ion beam implantation

DRUGER, Stephen D.; Department of Physics; Clarkson College of Technology; Potsdam NY 13676; USA; 315-268-6676; T; 2.1; 3.1; 4.1; Radiationless degradation of electronic excitation in large molecules; Raman and fluorescent scattering

DRULLINGER, Robert E.; Time and Frequency Division; National Bureau of Standards; Division 524; 325 S. Broadway; Boulder CO 80303; USA; 303-497-3183; E; 1.1; 3.2; 4.1; 5.6; Laser driven optial pumping as means of state selection and detection in; atomic beam, magnetic resonance spectrometer; Molecular development of visible and UV spectroscopic method to measure dimerization of atmospheric water vapor

DRUMMOND, David L.; U.S. Air Force Weapons Lab.; AFWL/ARAC; Kirkland AFB NM 87117; USA; 505-844-9836/3386; E; 2.3; 3.7; 3.8; Hydrogen fluoride reaction kinetics; Line broadening and anomalous dispersion effects in hydrogen fluoride lasers

DRUMMOND, Peter D.; Department of Physics; University of Auckland; Private Bag; Auckland; New Zealand; 09-73799x8837; T; 4.1; 1.4; 3.11; 3.6; 3.7; Quantum optics of cooperative effects; Quantum measurement; Pulse propagations in atomic or nonlinear media; Picosecond lasers

DRUZBICK, John; Physics and Astronomy; Virginia Military Institute; 210 Kent Street; Lexington VA 24450; USA; 703-463-0503; 703-373-6214; T; 1.1; 2.1

DUBE, Louis J.; Fakultaet Fuer Physik; Universitaet Freiburg; Hermann-Herder-Strasse 3; Freiburg 1BRW WG D-7800; West Germany; 049--761-203-3781; 049-76155-1784; T; 2.1; 2.2

DUBEN, Anthony J.; Department of Chemistry; South Dakota School of Mines and Tech.; Rapid City SD 57701; USA; 605-341-1776; T; 5.2; Solvent on molecular structures and spectral properties--effect

DUCAS, Theodore W.; Department of Physics; Wellesley College; Wellesley MA 02181; USA; 617-235-0320x429; E; 1.2; 3.1; Rydberg levels lifetimes as function of temperature of surroundings (influence of blackbody radiation); Photoionization of atomic levels

DUCOS, Jean-Pierre R.; 4 Rue Du Bief; Chilly-Mazarin

91380; France

DUFF, James W.; Aerodyne Research, Inc.; Crosby Drive; Bedford Research Park; Bedford MA 01730; USA; 617-275-9400; T; 2.1; High-energy energy transfer processes investigations

DUGAN, Charles H.; York University; 206 Petrie Building; Downsview Ontario; Canada

DUGGAN, Jerome L.; Department of Physics; North Texas State University; Denton TX 76203; USA; 817-788-2626; E/T; 4.2; Accelerator-based atomic physics

DUKE, Charles B.; Theoretical Physics and Chemistry; Webster Research Center; Xerox Corporation; 800 Phillips Road, 114/38D; Webster NY 14580; USA; 716-422-2109; T; 1.1; 3.1; 3.2; Electronic structure of organic molecules--calculations; Photoemission and UV absorption spectra--interpretation

DUNBAR, Robert C.; Department of Chemistry; Case Western Reserve University; Cleveland OH 44106; USA; 216-368-3712; E; 3.1; 4.4; Ion cyclotron resonance; Trapped ions--photodissociation

DUNCAN, M. M.; Department of Physics and Astronomy; University of Georgia; Athens GA 30602; USA; 404-542-2485; E; 2.1; 3.2; 4.2; 5.1; Electron loss of H-; Beam foil electron spectroscopy; Collisional ionization of projectiles and targets; Correlated systems

DUNCANSON JR., John A.; Center for Fast Kinetics Research; University of Texas; Pat-131; Austin TX 78712; USA; 512-471-7583; E; 3.8; 5.5; 5.3; 4.1; 3.6

DUNLAP, Brett I.; Code 6129; Chemistry Division; Naval Research Lab.; Washington DC 20375; USA; 202-767-3250; T; 1.1; 1.3; 2.3

DUNN, Gordon H.; Quantum Physics Division (NBS); Joint Inst. for Laboratory Astrophysics; National Bureau Standards; Campus Box 440; University of Colorado; Boulder CO 80309; USA; 303-492-7824; 303-497-3518; E; 2.1; 2.2; 2.3; 4.4; 5.4; Electron-ion collisions: recombination, excitation, ionization; Ion-molecule reactions at ultra-low temperatures; Trapped ion dynamics; Penning traps; Non-neutral plasmas; Physics of ions

DUNN, Terry S.; Beckton Dickinson Labware; 1950 Williams Drive; Oxnard CA 93030; USA

DUNNING, F. Barry; Department of Space Physics and Astronomy; Rice University; PO Box 1892; Houston TX 77251; USA; 713-527-8101; E; 1.2; 2.1; 3.9; 4.1; 5.1; Atomic and molecular properties in highly excited Rydberg states; Spin polarization measurements applications to surface studies; Spin dependencies in atom-atom and atom-surface interactions

DUNNING, Kenneth L.; 101-6 Highland Greens; Fort Ludlow WA 98365; USA

DUNNING JR., Thomas H.; Chemistry Division; Theoretical Chemistry Group; Argonne National Laboratory; 9700 S. Cass Avenue; Argonne IL 60439; USA; 312-972-3594; 312-972-3570; T; 1.1; 2.3; Ground and excited states of molecules and of potential energy surfaces for chemical reactions

DUQUETTE, David W.; Department of Physics; University of Wisconsin; Sterling Hall; Madison WI 53706; USA

DURUP, Jean; Atomiques Moleculaires; Laboratoire Des Collisions; Universite De Paris-Sud; Orsay 91405; France

DUTTA, Chizuko M.; Department of Physics; Rice University; PO Box 1892; Houston TX 77001; USA; 713-527-4945; T; 3.4; Atomic and molecular X-ray absorption spectra analysis

DUZY, Carolyn; AVCO Everett Research Lab.; 2385 Revere Beach Parkway; Everett MA 02149; USA; 617-389-3000-X734; T; 3.1; Photoexcitation and photoionization cross sections determination

DYKE, Thomas R.; Department of Chemistry; University of Oregon; Eugene OR 97403; USA; 503-686-4614; E; 1.3; 3.3; 3.9; 3.10; Molecular beam spectroscopy of van der Waals and hydrogen bonded complexes

DYLLA, H. F.; Plasma Physics Laboratory; Princeton University; PO Box 451; Princeton NJ 08544; USA; 609-683-3199; E; 5.1; 5.4; Plasma-wall interactions; Hydrogen, carbon, oxygen, etc.--low energy interactions

EAMES, David R.; GA Technologies; PO Box 85608; San Diego CA 92138; USA

EASO, Sajan; Department of Physics; Pennsylvania State University; 104 Davey Laboratory; University Park PA 16802; USA

EBERHARDT, William H.; Department of Chemistry; Georgia Institute of Technology; Atlanta GA 30332; USA; 404-894-4026; T; 1.1; Electron kinetic energy in bonding --role

EBERLY, Joseph H.; Department of Physics and Astronomy; University of Rochester; Rochester NY 14627; USA; 716-275-4576; T; 3.7; 3.8; 3.11; Quantum optics; Multiphoton processes

EBERT, Paul J.; L-Division; Lawrence Livermore National Lab.; P.O. Box 808; Livermore CA 94550; USA; 702-395-3954; 415-423-4736; E; 3.4; 4.2; 5.1; 5.4; 3.1; X-ray fluorescence; X-ray scattering and absorption; Photons, electrons and neutrons with matter-interaction of x-ray lasers

ECK, Thomas G.; Department of Physics; Case Western Reserve University; Cleveland OH 44106; USA; 216-368-4022; E/T; 1.1; 3.1; 3.9; 5.1

ECKBRETH, Alan C.; Chemical Physics Division; United Technologies Research Center; Silver Lane; East Hartford CT 06108; USA; 203-727-7269; E; 3.6; 4.1; 5.5; Coherent anti-Stokes Raman spectroscopy (CARS); Saturated laser-induced molecular fluorescence for flame and combustion diagnostics

ECKERT, Hans U.; 3901 Via Pavion; Palos Verdes CA 90274; USA; 213-375-1819; T; 5.3; 5.4; Excitation mechanisms in RT induction discharges

ECKHARDT, Craig J.; Department of Chemistry; Division of Physical Chemistry; University of Nebraska; Lincoln NE 68588; USA; 402-472-2734; 402-472-1429; E/T; 1.1; 1.3; 3.2; 3.3; 3.5; Energy transfer in molecular crystals; Lattice dynamics of molecular crystals; Triboluminescence; Piezomodulation spectroscopy of molecular crystals

ECKHOUSE, Shimon; Scientific Department; Ministry of Defense; PO Box 2250; Haifa; Israel

ECKSTROM, Donald J.; Molecular Physics Laboratory; SRI International; 333 Ravenswood Avenue; Menlo Park CA 94025; USA; 415-859-4398; E; 2.3; 3.1; 3.6; 4.1; 5.4; Excimer laser development; Laser physics; Plasma diagnostics; IR radiation--generation, transport and absorption

ECONOMOU, Nicholas P.; Group 87; Lincoln Lab.; MIT; 244 Wood Street; Lexington MA 02173; USA

EDDY, Thomas L.; Department of Mechanical Engineering; Georgia Institute of Technology; Atlanta GA 30332; USA; 404-894-3241; E/T; 5.4; Plasma spectroscopy

EDEN, James G.; 1801 Stratford Drive; Champaign IL 61821; USA

EDERER, David L.; Radiation Physics Division; Center for Radiation Research; National Measurement Laboratory; National Bureau of Standards; A251 Physics; Physics Building, Room A251; Gaithersburg MD 20899; USA; 301-921-2031; E; 1.1; 2.1; 3.2; 3.4; 3.6; Photo-electron spectroscopy from laser excited atoms; Electron spectroscopy from colliding laser excited atoms; Molecular ion fluorescence by synchrotron radiation; X-ray fluorescence spectroscopy excited by SR (Synchrotron Radiation)

EDGE, Ronald D.; Department of Physics; University of South Carolina; Columbia SC 29208; USA; 803-777-4121; E; 4.2; 4.3; 5.1; Ions in crystals--channeling; Rutherford backscattering--depth probability

EDIGHOFFER, John A.; TRW; 544 Forest Avenue; Palo Alto CA 94301; USA; 415-497-0121; 415-327-9209; E; 4.1; 4.2; 4.3

EDMONDS JR., Dean S.; 44 Tamarack Road; Weston MA 02193; USA

EDWARDS, Alan K.; Department of Physics and Astronomy; University of Georgia; Athens GA 30602; USA; 404-542-2485; E; 2.1; 4.2; Molecular dissociation produced by fast heavy ions

EDWARDS, Mark A.; Department of Physics; Johns Hopkins University; Homewood Campus; Baltimore MD 21218; USA

EDWARDS, Matthew E.; 5246 Cisco Court; Fayetteville NC 28303; USA

EDWARDS, Thomas H.; Department of Physics; Michigan State University; East Lansing MI 48824; USA; 517-349-2474 (Hm); 517-355-9708 (Wk); E; 3.3; 3.5; IR spectra of small molecules; Data processing of spectra

EERKENS, Jeff W.; Terra Nova, Inc.; PO Box 1178; Pacific Palisades CA 90272; USA; 213-973-2688; E/T; 4.1; High-energy molecular lasers

EFTEKHARI, Abbas; Department of Physics; University of Texas; PO Box 19059; Arlington TX 76019; USA

EGAN, Patrick O.; Department of Physics; L-296; LLNL; P.O. Box 808; Livermore CA 94550; USA; 415-423-3907; E; 1.4; 3.2; 5.4; 3.4; 4.2; Exotic atoms particle detector development

EGELHOFF JR., W. F.; National Bureau of Standards; Building 222-Room B248; Gaithersburg MD 20899; USA

EHLER, Arthur W.; Laser Fusion Division; Group L-4; Los Alamos National Lab.; PO Box 1663; Los Alamos NM 87545; USA; 505-667-4339; E; 3.6; 4.2; 5.2; 5.4; Multiply ionized atoms interactions produced by laser-matter interaction with gases, solids and plasmas

EHLERS, Kenneth W.; Magnetic Fusion Research Dept.; Accelerator and Fusion Research Div.; Lawrence Berkeley Laboratory; University of California; Bldg. 4, Rm. 214; 1 Cyclotron Road; Berkeley CA 94720; USA; 415-486-5011; 415-486-5314; E; 5.4; 5.3; 5.1; 4.3; 4.2; Negative hydrogen and deuterium ions by "surface" and "volume" production; Multi-ampere neutral hydrogen & deuterium beams; Acceleration and neutralization of both polarity ions; Multi-charged state heavy-ion generation

EHLERS, Vernon J.; 1600 Edgewood Avenue, SE; Grand Rapids MI 49506; USA

EHLERT, Ralph C.; 8225 W. Lorraine Place; Milwaukee WI 53222; USA

EHLOTZKY, Fritz; Institute for Theoretical Physics; University of Innsbruck; Sillgasse 8; Innsbruck A-6020; Austria; 0043 522 2724; T; 3.7; 3.8; 5.4; 3.6; 3.10

EHRLICH, Daniel J.; Lincoln Laboratory; MIT; C-128; Lexington MA 02173; USA; 617-862-2077; E; 3.6; 3.8; 4.1; Laser spectroscopy; Microscopic photochemical reactions; New laser development

EISELE, Fred L.; Electromagnetics Laboratory; Physical Sciences Division; Georgia Tech Research Institute; Georgia Institute of Technology; Atlanta GA 30332; USA; 404-894-3424; 404-894-3425; E; 5.6; 4.4; 5.3; 2.3; Tropospheric ion sampling and mass identification; Ion-transport; Ion-chemistry; Dissociative electron capture in dense gases

EISENSTADT, Maurice; Department of Medicine; Albert Einstein College of Medicine; 1300 Morris Park Avenue; Bronx NY 10461; USA; 212-430-2091; E; 1.3; 5.2; Macromolecules of biology, interaction with water, ions by NMR relaxation; Diamagnetic susceptibility of inhomogeneous biological tissue

ELANDER, Nils O.; Atomic Theory; Research Institute of Physics; S-104 05; Frescahvagen 24; Stockholm 50; Sweden; +46-8-15 0360; +46-8-768 8018; T; 1.1; 2.1; 1.2; 2.2; 4.2; Analytic collision theory, poles of s-matrix; Applications to resonances in ion-atom(ion)

and ion-electron collisions; Photofragment; Spectroscopy and molecular predissociation

ELIASSON, Baldur; Plasma Physics Department; Brown Boveri Research Center; Brown Boveri R CIE; Chilemattweg 5; Baden Aargau 5405; Switzerland; 056/848031; 056/848411; T; 5.3; 5.6; 2.3

ELIEZER, Isaac; Department of Chemistry; Oakland University; Rochester MI 48063; USA; 313-377-2142; T; 2.3; 3.2; 3.5; Spectroscopy and chemical physics pertaining to physical properties of atoms and simple molecules

ELIEZER, Shalom; Plasma Department; Soreq Nuclear Research Centre; Yavne 70600; Israel

ELKOMOSS, Sabry; 5 Rue De L'Universite; Strasbourg; France

ELKOWITZ, Allan B.; 66 Adams Street; Medfield MA 02052; USA

ELLIOTT, Greg; 5732 Bellevue; La Jolla CA 92037; USA

ELLIOTT, Robert B.; Physical Lab.; Trinity College; Dublin; Ireland

ELLIS, David G.; Department of Physics and Astronomy; University of Toledo; Toledo OH 43606; USA; 419-537-2276; 419-537-4634; T; 1.1; 1.2; 1.4; 3.1; 3.2; Spectra of highly ionized atoms; Plasma diagnostics with atomic spectroscopy; Polarization and quantum beats in optical emission; Correlation in atomic wave functions

ELLIS, Harry W.; Natural Science Collegium; Eckerd College; St. Petersburg FL 33733; USA; 813-867-1166x469; E; 5.3; Ion and neutral products created in electrical breakdown in dielectric gases--observations

ELLIS, John L.; 226 Jeffords Road; Rush NY 14543; USA

ELLIS, Walton P.; Group CHM-2; Los Alamos National Lab.; Box 1663 MSG738; Los Alamos NM 87545; USA; 505-667-4043; 505-667-4686; E; 3.1; 3.4; 5.1; Surface studies using Auger, LOSS, LEED, UV and synchrotron radiation; Low-energy ion scattering

ELLISON, Frank O.; Department of Chemistry; University of Pittsburgh; Pittsburgh PA 15260; USA; 412-624-5068; T; 1.1; 2.3; Valence--quantum theory; Molecules and ions in normal and excited states, and in course of reaction--electronic structure

ELLISON, G. B.; Department of Chemistry; University of Colorado; Boulder CO 80309; USA; 303-492-8603; E; 2.3; Organic molecules--chemical physics

ELLSWORTH, Louis D.; Department of Physics; Kansas State University; Manhattan KS 66502; USA

ELMQUIST, Randolph E.; Department of Physics; University of Virginia; McCormick Road; Charlottesville VA 22901; USA; E; 3.6; 3.1

ELSAYED-ALI, Hani E.; PO Box 2394; Station A; Champaign IL 61820; USA

ELSTON, Stuart B.; Department of Physics and Astronomy; University of Tennessee; Knoxville TN 37916; USA; 615-574-4794; E; 2.1; 4.2; Accelerator-based ion-atom collisions

ELTON, Dr. Raymond C.; Plasma Physics Division; U.S. Naval Research Laboratory; Code 4733; 4555 Overlook Avenue, SW; Washington DC 20375; USA; 202-767-2754; 202-767-3528; E; 5.4; 5.1; 4.1; 3.6; Plasma spectroscopy; X-ray laser research and development

EMERY, Guy T.; Department of Physics; Cyclotron Lab.; Indiana University; Bloomington IN 47405; USA; 812-335-2882; 812-335-9748; E/T; 1.1; 1.3; 3.1; 3.4; 4.2; Pionic atoms; Internal conversion

EMINYAN, Marcel E.; 85 BD Beaumarchais; Paris 75003; France

EMKEN, Walter C.; Department of Chemistry; Central Washington University; 317 Dean Hall; Ellensberg WA 98926; USA; 509-963-2902; 509-963-2811; E; 2.2; 3.3; Gaseous molecules--electron diffraction; IR spectroscopy including low temperature studies

EMMONS, Donald A.; Engineering Division; Frequency and Time Systems, Inc.; 34 Tozer Road; Beverly MA 01915; USA; 617-927-8220; E; 4.3; 4.1; Atomic beam (cesium) optics; Detector geometry and efficiency; Atomic beam source oven design and characterization; Hyperfine state resonance detection

ENDRES, Paul F.; Department of Chemistry; Bowling Green State University; Bowling Green OH 43403; USA; 419-372-0315; T; 1.3; 2.3; Microclusters--Molecular dynamics and Monte Carlo studies

ENGELHARDT, A. G.; Los Alamos National Laboratory; P.O. Box 1663; MS 554; Los Alamos NM 87545; USA; 505-667-7440; E; 5.1; Laser fusion

ENGELKEN, Robert D.; Department of Engineering; Electrical Division; Electronic-Photovoltaic Materials; Arkansas State University; P.O. Box 1080; Caraway; State University AR 72467; USA; 501-972-2088; 501-972-3421; E/T; 2.3; 3.1; 3.2; 3.8; Fundamental physics of photovoltaic processes; Physics of electrical transport in semiconductors; Optical absorption in semiconductors; Deposition of semiconductors thin films

ENGELKING, Paul C.; Department of Chemistry; University of Oregon; Eugene OR 97403; USA; 503-686-4656; 503-686-4001; E; 3.1; 3.5; 3.6; 3.7; Molecular radicals in a flowing afterglow--laser-induced fluorescence; Molecular ions--spectroscopy; Multiphoton dissociation and ionization; Atoms--photoionization

ENGELMAN JR., Rolf; Group CHM-1; Los Alamos National Laboratory; P.O. Box 1663 MS-G-740; Los Alamos NM 87545; USA; 505-667-4345; 505-662-7359; E; 1.1; 3.2; 3.1; Atomic and simple molecular optical spectra of actinide and radioactive elements--production and analysis; Isotope shifts in heavy elements

ENGLAND, Walter B.; Department of Chemistry; University of Wisconsin, Milwaukee; Milwaukee WI 53201; USA; 414-963-6731; 414-963-6153; T; 1.1; 3.5; Many-body perturbation theories--ab initio studies of spectroscopy and heavy alkaline earth monoxides--ab initio studies of spectroscopy and electronic structure; Carbon dioxide ions--ab initio studies

ENGLISH, Thomas C.; Efratom Div. of Ball Corporation; 18851 Bardeen Avenue; Irvine CA 92715; USA; 714-752-2891; E; 4.1; 4.3; 5.3; 3.3; 2.3; Atomic oscillators and clocks; Time and frequency; Engineering physics

ENNIS JR., Robert M.; Energy Systems Group; Oak Ridge Operation; TRW, Inc.; 800 Oak Ridge Turnpike, B-200; Oak Ridge TN 37830; USA; 615-482-9054; E; 5.3; Gas and liquid high voltage insulation and dielectric materials--applied research

ENO, Larry; Department of Chemistry; Frick Laboratory; Clarkson University; Potsdam NY 13676; USA; 315-268-6559; T; 2.1; 2.3; 3.8; Molecular collision problems within a quantum mechanical framework--analysis

ENSBERG, Earl S.; Fusion Project Division; General Atomic Co.; PO Box 81608; San Diego CA 92111; USA; 714-455-2267; E; 5.4; Applied atomic physics for fusion plasma diagnostics

ERBER, Thomas; Department of Physics; Illinois Institute of Technology; Chicago IL 60616; USA; 312-567-3382; T; 3.9; 4.2; 5.7; Atoms in super-strong magnetic fields

ERICKSON, Clifford W.; Honeywell; PO Box 524; Honeywell Plaza MN12 5219; Minneapolis MN 55440; USA

ERICKSON, Gary J.; Department of Chemistry; University of Iowa; Iowa City IA 52242; USA; 319-353-3788; T; 1.1; 2.3; Lineshape studies;

Molecular dynamics; Spectral lines--simulations

ERICKSON, Glen W.; Department of Physics; University of California, Davis; Davis CA 95616; USA; 916-752-1788/1500; T; 1.1; 1.4; One-electron atom energy levels; Quantum electrodynamics

ERIKSEN, Frederick J.; Department of Physics and Astronomy; University of Southern Mississippi; PO Box 9299; Southern Station; Hattiesburg MS 39401; USA; 601-266-7206; E/T

ERMLER, Walter C.; Department of Chemistry/Chemical Engineering; Stevens Institute of Technology; Castle Point Station; Hoboken NJ 07030; USA; 201-420-5520; T; 1.1; 2.3; 3.5; Molecular structure and spectra using ab initio formal procedures

ERMOLAEV, Alexei M.; Department of Physics; University of Durham; South Road; Durham DH13LE; England

ERNIE, Douglas W.; Department of Electrical Engineering; University of Minnesota; 123 Church Street, SE; Minneapolis MN 55455; USA; 612-373-3024; 612-376-9186; E/T; 5.3; 2.3; 2.1; 4.4; 4.1

EROS, Stephen; 1353 W. 83rd Street; Cleveland OH 44102; USA

ESCHRICH, Timothy C.; General Dynamics; PO Box 2507; Mail Zone 44-28; Pomona CA 91769; USA

ESHERICK, Peter; Laser Research and Development; Division 4216; Sandia Laboratories; PO Box 5800; Albuquerque NM 87185; USA; 505-844-5857; E; 4.1; High resolution stimulated Raman spectroscopy

ESTLER, Ron C.; Department of Chemistry; University of Southern California; University Park; Los Angeles CA 90007; USA; 213-743-7506; E; 3.6; 3.8; Laser-initiated reactions; Laser spectroscopy

ESTREICHER, Stefan K.; Depts. of Physics and Electrical Engineering; Rice University; PO Box 1892; Houston TX 77251; USA; 713-527-4938; 713-527-4020; T; 1.1; 1.3; Point defects in semiconductors (ab-initio cluster calculators); Rotronic and vibronic Jahn-Teller Effect

EUBANK, Harold P.; Research Department; Neutral Beam Division; Princeton Plasma Physics Lab.; Princeton University; P.O. Box 451; Princeton NJ 08544; USA; 609-683-3172; E; 5.4; Neutral beams for plasma heating; Diagnostic atom beams for plasma research

EVENSON, Kenneth M.; Time and Frequency Division; National Bureau of Standards; 325 S. Broadway; Boulder CO 80303; USA; 303-497-3397; E; 3.6; Laser spectroscopy

EWING, George E.; Department of Chemistry; Indiana University; Bloomington IN 47401; USA; 812-335-5754; E/T; 1.3; van der Waals molecules--structure and energy transfer

EWING, James J.; Vice President/Laser Division; Spectra Technology, Inc.; 2755 Northrup Way; Bellevue WA 98004; USA; 206-827-0460; E; 4.1; 5.3; 3.10; Direct multidiscipline R&D/engineering lasers and plasmas; Solid state laser development; High power laser development

EXTON, Reginald J.; Optical Spectroscopy Section; Instrument Research Division; Langley Research Center; NASA; MS 235A; Hampton VA 23665; USA; 804-865-2791; E; 3.2; 3.6; 3.10; 1.1; Spectral diagnostics-aeronautics; Line shape analysis; Coherent Raman spectroscopy; Environmental sensing (remote techniques)

EYLER, Edward E.; Physics Department; Yale University; 217 Prospect Street; New Haven CT 06511; USA; 203-436-3778; 203-436-4654; E; 1.2; 1.1; 3.1; 1.4; 3.5; Laser spectroscopy of He, H^2 and atmospheric molecules; Rydberg state structure and lifetimes

EZEKIEL, Shaoul; Department of Electrical Engineering; Massachusetts Institute of Technology; Room 33-113; Cambridge MA 02139; USA; 617-253-3783/2568; E; 3.6; 4.1; Laser spectroscopy--high resolution/high precision; Optical clocks; Optical gyroscopes

FABER, Roger J.; Department of Physics; Lake Forest College; Lake Forest IL 60045; USA

FADER, Walter J.; United Technologies Research Center; Silver Lane; East Hartford CT 06073; USA; 203-727-7158; T; 5.4; Impurities in plasma fusion devices; Neutral beam injection and charge exchange losses

FADLEY, Charles S.; Department of Chemistry; University of Hawaii; 2545 The Mall; Honolulu HI 96822; USA; 808-948-6401; 808-948-8112; E/T; 5.1; 5.2; 3.1; 2.2; 3.4; Surface physics and chemistry; Electron spectroscopy of solids and surfaces; Photoelectron diffraction

FAHEY, Albert J.; Department of Physics; Washington University; PO Box 1105; St. Louis MO 63130; USA

FAHEY, David W.; Aeronomy Laboratory; National Oceanic and Atmospheric Admin.; 325 S. Broadway; Boulder CO 80303; USA; 303-499-1000x3529; E; 2.3; 5.6; Ion-molecule reaction rate constants of atmospheric interest--measurement

FAIBIS, Aurel; Physics Division 203; Argonne National Lab.; 9700 S. Cass Avenue; Argonne IL 60439

FAINCHTEIN, Raul; Department of Physics; The Pennsylvania State University; 104 Davey Lab.; University Park PA 16802; USA

FAIRAND, Barry P.; 2554 Lane Road; Columbus OH 43220; USA

FAIRBANK, William M.; Department of Physics; Stanford University; Stanford CA 94305; USA

FAIRBANK JR., William M.; Department of Physics; Colorado State University; Fort Collins CO 80523; USA; 303-491-6660; E; 3.6; 1.3; 1.4; Single atom detection

FAIRCHILD, Clifford E.; Department of Physics; Oregon State University; Corvallis OR 97331; USA; 503-754-4631; E; 3.1; 3.6; 5.7

FALCONE, Roger W.; Department of Applied Physics; Edward L. Ginzton Laboratory; Stanford University; Stanford CA 94305; USA; 415-497-0189; E; 3.6; 3.8; Laser-induced atomic collision studies; Short wavelength light sources and spectroscopy

FALES, Norman J.; Analytical Section; Borg-Warner Chemical Tech. Center; PO Box 68; Washington WV 26181; USA; 304-863-7353; E; 1.1; 4.4; Chemical structures via their mass spectra--determination

FALK, Joel; Department of Electrical Engineering; University of Pittsburgh; 348 Benedum Hall; Pittsburgh PA 15261; USA; 412-624-5398; 412-624-6161; E/T; 3.10; 4.1

FALK, R. Aaron; Joint Institute for Laboratory Astrophysics; University of Colorado; Boulder CO 80309; USA; 303-492-7763; E; 2.2; Electron impact ionization and excitation of positive ions

FALLER, James; Quantum Physics Division (NBS); Joint Institute for Lab. Astrophysics; University of Colorado; Boulder CO 80309; USA; 303-492-3463 x8509; E; Experimental physics

FANG, Ta-Ming; Departments of Aerospace and Mechanical Engineering; Boston University; Boston MA 02215; USA; 617-353-2824; T; 3.6; 5.6; Atomic and molecular physics in atmospheric physics and laser interaction

FANG, Zuyun; Department of Physics; New York University; 4 Washington Place; New York NY 10003; USA

FANO, Ugo; Department of Physics and James Franck Institute; University of Chicago; 5640 S. Ellis Avenue; Chicago IL 60637; USA; 312-962-7010; T; 1.1; 2.1; 2.2; 2.3; General atomic and molecular theory; Collision dynamics

FANSLER, Todd D.; Engine Research Department; General Motors Research Lab.; Warren MI 48090; USA

FARLEY, John W.; Department of Physics; University of Oregon; Eugene OR 97403; USA; 503-686-4779; E; 3.3; 3.6; 1.2; 5.7; 1.1; Molecular ions; Blackbody radiation effects

FARNETH, William E.; Department of Chemistry; University of Minnesota; 139 Smith Hall; Minneapolis MN 55455; USA; 612-373-9929; E; 2.3; 3.1; 4.4; Small organic molecules/IR laser radiation interaction; Ion cyclotron resonance spectroscopy; Gas phase ion-molecule kinetics

FARRAR, James M.; Department of Chemistry; University of Rochester; Rochester NY 14627; USA; 716-275-5834; E; 2.3; 3.1; 4.3; 4.4; 5.5; Ion-neutral interactions--crossed beam studies; Solvation dynamics; Metal ion chemistry; Polyatomic cations--gas phase photodissociation

FARRAR, Thomas C.; Department of Chemistry; University of Wisconsin; Madison WI 53706; USA; 608-262-6158; T; Physical and analytical chemistry; Hydrides and other model compounds--molecular structure and dynamics; Relaxation phenomena and Nuclear magnetic resonance

FAULK, Jerry D.; PO Box 564; Wichita Falls TX 76307; USA; 817-766-0513; E/T

FAYER, Michael D.; Department of Chemistry; Stanford University; Stanford CA 94305; USA; 415-497-4446; E/T; 1.1; 2.3; 3.6; 3.10; 4.1; Intermolecular interactions; Molecular/optical radiation fields interactions

FEENEY, Robert K.; School of Electrical Engineering; Georgia Institute of Technology; Atlanta GA 30332; USA

FEHSENFELD, Frederick; Time and Frequency Division; National Bureau of Standards; 325 S. Broadway; Boulder CO 80303; USA; 303-497-5819; E; 1.1; 2.1; 2.3; Inelastic scattering of atoms and molecules; Chemical reactions and charge transfer; Ionization potentials and electron affinities; Molecular core binding energy

FEICHTNER, John D.; Applied Sciences Division; Westinghouse R&D Center; 1310 Beulah Road; Pittsburgh PA 15235; USA; 412-256-3629; E; 3.5; 3.6; Molecular spectra and laser interactions with matter

FEINBERG, Benedict; Accelerator and Fusion Research Div.; Lawrence Berkeley Laboratory; 1 Cyclotron Road; Berkeley CA 94720; USA; 415-486-5815; E; 4.3; Electron beam ion source (EBIS) research and development for production of highly stripped heavy ions

FEIOCK, Frank D.; W.J. Schafer Associates; 22222 Sherman Way; Canoga Park CA 91303; USA

FELD, Michael S.; Department of Physics; G.R. Harrison Spectroscopy Lab.; Massachusetts Institute of Technology; 77 Massachusetts Avenue; Cambridge MA 02139; USA; 617-253-7700; E/T; 1.1; 2.3; 4.1; 4.2; Lasers; Quantum optics; Coherent processes; Atomic and molecular structure and dynamics

FELDMAN, Barry J.; Group L-9; Basic Laser R&D Division; Los Alamos National Lab.; PO Bx 1663 MS535; Los Alamos NM 87545; USA; 505-667-5233; E/T; 2.1; Research in lasers; Nonlinear optics; Phase conjugation; Molecular collisions

FELDMAN, Donald W.; Applied Physics Division; Westinghouse R&D Center; 1310 Beulah Road; Pittsburgh PA 15235; USA; 412-256-3395; E; 3.6; 4.1; Gas discharge laser development; Atomic and molecular species--spectroscopy

FELDMAN, Henry R.; 4823 NE 42nd Street; Seattle WA 98105; USA

FELDMAN, Leonard C.; Physical Metallurgy and Ceramics Research; Materials Science Division; AT&T Bell Lab.; 600 Mountain Avenue; Murray Hill NJ 07974; USA; 201-582-5470; E; 5.1; 5.1; 5.2; Ion solid interactions in solids for study of surfaces

FELDMAN, Paul D.; Department of Physics and Astronomy; Johns Hopkins University; Baltimore MD 21218; USA; 301-338-7339; E; 5.6; 5.7; Planetary atmospheres; Astrophysics

FENN, John B.; Department of Engineering and Applied Science; Yale University; New Haven CT 06520; USA; 203-436-4812; E; 2.1; 5.1; Molecular beam methods development and application for studying: gas-gas and gas-surface collision processes

FENSKE, Richard F.; Department of Chemistry; University of Wisconsin; Madison WI 53706; USA; 608-262-0328; T; Inorganic chemistry; Ligand field theory and transition metal complexes; Molecular orbital theory; Photoelectron spectroscopy

FERGUSON, Stephen M.; Department of Physics; Western Michigan University; Kalamazoo MI 49008; USA; 616-383-4935; E; 3.4; 4.2; 5.1; Atomic collisions; Trace element analysis with particle-induced X-ray emission; Surface studies with Rutherford backscattering

FERRALLI, Michael W.; Physics and Chemistry Section; Corporate Research and Development; Lord Corporation; 2101 Peninsula Drive; Erie PA 16506; USA; 814-456-8511; E/T; 4.2; 5.1; Ion reactions on surfaces

FERRETT, Tricia A.; Room 1115; Department of Chemistry; MMRD; Lawrence Berkeley Lab.; Building 70A, Room 1115; Berkeley CA 94720; USA; 415-486-5666; E; 3.1; Photoionization

FESSENDEN, Richard W.; Department of Chemistry; Radiation Laboratory; University of Notre Dame; Notre Dame IN 46556; USA; 219-283-3446; E; 2.2; 4.2; Electron attachment to molecules and Van der Waals molecules

FETZER, Homer D.; Department of Physics; St. Mary's University; 1 Camino Santa Maria; San Antonio TX 78284; USA; 512-436-3233; 512-434-5475; E; 4.4; 5.7; 3.4; A&M ion spectrometry with space instrumentation; A&M instrumentation development and analysis; Characteristic X-rays--particle excitation

FEUERSANGER, Alfred E.; GTE Laboratories, Inc.; 40 Sylvan Road; Waltham MA 02154; USA

FIEGEL, Robert P.; Department of Physic and Astronomy; University of Oklahoma; 440 W. Brooks; Norman OK 73019; USA

FIELD, Lester M.; 12686 Kona Lane; Garden Grove CA 92641; USA

FIELD, Robert W.; Department of Chemistry; Physical Chemistry Division; Massachusetts Institute of Technology; Room 6-219; 77 Massachusetts Avenue; Cambridge MA 02139; USA; 617-253-1489; E/T; 1.1; 3.5; 3.6; 3.9; 3.2; Electronic structure of open-core and Rydberg states of diatomic molecules; Vibrationally excited polyatomic molecules--quantum ergodicity

FIGUEIRA, Joseph F.; CHM 5; Chemistry; Los Alamos National Laboratory; J566; Los Alamos NM 87545; USA; 505-667-5271; E; 3.6; 3.10; 4.1; 5.1; 5.3; Laser research and development; Interaction of laser radiation with matter; Non-linear optics

FILIPPELLI, Albert R.; Department of Physics; University of Wisconsin; 1150 University Avenue; Madison WI 53706; USA

FILSETH, Stephen V.; Department of Chemistry; York University; 4700 Keele Street; Downsview Ontario M3J1P3; Canada

FINCH, Eric C.; Department of Pure and Applied Physics; Trinity College; University of Dublin; Dublin 2; Ireland; (01) 772941; E; 1.4; Detectors: Interaction of fast heavy ions with radiation detectors

FINDLEY, Gary L.; Department of Chemistry; New York University; 4 Washington Place; New York NY 10003; USA; 212-598-2482; E/T; 1.1; 1.2; 3.1; 3.6; 3.7; Rydberg states in atoms and molecules; Photophysics and photochemistry in the vacuum ultraviolet; Quantum defect methods; Rydberg/valence interactions

FINEMAN, Morton A.; Department of Physics; Lycoming College; Williamsport PA 17701; USA; 717-326-1951; E; 2.3; 5.1; 5.4; Crossed beams for reactive collisions; Plasma-surface interactions by ions and neutrals

FINK, J. H.; 4023 East Avenue; Hayward CA 94542; USA

FINK, Manfred; Department of Physics; University of Texas; Austin TX 78712; USA; 512-471-5747; E; 2.2; Electron scattering

FINK, Richard D.; Department of Chemistry; Amherst College; Amherst MA 01002; USA; 413-542-2550/2345; E; 2.3; 4.3; Atomic and molecular beams; Chemical reactions --photochemical studies

FINK, Richard W.; School of Chemistry; Georgia Institute of Technology; Atlanta GA 30332; USA; 404-894-4030/4060; E; 3.4; 5.8; Inner-shell ionization phenomena; Laser spectroscopy of radioactive collinear beams (determination of nuclear properties)

FINK, William H.; Department of Chemistry; University of California, Davis; Davis CA 95616; USA; 916-752-0935/0937; T; 1.1; 5.1; Molecules--electronic structure calculations

FINKEL, Jack; 3802 Menlo Drive; Baltimore MD 21215; USA

FINKELSHTEIN, Eitan; 21 Liepos 10/10; Vilnius Lithuania SSR 23200; USSR

FINN, Edward J.; Department of Physics; Georgetown University; Washington DC 20057; USA; 202-625-6677; T; 1.1; 3.5; Ionic molecules; Modified Rittner potential used to describe physical properties of heterogeneous diatomics

FINZI, Jack; Hughes Aircraft Co.; Building 6 MS C-129; Culver City CA 90250; USA; 213-391-0711x3612; E; 2.1; 2.3; Vibrational energy transfer in small molecules; R-R transfer and temperature dependence

FIRESTONE, Richard F.; Department of Chemistry; Physical; Fast Kinetics; Ohio State University; 140 W. 18th Avenue; Columbus OH 43210; USA; 614-422-6819; E; 2.1; 2.3; 2.2; 3.6; 3.1; Kinetics of fast reactions of excited species; Excimer formation and decay in rare gases; Energy transfer mechanisms in gaseous systems; Laser probe fast reaction monitoring

FISANICK, Georgia J.; Chemical Kinetics Research Division; Bell Telephone Lab.; 600 Mountain Avenue; Murray Hill NJ 07974; USA; 201-582-2204; E/T; 3.7; Molecules with emphasis on fragmentation dynamics--multiphoton ionization spectroscopy

FISCHBECK, H. J.; Department of Physics and Astronomy; University of Oklahoma; Norman OK 73069; USA; 405-325-3962; E; 2.2; 4.2; 4.3; PIXE; Merging ion-electron beam to study radiation recombination

FISCHER, C. R.; Department of Physics; Queens College of CUNY; 65-30 Kissera Boulevard; Flushing NY 11367; USA; 212-520-7551; T; 1.1; 5.1; Chemisorption theory; Transition metal bonding using ab initio and semi-empirical theories

FISCHER, Charlotte F.; Department of Computer Science; Vanderbilt University; PO Box 6035-B; Nashville TN 37235; USA; 615-322-2926; T; 1.1; 3.1; Atomic transitions using the multi-configuration; Hartree-Fock method for isoelectronic sequences with relativistic effects included through the Breit-Pauli approximation

FISCHER, Traugott E.; Corporate Research Laboratories; Exxon Research and Engineering Co.; Route 22 East; Annandale NJ 08801; USA; 201-730-3045; E; 4.2; 5.1; Adsorption and reactions on solid surfaces; Applications to catalysis, metallurgy, tribology

FISHER, Edward R.; Chemistry and Chemical Engineering;

College of Engineering; Michigan Technological University; Houghton MI 49931; USA; 906-487-2047; T/E; 4.1; 4.2; 5.3; 5.4; Plasma chemistry and excited state processes measurements in molecular discharges; Nuclear-driven and high-energy beam driven lasers calculations

FISHER, Galen B.; Physical Chemistry Division; General Motors Research Lab.; General Motors Technical Center; 12 Mile and Mound Roads; Warren MI 48090; USA; 313-575-7192; E; 2.3; 5.1; Catalytically interesting materials--spectroscopy and chemical reactions involving molecules at surfaces; Inorganic molecular analog experiments

FISHER, George P.; R&D Associates; PO Box 9695; Marina Del Rey CA 90295; USA

FISHER, H. Leonard; Technical Information; Library; Lawrence Livermore National Lab.; PO Box 5500; Livermore CA 94550; USA; 415-422-5291; 5.4; 2.1; 3.6; 3.1; 3.8; Information retrieval; Literature searching; Database development

FISHER, Leon H.; 102 Encinal Avenue; Atherton CA 94025; USA

FISHER, Robert A.; Group CHM-6; Los Alamos National Lab.; Mail Stop E-535; Los Alamos NM 87545; USA; 505-667-6555; E/T; 3.6; 4.1; 3.10; Nonlinear optics; Laser physics

FISK, George A.; Laser Research and Development; Division 4218; Sandia National Laboratories; Albuquerque NM 87185; USA; 505-844-5164; E; 2.1; Rate constants governing transfer of energy between excited states of oxygen and iodine--determination

FITE, Wade L.; Department of Physics and Astronomy; University of Pittsburgh; Pittsburgh PA 15260; USA; 412-624-4356; E; 2.1; 2.2; 2.3; 4.4; 5.6; Ionization processes (associative ionization); Airborne particulate detection; Mass spectrometry

FITZHUGH, Raymond L.; Department 469B; Goodyear Plant I; 1144 E. Market Street; Akron OH 44316; USA

FITZSIMMONS, William A.; Department of Physics; University of Wisconsin; 475 N. Charter Street; Madison WI 53706; USA; 608-262-8962

FLAMM, Daniel L.; Chemical Engineering Division; Bell Telephone Laboratories; Room 6E-216; 600 Mountain Avenue; Murray Hill NJ 07974; USA; 201-582-3000; E; 5.1; 5.3; Plasma discharges and their physical and chemical interactions with solids

FLANNERY, Martin R.; School of Physics; Georgia Institute of Technology; Atlanta GA 30332; USA; 404-894-5263; T; 1.1; 1.2; 2.1; 2.3; Development of basic theoretical foundations of ion-ion recombination and of Rydberg atom-atom, molecule collisions

FLEISCHMANN, Hans H.; Department of Applied Physics; Cornell University; Clark Hall; Ithaca NY 14853; USA; 607-256-4910; T; 2.1; 5.4; Atomic collisions; Atomic processes in plasmas

FLEMING, Richard G.; Department of Physics; Midwestern State University; Wichita Falls TX 76308; USA

FLEMING, Ronald H.; Charles Evans and Associates; 301 Chesapeake Drive; Redwood City CA 94063; USA; 415-369-4567; E; 4.4

FLETCHER, Gary D.; Department of Physics; University of Virginia; McCormick Road; Charlottesville VA 22901; USA; 804-924-7591; E; 3.9; 3.6; 3.3; 3.1; 3.7

FLETCHER, Paul C.; 11359 Fuerte Drive; El Cajon CA 92020; USA

FLETCHER, Paul M.; 59 Neillian Way; Bedford MA 01730; USA

FLETCHER, William H.; Department of Chemistry; University of Tennessee; Knoxville TN 37916; USA; 615-974-3454; 615-588-1917; E; 4.1; 3.6; High resolution Raman spectroscopy

FLEURY, Paul A.; Physical Research Laboratory; AT&T Bell Laboratories; Room 1D-269; 600 Mountain Avenue; Murray Hill NJ 07974; USA; 201-582-2276; E; 3.10; 3.5; 3.6; Macromolecules --laser spectroscopy

FLIFLET, Arne W.; Code 4740; Naval Research Lab.; Washington DC 20375; USA

FLUSBERG, Allen; Avco Everett Research Laboratories; 2385 Revere Beach Parkway; Everett MA 02149; USA; 617-389-3000

FLYNN, G. W.; Department of Chemistry; Columbia University; 315 Havemeyer; New York NY 10027; USA; 212-280-4162; E; 3.6; 3.8; Energy transfer--laser studies

FOHL, Timothy; Engineering Development; Engineering Center; GTE Products Corp., Lighting Group; 60 Boston Street; Salem MA 01970; USA; 617-777-1900; E; 5.4; 5.3; 5.1; 4.1; Light source development based on radiating atoms and molecules; Fluid dynamics of arcs

FOLEY, Charles K.; Department of Chemistry; Duke University; Durham NC 27706; USA; 919-684-2849; 919-684-2414; T; 1.1; 5.2; Nuclear magnetic resonance-theoretical; Chemical shifts-calculation; Basis set dependence of chemical shift; Geometry dependence of chemical shift

FOLTZ, Greg W.; Chemical Physics Section; Health and Safety Research Division; Oak Ridge National Laboratory; Building 5500; PO Box X MS 202; Oak Ridge TN 37830; USA; 615-574-6322; E; 3.1; 3.8; Absolute cross sections for photoionization of excited cesium states using resonance ionization spectroscopy; Collisional broadening of excited cesium states

FOLTZ, Nevin D.; Department Physics; Molecular Physics; Laser Physics; Memorial University, Newfoundland; Elizabeth Avenue; St John's NF A1B3X7; Canada; 709-737-8834; 709-737-8736; E; 3.10; 4.1; 5.6; 3.6

FONCK, Raymond J.; Plasma Physics Laboratory; Princeton University; PO Box 451; Princeton NJ 08544; USA; 609-683-3281; E; 3.6; 5.4; Atomic and ionic emissions used in diagnostics of tokomak plasmas; Resonance and forbidden measurements for transitions of highly ionized atoms

FONDER, Edward F.; 717 Arlington Avenue Apt. 5D; Plainfield NJ 07060; USA

FONER, Samuel N.; Applied Physics Laboratory; Johns Hopkins University; Johns Hopkins Road; Laurel MD 20707; USA; 301-953-5000; E; 2.3; 4.4; 1.1; 5.1; 2.2; Free radicals and excited state molecules produced in chemical reactions; Modulated molecular beam mass spectrometry of highly reactive species; Ionization potentials and bond-dissociation energies by electron impact; Energy accommodation and chemical reactions on heated surfaces

FONTANA, Peter R.; Department of Physics; Oregon State University; Corvallis OR 97330; USA; 503-754-4631; T; 1.1; Atomic radiative processes

FORBRICH JR., Lt. Col. Carl A.; Munitions Division; Air Force Armament Laboratory; AFATL/DLJ; Eglin AFB FL 32542; USA; 904-882-2200; T; 2.3; 3.6; Chemical physics related to laser physics

FORD, A. L.; Department of Physics; Texas A&M University; College Station TX 77843; USA; 713-845-3337/7717; T; 1.1; 2.1; 2.3; Inner-shell ionization and charge transfer in ion-atom collisions; Structure and properties of small diatomic molecules; Born-Oppenheimer breakdown

FOREMAN, Larry R.; Dept. of Chemical Engineering and Math'l Science; University of Minnesota; Minneapolis MN 55455; USA; 612-373-5688; E/T; 5.3; Electron drift in rare gases and molecular gas-rare gas mixtures

FORK, Richard L.; Bell Laboratories 4D-417; Holmdel NJ 07733; USA

FORNARI, Luigi S.; Department of Physics; University of Texas, Arlington; PO Box

19059; Arlington TX 76019; USA; 817-273-2266; E; 1.4; 2.1; 2.2; 4.2; 4.3; Positron inelastic scattering in gases; Correlations and coherence in ion-atom-molecule scattering; Superposition principle studies

FORRESTER, A. T.; Department of Physics; University of California, Los Angeles; 7731 Boelter Hall; Los Angeles CA 90024; USA; 213-825-3311; E; 2.3; 4.3; Ion sources, positive and negative-improved; Positive to negative deuterium ions-double charge exchange

FORSEN, Harold K.; Laser Enrichment Division; Exxon Nuclear Co., Inc.; 777 106th Avenue, NE; Bellevue WA 98008; USA; 206-453-4374; E; 2.3; 3.2; 3.6; 4.3; Uranium atomic spectroscopy; Laser interactions; Ion acceleration; Kinetic collisions

FORSLEY, Lawrence P.; Laboratory for Laser Energetics; University of Rochester; 250 E River Road; Rochester NY 14623; USA; 716-2755-570/5101; E; 3.6; 4.3; Short-lived phenomena on picosecond and sub-picosecond time scales--development of instrumentation for explaining

FORSYTH, James M.; Laboratory for Laser Energetics; University of Rochester; 250 E. River Road; Rochester NY 14623; USA; 716-275-5659; E; 3.4; 4.1; 5.4; X-rays produced by laser-heated plasmas--applications

FORTNA, John D.; 2104 N. Troy Street; Arlington VA 22201; USA

FORTNER, Richard J.; Physics Division 401; Lawrence Livermore National Lab.; PO Box 808 L330; Livermore CA 94550; USA; 415-443-7957; E/T; 3.4; 4.2; 5.4; Plasma spectroscopy; Inner-shell vacancies in ion-atom collisions

FORTSON, E. Norval; Department of Physics FM-15; University of Washington; Seattle WA 98195; USA; 206-543-2665; 206-543-2770; E; 1.4; 3.3; 3.6; 4.1; Parity and time reversal symmetry in atoms

FOSSUM, Eric R.; Department of Electrical Engineering; Columbia University; Seeley Mudd Building; New York NY 10027; USA

FOSTER, Gershom C.; Department of Physics; University of Connecticut; Torrington Campus; Torrington CT 06790; USA; 203-482-7635x33; T; 2.2; Atomic collisions; Glauber/Eikonal methods applied to charged particle-atom ion collisions

FOU, Cheng-Ming; Department of Physics; College of Arts and Science; University of Delaware; Newark DE 19716; USA; 302-451-2677; 302-451-6542; E; 2.1; 4.2; 2.2

FOURNIER, Paul G.; Physique; Spectroscopie de Translation; Universite Paris Sud; Bat 478; Orsay 91400; France; 6.9417486; E/T; 1.1; 1.1; 1.1; 3.1; KeV atomic and molecular collisions; Doubly charged states; Ions-surface interactions CSIMS

FOWLER, Bruce W.; Close Combat Team; Advanced Systems Concepts Office; U.S. Army Missile Command; Redstone Arsenal AL 35898; USA; 205-876-1245; 205-876-1442; E/T; 1.1; 3.1; 3.3; 3.7; 3.11; Propagation of radiation through the turbulent boundary layer and its constituents, overt and natural sources and detectors of that radiation; Estimation of the effects of these phenomena on the rate of attrition; Development of parametric representations of boundary layer effects

FOWLER, Howland A.; 5413 Albemarle Street; Bethesda MD 20016; USA

FOWLER, Richard G.; Department of Physics and Astronomy; University of Oklahoma; Room 131; 440 W Brooks Street; Norman OK 73019; USA; 405-325-3961; 405-329-6529; E/T; 5.3; 3.1; 2.2; 3.3; 5.4; Discharge models--lighting mechanisms; Atomic and molecular lifetimes; Collision cross sections; Infra red determination of organic structure

FOX, Kenneth; Department of Physics and Astronomy; University of Tennessee; Knoxville TN 37996; USA; 615-974-3342; 301-344-8533; T/E; 3.3; 3.5; 3.6; 1.1; 1.3; High-resolution IR spectroscopy of simple molecules in gas phase; Laser spectroscopy; Spectral frequencies and intensities--isotopic dependence; Atmospheric and environmental applications of A&M physics

FOX, Robert A.; Department of Physics; Natural Sciences Division; University of Hawaii, Hilo; 1400 Kapiolani Street; Hilo HI 96720; USA; 808-961-9379; 808-961-9383; E; 3.1; 3.6; 2.2

FOX, Russell E.; 215 Woodside Road; Pittsburgh PA 15221; USA

FRADKIN, David M.; Department of Physics and Astronomy; Wayne State University; Detroit MI 48202; USA; 313-577-2792; 313-577-2720; T; 3.7; 2.2; Radiation reaction in intense laser fields; Scattering of charged particles

FRANCIS, Kevin J.; 6733 W 115th Street; Worth IL 60482; USA

FRANCK, Carl P.; Department of Physics; Atomic and Solid State Phys.; Cornell University; Clark Hall; Ithaca NY 14853; USA; 607-256-3562; 607-256-5215; E; 3.10; 1.1; 4.2; 3.4; Scattering, inelastic X-ray; Excitation dynamics ground state correlations; Synchrotron radiation; K-shell excitations

FRANCKE, Ricardo E.; Institute De Fisica Da Ufrgs; Predio M-117; Ave Bento Goncalves 9500; Porta Alegre 90000; Brazil; 0512-36-4677; E; 1.1; 2.1; 3.6; 4.1

FRANCO, Victor; Department of Physics; Brooklyn College; Brooklyn NY 11210; USA; 212-780-5806; 212-780-5418; T; 2.1; 2.2; Intermediate and high energy atomic collision theory including ion-atom, atom-atom, charged particle-atom collisions; Glauber theory and eikonal exchange calculations

FRANKEL JR., Donald S.; Aerodyne Research, Inc.; Bedford Research Pk.; Crosby Dr.; Bedford MA 01730; USA; 617-275-9400; E; 3.8; 5.5; 5.6; High altitude rocket plane phenomenology; Laser induced chemistry; Environmental/atmospheric chemistry

FRANKEL, Robert D.; Laboratory for Laser Energetics; University of Rochester; 250 E. River Road; Rochester NY 14623; USA; 716-275-4548; E; 1.3; Biological molecules--kinetic X-ray diffraction

FRANKEN, Peter; Optical Sciences Center; University of Arizona; Tucson AZ 85721; USA

FRANKLIN, Joseph L.; Department of Chemistry; Rice University; PO Box 1892; Houston TX 77001; USA; 713-527-4845; E; 2.3; Energy and angular distribution of products of ionic dissociation; Ion-molecule reactions

FRANZ, Frank A.; Department of Physics; Indiana University; Bloomington IN 47405; USA; 812-337-3735; E; 2.1; 3.6; Atomic collisions; Optical pumping; Laser spectroscopy; Relaxation phenomena

FRANZEN, Wolfgang; Department of Physics; Boston University; Boston MA 02215; USA; 617-353-2615; E/T; 1.2; 2.2; Excitation function measurement for production of high Rydberg states in helium by bombardment with energy-analyzed electrons

FRAZIER, George B.; Physics International Company; 2700 Merced Street; San Leandro CA 94577; USA

FRECHETTE, Michel F.; Institute De Recherche; D'Hydro-Quebec; Case Postale 10000 Local P-70; Varennes Que J0L2P0; Canada

FRED, Mark; Chemistry Division; Argonne National Laboratory; 9700 S. Cass Avenue; Argonne IL 60439; USA; 312-972-3637; E/T; 3.2; Atomic spectroscopy; Actinide elements--term analysis

FREDERICK, Jonathan W.; PO Box 392; Enosburg Falls VT 05450; USA

FREED, Charles; Lincoln Laboratory; MIT; P.O. Box 73; 244 Wood Street; Lexington MA 02173; USA; 617-863-5500 x3639; 617-259-9338 (H); E; 1.1; 2.1; 3.3; 4.1; 5.3; Vibrational rotational constants of carbon dioxide isotopes--determination; High resolution IR spectroscopy

FREED, Jack H.; Department of Chemistry; Baker Laboratory of Chemistry; Cornell University; Ithaca NY 14853; USA; 607-256-3647; E/T; 5.2; Molecular dynamics and magnetic resonance in fluids

FREED, Karl F.; James Franck Institute; Department of Chemistry; University of Chicago; 5640 Ellis Avenue; Chicago IL 60637; USA; 312-962-7202; T; 1.1; 1.3; 2.3; 3.1; 3.8; Molecular and atomic electronic structure; Photodissociation and molecular relaxation processes; Statistical mechanics of polymers; Collisional energy transfer in molecules

FREEDMAN, Andrew; Center for Chemical and Environmental Physics; Research Group; Aerodyne Research, Incorporation; 45 Manning Road; Billerica MA 01821; USA; 617-663-9500; E; 2.3; 3.1; 3.6; 5.6; 5.5; Photo-chemistry and spectroscopy; Chemical kinetics; Spectral diagnostics

FREEDMAN, Stuart; Physics Division-Building 203; Argonne National Lab.; 9700 South Cass Avenue; Argonne IL 60439; USA

FREEMAN, Daryl E.; Harvard College Observatory; 60 Garden Street; Cambridge MA 02138; USA; 617-495-2783; E; 3.2; 3.5; High resolution vacuum UV spectroscopy of atoms and small molecules

FREEMAN, Gordon R.; Department of Chemistry; University of Alberta; Edmonton AB T6G2G2; Canada; 403-432-3468; E; 5.2; 5.3; 4.2; 2.2; Electron transport in fluids; Optical absorption spectra of electrons in fluids; Electrons in fluids, energy states; Electrons in fluids, molecular structure effects

FREEMAN, Mark P.; Dorr-Oliver Company; Stamford CT 06904; USA

FREEMAN, Richard R.; Research Division; Bell Telephone Laboratories; Holmdel NJ 07733; USA; 201-949-4933; E; 3.6; 4.1; Spectroscopy; Nonlinear optics

FREEMAN, Robert D.; Department of Chemistry; Oklahoma State University; Stillwater OK 74078; USA; 405-624-6676; T; 1.1; Atomic and molecular data in statistical thermodynamic calculations

FREUND, Hans J.; Physical Chemistry; Inst Fur Physikalische Und; Theoretische Chemie Der Univ.; University Erlangen; Egerland Str. 3; Erlangen D-8520; FRG; 0913-1-857310; 09131-857342; E/T; 1.1; 3.1; 4.2; 5.1; 5.2; Electron spectroscopy; Molecular adsorption on surfaces; Molecular solids

FREUND, Robert S.; Chemical Kinetics Research Division; Bell Telephone Laboratories; Room 1D-256; 600 Mountain Avenue; Murray Hill NJ 07974; USA; 201-582-3865; E; 1.2; 2.2; 3.6; High-Rydberg molecules; Electron-molecule collisions; Spectroscopy

FRIBERG, Stephen R.; Department of Physics; Bausch and Lomb Hall; University of Rochester; Rochester NY 14627; USA

FRIDDEL, Kenneth D.; 17044 16th Avenue, SW; Seattle WA 98166; USA

FRIED, Zoltan; Department of Physics and Applied Science; University of Lowell; Lowell MA 01854; USA; 617-452-5000 x2574; T; 1.3; 3.7; 3.9; Multiphoton emission in exotic atom cascades; Atoms in external fields

FRIEDMAN, Helen L.; 242-23 54th Avenue; Douglaston NY 11362; USA

FRIEDMAN, Herbert; Code 4190; Space Science Division; US Naval Research Laboratory; 4555 Overlook Avenue, SW; Washington DC 20375; USA; 202-767-3363; E; 3.4; 5.7; X-Ray astronomy; X-Ray spectroscopy; Observational astronomy

FRIEDMAN, Joel M.; Bell Telephone Laboratories; AT&T Bell Laboratories; 600 Mountain Avenue; Murray Hill NJ 07974; USA; 201-582-3894; E; 1.3; 3.5; 4.1; Macromolecular systems --spectroscopy (resonance Raman); Biomolecular physics; Ultrafast macromolecular dynamics; Time resolved Raman scattering

FRIEDRICH, Donald M.; Department of Chemistry; Hope College; Holland MI 49423; USA; 616-392-5111x3222; 616-392-5111x3213; E; 3.1; 3.7; Two-photon spectroscopy and photoselection spectroscopy

FRIEDRICH, Harald S.; Physik-Department; Theoretische Physik; T. U. Muenchen; James-Franck-Strasse; Garching D-8046; West Germany; 089-3209-2358; 089-3209-2354; T; 3.9; 1.1; 1.2; 1.3; 2.1

FRIND, Gerhard; Corporate Research and Development Center; General Electric Co.; 1 River Road; Schenectady NY 12345; USA; 518-861-5392; E; 5.4; Arc plasmas temperature measurements

FRITSCH, Wolfgang; Bereich P; Hahn-Meitner-Institut; Glienicker Str 100; Berlin 39 D-1000; West Germany; T; 2.1; 4.2; 5.4

FROELICH, David V.; Physics Department; University of Notre Dame; Notre Dame IN 46556; USA; 219-239-5662; 219-239-5673; T; 5.2; 1.3; Deep levels in semiconductors in bulk and at surface

FROMMHOLD, Lothar W.; Department of Physics; University of Texas; Austin TX 78712; USA; 512-471-5100; E; 3.8; Atoms colliding--Raman and IR spectroscopy

FROST, Lindsay R.; 25 Commercial Road; Port Noarlunga 5167S; Australia

FRUEHOLZ, Robert P.; Atomic Physics Section; Chemistry and Physics Laboratory; The Aerospace Corporation; 92957; Los Angeles CA 90009; USA; 213-648-6975; E/T; 3.3; 3.1

FRY, Edward S.; Department of Physics; Texas A&M University; College Station TX 77843; USA; 713-845-5125; E; 1.1; 2.2; 5.1; Low energy electron-hydrogen atom scattering-differential cross sections; Hydrogen atom scattering by surfaces using two photo laser excitation; to the 2S state; Atomic cascade coincidence experiments

FRYE, Daniel D.; Physics Department; University of Virginia; McCormick Road; Charlottesville VA 22901; USA; T; 1.1; 3.1

FUHR, Jeffrey R.; Atomic and Plasma Radiation Division; National Bureau of Standards; Gaithersburg MD 20899; USA; 301-921-2071; E; 1.1; 3.1; 5.7; Atomic transition probabilities--critical evaluation of; experimental and theoretical data

FUJISAKI, Iatsuo; Toyotama Kami 1-Chome; 26 Banchi; Nerima-Ku Tokyo; Japan

FUJIWARA, Izuru; Department of Math Sciences; Osaka Furitsudai Surikogaku; Univ. of Osaka Pref; Sakai 591; Japan

FUKUI, Katsura; Air Force Geophysics Lab.; Hanscom AFB MA 01731; USA; 617-861-4281; E; 5.6; 5.7; Solar UV radiation flux analysis; Aeronomy applications; Upper atmospheric condition analysis

FULLER, Kirk; Department of Physics; Texas A&M University; College Station TX 77843; USA

FULTON, Robert L.; Department of Chemistry; Dittmer Laboratory of Chemistry; Florida State University; Tallahassee FL 32306; USA; 904-644-6449; T; 2.1; Rotational relaxation--theory; Di-electric friction--theory; Fluctuation phenomena--theory

FURST, Mitchell L.; Department of Physics U-46; University of Connecticut; Storrs CT 06268; USA; 203-486-5112; 203-486-4915; E; 2.1; 3.2; 3.4; 4.2; Ion-atom collision spectroscopy of Looke V-Z ions incident on Ar gas; High-resolution soft X-ray and VUV spectra from multi-L vacancy transitions

FURUMOTO, Horace; Candela Corporation; 19 Strathmore Road; Natick MA 01760; USA; 617-653-7373; E; 4.1; 3.1; Dye laser research and development; Chemical kinetics of dyes

FUSON, Nelson; Department of Physics (retired); Molecular Spectroscopy; Fisk University; 1803 Morena St; Nashville TN 37208; USA; 615-329-0823; E; 3.3; 5.2; Infrared and Raman spectra

FUTRELL, J. H.; Department of Chemistry; University of Utah; Salt Lake City UT 84112; USA; 801-581-7307; E/T; 2.3; 4.4; Gas phase reaction dynamics; Ion-molecule reactions and energy transfer; Uni-molecular and collision-induced dissociation of ions; Mass spectrometry;

GABBARD, Fletcher; Department of Physics and Astronomy; University of Kentucky; Lexington KY 40506; USA; 606-259-2960; E; 3.4; 4.2; X-ray spectra applied to elemental analysis

GABRIEL, Oscar V.; 345 Poplar Ave. Apt. 0-507; Devon PA 19333; USA

GABRIELSE, Gerald; Department of Physics FM-15; University of Washington; Seattle WA 98195; USA; 206-543-1372; 206-543-2770; T; 1.4; 4.4; Tests of fundamental symmetries and theories (QED, CPT, etc.); Relativistic synchrocyclotron at sub-eV energies; Precision comparison of proton and antiproton masses; Particle trapping and trap improvement

GAILY, T. D.; Department of Physics; University of Western Ontario; London Ontario N6A3K7; Canada; 519-679-2568; E; 3.6; 1.4; 2.1

GAISER, James E.; Department of Physics; East Carolina University; Greenville NC 27834; USA; 919-757-6894; 919-757-6379; E; 2.1; 4.2

GALANTOWICZ, Thomas A.; Department of Electrical Engineering; Union College; Schenectady NY 12308; USA; 518-370-6270; E/T; 4.3; Submillimeter wave sources, detectors and signal processing components

GALBRAITH, Harold W.; Theoretical Division; Group T-12; Los Alamos National Laboratory; PO Box 1663; Los Alamos NM 87545; USA; 505-667-6198; T; 3.6; 3.7; 4.1; Laser spectroscopy; Multiphoton excitation; Quantum electronics

GALLAGHER, Alan C.; Joint Institute for Laboratory Astrophysics; University of Colorado; PO Box 440; Boulder CO 80309; USA; 303497-3936; E; 2.2; 2.1; 5.3; Collisional energy transfer and line-shapes in gases; Electron collisional excitation; Gas discharges; Gas discharge chemistry

GALLAGHER, Jean W.; Atomic Collision Data Center; Joint Institute for Laboratory Astrophysics; University of Colorado; Campus Box 440; Boulder CO 80309; USA; 303-492-8089; FTS 320-3181; E/T; 2.1; 2.2; 2.3; 3.1; 5.3; Compilation and review of data in specialty areas

GALLAGHER, Thomas F.; Physics Department; University of Virginia; McCormick Road; Charlottesville VA 22901; USA; 804-924-6817; E; 1.2; 2.1; 3.6; 3.8; 4.1; Rydberg states--bound and autoionizing--using IR and optical (laser) spectroscopy--studies of structures; Highly-excited atomic states--collision properties; Frequency modulation spectroscopy

GALLO JR., Charles F.; Research and Engineering Division; Xerox Corporation; Building 129; Rochester NY 14644; USA; 716-422-5912; E/T; 5.3; Gas discharges, coronas, sparks, arcs, and glow discharges--radiative and electrical properties

GALLUP, Gordon A.; Department of Chemistry; University of Nebraska; Lincoln NE 68588; USA; 402-472-2740; T; 1.1; 2.2; 2.1; Energy surfaces for polyatomic molecules--calculation; Vibration states of temporary negative ion resonances--properties

GAMO, Hideya; School of Engineering; University of California, Irvine; Irvine CA 92717; USA; 714-833-5819; E/T; 3.8; 4.1; Atomic and molecular collisions--pressure broadening and phase shift; Gain, saturation parameter and collision effect of atoms and molecules in gas lasers

GAN, Zi-Zhao; Department of Physics; Princeton University; PO Box 708; Princeton NJ 08544; USA

GANAS, Perry S.; Department of Physics; California State University, LA; Los Angeles CA 90032; USA; 213-224-3607; T; 1.1; 2.2; 3.1; Electron scattering from atoms and molecules; Optical oscillator strength calculation; Excitation, ionization, photoionization and elastic cross sections

GANGOPADHYAY, Pradip; Department of Physics and Astronomy; University of Pittsburgh; Allen Hall; Pittsburgh PA 15260; USA

GAO, Ru-Shan; Department of Physics; Rice University; PO Box 1892; Houston TX 77251; USA

GARA, Aaron D.; Newport Corporation; PO Box 8020; 18235 Mt. Baldy Circle; Fountain Valley CA 92728; USA

GARBUNY, Max; Applied Physics Division; Westinghouse Research and Development Center; Building 401; 1310 Beulah Road; Pittsburgh PA 15235; USA; 412-256-3267; E/T; 3.8; Molecular energy transfer processes induced by laser radiation; Laser-induced chemistry

GARCIA, Jorge R.; Reinoldus-Gymnasium; Hoekerstrasse 7; Dortmund-Dorstfeld 4600; West Germany

GARCIA JR., Jose D.; Department of Physics; University of Arizona; Tucson AZ 85721; USA; 602-626-1114; T; 2.1; 2.3; Charge exchange and excitation collision theory

GARDINER, Crispin W.; Department of Physics; University of Waikato; Hamilton; New Zealand

GARDINER JR., W. C.; Department of Chemistry; University of Texas; Austin TX 78712; USA; 512-471-7166; E; 2.3; Hydrocarbons--high temperature reactions

GARDNER, F. M.; 111 Stockade Road; S. Glastonbury CT 06073; USA

GARDNER, J. H.; Brigham Young University; Physics 296 Eyring Science; Provo UT 84601; USA

GARDNER, Larry D.; Atomic and Molecular Physics Division; Mail Stop 50; Harvard-Smithsonian Center for Astrophysics; 60 Garden Street; Cambridge MA 02138; USA; 617-495-7286; 617-495-5475; E; 2.2; 3.1; 3.6; Dielectronic recombination electron impact excitation of ions; Neutral free radicals via lasers and fast molecular/ionic beams technique; Photodissociation and photodetachment

GARETZ, Bruce A.; Department of Chemistry; Polytechnic Institute of NY; 333 Jay Street; Brooklyn NY 11201; USA; 212-643-3565; E/T; 2.3; 3.5; 4.1; Molecular spectroscopy using lasers; Molecular dynamics via computer modeling; Nonlinear optics

GARING, John S.; Optical Physics Division; USAF Geophysics Laboratory; Hanscom AFB MA 01731; USA; 617-861-2951; 617-862-7592; E; 5.6; 3.5; 1.1; 3.1; Atomic and molecular science research; Laboratory director

GARRETT, William R.; Chemical Physics Department; Chemical Physics Division; Oak Ridge National Laboratory; PO Box X; Oak Ridge TN 37831; USA; 615-574-6231; E/T; 3.10; 3.7; 3.6; 1.1; 2.2; Nonlinear optics; Multiphoton ionization phenomena; Electron collision phenomena; Negative ions

GARSCADDEN, Alan; Plasma Physics; Power Division; Air Force Wright Aeronautical Lab.; USAF; AWALPOOC-3; Wright-Patterson AFB; Dayton OH 45433; USA; 513-255-2923; 513-255-3835; E; 1.1; 1.2; 2.2; 2.3; 3.9; Electric discharge research; electron transport theory; Plasma enhanced chemistry and thin films; Non-linear plasma diagnostics, combustion studies; Plasma instability, light sources

GARSTANG, Roy H.; Joint Institute for Laboratory Astrophysics; University of Colorado; Boulder CO 80309; USA; 303-492-7795; T; 1.1; 3.9; 3.1; 5.7; Atomic transition probabilities and atomic energy levels--calculations; Atoms in high magnetic fields--behavior

GARTH, J. C.; RADC/ESR; Rome Air Development Center; Hanscom AFB MA 01731; USA

GATLAND, Ian R.; School of Physics; Georgia Institute of Technology; Atlanta GA 30332; USA; 404-894-5261; T; 1.1; 2.3; 5.3; Ions with atoms and molecules--interaction potential; Ion drift, diffusion and chemical reactions

GAVIN, Basil; Accelerator and Fusion Division; Lawrence Berkeley Lab.; University of California; Hearst Avenue; Berkeley CA 94720; USA; 415-486-6159; E; 4.3; Ion source development; Heavy elements

GAY, Timothy J.; Department of Physics; University of Missouri, Rolla; Rolla MO 65401; USA; 314-341-4797; 314-341-4702; E/T; 2.1; 2.2; 2.3; 4.2; 5.1

GEBALLE, Ronald; Dean, Graduate School; Vice Provost for Research; University of Washington; AG-10; 201 Administration Building; Seattle WA 98195; USA; 206-543-5900; E; 3.1; 5.4; Negative alkali ions--photodetachment; Slow, highly charged ion beams from laser-induced plasmas

GELBART, William M.; Department of Chemistry; University of California, LA; 405 Hilgard Avenue; Los Angeles CA 90024; USA; 213-825-2005; T; 3.5; 3.8; 5.1; Light scattering and optical properties of simple fluids; Molecular radiationless transitions and gas phase photochemistry; Spectra and dynamics of vibrationally excited molecules

GELBWACHS, Jerry A.; Chemistry and Physics Lab.; The Aerospace Corporation; PO Box 92957; Los Angeles CA 90009; USA; 213-648-5949; E; 1.2; 3.2; 3.5; Atomic and molecular spectroscopy; Photoacoustic spectroscopy; Rydberg atom spectroscopy

GELFAND, Jack J.; Dept. of Mechanical and Aerospace Engineering; Princeton University; D-426 Engineering Quadrangle; Princeton NJ 08544; USA; 609-452-4745; E; 2.1; 2.3; 3.5; Molecular collision dynamics; Gas phase vibrational and rotational relaxation; Chemical dynamics; Molecular spectroscopy

GELINAS, Robert J.; Advanced Laser Division; Lawrence Livermore National Lab.; P.O. Box 808; Livermore CA 94550; USA; 415-423-2267; T; 4.1; 5.2; 5.6; Computational physics; Non-equilibrium physics; Thermoelasticity/stress-birefringence; Solid state laser design

GELLER, M.; Cometary and Molecular Physics; Space Sciences; Jet Propulsion Laboratory; 4009 Ventura Cyn; Sherman Oaks CA 91423; USA; 818-354-2593; T; 1.1; 3.3; 3.5; 3.9; 5.6

GELTMAN, Sydney; Quantum Physics Division (NBS); Joint Institute for Laboratory Astrophysics; University of Colorado; Boulder CO 80309; USA; 303-492-7853; T; 3.7; 3.8; 3.9; 2.1; 2.2; Atomic ionization by coherent radiation fields

GEMMELL, Donald S.; Physics Division; Argonne National Laboratory; Building 203; 9700 S. Cass Avenue; Argonne IL 60439; USA; 312-972-4044; E; 4.2; Fast (MeV) molecular-ion beams

GENACK, Azriel; Physical and Materials Sciences Laboratory; Corporate Research Laboratories; Exxon Research and Engineering Co.; PO Box 45; Linden NJ 07036; USA; 201-474-2377; E; 2.3; 3.8; 4.1; Optical coherent transient experiments in atoms and molecules; Time scale measurements of dephasing energy relaxation, laser-induced chemistry measurements

GENTILE, Thomas R.; Department of Physics; 26-229, MIT; Cambridge MA 02139; USA; 617-253-6813; 617-253-0883; E; 1.2; 3.11

GENTRY, W. R.; Department of Chemistry and Chemical Physics; University of Minnesota; 207 Pleasant Street, SE; Minneapolis MN 55455; USA; 612-373-0174; E; 2.3; Molecular collisions--chemical dynamics

GEOHEGAN, David B.; Department of Electrical Engineering; Gaseous Electronics Lab.; 607 E. Healy Street; Champaign IL 61820; USA

GEORGE, James; 15 Oakledge Road; Swampscott MA 01907; USA

GEORGE, Patricia M.; Materials Science; Solid State Research; MS 53-8241; Aerojet Electrosystems; P.O. Box 296; 1100 W. Hollyvale; Azusa CA 91702; USA; 818-812-2854; 818-812-2717; E; 5.1

GEORGE, Simon; Department of Physics and Astronomy; California State University; Long Beach CA 90840; USA; 213-498-4924; E; 3.3; Antimony and tellurium--IR atomic emission spectra

GEORGE, T. V.; Plasma Technologies; Development and Technology; Office of Fusion Energy; U.S. Department of Energy; Washington DC 20005; USA; 301-353-4965; E; 1; 2.6; 2.6; 1; 3.4; Scattering of optical radiation

GEORGE, Thomas F.; Department of Physics and Astronomy; State University of NY, Buffalo; 239 Fronczak Hall; Buffalo NY 14620; USA; 716-636-2531; T; 3.8; 5.1; Laser-stimulated molecular rate processes both in the gas phase and at a solid surface

GEORGES, A. Thomas; Department of Physics; University of Crete; Herakleion Crete; Greece

GERARDI, Gary J.; Department of Chemistry; William Patterson College; 300 Pompton Road; Wayne NJ 07470; USA; 201-595-2438; E/T; 5.2; Dynamic nuclear polarization and other magnetic resonance measurements used to study molecular motions on liquids

GERARDO, James B.; Laser Research and Development; Division 4210; Sandia National Laboratories; PO Box 5800; Albuquerque NM 87185; USA; 505-844-3871; E; 3.6; 5.3; Gas-laser kinetics; Laser spectroscopy

GERBER, Andrew; 412 Whitney Ave. No. 1; New Haven CT 06511; USA

GERJUOY, Edward; Department of Physics and Astronomy; University of Pittsburgh; Pittsburgh PA 15260; USA; 412-624-4335; 412-243-5774 (H); T; 2.1; 2.2; 5.6; Atmospheric pollution by particulates; Variational methods; Formal theory of collisional processes

GERMAN, Kenneth R.; Burleigh Instruments Incorporation; Burleigh Park; 28 Chestham Way; Fishers NY 14453; USA; 716-924-9355; E; 4.1; 1.1

GERSH, Michael E.; Aerodyne Research, Incorporation; Crosby Drive; Bedford MA 01730; USA; 617-275-9400; E/T; 2.3; Kinetics and spectroscopy--experimental and analytical studies

GERSHON, Nahum D.; DCRT; National Institutes of Health; Building 12A Rm 2007; Bethesda MD 20205; USA; 301-496-1135; T; 5.1; Light scattering from molecular fluids

GERSTEN, Joel I.; Department of Physics; City College of New York; New York NY 10031; USA; 212-690-6908; T; 5.1; Molecules with solids--interactions

GESCHWIND, Stanley; Bell Laboratories 1D-330; PO Box 261; Murray Hill NJ 07974; USA

GETHNER, Jon S.; Science Laboratories; Corporate Research Laboratories; Exxon Research and Engineering Company; PO Box 45; Linden NJ 07036; USA; 201-474-2662; E; 5.1; 5.2; Absorbed molecules--spectroscopy; Reactions in dense media--kinetics

GEVA, Michael; 2A-444; Bell Lab.; 600 Mountain Avenue; Murray Hill NJ 07974; USA

GHANBARI, Ebrahim; Varian /Extrian; Blackburn Industrial Park; Gloucester MA 01930; USA

GIANNELLA, Ruggero; Cre Enea Di Frascati; C P 65; Frascati Roma 00044; Italy

GIANTURCO, Franco A.; Department of Chemistry; Theoretical Chemistry; University of Rome; Citta Universitaria; Roma 0018500044; Italy; (06)493202x30; T; 1.1; 2.1; 2.2; 2.3; 5.7; Interaction dynamics in Van der Waals molecules; Slow-energy electron-polyatomics collisions; Proto-molecule charge-transfer collisions; Collisional vibrational excitations

GIBBONS, Patrick C.; Department of Physics; Washington University; Campus Box 1105; St. Louis MO 63130; USA; 314-889-6271; E; 2.2; 5.2; 3.4; 1.1; Collective excitations in atoms

GIBBS, Hyatt M.; Bell Telephone Laboratories; 600 Mountain Avenue; Murray Hill NJ 07974; USA; 201-582-5370; E; 3.6; 4.1; Laser beams with absorbers--interactions; Superfluorescence; Superradiance; Optical bistability

GIBSON, F. P.; 6523 Marsh Avenue, NW; Huntsville AL 35806; USA

GIDLEY, David W.; Department of Physics; Randall Laboratory; University of Michigan; Ann Arbor MI 48109; USA; 313-763-3464; E; 1.4; 5.1; 1.3; 2.2; QED fundamental tests using positronium positron physics

GIEN, Tran T.; Department of Physics; Memorial University of Newfoundland; 11 Parliament Street; St. Johns A1A2Y6; Canada; 709-753-3386; 709-739-5799; T; 2.2; 2.1; Atomic and molecular collisions

GIESE, Clayton F.; Department of Physics; University of Minnesota; Minneapolis MN 55455; USA; 612-373-5469/5807; E; 2.1; 2.3; 4.4; Mass spectroscopy; Ion-molecule interactions; Ionization-dissociative; Ion-molecule collisions-rotational and vibrational excitations

GIESE, John P.; Department of Physics; James R. Macdonald Laboratory; Kansas State University; Cardwell Hall; Manhattan KS 66502; USA; 913-532-6786; 913-532-6777; E; 2.1; 4.2; 2.3; 1.1; Energy gain spectroscopy in electron capture collisions

GILBERT, Sarah L.; Time and Frequency-524.10; National Bureau of Standards; 325 Broadway; Boulder CO 80303; USA; E; 1.1; 1.4; 3.6; 3.9; 4.1

GILBODY, Henry B.; Pure and Applied Physics; Queens University of Belfast; Belfast BT71NN; United Kingdom; (0232)245133; E; 2.1; 2.3; 4.2; 4.3; 5.4

GILLEN, Keith T.; Molecular Physics Dept.; Chemical Physics Laboratory; SRI International; 333 Ravenswood Avenue; Menlo Park CA 94025; USA; 415-859-3250; E; 2.1; 2.3; 3.6; 4.4; 5.1; Collisions of metastable excited atoms and ions; Nonresonant multiphoton ionization; Surface analysis techniques

GILLES, Paul W.; Department of Chemistry; University of Kansas; Lawrence KS 66045; USA; 913-864-3829; E; 1.1; 4.4; Energies-dissociation; Atoms and molecules produced in high temperature vaporizing systems--mass spectrometric study

GILLESPIE, George H.; Physical Dynamics, Inc.; PO Box 1883; La Jolla CA 92038; USA; 619-457-3201x321; T; 2.1; Atomic collisions at very high energies

GILLESPIE, Sherry J.; 6 Nahma Avenue; Essex Junction VT 05452; USA

GILLISPIE, Gregory D.; Department of Chemistry; North Dakota State University; Fargo ND 58105; USA; 701-237-8244; E; 3.6; 2.3; 4.3; 5.2; Proton tunneling in intramolecular hydrogen bonds; Excited state proton transfer; Supersonic jet spectroscopy; Shpol'skii matrix spectroscopy

GILMORE, Forrest R.; Nuclear Effects Division; R&D Associates; PO Box 9695; Marina Del Rey CA 90295; USA; 213-822-1715; T; 5.6; 5.5; 2.3; 3.1; Atomic and molecular cross sections, reaction rates, and radiation rates to the upper atmosphere, to atmospheric nuclear bursts, and to flames--application

GILMORE, John; Vice-President; Minuteman Laboratories, Inc.; 912-920 Main Street; Acton MA 01720; USA; 617-263-2632/8888; E; 4.3; XUV and UV monochromators--design and manufacture; Polychromators plus sources and detectors in the 2-1100 angstrom region

GILSON, Greyson; PO Box 237; Mt. Sunapee NH 03772; USA

GIMARC, Benjamin M.; Department of Chemistry; University of South Carolina; Columbia SC 29208; USA; 803-777-2677; T; 1.1; 2.3; Applications of quantum mechanical methods to chemical; Problems including molecular structures, relative molecular stabilities, barriers to inversion and internal rotation, hydrogen bonding, and reaction mechanisms

GINELL, Robert; Department of Chemistry; Brooklyn College of CU, New York; Bedford Ave and Ave H; Brooklyn NY 11210; USA; 516-868-6943; 718-780-5753; T; 1.1; Association theory: refers to the structure and forces of associated molecules

GINGERICH, Karl; Department of Chemistry; Texas A&M University; College Station TX 77843; USA; 409-845-2431; E; 3.5; 4.4; 5.2; 3.3; High temperature molecules --mass spectrometry and matrix isolation spectroscopy

GINSBURG, Charles A.; 1170 Genesee Street; Rochester NY 14611; USA

GINTER, Marshall L.; Institute for Physical Science and Technology; University of Maryland; College Park MD 20742; USA; 301-454-4411; E/T; 1.2; 1.1; 3.2; 4.2; 3.1; High resolution (RP> 150,000) spectroscopy, 1-1000 nm; Channel couplings in atoms and small molecules; Optical systems in high and ultrahigh vacuum; VUV-soft X-ray light sources and applications

GIORDMAINE, Joseph A.; Advanced Technology Development Lab.; AT&T Bell Laboratories; 600 Mountain Avenue; Murray Hill NJ 07974; USA; 201-582-2173; E; 3.10; 3.11; 5.1; Atomic and molecular beams with surfaces--interaction; Molecular beam epitaxial crystal growth

GIRARD, Thomas A.; Institute De Physique; Univ. De Louvain; Phemin Du Cyclotron Z B-1348; Louvain La Neuve; Belgium

GIRARDEAU JR., Marvin D.; Department of Physics; University of Oregon; Eugene OR 97403; USA; 503-686-5210; T; 2.3; Quantum field theory of composite particles and its application to reactive atomic collision theory and chemical reaction theory foundation

GISLASON, Eric A.; Department of Chemistry; University of Illinois, Chicago; PO Box 4348; Chicago IL 60680; USA; 312-996-5423; E/T; 2.1; 1.3; 2.3; Vibronic energy transfer in collisions; Negative-ion resonance calculations; Ion-molecule collisions

GLAB, Wallace L.; Loomis Laboratory of Physics; University of Illinois; Champaign-Urbana Campus; Urbana IL 61801; USA

GLADISCH, Michael W.; Physics Institute; Philosophenweg 12; Heidelberg 6900; West Germany; 6221-569371; E; 1.4; 1.3; 3.3; 3.6; 4.2

GLASS, Alexander J.; 2025 Hill Street; Ann Arbor MI 48104; USA

GLASS, Solomon J.; Department of Physics; Howard University; Washington DC 20059; USA

GLASSGOLD, A. E.; Department of Physics; New York University; 4 Washington Place; New York NY 10003; USA; 212-598-2020; T; 5.7; Astrophysical applications

GLAUBER, Roy J.; Department of Physics; Lyman Laboratory; Harvard University; Cambridge MA 02138; USA; 617-495-2869; T; 2.1; 3.8; Atoms with radiation fields-collective interactions; Charged particles--scattering

GLICK, Madeleine S.; Institute of Applied Physics EPFL; PHB Ecublens; Lausanne CH CH1015; Switzerland

GLICKSTEIN, Stanley S.; Welding Engineering Division; Westinghouse-Bettis Atomic Power La; PO Box 79; West Mifflin PA 15122; USA; 412-462-5000x489; E/T; 5.4; Welding arc physics

GOBLE, Alfred T.; Department of Physics; Union College; Schenectady NY 12308; USA; 518-390-6254; E; 3.2; Optical isotope shifts by pressure-scanned Fabry-Perot; interferometer-measurement

GODDARD III, William A.; Department of Chemistry; Division of Chemistry and Chemical Engineering; Arthur Amos Noyes Lab. of Chemical Physics; California Institute of Technology; Pasadena CA 91125; USA; 818-356-6544; 818-356-6547; T; 2.3; 1.1; 5.1; Studies of reaction processes using ab initio and simulation methods; Ab initio electronic wavelengths including electron correlation effects; Gas phase reactions and reactions on surfaces--primary applications

GOFFE, Thomas V.; Department of Physics and Astronomy; University of Nebraska, Lincoln; Lincoln NE 68588; USA

GOGOL, Carl; Inficon Leybold-Hereaus; 6500 Fly Road; East Syracuse NY 13057; USA

GOLD, L. Peter; Department of Chemistry; Pennsylvania State University; 152 Davey Laboratory; University Park PA 16801; USA; 814-865-7694; E; 3.1; 3.2; 3.3; 3.6; 3.7; Electronic laser spectroscopy of small molecules

GOLDBERG, Ira B.; Physics and Chemistry Division; Rockwell Internat'l Science Center; PO Box 1085; Thousand Oaks CA 91360; USA; 805-498-4545x276; E; 2.3; 3.8; Electron parametric resonance of gas phase and solid state species; Kinetic determinations; Laser chemistry

GOLDBERG, Lawrence S.; U.S. Naval Research Lab.; Code 6510; 4555 Overlook Ave., SW; Washington DC 20375; USA; 202-767-2499; E; 2.3; 3.6; 5.2; Laser picosecond pulse studies of atoms, molecules and solids

GOLDBERG, Leo; Kitt Peak National Observatory; PO Box 26732; Tucson AZ 85726; USA; 602-327-5511; E/T; 5.7; Stellar spectroscopy

GOLDBERG, Leon P.; Physical Sciences; Liberal Arts and Science; Glassboro State College; Route 322; Glassboro NJ 08028; USA; 609-863-6333; 609-429-6393; E/T; 5.4; 5.7; 3.9

GOLDBERGER, Arthur L.; Department of Physics; Behlen Laboratory; University of Nebraska; SB-66; Lincoln NE 68588; USA; 402-472-1230; E; 2.1; Radiation from atomic states measured in coincidence with the scattered projectile in ion-molecule collisions at low energy levels--linear and circular polarization

GOLDE, Michael F.; Department of Chemistry; University of Pittsburgh; Pittsburgh PA 15260; USA; 412-624-5084; E; 2.1; 2.3; 3.6; Spectroscopy and gas phase chemical kinetics; Energy transfer; Chemiluminescence

GOLDEN, David E.; Department of Physics and Astronomy; University of Oklahoma; Norman OK 73019; USA; 405-325-3961; E; 2.2; 3.6; 4.2; Electron scattering; Electron photon correlation; Lasers

GOLDEN, David M.; Chemical Kinetics Division; SRI International; 333 Ravenswood Avenue; Menlo Park CA 94025; USA; 415-326-6200; E; 2.3; 3.8; 5.1; Reactivity; Energy transfer; Laser induced processes and thermal energization; Gas-surface interactions

GOLDEN, Lawrence B.; Department of Physics; Pennsylvania State University; Scranton Campus; 120 Ridgeview Drive; Dunmore PA 18512; USA; 717-961-4752; T; 2.2; Electron collisional excitation; Highly charged ions--ionization

GOLDEN, Sidney; Department of Chemistry; Brandeis University; Waltham MA 02254; USA; 617-647-2803; T; 1.1; 2.3; 3.2; Hartree-Fock energies--upper and lower bounds; Solvated electrons--optical absorption; Quantum effects on mutual bonding of particles

GOLDFIELD, Evelyn M.; Center for Astrophysics; 60 Garden Street B-320; Cambridge MA 02138; USA

GOLDFLAM, Rudolf; Department of Physics; University of Washington; Seattle WA 98195; USA; 206-543-6896; E/T; 1.1; 2.1; Molecular collisions--closure approximations

GOLDMAN, Leonard M.; Department of Mechanical and Aerospace Science; Laboratory for Laser Energetics; University of Rochester; 250 E. River Road; Rochester NY 14623; USA; 716-275-5285; 716-275-2048; E; 3.6; 5.4; 5.1; Lasers with plasmas--spectroscopic techniques in study of interactions

GOLDSMITH, John E. M.; Combustion Sciences Department; Diagnostics Research Division 8354; Sandia National Laboratories; P.O. Box 969; Livermore CA 94550; USA; 415-422-2432; E; 3.7; 3.6; 5.5; Multiphoton excitation techniques for combustion diagnostics; Nonlinear laser spectroscopy

GOLDSMITH, Samuel; Lab. for Plasma and Fusion Energy; University of Maryland; College Park MD 20742; USA

GOLDSTEIN, John C.; Group X-1; Los Alamos National Lab.; PO Box 1663 MS531; Los Alamos NM 87545; USA; 505-667-7281; T; 2.1; 3.6; 4.1; Nanosecond pulses with molecules--interaction of; Molecular systems--rotational relaxation in saturation; Molecular laser systems--multilevel effects contributing to gain

GOLDSTEIN, Raymond; Mail Code 169-506; Jet Propulsion Laboratory; 4800 Oak Grove Dr.; Pasadena CA 91109; USA; 818-354-0241; E; 4.4; 5.3; 5.4; 5.7

GOLDWIRE JR., Henry C.; LGF Group; L-451; Lawrence Livermore National Laboratory; 808; 5340 Crown Court; Livermore CA 94550; USA; 415-423-0160; 415-422-7381; T; 3.1; 1.1; 5.7; 1.4; 3.7

GOLE, James L.; Department of Chemistry; Georgia Institute of Technology; 225 North Avenue; Atlanta GA 30332; USA; 404-894-4029/4011; E; 1.1; 2.3; 3.6; Diatomic metal oxides or halides small metal aggregates, and molecular ions--high temperature chemistry focusing on molecular electronic structure and basic parameters; Spectroscopy and lasers and quantum chemistry

GOLIGHTLY, Danold W.; Department of Interior; Geologic Division; Branch of Analytical Chemistry; U.S. Geological Survey; 957 National Center; Reston VA 22092; USA; 703-860-7652; E; 3.2; 5.4; Temperatures and electron concentrations in laboratory plasmas--spectroscopic measurement; UV-visible spectral wavelengths--measurements and spectral line intensities for free atom species in laboratory plasmas

GOLOVCHENKO, J. A.; Radiation Physics Division; Bell Telephone Laboratories; 600 Mountain Avenue; Murray Hill NJ 07974; USA; 201-582-4407; E; 4.2; 5.1; Charged particles with solids--interaction

GOLUB, John E.; Dept. of Physics; Jefferson Lab.; Harvard University; Oxford St.; Cambridge MA 02138; USA; 617-495-2788; E; 3.10; 2.1; 3.7; 3.6; 3.11; Fast coherent transient studies of relaxation processes; Time domain spectroscopy of atoms and solids; Aspects of the interaction of atoms with stormy laser fields

GONCALVES, Antonio M.; Department of Chemistry; Temple University; Philadelphia PA 19122; USA; 215-787-7968; E; 2.3; Photoexcited organic triplet states--fast time--resolved magnetic resonance

GOOD JR., Roland H.; Department of Physics; Pennsylvania State University; 104 Davey Laboratory; University Park PA 16802; USA; 814-863-2055; 814-238-5818; T; 1.1; 5.1; 3.9; 3.1; Metallic binding; field electron and ion emission; Stark effect in hydrogen; Ortho-para transition in hydrogen

GOOD JR., William E.; Department of Chemistry; Suffolk University; Beacon Hill; Boston MA 02114; USA; 617-723-4700x255; E/T; 3.6; 3.10; Raman scattering spectroscopy

GOODMAN, Irwin H.; 2629 W. Jarlath Street; Chicago IL 60645; USA

GOODMAN, Leonard S.; Physics Division; Argonne National

Laboratory; Bldg 203 F-117; 9700 S. Cass Avenue; Argonne IL 60439; USA; 312-972-4040 /4004; E; 1.1; 3.6; 5.6; Laser and laser-rf-double resonance spectroscopy; Atomic and molecular structure

GOODMAN, Zoe M.; Department of Physics; Yale University; PO Box 6666; 217 Prospect Street; New Haven CT 06511; USA

GOODSON, Donald W.; 2210 Crestwood Drive; Tyler TX 75701; USA

GOORVITCH, David; Astrophysics Experiments Branch; Space Sciences Division; Ames Research Center; NASA; N245-6; Moffett Field CA 94035; USA; 415-694-5502; FTS 464-5502; E; 3.3; 5.6; 5.7; 1.1; Laboratory IR spectroscopy related to planetary atmospheres

GORDON, James P.; Bell Telephone Laboratories; Crawfords Corner Road; Holmdel NJ 07733; USA; 201-949-2227; T; 3.6; 4.1; Quantum electronics; Maser and laser operation--quantal and semiclassical theories; Light and matter interaction

GORDON, Mark S.; Department of Chemistry; North Dakota State University; Fargo ND 58105; USA; 701-237-8829; T; 1.1; 2.3; Quantum chemistry on excited state potential energy surfaces; Excited state reaction dynamics; Franck-Condon factors

GORDON, Robert J.; Department of Chemistry; University of Illinois, Chicago; PO Box 4348; Chicago IL 60680; USA; 312-996-3280; 312-996-5430; E/T; 2.1; 2.3; 3.8; Gas phase kinetics; Laser induced chemical reactions; Energy transfer scattering theory

GORDON, Roy G.; Department of Chemistry; Harvard University; 12 Oxford Street; Cambridge MA 02138; USA; 617-495-4017; T; 1.1; 2.1; 3.5; Intermolecular forces; Molecular collisions dynamics; Molecular spectral line shapes

GORMAN, Joseph G.; Westinghouse Research Lab.; Beulah Road 301-3B5; Pittsburgh PA 15235; USA

GOSS, Larry P.; Research Applications Div.; Systems Research Laboratory, Inc.; 2800 Indian Ripple Rd.; Dayton OH 45440; USA; 513-252-2706; 513-426-6000; E; 3.1; 4.1; 3.9; 5.5; 1.2

GOULD, Harvey A.; Lawrence Berkeley Laboratory; University of California; Building 71, Room 257; Berkeley CA 94720; USA; 415-486-4976; E; 1.4; 4.2; 2.1; 4.2; QED in few-electron very high-Z atoms; Relativistic and ultrarelativistic charge-changing collisions; CP violation (electric-dipole moments) in atoms

GOULD, Phillip L.; Physics Department; Room 26-247; Research Laboratory of Electronics; Massachusetts Institute of Technology; Room 26-247; 77 Massachusetts Avenue; Cambridge MA 02139; USA; 617-253-4167; E; 3.6; 3.7; 3.11; Radiation pressure

GOUTERMAN, Martin; Chemistry; University of Washington; Seattle WA 98195; USA; 206-543-1645; E/T; 1.3; 3.2; 3.8; 5.9; Properties of metal free and metalloporphyrins; Optical spectra (absorption & emission) of porphyrins; Photochemistry and electron transfer in porphyrins; Applications of A&M physics to chemical sensors

GOWDA, Ramakrishna; Department of Physics; Baylor University; Waco TX 76703; USA

GRABOSKE JR., Harold C.; Physics Department; V-Atomic /Dense Plasma Division; Lawrence Livermore National Lab.; Department of Energy; Box 808, L-296; Livermore CA 94550; USA; 415-422-7264; T; 5.7; 5.4; Dense partially ionized plasmas using approximate quantum perturbation techniques--modeling; Quantum mechanics of dense plasmas; State, transport properties of astrophysical mixtures

GRACE, Dennis R.; 3123 1/2 Byron; San Diego CA 92106; USA

GRAFF, Margaret M.; Department of Physics; University of Oregon; 60 Garden Street; Eugene OR 97403; USA; 503-686-4751; 503-485-8541; E/T; 5.7; 2.3; 3.1

GRAFT, Ronald D.; 5006 Gainesborough Drive; Fairfax VA 22032; USA

GRAHAM, Gerald; Department of Physics; University of Delaware; Newark DE 19711; USA

GRAHAM, W. J.; 2500 Colonial Road; Accokeek MD 20607; USA

GRAHAM, William G.; Department of Physics; New University of Ulster; Coleraine BT52 1SA; N. Ireland; 0265-4141 x471; 0165-85-696; E; 2.1; 4.2; 5.1; 5.4

GRAM, Peter A. M.; Group MP-4; Los Alamos National Laboratory; PO Box 1663; Los Alamos NM 87545; USA; 505-667-5562; E; 3.1; 4.2; Negative hydrogen ions using relativistic Doppler shift-photodetachment

GRANNEMAN, Ernst H.; 5 J C Ritsemalaan; Kortenhoef 1241AP; Holland

GRANT, David E.; Applied Research Laboratory; University of Texas; PO Box 8029; Austin TX 78712; USA

GRANT, Edward R.; Department of Chemistry; Cornell University; Ithaca NY 14853; USA; 607-256-4388; E; 2.3; 3.6; 3.8; 3.7; 5.5; Dynamics of elementary chemical reactions--investigations; Unimolecular decomposition-dynamics; Laser chemistry and spectroscopy

GRASSO, Robert P.; 1424 Richmond Rd-Apt. E3; Lyndhurst OH 44124; USA

GRATTIDGE, W.; 1 Washington Avenue; Schenectady NY 12305; USA

GRAVES, John L.; Department of Chemistry; Univ. of North Carolina, Greensboro; Greensboro NC 27412; USA; 919-379-5475; T; 1.1; 2.3; Atomic and molecular quantum chemistry; Empirical approaches to atomic-molecular energies; Energetics of kinetics of small to large molecules

GRAY, Eoin W.; Fusion Research; Power Flow Research; Pulsed Power Sciences; Sandia National Laboratory; 6200 East Broad Street; Albuquerque NM 87111; USA; 505-844-6845; E; 5.1; 5.3; 5.4; 2.3; Surface reactions in plasmas; Ion-molecule reactions in discharges; Insulator flashover in fusion accelerators

GRAY, Tom J.; Department of Physics; James R. Macdonald Laboratory; Kansas State University; Cardwell Hall; Manhattan KS 66506; USA; 913-532-6029; 913-532-6777; E; 2.3; 1.3; 2.1; 4.2; 4.3; Production of highly charged molecular ions by hard collisions; Study of lifetimes for metastable highly charged molecular ions; Spectroscopy of molecular fragments produced by hard collisions; Superconducting linear accelerator development and construction

GRAYBEAL, Jack D.; Department of Chemistry; Virginia Polytechnic Institute of State University; Davidson Hall; Blacksburg VA 24061; USA; 703-961-5406; E; 3.3; 1.1; Microwave spectroscopy; Nuclear quadrupole resonance

GREEN, Alex E.; Dept. of Physics and Nuclear Engineering Sciences; University of Florida; Gainesville FL 32611; USA; 904-392-2001; T/E; 2.2; 5.6; Charged particle energy deposition using fundamental atomic and molecular cross sections as inputs; Atomic and molecular cross sections and properties from first principles and measurement of solar and atmospheric UV in UV B region

GREEN, Byron D.; Physical Sciences, Inc.; 30 Commerce Way; Woburn MA 01801; USA; 617-933-8500; E/T; 2.1; 2.2; 3.3; 5.6; Energy deposition by electrons in atmospheric gases; IR fluorescence; Energy transfer

GREEN, Joseph M.; Physics Division; Research and Development Associates; 4640 Admiralty Way; Marina del Rey CA 90291; USA; 213-822-

1715; T; 5.4; Radiative properties of dense plasmas --calculation

GREEN, Sheldon; NASA Goddard Institute for Space Studies; 2880 Broadway; New York NY 10025; USA; 212-678-5562; T; 1.1; 2.1; 5.7; Molecular structure and scattering

GREEN, Thomas A.; Laser Physics Division; Sandia National Laboratories; PO Box 5800; Albuquerque NM 87185; USA; 505-844-4395; T; 2.3; 5.3; Gas discharge modeling for H.V. switch; Charge exchange for highly stripped ions

GREENBERG, Adam E.; 51 Gore Street; Cambridge MA 02141; USA

GREENBERG, Jack S.; Department of Physics; Sloane Physics Laboratory; Yale University; New Haven CT 06520; USA; 203-436-1649; E; 1.4; 2.1; 3.5; 4.2; Ion-atom collisions in high energy heavy-particle systems; Quasimolecular spectroscopy; Quantum electrodynamics of strong fields

GREENE, Arthur E.; Theoretical Chemistry and Molecular Physics Division; Group T-12; Los Alamos National Lab.; P.O. Box 1663 MS569; Los Alamos NM 87545; USA; 505-667-7799; T; 2.2; 5.3; Electron impact cross sections for modeling laser plasmas--determination

GREENE, Chris H.; Department of Physics and Astronomy; Louisiana State University; Baton Rouge LA 70803; USA; 504-388-6845; 504-388-2262; T; 3.1; 1.1; 3.9; 1.2; 2.2; Electron interactions with a charged or neutral target; Photoionization and photodissociation, polarization of photofragments

GREENE, Edward F.; Department of Chemistry; Brown University; Providence RI 02912; USA; 401-863-1193; E; 2.3; 5.1; 2.1; Atomic and molecular scattering; Molecules with surfaces--interactions; Molecules with electron beams--interactions

GREENE, Thomas J.; UPM #193; Dhahran; Saudi Arabia

GREENEBAUM, B.; 16 Illinois Street; Racine WI 53405; USA

GREENLEES, G. W.; Department of Physics; University of Minnesota; Minneapolis MN 55455; USA; 612-373-3331; E; 1.1; 3.6; 4.2; 4.4; 4.1; Isotope shift and hyperfine splittings on radioactive atoms

GREGORY, Donald C.; Physics Division; Oak Ridge National Laboratory; P.O. Box X; Building 6003; Oak Ridge TN 37831; USA; 615-574-4706; FTS-624-4706; E; 2.2; 5.4; 4.2; Ionization and excitation of multicharged ions by electron impact; Highly stripped heavy ion experiments; Crossed electron-ion beams; Synchrotron experiments

GREVE, Peter; L-Flab; Central R&D; Carl Zeiss; 1369180; Oberkochen FR Germany 7082; West Germany; E; 3.6; 4.1; 3.1; 3.8; 5.7

GREYBER, H. D.; 10123 Falls Road; Potomac MD 20854; USA

GREYTAK, Thomas J.; Department of Physics; Massachusetts Institute of Technology; Room 13-2074; Cambridge MA 02139; USA; 617-253-6818; E; 1.1; 2.1; Spin-polarized atomic hydrogen--creation and study

GRIEM, Hans R.; Laboratory for Plasma and Fusion Energy Studies; Mathematical, Physical Science and Engineering; University of Maryland; College Park MD 20742; USA; 301-454-3516; E; 5.4; Plasma spectroscopy; Line broadening by charged particles

GRIFFIN, Donald C.; Department of Physics; Rollins College; Winter Park FL 32789; USA; 305-646-2664; T; 1.1; 2.2; Atomic structure calculations; Electron scattering from atoms

GRIFFIN, P. M.; 107 Wood Ridge Lane; Oak Ridge TN 37830; USA

GRIFFIN, Richard D.; Code 6513; Naval Research Lab.; Washington DC 20375; USA

GRIFFITHS, James E.; Physical and Inorganic; Chemical Research Division; Bell Telephone Laboratories; 600 Mountain Avenue; Murray Hill NJ 07974; USA; 201-582-3034; E; 4.1; Raman scattering--properties and structure of materials

GRISCHKOWSKY, Daniel R.; Physical Sciences Division; IBM Thomas J. Watson Research Center; PO Box 218; Yorktown Heights NY 10598; USA; 914-945-2057; E/T; 3.6; Laser light and atomic vapors--near-resonant interaction

GROENEVELD, Karl-Ontjese E.; Institut F. Kernphysik; Physics; J.W. Goethe Universitaet; August-Euler Str 6; Frankfurt 90 Germany D6000; West Germany; 069-798-4238; 06039-3225; E; 2.1; 4.2; 5.2; 1.1; 5.1

GROSJEAN, Dennis F.; Research Applications Division; Systems Research Laboratories; 2800 Indian Ripple Road; Dayton OH 45440; USA; 513-252-4264; 513-426-6000; E; 5.3; 5.4; 4.1; 3.6; 3.2

GROSSMAN, Arthur A.; Columbia University; Room 202, SW Mudd Building; New York NY 10027; USA

GROSSMAN, M. W.; 249 Rutledge Road; Belmont MA 02178; USA

GROTCH, Howard; Department of Physics; Pennsylvania State University; 104 Davey Laboratory; University Park PA 16802; USA; 814-863-0007; T; 1.4; Quantum electrodynamics of simple atoms

GROVER, James R.; Department of Chemistry; Brookhaven National Laboratory; Building 555; Upton NY 11973; USA; 516-282-4348; 516-751-0636; E; 1.3; 2.3; 3.1; 4.2; 4.4; Photoionization mass spectrometry in molecular beams with synchrotron light; Reactive scattering in crossed molecular beams

GRUBER, Carl L.; Department of Electrical Engineering; South Dakota School of Mines and Technology; Rapid City SD 57701; USA; 605-394-2459; E; 4.1; 5.1; 5.3; High efficiency RF discharge hydrogen dissociators including surface atomic recombination mechanisms; Transverse RF discharge carbon dioxide waveguide lasers--development

GRUEN, Dieter M.; Materials Science and Technology Division; Argonne National Laboratory; Building 200; 9700 S. Cass Avenue; Argonne IL 60439; USA; 312-972-3513; E; 3.6; 3.7; 3.10; 4.1; 5.1; Laser fluorescence for neutral atom detection; Multiphoton-resonance ionization of sputtered atoms; Laser spectroscopy of matrix isolated metal clusters; Second harmonic generation at surfaces

GRUENEBAUM, J.; 1746 47th Street; Brooklyn NY 11204; USA

GUBERMAN, Steven L.; Institute for Scientific Research; 271 Main Street, Suite 302; Stoneham MA 02180; USA; 617-438-7894; 617-729-4421; T; 1.1; 2.2; 3.2; 3.1; 5.3

GUCH JR., Steve; Electro-Optical Organization; Research and Development Division; GTE-Sylvania/Western Division; 100 Ferguson Drive; Mountain View CA 94042; USA; 415-966-2261; E; 4.1; Ionic molecular laser media--advanced research and development

GUDMUNDSEN, Richard A.; Electronic Research Center; Rockwell International; 3370 Mira Loma Avenue; Anaheim CA 92806; USA; 714-632-3614; T; 3.6; Laser physics

GUENTHER, Arthur H.; US Air Force Weapons Laboratory; AFWL/CCN; Kirkland AFB NM 87117; USA; 505-844-9856; E/T; Discharge research relative to high power switching and electrical failure of dielectrics, gaseous, liquid, solid including optical materials

GUEST, Gareth; Fusion Plasma Theory Division; General Atomic Co.; PO Box 81608; San Diego CA 92138; USA; 714-455-3466; T; 5.4; Heavy and medium-A impurities--effects on properties of tokamak plasmas

GUILLORY, William A.; Department of Chemistry; University of Utah; Salt Lake City UT 84112; USA; 801-581-7832; E; 2.3; 3.2; 3.3; 3.8; Photochemical processes; IR and

UV spectroscopy of very reactive species; Time-resolved spectroscopy

GUIRAGOSSIAN, Zaven G.; Plasma Physics Div. R1-1070; TRW, Inc.; 1 Space Park; Redondo Beach CA 90278; USA; 213-536-1106; E; 4.2; 4.3; Intense ion source development; Charged and molecular beams

GUMNICK, J. L.; 5510 Yarwell; Houston TX 77096; USA

GUNDERSEN, Martin A.; Department of Electrical Engineering; Texas Tech University; Lubbock TX 79409; USA; 806-742-3501; E; 3.6; 4.1; Electro-optics; Lasers; Light with matter-interaction

GUNDERSEN, Roy; University Library of Trondheim; Erling Skakkesgt 47c; Trondheim N-7000; Norway; 47759227; T; 2.1

GUNTER JR., William D.; 5290 Dellwood Way; San Jose CA 95118; USA

GUNTON, Robert C.; Division 52-12, Bldg 205; Lockheed Palo Alto Research Lab.; 3251 Hanover Street; Palo Alto CA 94304; USA; 415-493-4411x5237; T; 2.3; 5.6; Neutral and ion chemistry of the atmosphere

GUO, Hong; 5707 Walnut Street; Pittsburgh PA 15232; USA

GUO, Theodore C.; 10618 Tanager Lane; Potomac MD 20854; USA

GUPTA, Om P.; 145 Talaram Bagh; Allahabad 211006; India

GUPTA, Rajendra; Department of Physics; University of Arkansas; Fayetteville AR 72701; USA; 501-575-5933; 501-575-2506/2507; E; 3.6; 4.1; Laser spectroscopy; Optical techniques

GUSINOW, Michael A.; Combustion Science Division; Sandia National Laboratories; Livermore CA 94550; USA; 415-422-3411; E; 2.3; 5.5; Combustion environments--fundamental chemical physics studies of basic processes

GUSTAFSON, Ture K.; Dept. of Electrical Engineering and Computer Science; University of California; Berkeley CA 94720; USA; 415-642-3139; E/T; 4.1; Atoms and molecules --nonlinear optical response

GUSTAFSON, Walter R.; PO Box 233; Route #1; Furlong PA 18925; USA

GUTCHECK, Robert A.; Physical Electronics Division; SRI International; Building 410A; 333 Ravenswood Avenue; Menlo Park CA 94025; USA; 415-326-6200x5598; E; 3.1; 3.8; 4.2; Synchrotron radiation used to excite rare gas-halide mixtures; Excimer formation --study of kinetics involved

GUTMAN, William M.; Opti-Metrics, Inc.; P.O. Drawer E; White Sands NM 88002; USA; 505-523-5114; 505-678-3588; E; 5.6; 3.6; Atmospheric propagation of radiation

GUTZWILLER, Martin C.; Physical Sciences Division; IBM Thomas J. Watson Research Center; PO Box 218; Yorktown Heights NY 10598; USA; 914-945-1042; T; 1.1; Classical mechanics with quantum mechanics in atomic and molecular systems --relation

GWINN, William D.; Department of Chemistry; University of California; Berkeley CA 94720; USA; 415-642-1047; T; 1.1; 3.3; Molecular structure; Microwave spectroscopy

HAAN, Stanley L.; Department of Physics; Calvin College; Grand Rapids MI 49506; USA; 616-957-6339; T; 3.1; 3.7; 3.10

HACKEL, Lloyd A.; Y Division; Lawrence Livermore National Lab.; PO Box 808; Livermore CA 94550; USA; 415-422-6225; E; 3.6; Pulse propagation; Spectroscopy

HADEISHI, Tetsuo; Lawrence Berkeley Laboratory; University of California; Berkeley CA 94720; USA; 415-486-5734; E; 4.3; Molecular detection research and detector development

HAENSEL, R.; Institute Experimental Physik; Universitaet Kiel; Olshausenstrasse 40-60; Kiel 23; West Germany

HAFF, Peter K.; Kellogg Radiation Lab. 106-38; Cal Tech; Pasadena CA 91125; USA

HAGAN, Lucy B.; 6212 Redwing Road; Bethesda MD 20034; USA

HAGEN, Gunter; Electron Device Physics Division; Hughes Research Laboratories; 3011 Malibu Canyon Road; Malibu CA 90265; USA; 213-456-6411; E; 5.1; Atomic, ionic, free radical and excited molecular fragments; with surfaces--interaction

HAGLUND JR., Richard F.; Department of Physics and Astronomy; Vanderbilt University; Nashville TN 37027; USA; 615-322-7964; 615-322-2828; E; 5.1; 3.6; 4.1; 3.2; 5.3; Interactions of spin-polarized nuclei with surfaces

HAGMANN, Siegbert J.; Department of Physics; Kansas State University; Manhattan KS 66506; USA; 913-532-6776; E; 2.1; 4.2; Dynamics of different inner-shell vacancy production mechanisms; Excitation patterns and alignment of collisionally excited states

HAGSTROM, Stanley; Department of Chemistry; Indiana University; Bloomington IN 47401; USA; 812-337-0938; T; 1.1; Ab initio calculations on atoms and diatomic molecules by configuration interaction and Hylleraas techniques; Transition probabilities and polarizabilities; Relativistic effects in light atoms

HAGSTROM, Stig B.; Xerox Park; 3333 Coyote Hill Road; Palo Alto CA 94304; USA

HAGSTRUM, Homer D.; Surface Physics Research Division; Bell Telephone Laboratories; 600 Mountain Avenue; Murray Hill NJ 07974; USA; 201-582-3933; E; 5.1; Electronic interaction with surfaces of atoms carrying potential energy

HAHN, Yukap; Department of Physics; University of Connecticut; Storrs CT 06268; USA; 203-486-4469; 203-486-4915; T; 1.2; 2.1; 2.2; 5.4; 5.7; Electron-ion interactions, leading to electron capture, ionization and their inverse processes; Atomic processes in plasma; Ion-atom collisions at high energies and high Rydberg states

HAHNE, Gerhard E.; Computational Chemistry Branch; Thermosciences Division; Ames Research Center; NASA; Mail Stop 230-3; NASA/AMES; Moffett Field CA 94035; USA; 415-694-6140; T; 2.1; 3.8; Elastic and vibration rotation inelastic scattering of small molecules; Vibration rotation lines in di-atomic molecules --collision broadening

HAKE JR., Richard D.; Electromagnetic Sciences Laboratory; SRI International; 333 Ravenswood Avenue; Menlo Park CA 94025; USA; 415-326-6200; T/E; 2.3; Photoemissions--use of reaction rates for calculating

HALL, Frederick G.; Bariles Mobile Village Lot A2; Rome NY 13440; USA

HALL, Gregory; Smith Hall; University of Minnesota; 207 Pleasant Street, SE; Minneapolis MN 55455; USA

HALL, James M.; Department of Physics; Atomic, Molecular, and Optical; James R. Macdonald Laboratory; Kansas State University; Cardwell Hall; Manhattan KS 66506; USA; 913-532-6876; 913-532-6975; E; 2.1; 4.2; Atomic and molecular collisions (electron capture and ionization); Accelerator-based atomic and molecular physics

HALL, John L.; Quantum Physics Division [NBS]; Joint Institute for Lab. Astrophysics; University of Colorado; Boulder CO 80309; USA; 303-492-7843; E; 1.4; 3.6; Laser spectroscopy; Precision measurement

HALL, Laurence S.; Lawrence Livermore Laboratory; University of California; PO Box 808; L-630; Livermore CA 94550; USA

HALL, Richard B.; Surface Chemical Dynamics; Physical and Materials Laboratory; Exxon Research and Engineering Co.; Route 22E; Annandale NJ 08801; USA; 201-730-2733; E/T; 3.8; 5.1; 3.6; 3.10; Laser chemistry; Radiation with surfaces/surface adsorbates-interaction; Surface reaction dynamics and kinetics

HALL, Richard I.; LPOC T12 E5; Univ. P. et M. Curie; 4 Place Jussieu; Paris Cedex 05 75230; France; 4336 2525x4313; E; 2.2; 2.1; 2.3; Low energy electron collisions-resonance spectroscopy; Negative ion collisions -electron detachment; Dissociative electron attachment

HALL, Robert J.; Combustion Sciences; Power and Industrial Systems; United Technologies Research Center; Silver Lane; East Hartford CT 06108; USA; 203-727-7349; T; 3.6; 3.10; 3.3; 3.11; 5.5; Non-linear optical processes (CARS) to combustion diagnosis-application

HALL, Rodney B.; Naval Air Development Center; Code 3012; Warminster PA 18974; USA

HALLBERG, Bengt O.; BOH Optical Lab.; PO Box 3022; Solnan S17103; Sweden

HALPERN, Alvin M.; Department of Physics; Brooklyn College of CUNY; Brooklyn NY 11210; USA; 718-780-5418; T; 2.1; 2.2; Atomic collision theory

HAM, Mooyoung; Advance Development Laboratory; General Electric Co.; 48 Helen Drive; Glens Falls NY 12801; USA; 518-761-7432; 518-798-3864; E; 1.3; 3.9; 4.3; 5.4; Space charge source design; Ion and electron gun design; Plasma source design; Sputtering and ion plating (and implantation)

HAMADTO, S. A.; The Physics Department; University of Riyadh; PO Box 2455; Riyadh; Saudi Arabia

HAMEKA, Hendrik F.; Department of Chemistry; University of Pennsylvania; Philadelphia PA 19104; USA; 215-898-0303; 215-898-8317; T; 1.4; 3.10; 3.9; 1.1; Quantum theory applied to molecules; Perturbation theory

HAMERMESH, Morton; Department of Physics; University of Minnesota; Minneapolis MN 55455; USA; 612-376-4612; 612-933-5708; T; 1.1; 3.10; 5.8

HAMILTON, Peter A.; Chemistry; Queen Mary College; Mile End Road; London E1 4NS; ENGLAND; 01-980-4811; E; 1.1; 2.3; 3.2; 3.6; Laser induced fluorescence of cations and free radicals; Kinetics of atmospheric reactions

HAMM, Robert W.; Group AT-1; Los Alamos National Laboratory; PO Box 1663 MS817; Los Alamos NM 87545; USA; 505-667-5521; E; 3.1; 4.2; 800 MeV H- and Ho beams--using photodetachment

HAMMER, J. M.; RCA Lab.; Princeton NJ 08540; USA; E/T; 4.1; 3.10; 3.9; 3.11; Optical waveguides; Diode lasers; Optical modulation

HAMMERMAN, I.; 33 Eisenberg Street; Rehovot Rehovot 76287; Israel

HAMMOND, Gordon L.; Explosion Dynamics Branch; Energetic Materials Division; White Oak Laboratory; U.S. Naval Surface Weapons Center; Silver Spring MD 20903; USA; 301-394-1354; E; 2.1; 3.1; 5.4; 5.7; Collisional line broadening; Plasma diagnostics in explosion phenomena; Stellar atmosphere models

HAMMOND, Ray L.; 1050 N.

HAMO-LEILA, Mustafa A.;
Atomic Energy Commission;
PO Box 6091; Damascus; Syria

HAMPSON JR., Robert F.;
Chemical Kinetics Data Center;
National Bureau of Standards;
Building 222, Room A165;
Gaithersburg MD 20899; USA;
301-921-2565; T; 2.3; 5.5;
5.6; Chemical Kinetics Data
Center Director

HANCE, Robert L.; Department
of Chemistry; Abilene Christian
University; PO Box 8127;
Station ACU; Abilene TX
79699; USA; 915-677-1911x2157;
E; 2.1; 5.1; Ions and atoms
(100-2000 eV) from small
molecules--small angle scattering; Small molecules on
metal surfaces--reactions

HANCOCK, Kent J.; Office of
Uranium Enrichment; Advanced
Technology Projects; U.S.
Department of Energy; NE-35;
Washington DC 20545; USA;
301-353-5054; E; 3.8; 4.1;
Isotope separation R&D-program
administrator

HANNA, Stanley S.; Department
of Physics; Stanford University; Stanford CA 94305;
USA; 415-497-4612; E; 1.1;
2.1; 4.2; 5.2; Ions produced
in atomic collisions--hyperfine
interactions; Atoms and
ions in solids (metals and
compounds) and as impurities
produced by ion implantation-
hyperfine interactions

HANRAHAN, Robert J.; Department
of Chemistry; 406 Nuclear
Science Center; Nuclear-Radi-
ation Chemistry; University
of Florida; Gainesville FL
32611; USA; 904-392-1442;
904-376-7754; E; 2.3; 3.8;
4.2; 4.4; Pulse radiolysis
of gases and gas mixtures;
Steady-state radiolysis;
Flash and steady-state photochemistry; Ion-molecule
reactions--mass spectroscopy

HANSCH, Theodor W.; Department
of Physics; Stanford University; Stanford CA 94305;
USA; 415-497-3571; E; 3.6;
Laser spectroscopy of atoms
and molecules

HANSEN, Lorin K.; Rasor
Associates, Inc.; 253 Humboldt
Court; Sunnyvale CA 94086;
USA; 408-734-1622; E/T;
5.4; Cesium plasma diodes
and negative ions in plasmas

HANSON, David M.; Department
of Chemistry; State University
of NY, Stony Brook; Stony
Brook NY 11794; USA; 516-
246-7660; 514-246-5050; E;
3.4; 3.6; 3.9; 5.2; Molecular
solids--stark spectroscopy;
Energy and electron transfer
processes

HANSON, Frank E.; Naval
Ocean Systems Center; Code
8113; San Diego CA 92152;
USA

HANSON, Gary R.; Submicron-
Knight Lab.; Cornell University; Ithaca NY 14853; USA

HANSON, Harold P.; Committee
on Science and Technology;
2321 Rayburn House Office
Building; Washington DC
20515; USA; 202-225-6375;
202-546-3698; E/T; 1.1; 3.4

HANSON, Ronald K.; Department
of Mechanical Engineering;
High Temperature Gasdynamics
Lab.; Stanford University;
Stanford CA 94305; USA;
415-497-1745; E; 2.3; 3.1;
3.6; 5.5; Spectroscopic
parameters (line strengths
and collision widths) of
combustion gases using tunable
laser; Laser fluorescence
flowfield imaging

HANSTEEN, Johannes M.; Department of Physics; University
of Bergen; Allegt. 55; Bergen
N-5000; Norway; 21 28 68;
095 47 5/21 27 61; T; 1.4;
2.1; 3.4; 4.2; 5.8.; Inner-
shell ionization by heavy
charged projectiles; Semiclassical collision theory
development (SCA); SCA theory
application to ionization,
delta-electron emission,
pair-production, chanelling

HAPPER, William; Department
of Physics; Princeton University; Princeton\NJ 08544;
USA; 609-452-5584; 608-921-
1487; E; 3.6; Laser spectroscopy; Optical pumping; Magnetic
resonance

HAQUE, M. A.; 2 Pence Court;
Greensboro NC 27408; USA

HARDCASTLE, Donald L.; Department of Physics; Baylor
University; Waco TX 76703;
USA; 817-755-2511; T; 1.1;
Atomic orbital calculations
(perturbation, finite difference techniques)

HARDIS, Jonathan E.; Far
Ultraviolet Physics; Radiation
Physics Division; National
Measurement Laboratory;
National Bureau of Standards;
Bldg 221/Room A251; Gaithers-
burg MD 20899; USA; 301-
921-2031; E; 1.1; 3.1;
3.2; 4.2; 3.6; Synchrotron
light source

HARDISSON, Arturo; La Laguna;
22 Av Lucas Vega; Tenerife;
Spain

HARGIS JR., Philip J.;
Laser Analytic Spectroscopy
Div.; Sandia National La-
boratories; PO Box 5800;
Albuquerque NM 87185; USA;
505-844-2821; E; 3.6; 3.8;
5.1; Spectroscopic diagnostics
and photochemistry for
molecules of interest in
chemical vapor deposition
reactions

HARMIN, David A.; Department
of Physics and Astronomy;
University of Kentucky;
Sort 0055; Lexington KY
40506; USA; 606-257-6722;
T; 1.2; 2.2; 3.9; DC stark
effect on Rydberg states;
Dielectronic recombination;
Quantum-defect theory;
Microwave ionization

HARMONY, Marlin D.; Department
of Chemistry; University
of Kansas; Lawrence KS
66045; USA; 913-864-4673;
913-864-3980; E; 1.1; 3.3;
3.5; 3.6; 3.9; Molecular
structure determination;
Studies of transient molecules

HARNEY, Robert C.; Optics
Division; Lincoln Laboratory;
MIT; PO Box 73; Lexington
MA 02173; USA; 617-862-5500
x7170; T; 4.1; Electric
dipole forbidden Raman scattering processes-studies

HARRELL II, Evans M.; Department of Mathematics; Georgia
Institute of Technology;
Atlanta GA 30332; USA;
404-894-2715; 404-233-3381;
T; 1.1; Eigenvalue problems
arising in atomic and mo-
lecular physics--mathematical
aspects

HARRELL JR., Joe J.; 1430
Lemhurst Drive; Pensacola
FL 32507; USA

HARRIMAN, John E.; Department
of Chemistry; Theoretical
Chemistry Institute; Uni-
versity of Wisconsin; 1101
University Avenue; Madison
WI 53706; USA; 608-262-1511;
T; 1.1; 5.2; Spin densities
calculation in atoms and
molecules; Molecular radicals
ESR in host single currents;
Reduced density matrices

HARRIS, Bernard; Instrumen-
tation Group; Corporate
Research; Cabot Corporation;
Concord Road; Billerica MA
01821; USA; 617-663-3455;
E/T; 3.4; 3.2; 3.1

HARRIS, Charles B.; Department
of Chemistry; University
of California; D-87 Hildebrand
Hall; Berkeley CA 94720;
USA; 415-642-2814; E/T;
2.3; 3.6; Chemical physics;
Matter and radiation--coherent
properties

HARRIS, David O.; Department
of Chemistry; Univ. of
California, Santa Barbara;
Santa Barbara CA 93106;
USA; 805-961-2534; E; 3.5;
3.6; Small transient mole-
cules--laser spectroscopy

HARRIS, Frank E.; Department
of Physics; University of
Utah; Salt Lake City UT
84112; USA; 801-581-8445;
T; 1.1; Electronic structure
computation--development of
methods

HARRIS, Harold H.; Department
of Chemistry; University of
Missouri; St Louis MO 63121;
USA; 314-553-5344; 314-993-
5422; E/T; 2.1; 2.3; 3.2;
3.6; Trajectory calculations
on multiple potential energy
surfaces; Ion-molecule reactions in beam-gas experiments

HARRIS, L. P.; GE R&D Center;
PO Box 43; Schenectady NY
12345; USA

HARRIS, Stephen E.; Dept.
of Elec'l Engineering and
Applied Physics; Edward L
Ginzton Laboratory; Stanford
University; Stanford CA
94305; USA; 415-497-0224;
E; 1.1; 2.1; 4.1; Collisions;
Autoionizing states; Laser
devices

HARRIS, Stephen J.; Physical
Chemistry Division; General
Motors Research Laboratory;

General Motors Technical Center; 12 Mile and Mound Roads; Warren MI 48090; USA; 313-575-2595; E; 5.5; Intracavity spectroscopy--details; Kinetics relating to combustion

HARRISON, Allen; Electrical Engineering; Sensor Product Development; Schlum Berger Well Services; 4002 Byron; Houston TX 77005; USA; 713-668-4794; 713-928-4390; E/T; Electromagnetic radiation sensors; Oil well logging sensor development; Magnetic materials analysis

HARRISON JR., Don E.; Department of Physics; U.S. Naval Postgraduate School; Monterey CA 93943; USA; 408-646-2877; T; 5.1; 5.2; 5.3; Atom and molecule ejection from clean and chemically reacted single crystal metal targets by ions --computer simulation

HARSHAW, Robert C.; University of Texas of Dallas; PO Box 688; MS NB1-101; Richardson TX 75080; USA

HART, Mark W.; Department of Physics; Rice University; PO Box 1892; Houston TX 77251; USA

HART, R. R.; PO Box 124; Porter Square Branch; Cambridge-B MA 02140; USA

HART, Raymond K.; Pasat Research Associates, Inc.; 585 Royervista Drive, NE; Atlanta GA 30342; USA; 404-256-0410; E; 4.3; 3.3

HARTER, William G.; Department of Physics; Georgia Institute of Technology; Atlanta GA 30332; USA; 404-894-5238; T; 3.5; 3.8; High resolution molecular spectra and related effects in quantum electronic; Symmetry analysis of spectra; Unitary group methods for multi-electron and multi-nucleon configurations

HARTFORD JR., Allen; Group AP-4; Applied Photochemistry Division; Los Alamos National Laboratory; PO Box 1663 MS567; Los Alamos NM 87545; USA; 505-667-7121; E; Laser-induced chemistry--technical management of programs

HARTMAN, Paul L.; Dept. of Physics and Applied Engineering Physics; Cornell University; Clark Hall; Ithaca NY 14853; USA; 607-256-5205; E; 3.1; 3.2; 4.2; Spectroscopy; Physical optics; Accelerator related atomic physics

HARTMANN, Sven R.; Department of Physics; Columbia University; New York NY 10027; USA; 212-280-3272; E; 3.6; 3.10; 2.1; 1.1; Optical coherent transient techniques to study relaxation and spectroscopy in solid state and atomic physics

HARVEY, James F.; Lawrence Livermore National Lab.; PO Box 808; Livermore CA 94550; USA; 415-443-1388; T; 3.1; Krypton excimer--photoionization

HARVEY, Kenneth C.; Center for Absolute Physical Quantities; National Bureau of Standards; Building 221, Room A141; Gaithersburg MD 20899; USA; 301-921-2061; E; 1.4; 3.2; 4.3; One- and two-electron atoms--high resolution spectroscopy; Slow atomic beams and atomic traps for precision spectroscopy--development

HARVEY, Nancy M.; Chemical Physics Division; Chemistry and Physics Laboratory; The Aerospace Corporation; PO Box 92957; Los Angeles CA 90009; USA; 213-648-7416; T; 2.1; 2.3; Molecular scattering theory for inelastic and reactive scattering

HASE, William L.; Department of Chemistry; Physical; Wayne State University; 435 Chemistry Building; Detroit MI 48202; USA; 313-577-2694; T; 2.1; 2.3; Intermolecular and intramolecular dynamics by classical and semiclassical procedures--simulation; Unimolecular reaction rate theory; Computer simulation of chemical reaction dynamics

HASKELL, Hugh B.; Sciences Department; N.C. School of Science and Mathematics; P.O. Box 2418; Durham NC 27705; USA; 919-683-6585; T; 2.2; Low energy electron-atom and electron-ion scattering

HASKINS, C. P.; 1545 18th Street, N.W.; Washington DC 20036; USA

HASS, Michael; Nuclear Physics Department; Wetzmann Institute; Rehovot; Israel

HASSAN, H. A.; Department of Mechanical and Aerospace Engineering; North Carolina State University; P.O. Box 7910; Raleigh NC 27695; USA; 919-787-1806; T; 4.1; Nuclear pumped lasers for atomic and molecular systems

HASTIE, John W.; High Temperature Chemistry; Department of Ceramics; Institute for Materials Science and Engr.; National Bureau of Standards; Gaithersburg MD 20899; USA; 301-921-2859; 301-921-3618; E; 1.1; 2.3; 6,8; 3,4; 5; Molecular species at high temperature --structure, stability and kinetics

HASTINGS, Julius M.; Chemistry Division; Brookhaven National Laboratory; Upton NY 11973; USA; 516-282-4377; E; Magnetic systems--critical phenomena; Magnetic systems--phase transformations

HATFIELD, Lynn L.; Department of Physics and Engineering Physics; Texas Tech University; PO Box 4180; Lubbock TX 79409; USA; 806-742-3767; E; 5.1; 5.3; 5.4; Areas on surface; Pulsed power physics

HAUER, Allan; Los Alamos National Laboratory; L-A MS 554; Los Alamos NM 87545; USA

HAUGH, John J.; 6965 Maiden Lane; San Jose CA 95120; USA

HAUGSJAA, Paul O.; Solid State Technology; Advanced Component Technology Ctr.; Component Research Laboratory; GTE Laboratories, Inc.; 40 Sylvan Road; Waltham MA 02154; USA; 617-890-8460; 617-466-2512; E/T; 2.3; 4.2; 4.3; 5.1; 5.3; Discharge physics applied to light sources; A&M physics applied to plasma processing of materials and devices; Montecarlo modeling; RF/microwave power coupling to discharge plasmas

HAUS, Hermann A.; Dept. of Elect'l Engineering and Computer Science; Research Lab. of Electronics; Massachusetts Institute of Technology; Room 36-351; Cambridge MA 02139; USA; 617-253-2585; E/T; 3.8; Carbon dioxide isotopes--theory and measurement of pressure shift and line broadening

HAUSER, Ulrich A.; Physics; National Science; Physik Institute; University of Cologne; Universitaets Str. 14; D-5000 Koeln 41; F.R. Germany; 4703546; 4703499; E; 1.1; 2.1; 3.1; 4.2; 3.6; Doppler shift analysis; Metastable molecular states

HAVENER, Charles C.; Physics Division; Oak Ridge National Lab.; P.O. Box X; Oak Ridge TN 37831; USA; 615-574-4704; E; 2.1; 4.3; Electron capture of multi-charged ions on gases and H; Ion-atom merged beam apparatus

HAVEY, Mark D.; Physics Department; Old Dominion University; 1040 W. 45 Street; Norfolk VA 23508; USA; 804-440-4612; 804-440-4610; E; 3.8; 3.6; 2.1

HAWK, James F.; Physics; Natural Sciences and Mathematics; University of Ala. at Birmingham; University Station; Birmingham AL 35294; USA; 205-934-8088; E; 4.1

HAWKE, Ronald S.; 2369 Westminster Way; Livermore CA 94550; USA

HAY, Philip J.; Theoretical Division; T-12; Los Alamos National Laboratory; PO Bx 1663 J569; Los Alamos NM 87545; USA; 505-667-2097; T; 1.1; 2.3; Molecules--electronic structure; Potential energy surfaces; Relativistic effects and excited states; Transition-metal and actinide chemistry

HAYASHI, Shigeo; Department of Materials Sciences; University of Electrocommunications; 1-5-1 Chofugaoka; Chofu Tokyo 182; Japan; 0424-83-2161; 0424-81-0208; T; 2.2; Spin polarization of electrons due to multiple scattering by molecules

HAYDEN, Howard C.; Department of Physics U-46; University of Connecticut; Storrs CT 06268; USA; 203-486-3766; E; 3.8; 4.2; 5.2; Ion implantation; Ion-atom collision spectroscopy

HAYES, Edward F.; Chemistry Division; National Science Foundation; Washington DC 20550; USA; 202-357-7501; T; 1.1.; 2.1; Scattering; Electron structure

HAYHURST, Thomas L.; Infrared Sciences; Chemistry and Physics Laboratory; The Aerospace Corporation; P.O. Box 92957; Mail Station M2-251; Los Angeles CA 90009; USA; 213-648-5996; 213-648-5686; E/T; 1.1; 3.2; 3.8; 3.1; 3.2; Semi-empirical atomic theory; Hartree-Fock; Visible and near IR spectroscopy

HAYNES, Sherwood K.; Department of Physics; Michigan State University; East Lansing MI 48824; USA; 517-332-4893; E; Radioactive substances--Auger spectra

HAYS, Gerald N.; Division 1128; Sandia National Laboratories; P.O. Box 5800; Albuquerque NM 87185; USA; 505-844-8898; E; 5.3; 2.3; 5.4

HAZI, Andrew U.; Physics Department; High Temperature Physics Division; Lawrence Livermore National Lab.; P.O. Box 808; Livermore CA 94550; USA; 415-422-6195; T; 2.2; 1.1; 3.1; Atomic and molecular physics; Advanced UV and XUV lasers; Atomic processes in hot dense plasmas

HEAD, Charles E.; Department of Physics; University of New Orleans; New Orleans LA 70122; USA; 504-283-0341; E; 2.1; 3.6; 4.2; Low-energy accelerator (<50kV) based atomic and molecular physics; Dye-laser based spectroscopy

HEAD, Martha E.; University of New Orleans; PO Box 1362; Lake Front; New Orleans LA 70148; USA

HEAPS, William S.; Laboratory for Planetary Atmosphere; NASA Goddard Space Flight Center; Code 963; Greenbelt MD 20771; USA; 301-344-5106; E; 3.6; 5.6; Spectroscopy on atmospheric species to develop measurement techniques

HEATON, Marie M.; Department of Chemistry; New Mexico State University; PO Box 3C; Las Cruces NM 88003; USA; 505-646-1826; T; 1.1; Ab initio calculations to elucidate molecular energetics, structure and properties

HEBERLEIN, Joachim V. R.; Plasma and Nuclear Sciences; Applied Sciences; Westinghouse R&D Center; 1310 Beulah Road; Pittsburgh PA 15235; USA; 412-256-1581; E; 5.4; 5.3; Plasma heat transfer; Plasma synthesis of materials

HECKEL, Blayne R.; Physics; University of Washington; FM-15; Seattle WA 98195; USA; 206-543-8785; 206-543-2770; E; 1.4; 3.9; 4.1; 5.8; Measurements of atomic electric dipole moments; Interactions of gravitational, axion, fields with atomic spins; High frequency shift resolution with optical pumping; Nuclear polarization through optical pumping

HECKMANN, Paul H.; Department of Physics and Astronomy; Experimentalphysik III; Dynamitron-Tandem Laboratory; Ruhr-Universitaet Bochum; PO Box 102148; Bochum Nordrhein-Westf D-4630; West Germany; (0234)7003600; (0234)7006205; E; 3.2; 4.2; 1.1; 1.4; 2.1; Lifetimes and transition probabilities; Spectroscopy of highly charged ions

HEDBERG, Kenneth W.; Department of Chemistry; Oregon State University; Corvallis OR 97331; USA; 503-754-2081; E; 1.1; Gas molecules--molecular structure and motions

HEDIN, Lars T.; Department of Theoretical Physics; University of Lund; Solvegation 14A; Lund S22362; Sweden; 0461-09080; T; 5.2; Inter shell excitations of atoms in solids

HEER, Clifford V.; Department of Physics; Ohio State University; 174 West 18th Avenue; Columbus OH 43210; USA; 614-422-4275; E/T; 3.6; 3.10; 4.1; Optics; Laser gyro; Photon echoes and phase conjugation; Trapping neutral atoms and molecules

HEFFERLIN, Ray A.; Department of Physics; Science; Southern College; PO Box H; Collegedale TN 37315; USA; 615-238-2869; T; 1.1; 1.3; Periodic law of molecules and its graphical representations; Systematic study and prediction of molecular data

HEGSTROM, Roger; Department of Chemistry; Wake Forest University; Winston-Salem NC 27109; USA

HEICKLEN, Julian P.; Department of Chemistry; 152 Davey Laboratory; Pennsylvania State University; University Park PA 16802; USA; 814-865-9621; E; 2.3; 3.8; Gas phase kinetics and photochemistry

HEIDNER III, R. F.; Aerophysics Laboratory; The Aerospace Corporation; PO Box 92957; Building 130, Room 145; Los Angeles CA 90009; USA; 213-648-5610; E; 2.3; 3.8; Atoms and small molecules--gas phase kinetics; Energy transfer studies related to chemical lasers

HEIL, Hans; 3880 Rambla Orienta; Malibu CA 90265; USA

HEIL, Timothy G.; Department of Physics and Astronomy; University of Georgia; Athens GA 30602; USA; 404-542-2485; T; 2.1; 1.1; 5.7

HEINZEN, Daniel J.; Physics; Spectroscopy Laboratory; MIT; 305 Memorial Drive-210A; Cambridge MA 02139; USA; 617-253-5077; E/T; 3.11

HELBIG, Herbert F.; Department of Physics; Clarkson University; Potsdam NY 13676; USA; 315-268-2348; E/T; 5.1; Atoms in solid and liquid surfaces--ion-atom collisions

HELLER, Donald F.; Laser Research and Development; Corporate Research and Development; Materials Laboratory; Allied Corporation; 7 Powderhorn Drive; Mt. Bethel NJ 07060; USA; 201-560-1750; 201-469-9467; E/T; 4.1; 3.10; 3.8; 3.6; 3.7; Laser physics, solid state and molecular lasers, laser development; Intramolecular dynamics, relaxation processes, energy redistribution; Nonradiative processes, radiationless transitions, ultrafast processes; Novel optical diagnostics moire techniques, laser diagnostic methods

HELLER, Eric J.; Department of Chemistry; University of California, LA; 405 Hilgrad Avenue; Los Angeles CA 90024; USA; 213-825-2260; T; 1.1; 2.1; 3.1; Scattering theories--quantum and semiclassical; Molecular collisions and photodissociation; Molecular bound states; Intramolecular energy transfer

HELLWARTH, Robert W.; Department of Physics; University of Southern California; Los Angeles CA 90089; USA; 213-743-6390; E; 3.6; 4.1; Lasers and light scattering; Nonlinear optics

HELLWIG, Helmut W.; Frequency and Time Systems, Inc.; 34 Tozer Road; Beverly MA 01915; USA; 617-927-8220; E; 4.3; 4.1; 3.9; 2.2; Atomic clocks; Optical pumping

HELM, Hanspeter; Molecular Physics Laboratory; SRI International; 333 Ravenswood Avenue; Menlo Park CA 94025; USA; 415-326-6200x4637; E; 3.5; Molecular ions and neutrals--photofragment spectroscopy

HELMAN, William P.; Radiation Laboratory; University of Notre Dame; Notre Dame IN 46556; USA; 219-283-6527; E; Radiation chemistry and photochemistry

HELMS, David A.; Department of Physics; Rice University; 6100 S. Main Street; Houston TX 77005; USA; 713-527-4896; E; 2.2; Dense rare gas mixtures--electron beam excitation

HENCHMAN, Michael J.; Chemistry; Brandeis University; Waltham MA 02254; USA; 617-647-2821; 617-861-4028; E/T; 2.3; 4.4; 5.6; 5.8; 1.3; Chemical reactions of gaseous ions; Properties of solvated ions: relationship to solutions

HENDERSON, George A.; Department of Physics; Southern Illinois University; Edwardsville IL 62026; USA; 618-692-

2472; T; 1.1; Atoms and molecules with emphasis on one-electron properties and density matrix methods--electronic structure; Density functional theory

HENLEY, Ernest M.; Department of Physics F7-15; University of Washington; Seattle WA 98195; USA; 206-543-2995; T; 1.1; 1.4; Atoms--parity violation and weak neutral current effects; Hartree-Fock schemes--extended

HENNEBERGER, Walter C.; Department of Physics and Astronomy; Southern Illinois University; Carbondale IL 62901; USA; 618-453-3659; T; 1.1; Aharonov-Bohm effect

HENRICHS, P. M.; Chemistry Division; Organic Photo Science; Eastman Kodak Research Laboratories; 1999 Lake Avenue; Rochester NY 14650; USA; 716-477-6229; E/T; 5.2; Molecular characterization and dynamic studies involving NMR spectroscopy; Structure and dynamic studies involving NMR spectroscopy

HENRY, Hugh F.; 404 Linwood Drive; Greencastle IN 46135; USA

HENRY, Lucien R.; 5 Rue Lagarde; Paris 5EME; France

HENRY, Ronald J.; Department of Physics and Astronomy; Louisiana State University; Baton Rouge LA 70803; USA; 504-388-2261; T; 2.2; Electron-atom, -ion, and -molecule collisions; Atoms in electric and/or magnetic fields

HENSLEY, Eugene B.; Department of Physics; University of Missouri; Columbia MO 65201; USA

HENSON, Bob L.; Department of Physics; University of Missouri, St. Louis; 8001 Natural Bridge Road; St Louis MO 63121; USA; 314-553-5931; E/T; 5.3; Ion transport phenomena in gases; Corona discharge

HERBERT, Carroll; Quantum Technical Laboratories; 1732 Fourth Street; Harvey LA 70058; USA; 504-368-6992; E; 3.5; Heterocyclic hydrocarbons to determine source and quantity for oil and gas exploration--fluorescence spectroscopy

HERBST, Eric; Department of Physics; Duke University; Durham NC 27706; USA; 919-684-8180; 919-684-8140; T/E; 2.1; 3.3; 3.5; 5.7; 2.3; Ion-molecule scattering calculations; Molecular astrochemistry; Astrochemistry: organic molecules in space

HERBST, Jan F.; Department of Physics; Gen'l Motors Research Laboratories; General Motors Technical Center; 12 Mile and Mound Roads; Warren MI 48090; USA; 313-575-3382; T; 1.1; 5.2; Electronic energy levels--calculation; Localized atomic-like states in rare earth materials

HERGENROTHER, Rudolf C.; 45 Hidden River Drive-Route #2; Sarasota FL 33582; USA

HERGLOTZ, Heribert K.; Engineering Division; E.I. Dupont de Nemours and Co., Inc.; Experimental Station, Building 357; Wilmington DE 19898; USA; 302-772-3050; E; 1.1; 3.4; 5.1; Polymer structure; X-ray structure; Surface analysis by ESCA

HERITAGE, Jonathan P.; Bell Telephone Laboratories; Crawfords Corner Road; Holmdel NJ 07733; USA; 201-842-6975; E; 5.1; Surface adsorbed molecular species using optical techniques--vibrational spectroscopy

HERM, Ronald R.; Infrared Sciences Department; The Aerospace Corporation; PO Box 92957; Los Angeles CA 90009; USA; 213-648-7010; E; 5.6; 2.3; 3.3; 3.10

HERMAN, Frank; Surface Physics Division; IBM Research Laboratory; 5600 Cottle Road; San Jose CA 95193; USA; 408-256-6254; T; 1.1; 5.1; Atomic structure calculations; Chemisorption of small molecules on solid surfaces

HERMAN, Irving P.; Special Studies Group; Physics Division; Lawrence Livermore National Lab.; PO Box 808; Livermore CA 94550; USA; 415-422-1132; E; 5.7; Laser-assisted light isotope separation; Laser-initiated chemistry and laser spectroscopy

HERMAN, Robert; Department of Civil Engineering; Center for Studies-Statistical Mechanics; University of Texas, Austin; 6.806 E. Cockrell, Jr. Hall; Austin TX 78712; USA; 512-471-4379; 512-471-7253/7254; E; 1.1; Electron in the field of a fixed electric dipole in the presence of a magnetic field--stability and optical properties; Electron in the field of a multipole--studies

HERMAN, Roger M.; Department of Physics; Pennsylvania State University; 104 Davey Laboratory; University Park PA 16802; USA; 814-865-6092; T; 1.1; 3.6; 5.6; Collisional line broadening; Collisional relaxation of atomic/molecular states

HERMSMEIER, Brent D.; Department of Chemistry; University of Hawaii; 2545 The Mall; Honolulu HI 96822; USA

HERNANDEZ-RIVERA, Samuel P.; Department of Chemistry; University of Puerto Rico; Mayaguez 00708; Puerto Rico

HERNQUIST, Karl G.; RCA Laboratories; David Sarnoff Research Center; Princeton NJ 08540; USA; 609-734-2932; E/T; 5.3; 5.4; 5.1; 4.1

HEROUX, Leon J.; UV Surveillance and Remote Sensing Branch (LIU); Ionospheric Physics Division; U.S. Air Force Geophysics Laboratory; Hanscom AFB MA 01731; USA; 617-8613311; E; 3.2; 5.6; Rocket and satellite spectrometry of effects of solar UV radiation on the upper atmosphere and ionosphere; Photoionization rates in the atmosphere; Atmospheric UV horizon radiation

HERRICK, David R.; Department of Chemistry; University of Oregon; Eugene OR 97403; USA; 503-686-5208; T; 1.1; 1.2; Electron correlation effects in excited states of atoms and molecules; including multiply excited atoms--Lie algebraic investigations; Stark mixing at Rydberg levels; Conjugated polyenes--valence excited spectra

HERRING, Conyers; Department of Applied Physics; Stanford University; Stanford CA 94305; USA; 415-497-0686; T; 1.1; 5.2; Ground-state energy of a many-electron system in terms of its electron density distribution --search for improved ways; Interatomic forces in solids

HERSCHBACH, Dudley R.; Department of Chemistry; Harvard University; 12 Oxford Street; Cambridge MA 02138; USA; 617-495-3218; 617-498-5775; E/T; 2.3; 1.1; 1.3; 5.1; Chemical reactions and energy transfer processes studied by molecular beam and spectroscopic methods--dynamics; Weakly-bound molecular clusters, especially electron attachment--property; Ionization and reactions on surfaces

HERWIG, Lloyd O.; 2309 N. Stafford Street; Arlington VA 22207; USA

HERZBERG, Gerhard; Herzberg Institute of Astrophysics; National Research Council of Canada; 100 Sussex Drive; Ottawa Ontario K1A0R6; Canada; 613-990-0917; 613-746-4126

HERZENBERG, Arvid; Applied Physics; Yale University; PO Box 2157; Yale Station; New Haven CT 06520; USA

HERZFIELD, Judith; Biophysical Laboratory; Harvard Medical School; 25 Shattuck Street; Boston MA 02115; USA; 617-732-1956; E/T; Proteins--structure and function

HESS JR., Doren W.; Engineering Department 28I; Atlanta Instrumentation Division; Scientific-Atlanta, Inc.; PO Box 105027; Atlanta GA 30348; USA; 404-925-5000; 404-448-1850; E; 3.3; 3.6; 3.9; 3.11; 1.1; Microwave spectroscopy of ions in solids; Interaction of coherent radiation with atomic electrons; Crystal-field splittings of electronic energy levels; Instrumentation for microwave and laser spectroscopy

HESS, Roger A.; Lab. for Plasma and Fusion; University of Maryland; College Park MD 20742; USA; 301-454-3536; E; 3.2; 5.4; Plasma spectroscopy; Impurity studies; Ions in plasmas

HESSEL, Merrill M.; Molecular Spectroscopy Division; National Bureau of Standards; Gaithersburg MD 20899; USA; 301-921-2021; E; 5.7; Laser spectroscopy of molecular species; Laser measurement techniques for obtaining molecular properties--development

HESSEL, Norman B.; 1483 First Avenue; New York NY 10021; USA

HESSLER, Jan P.; Chemistry Division; Argonne National Laboratory; 9700 S. Cass Avenue; Argonne IL 60439; USA; 312-972-3717; E; 2.3; 3.6; 3.7; 3.10; 3.1; Nonlinear processes for generation of VUV light; Multi-photon excitation for analytical applications; VUV excitation of polyatomic molecules; Chemical kinetics of hydrocarbon reactions

HETTEMA, John M.; Department of Physics; Sloane Physics Lab.; Yale University; 217 Prospect Street; New Haven CT 06520; USA; 203-436-3567; 203-436-8671; E; 1.4

HEXTER, Robert M.; Department of Chemistry; University of Minnesota; 207 Pleasant Street, SE; Minneapolis MN 55455; USA; 612-373-2236; E/T; 5.1; Molecules adsorbed on clean metal surfaces--Raman spectra; Molecules adsorbed on clean metal surfaces--secondary ion mass--spectroscopy

HICHWA, Richard D.; Univ. of Michigan Medical School; Box 56; 3480 Kresge I; Ann Arbor MI 48109; USA

HICKMAN, Albert P.; Molecular Physics Department; Molecular Physics Laboratory; SRI International; 333 Ravenswood Avenue; Menlo Park CA 94025; USA; 415-859-4478; T; 2.1; Two-body and non-reactive three-body atomic and molecular collision processes--calculations

HICKS, Ross G.; NASA; 2880 Broadway; New York NY 10025; USA

HIERL, Peter M.; Department of Chemistry; University of Kansas; Lawrence KS 66045; USA; 913-864-3019; E; 2.3; Dynamics of ion-molecule reactions at relative collision energies from; 0.1 to 100 eV--ion beam studies

HIGAKI, Hajime; 7-4-81 Asahi-Cho; Kashiwa-Shi; Meteorological College; Chiba-Ken 277 6045; Japan

HIJAB, Raif S.; PO Box 9466; Berkeley CA 94709; USA

HILBORN, Robert C.; Department of Physics; Oberlin College; Oberlin OH 44074; USA; 216-775-8330; E; 1.1; 3.6; 3.7; Collisional effects in near-resonance laser light scattering; Free radicals-high-resolution spectroscopy; Excited-state lifetime studies of atoms and molecules

HILDEBRAND, Joel H.; Department of Chemistry; University of California; Berkeley CA 94720; USA; 415-525-2131; T/E; Professor emeritus

HILDUM, Edward A.; 100 University; Menlo Park CA 94025; USA

HILL, Dennis; 2801 N Redmond; Oklahoma City OK 73127; USA

HILL, John C.; Department of Physics; Physics Division; General Motors Research Labs.; General Motors Technical Center; 12 Mile and Mound Roads; Warren MI 48090; USA; 313-575-2838; E; 3.3; 3.6; 5.5; 5.6; Vehicle emissions; Semiconductor lasers

HILL, R. A.; Neutron Devices and Technology Div.; Sandia National Laboratories; PO Box 5800; Albuquerque NM 87185; USA; 505-844-1675; E; Plasma physics; Plasma focus device used as a neutron generator

HILL, R.; Sharp Physical Lab.; University of Delaware; Newark DE 19716; USA

HILL JR., Ralph H.; Chemical Lasers Division; U.S. Air Force Weapons Lab.; AFWL/ARAC; Albuquerque NM 87117; USA; 505-844-9836/3386; E; 4.1; Chemical lasers

HILL, Robert M.; Molecular Physics Laboratory; SRI International; 333 Ravenswood Avenue; Menlo Park CA 94025; USA; 415-326-6200x3331; E; 1.2; 2.1; 2.3; Rydberg atoms--reactive collisions; Gaseous excimer lasers--kinetics; Air subject to intense particle beams--kinetics

HILL, Robert N.; Department of Physics; University of Delaware; Newark DE 19716; USA; 302-451-8787; 302-451-2661; T; 1.1; Mathematical physics of atoms and molecules; Numerical methods-numerical analysis for atomic and molecular problems; Rigorous bounds to bound state properties

HILL III, Wendell T.; Institute for Physical Science and Tech.; University of Maryland; IPST Building; College Park MD 20742; USA

HILLARD, Grover B.; 25801 Lakeshore Boulevard #34; Euclid OH 44117; USA

HILLIER, J.; 22 Arreton Road; Princeton NJ 08540; USA

HILLMAN, John J.; Molecular Astrophysics Section; Code 693.1; NASA Goddard Space Flight Center; Greenbelt MD 20771; USA; 301-344-7974; E/T; 3.3; Ultra-high resolution IR molecular spectroscopy; IR astronomy; Planetary atmospheric constituents and molecular absorption/emission from other; IR astrophysical sources--heterodyne detection

HINCHEN, John J.; United Technologies Research Center; Silver Lane; East Hartford CT 06108; USA; 203-727-7525; E; 2.1; 2.3; 3.2; 3.5; 3.6

HINDS, Edward A.; Department of Physics; Sloan Physics Laboratory; Yale University; 217 Prospect Street; New Haven CT 06520; USA; 203-436-8671; E; 1.1; 1.2; 1.4; 3.2; 3.3; Simple systems--fundamental high precision studies; Symmetries

HINNOV, Einar; Plasma Physics Laboratory; Princeton University; Princeton NJ 08544; USA; 609-683-3154; E/T; 5.4; Spectrum lines of highly-ionized atoms in high temperature (>1 keV) plasmas--identification and measurement; Plasma diagnostics--applications of studies

HINRICHS, C. K.; EG&G Energy Measurements; P.O. Box 204; 2801 Old Crow Canyon Road; San Ramon CA 94583; USA; 415-838-3295; 415-837-9074; E; 5.1; 5.4; 4.1; 3.4; 3.1; High speed diagnostics-electronic streak tube development; Inertial fusion-magnetic fusion diagnostics employing electro-optics; X-ray absorption and emission, fluorescence, radiation transport; Electron-ion optics

HINTERLONG, Stephen J.; 9 Kimberly Court Apt. 28; Red Bank NJ 07701; USA

HIPPS, Kerry W.; Chemical Physics; Chemical Physics Program; Washington State University; Pullman WA 99164; USA; 509-335-5822; 509-335-3033; E/T; 1.1; 2.2; 3.2; 3.3; 3.5; Inelastic electron tunneling spectroscopy; Excited state molecular geometry; Vibrational spectroscopy of surfaces; Optical spectroscopy of surfaces

HIRD, Brian; Department of Physics; University of Ottawa; 375 Nicholas; Ottawa Ontario K1N6N5; Canada; 613-564-3356; E; 2.1; 4.2; Negative ions

HIRSCH, John M.; 1502 Cedarbrook; Houston TX 77055; USA

HIRSCH, Robert G.; Engineering Research and Development; E.I. Dupont de Nemours and Co., Inc.; Experimental Station E304; Wilmington DE 19898; USA; 302-772-2535; E/T; 2.3; 5.4; Ion molecule reactions; Trace gas analysis; Electron capture detection

HIRSCHFELDER, Joseph O.; Theoretical Chemistry Institute; University of Wisconsin; 1101 University Avenue; Madison WI 53706; USA; 608-262-1511; T; 1.1; 3.6; Molecular quantum mechanics; Intermolecular forces; Statistical mechanics; Molecular properties in electromagnetic fields

HIRSH, Merle N.; Laboratory for Laser Energetics; University of Rochester; Rochester NY 14627; USA; 716-275-2314; E; 4.2; 5.3; 5.4;

Gaseous electronics and corona discharge; Plasma chemistry

HIRSHFIELD, Jay L.; Dept. of Engineering and Applied Science; Mason Laboratory; Yale University; 9 Hillhouse Avenue; New Haven CT 06520; USA; 203-432-4553; E; 4.1; 5.3; 5.4; Plasma physics; Ion confinement in laser-initiated vacuum arcs; Free electron lasers

HISATUNE, I. C.; Department of Chemistry; Pennsylvania State University; 152 Davey Laboratory; University Park PA 16802; USA; 814-865-5761; E; 2.3; Transient molecules appearing during chemical reactions--spectroscopic studies

HISKES, John R.; Lawrence Livermore National Lab.; PO Box 808; Livermore CA 94550; USA; 415-422-9834; T; 2.3; Negative hydrogen and deuterium ions--processes leading to production

HLOUSEK, Louis E.; 3103 E. Locust; Ft. Collins CO 80524; USA

HO, Tak-San; Department of Chemistry; University of Kansas; Lawrence KS 66045; USA

HO, William W.; Physics and Chemistry Division; Rockwell Internat'l Science Center; 1049 Camino Dos Rios; Thousand Oaks CA 91360; USA; 805-498-4545; E; 3.3; Gases, liquids and solids at microwave and millimeter-wave frequencies --di-electric properties

HO, Yew K.; Department of Physics and Astronomy; Louisiana State University; Baton Rouge LA 70803; USA; 504-388-6855; E; 2.2; 1.1; 3.1; 3.9; 5.4

HOBBS, Robert; Electromagnetics and Physics Division; United Technologies Research Ctr.; Silver Lane; East Hartford CT 06108; USA; 203-727-7421; T; 1.1; 2.1; 3.8; Molecular structure; Molecular collision dynamics; Laser-induced chemistry

HOCHSTRASSER, Robin M.; Department of Chemistry; University of Pennsylvania; 34th and Spruce Streets; Philadelphia PA 19104; USA; 215-898-8410; E; 2.3; 3.5; 3.6; 3.9; Molecules--spectroscopy; Dye lasers and picosecond lasers for studies of molecular spectroscopy and relaxation processes in molecules

HOCHULI, Urs E.; Department of Electrical Engineering; University of Maryland; College Park MD 20742; USA; 301-454-6855; E; 4.1; Gas laser technology

HOCKER, L. O.; PO Box 1016; 199 Main Street; North Falmouth MA 02556; USA

HODGE JR., William L.; Fusion Research Center; University of Texas; RLM 11.222; Austin TX 78712; USA

HODGES JR., Dean T.; Research and Development; Lasers and Optics Division; Newport Corporation; P.O. Box 8020; 18235 Mt. Baldy Circle; Fountaine Valley CA 92728; USA; 714-963-9811; E; 4.1; 5.3; Polar molecules for generating coherent IR radiation--IR pumping

HODGES, Ronald V.; Chemistry Department; Materials Sciences Directorate; Palo Alto Research Laboratory; Lockheed Missiles and Space Co.; 3251 Hanover; Palo Alto CA 94304; USA; 415-424-2518; E; 5.3; 2.3; Cross sections of processes occurring in collisions of ionic and metastable or ground state neutral atoms with various gases--measurements; Electrical breakdown of gases; Electrical discharges

HODGSON, Rodney T.; Physical Sciences Division; IBM Thomas J. Watson Research Center; PO Box 218; Yorktown Heights NY 10598; USA; 914-945-2886; E; 5.2; Ion implantation using high power pulsed beams

HOERLIN, Herman W.; Los Alamos National Laboratory; PO Box 1663 MS672; Los Alamos NM 87545; USA; 505-667-4997; E; 4.3; 5.6; Atmospheric and high altitude atomic test data for interaction of gamma rays, X-rays, neutrons and debris with environment --review

HOFFMAN, J. M.; 4908 Glenwood Hills Drive, NE; Albuquerque NM 87111; USA

HOFFMAN, Nelson M.; Inertial Fusion and Plasma Theory; Applied Theoretical Physics; Los Alamos National Laboratory; University of California; Mail Stop E 531; Los Alamos NM 87545; USA; 505-667-3417; 505-662-9337; T; 5.1; 5.4

HOFFMANN, Roald; Department of Chemistry; Cornell University; 222 Baker Laboratory; Ithaca NY 14853; USA; 607-256-3419; T; 2.3; Applied quantum chemistry

HOFMANN, G. A.; 3750 Riviera Drive NR G; San Diego CA 92109; USA

HOFSTADTER, Robert; Department of Physics; Stanford University; Stanford CA 94305; USA

HOLIEN, Erling; Institute of Physics; Theoretical Group; University of Oslo; 1048; Sverrestien 1; Oslo N-3 Norway 1310; Norway; 456010; T; 1.1; 1.2; 1.3; 1.4; 2.1; General structure and properties of atoms and molecules, energy levels, wave functions, interactions (potentials); Atomic and molecular collisions excluding electron collisions; Electron and positron collisions with atoms and molecules, e.g. energy loss

HOLLAND, Orin W.; Solid State Division; Oak Ridge National Laboratory; Building 3003 (X-10); Oak Ridge TN 37830; USA; 615-574-7271; E; 4.2; 5.1; Positive ion channeling spectroscopy to study reordering of surfaces

HOLLBERG, Leo W.; 572 Arapahoe; Boulder CO 80302; USA

HOLMES, Charles P.; 10017 Herding Row; Columbia MD 21046; USA

HOLMES, John R.; Department of Physics and Astronomy; University of Hawaii, Manoa; 2505 Correa Road; Honolulu HI 96822; USA; 808-948-7087; E; 1.1; Atoms--transition probabilities

HOLROYD, Richard; Chemistry Division; Brookhaven National Laboratory; Bldg 555; Upton NY 11973; USA; 516-345-4329; E; 3.1; 3.8; 4.2; 5.2; Photochemistry; Photoionization in liquids; Laser photodetachment

HOLSTEIN, T. D.; Department of Physics; University of California, LA; Los Angeles CA 90024; USA; 213-825-1796; T; 5.2; Atoms and electrons in solids--interactions

HOLT, Helen K.; National Bureau of Standards; Gaithersburg MD 20899; USA; 301-921-2061; T; 2.2; 2.3; 3.1; 3.9; Light with atoms--interaction; Electron-atom scattering; Stark effect and resonance trapping

HOLT, Richard A.; Department of Physics; University of Western Ontario; London Ontario N6A3K7; Canada; 519-679-6168; 519-679-2568; E; 3.6; 1.4; 4.1; 3.3; 4.2

HOLTON, Gerald J.; Department of Physics; Jefferson Laboratory; Harvard University; Cambridge MA 02138; USA; 617-495-4474; E; 2.1; 1.3; Molecular relaxation at high pressures

HOLWAY JR., L. H.; 4 Everett Street; Nattick MA 01760; USA

HOLZBERLEIN, T. M.; Department of Physics; Principia College; Elsaa IL 62028; USA

HOLZSCHEITER, Mike H.; Department of Physics; Texas A&M University; College Station TX 77843; USA

HONG, Siu-Ping; Bell Telephone Laboratories; AT&T Bell Laboratories; 555 Union Boulevard; Allentown PA 18103; USA; 215-439-7610; E; 5.3; Gas breakdown in very small gaps

HONIG, Richard E.; Materials Characterization Research Div.; RCA Laboratories; David Sarnoff Research Center; Princeton NJ 08540; USA; 609-734-3241; E; 5.1; Surface studies and analysis by various ion, electron and X-ray spectroscopies

HOOD, Robin J.; Department of Chemistry; Marygrove College; 8425 W. McNichols

Road; Detroit MI 48221; USA; 313-862-8000; E/T

HOOPER JR., E. B.; M-Division; Lawrence Livermore National Lab.; PO Bx 808 L-433; Livermore CA 94550; USA; 415-422-6792; E/T; 5.4; Plasma physics and neutral beam physics

HOOPER, Henry O.; Department of Physics; University of Maine; Orono ME 04469; USA; 207-581-7546; E; Magnetic resonance (NMR and NQR) of solids including glasses

HOPKINS, Jeffrey L.; Department of Physics; University of Georgia; Athens GA 30602; USA

HOPPE, John C.; Photo-Optical Instrument Section; Instrument Research; Langley Research Center; NASA; MS 236; Hampton VA 23665; USA; 804-865-3234; E; 4.1; 3.6; Aerodynamic and fluid dynamic properties in wind tunnels--lasers used to excite atomic and molecular species; Visualize the flow fields around models; Conduct non-intrusive measurements

HOPPER, Darrel G.; Theoretical Chemistry Division; JAYCOR; PO Box 3126; Dayton OH 45431; USA; 513-429-9909; 513-873-2202; T/E; 1.1; 2.1; Molecular structure and spectra with advanced state-of-the-art ab initio methods--apriori calculations; Ion-molecule and neutral-neutral scattering processes and fundamental research on state-to-state processes--collisional studies

HORNBECK, J. A.; 5196 Crystal Drive; Beulah MI 49617; USA

HORSLEY, John A.; Corporate Research Laboratories; Exxon Research and Engineering Co.; PO Box 45; Linden NJ 07036; USA; 201-474-2585; T; 3.1; 3.8; IR laser-induced dissociation of complex molecules --modeling; Laser isotope separation

HORVITZ, Samuel; PO Drawer 303; Waterford CT 06385; USA

HORWITZ, Alexander B.; Chemistry Division; Brookhaven National Laboratory; Building 555A; Upton NY 11973; USA; 516-345-4375/4372; E; 3.1; 3.7; IR multiple photon dissociation characterization of the mechanisms and dynamics of dissociation--absorption process

HOSE, Gabriel; Department of Chemistry; University of Southern California; University Park; Los Angeles CA 90089; USA

HOSOYA, Haruo; Department of Chemistry; Ochanomizu University; Bunkyo-Ku Tokyo 112; Japan

HOUGEN, Jon T.; Molecular Spectroscopy Division; National Bureau of Standards; Gaithersburg MD 20899; USA; 301-921-2021; T; 3.5; Molecular spectroscopy

HOUK, Thomas W.; Department of Physics; Miami University; Oxford OH 45056; USA; 513-529-2713; E; Rare earths and organics--fluorescence spectroscopy; Cotton Mouton Effect

HOUSTON, John M.; Electronic Power Systems Branch; General Electric Corporate R&D Center; PO Box 8; Schenectady NY 12301; USA; 518-385-8591; E; 3.4; 4.1; 5.3; Discharge lamps for lighting; Xenon X-ray detectors for computer tomography; Novel flashlamps for laser pumping

HOUSTON, Paul L.; Department of Chemistry; Baker Laboratory; Cornell University; Ithaca NY 14853; USA; 607256-4303; E; 2.1; 3.1; 5.1; Energy transfer; Photodissociation dynamics; Gas-surface interactions

HOWALD, Arthur M.; 9173 Harlaxton Court; Knoxville TN 37923; USA

HOWARD, Carleton J.; Aeronomy Laboratory; Nat'l Oceanic and Atmospheric Admin.; R/E/AL2; 325 S. Broadway; Boulder CO 80303; USA; 303-497-5820; E; 2.3; 5.6; 5.5; 3.6; 3.8; Atmospheric chemistry; Chemical kinetics; Small molecules and atoms--reactions

HOWARD, John S.; Aztec Enterprises; 9145 East Kenyon Ave. Suite 101; Denver CO 80237; USA

HOWARD, Robert E.; Department of Chemistry; University of Tulsa; 600 South College; Tulsa OK 74104; USA; 918-592-6000x2686; 918-592-6000x3025; T; 2.3; 3.8; 5.6

HOWGATE, David W.; Research Directorate (DRSMI-RRD); U.S. Army Missile Command; Redstone Arsenal AL 35809; USA; 205-876-4805; T; 3.6; 4.1; Superradiance; Optical bistability; Laser physics

HOWORKA, Franz; Institut Fuer Atomphysik; Karl Schoenherrstrasse 3; Innsbruck A6020; Austria

HRUBESH, Lawrence W.; Chemistry and Material Science Div.; Lawrence Livermore National Lab.; PO Box 808; Livermore CA 94550; USA; 415-423-1691; E; 3.3; 3.6; 3.8; 3.9; Pure rotational spectroscopy at microwave and far IR wavelengths

HSIAO, Yu-Yuan R.; 4715 193rd, SE; Issaquah WA 98027; USA

HSIEH, You-Fong; Department of Physics; 666 W. Hancock; Detroit MI 48202; USA

HSU, Donald K.; Department of Physics and Computer Science; Saint Peter's College; Kennedy Boulevard; Jersey City NJ 07306; USA; 201-333-4400x489; E/T; 1.1; 3.1; 3.6; Laser fluorescence and electron beam lifetime measurements; High resolution absorption; Magnetic rotation spectroscopic methods and Franck Condon pre-dissociation; F-values and term values of energy level

HSU, Hsiung; Department of Electrical Engineering; College of Engineering; Dreese Laboratory; Ohio State University; 2015 Neil Avenue; Columbus OH 43210; USA; 614-422-7251; E/T; 3.10; 4.1; Phase conjugation; Nonlinear optics; Quantum electronics; Solid state circuits and devices

HSU, Shaw L.; Department of Polymer Science and Engineering; Graduate Research Center; University of Massachusetts; Amherst MA 01003; USA; 413545-0433; 413-545-2678; E; 1.3; 3.5; 3.3; High polymers-spectroscopic characterization

HSUEH, Ching-Yu; 2924B Sandage; Fort Worth TX 76109; USA

HUANG, Chour-Yih; Energy Conversion Devices; 1675 W. Maple Road; Troy MI 48084; USA

HUANG, Keh-Ning; Department of Physics; University of Notre Dame; Notre Dame IN 46556; USA; 219-283-7778; T; 1.1; 1.3; 1.4; Quantum electrodynamic theory; Polarization correlation in particle scatterings; Relativistic and correlation effects in atoms; Exotic atoms

HUBBARD, Paul S.; Department of Physics; University of North Carolina; Chapel Hill NC 27514; USA; 919-933-3022; 919-929-5673; T; 5.2; Molecular motion and nuclear magnetic relaxation in liquids and gases

HUBBELL, John H.; Radiation Physics Division; Center for Radiation Research; National Bureau of Standards; Gaithersburg MD 20899; USA; 301-921-2685; T; 3.1; 3.4; Staff member-Photon and Charged Particle Data Center; X-Ray mass attenuation coefficients 10 ev to 100 gev; Atomic form factors and incoherent scattering functions; Radiation fields from extended sources in absorbing and scattering media

HUBERT, Jay M.; Analytical Research and Services Div.; Chevron Research Company; P.O. Box 1627; Richmond CA 94802; USA; 415-620-4675; 415-620-4010; E; 5.1; 4.4

HUCHITAL, David A.; Perkin-Elmer Corporation; MS 240; Main Ave.; Norwalk CT 06856; USA

HUDDLESTON, Rodney K.; Chemistry Division; Argonne National Laboratory; 9700 S. Cass Avenue; Argonne IL 60439; USA; 312-972-3485; E; 3.8; 4.2; 5.2; Electron transfer reactions in low temperature glasses; Radiation chemistry

HUDGENS, Jeffrey W.; National Bureau of Standards; Gaithersburg, MD 20899; USA; 301-

921-1000; E; 3.7; 4.3; 4.4; IR and UV multiphoton ionization mass spectrometry; Radiationless processes in molecular beams

HUDSON, David F.; U.S. Naval Surface Weapons Center; F 43 Division; White Oak Laboratory; 10901 N.H. Ave.; Building 4, Room 237; Silver Spring MD 20903; USA; 202-394-2309; 202-394-1248; E/T; 1.1; 1.2; 5.3; High Rydberg atoms--collisional properties; Autoionizing states and planetary atoms; Discharge instabilities, negative ions

HUDSON, G. Martin; Department of Physics; University of Central Florida; PO Box 25000; Orlando FL 32816; USA; 305-275-2333/2325; E; 4.2; 4.3; 5.6; Ion beam analysis techniques to air pollution--development; PIXE proton induced X-ray emission analysis and PESA proton elastic scattering analysis

HUDSON, George E.; Electro-Optics Branch; Research Department; White Oak Laboratory; US Naval Surface Weapons Center; Silver Spring MD 20910; USA; 301-394-2751; T; 2.2; 3.6; 4.1; High energy electron beams; Physical and geometric optics; Radiation

HUDSON, Robert D.; Stratosphere Chemistry and Physics Branch; NASA Goddard Space Flight Center; Greenbelt MD 20771; USA; 301-344-6358; E; 3.5; 3.6; 3.7; 5.6; Atoms and molecules in UV of atmospheric interest--laser spectroscopy; Two photon spectroscopy

HUEBNER, Russell H.; Biomedical and Environmental Research; Office of the Director; Argonne National Laboratory; University of Chicago/USDOE; Bldg. 202; 9700 S. Cass Avenue; Argonne IL 60439; USA; 312-972-3807; 312-972-3804; E; 1.1; 2.2; 3.1; 4.1; 5.6; Research management; Science policy; Radiation physics; Radiological protection

HUEBNER, Walter F.; Equation of State and Opacity; Group T-4; Theoretical Division; Los Alamos National Laboratory; PO Box 1663 MB212; Los Alamos NM 87545; USA; 505-667-5751; T; 1.1; 3.1; 5.7; 5.6; Direct efforts in equation of state, opacity, astrophysics; Photon cross sections with atoms, ions and molecules; Particle collision cross sections

HUEHNERMANN, Harry; Lindenweg 6A; Marburg 355 7545; Germany

HUENNEKENS, John P.; 3V Magie; Faculty Road; Princeton NJ 08540; USA

HUERTA, Manuel A.; Department of Physics; University of Miami; Coral Gables FL 33124; USA; 305-284-2323; T; 5.3; 5.4

HUESTIS, David L.; Molecular Physics Department; Chemical Physics Laboratory; SRI International; 333 Ravenswood Avenue; Menlo Park CA 94025; USA; 415-859-3464; 415-859-3122; E/T; 1.1; 2.3; 3.2; 3.6; 3.8; Kinetic and optical processes in excimer laser media; Applied mathematics and mathematical physics; Scattering theory, especially resonances; Solids--electronic structure

HUFFAKER, James N.; Department of Physics and Astronomy; University of Oklahoma; 440 W. Brooks Street; Norman OK 73019; USA; 405-325-3962; T; 1.1; Analytic potential function, eigenfunctions, etc., from spectroscopic measurements and analyses--calculation

HUFFMAN, Robert E.; Ultraviolet Surveillance and Remote Sensing; Ionospheric Physics; US Air Force Geophysics Lab.; Hanscom AFB MA 01730; USA; 617-861-3043; E; 3.2; 5.6; Atmospheric spectroscopy in UV

HUG, William F.; Light Source Research Department; Xerox Electro-Optical Systems, Inc.; 300 N. Halstead Street; Pasadena CA 91107; USA; 213-351-2351; E; 4.1; 5.3; Optical and gaseous discharge properties of gas and metal vapor lasers

HUGHES, Raymond H.; Department of Physics; University of Arkansas; Fayetteville AR 72701; USA; 501-575-6571; E; 5.1; 4.3; 3.2; Ions generated by lasers interacting with surfaces; Optical emissions from relativistic e-beams on gases

HUGHES, Vernon W.; Department of Physics; Yale University; PO Box 6666; 217 Prospect Street; New Haven CT 06511; USA; 203-436-3566-8673; E; 1.3; 1.4; 2.2; 3.9; 5.8; Nucleons--spin structure; Parity violating fundamental interactions; Muonium, positronium and helium fine structure; Zeeman effect

HUGHES, William M.; Physics Division; Los Alamos National Lab.; PO Box 1663; Los Alamos NM 87545; USA; 505-667-7535-4415; E; 1.1; 2.2; 2.3; 4.2; Chemical kinetics for laser application processes for atoms, molecules, ions, recombination, association attachment, radiative lifetimes, photolysis--evaluation

HUHNERMANN, Harry; Physics Department; University of Marburg; Renthof 5; Marburg 3550; Germany; 6421-284153; 6421-32163; E; 1.1; 3.6; 4.2; 4.4; CLIBS (Collinear Laser-Ion Beam Spectroscopy)

HULBERT, Steven L.; Department of Physics; Brookhaven National Lab.; Building 510-B; Upton NY 11973; USA

HULET, Randall G.; Massachusetts Institute of Tech.; Room 26-214; Cambridge MA 02139; USA

HULIS, Malcom E.; 2108 W. Surrey Drive; Muncie IN 47304; USA

HUMMER, Charles R.; Department of Physics; Behlen Laboratory; University of Nebraska; Lincoln NE 68588; USA; 402-472-2770

HUMMER, David G.; Quantum Physics Div. (NBS); Joint Institute for Lab. Astrophysics; University of Colorado; Boulder CO 80309; USA; 303-492-7837; FTS-323-3760; T; 2.2; 4.1; 5.7; Electron-collision cross sections for light atoms and positive ions--calculation; Radiative transfer theory for laboratory and astrophysical systems; Saturation spectroscopy and nonlinear optics--contributions to theory

HUMPHREY, Charles H.; 219 Follen Road; Lexington MA 02173; USA

HUMPHRIES, Kent C.; Far West Technology, Inc.; 330-D South Kellogg; Goleta CA 93117; USA; 805-964-3615; E; 3.1; 3.8; 2.2; Dosimetry and radiation measurements; Radiation diagnostics

HUNEKE, John C.; Charles Evans and Associates; 301 Chesapeake Drive; Redwood City CA 94063; USA; 415-369-4567; E; 4.4; 4.2

HUNT, Angus L.; Magnetic Fusion Energy Division; Lawrence Livermore National Lab.; PO Bx 808 L-437; Livermore CA 94550; USA; 415-422-9802; FTS-532-9802; E; 5.4; Atomic and molecular physics used to diagnose high temperature plasma

HUNTER, Lawrence W.; Applied Physics Laboratory; Johns Hopkins University; Johns Hopkins Road; Laurel MD 20810; USA; 301-953-7100; E

HUNTER, Scott R.; Atomic, Molecular and High Voltage Physics Grp.; Health and Safety Research Division; Oak Ridge National Lab.; Department of Energy; PO Box X; Oak Ridge TN 37831; USA; 615574-4948; 615-574-4662; E; 5.3; 2.2; 2.36; 4.4; Electron transport parameters in low and high gas discharges; Electron attachment studies in low energy swarm and beam experiments; Laser modified electron attachment and ionization processes in low pressure gas discharges

HUNTLEY, Wright H.; HOLOGRAF; 2255 G Martin Avenue; Santa Clara CA 95050; USA; 408-727-6657; E; 3.8; 4.1; Photochemical phenomena and physical optics

HUNTOON, Andrew E.; 34 Mahler Court; Appleton WI 54915; USA

HUNTRESS JR., Wesley T.; Molecular Physics and Chemistry Div.; Jet Propulsion Laboratory; 4800 Oak Grove Drive; Pasadena CA 91103;

USA; 213-354-2140

HUO, Winifred M.; Physical Sciences Branch; NASA Ames Research Center; Moffett Field CA 94035; USA; 408-965-6189; T; 2.2; 3.7; Multiphoton excitation processes in small molecules--calculation of transition probabilities; Electron attachment of Van der Waals molecules

HURST, George S.; Institute of RIS; 22238; One Pellissippi Center; Knoxville TN 37933; USA; 615-966-0689; FTS-624-5893; E; 2.3; 3.6; 5.6; 5.8; Resonance ionization spectroscopy and one-atom detection; Laser techniques for ultrasensitive detection; Nuclear physics, chemical physics, photophysics and environmental sciences--applications

HURST, Robert P.; Department of Physics and Astronomy; State University of NY, Buffalo; 337 Fronczak Hall; Amherst NY 14260; USA; 716-636-2530; T; 1.1; Atomic polarization in ionic crystals; Momentum wavefunctions for atoms

HUSMANN, Otto K.; RA-1; Space Division; Messerschmitt Bolkow-Blohm; PO Box 801169; 8000 Munich 80; 800 Munich 80 Bavaria 8000; West Germany; 089-60007293; 089-6122871; E; 3.1; 3.2; 2.1; 2.2; 5.1; Materials degradation in simulated space environment studies

HUTCHINSON JR., Clyde A.; Department of Chemistry; University of Chicago; Chicago IL 60637; USA; 312-753-8618; E; 1.1; Hydrogen atom coordinates in large molecules-structure and precise determination

HUTCHISON, Sheldon B.; Research and Development; Laser Physics Laboratory; XMR, Inc.; 5403 Betsy Ross Drive; Santa Clara CA 95054; USA; 408-988-2426; 408-243-9489; E; 3.6; 3.2; 4.1; 5.3; 3.8; Laser transition spectroscopy for new laser systems; Laser-materials interactions research; Gas discharge research for laser device development; Optical laser beam clean up research

HUTSON, Richard L.; Medium Energy Physics Division; Los Alamos National Lab.; Box 1663 MSH844; Los Alamos NM 87545; USA; 505-667-8293; E; 5.2; 4.3; Muonic atoms; Muonic X-rays; Muon spin relaxation

HUTTER, Edwin C.; 54 Van Dyke Road; Princeton NJ 08540; USA

HUWEL, Lutz; Department of Physics; Wesleyan University; Middletown CT 06457; USA; 203-347-9411x2815; 203-347-9411 x3035; E; 1.2; 2.1; 2.3; 3.6; 3.8; Charge transfer, rearrangement processes and negative ion formation in low energy collisions involving Rydberg atoms

HWANG, Dah-Min D.; Department of Physics; Univ. of Illinois, Chicago Circle; PO Box 4348; Chicago IL 60680; USA; 312-996-3446/3400; E; 4.1; Raman spectroscopy on molecules

HWANG, Wei; Department of Electrical Engineering; Columbia University; 520 W. 120th Street; New York NY 10027; USA

HYLTON, Derrick J.; 1012 Bay Street; Staten Island NY 10305; USA

HYMAN, Howard A.; Avco-Everett Research Laboratory; 2385 Revere Beach Parkway; Everett MA 02149; USA; 617-389-3000 x211; T; 2.1; 2.2; 3.1; 3.10; 4.1; Atomic and molecular spectroscopy; Electron scattering processes; Kinetics; Laser physics

ICHIMURA, Hideo; Department of Physics; University of Windsor; Windsor Ontario N9B3P4; Canada

IKUTA, Takashi; Department of Applied Electronics; Osaka Electro-Commun Univ.; Neyagawa Osaka; Japan

INGOLD, John H.; Lighting Research & Technical Services Division; General Electric Co.; Nela Park; Cleveland OH 44112; USA; 346-266-3137; T; 5.3; 5.4; Gas discharge light source; Low energy plasma physics

INNES, Frederick R.; Space Physics Division; US Air Force Geophysics Lab.; Hanscom AFB MA 01730; USA; 617-861-3220; T; 1.1; 2.3; 5.4; Atomic and molecular structure in ion-molecule reactions; Transition moments in atoms and molecules; Phenomenology of beams in plasmas

INNES, K. K.; Department of Chemistry; State University of NY, Binghamton; Binghamton NY 13901; USA; 607-777-2269-4246; E/T; 1.1; 3.5; Molecular spectroscopy and structure

INOKUTI, Mitio; Environmental Research Division; Argonne National Laboratory; 9700 S. Cass Avenue; Argonne IL 60439; USA; 312-972-4186; 312-972-4185; T; 2.3; 3.1; 5.6; Atomic collision processes pertinent to action of ionizing radiation in matter

INTEMANN, Robert L.; Department of Physics; Temple University; Philadelphia PA 19122; USA; 215-787-7697; 215-787-1398; T; 5.8; 2.2; 1.1; Inner shell physics

INUISHI, Yoshio; Faculty of Engineering; Osaka University; Yamada-Oka Suita; Osaka; Japan

IOUP, George E.; Department of Physics; University of New Orleans; New Orleans LA 70148; USA; 504-286-6709; 504-283-6820; T; 4.4; 2.1; 2.3; Deconvolution and mathematical digital filtering; Inverse theory

IOUP, Juliette W.; Physics; University of New Orleans; New Orleans LA 70148; USA; 504-286-6715; T; 4.4; 2.1; Deconvolution and mathematical digital filtering; Inverse theory; Computer model of the quadropole mass spectrometer 127-degree cylindrical electrostatic analyzer--transmission

ISHIDA, Takanobu; Department of Chemistry; State University of NY, Stony Brook; Stony Brook NY 11794; USA; 516-246-7165; E/T; 1.1; Elucidation of molecular forces, both intra- and intermolecular isotope effects, M.O. on clusters, zero-point energy; Fractionation of stable isotopes--developmental studies

ISHIHARA, Takeshi; Institute of Applied Physics; University of Tsukuba; Sakura Mura Ibaraki 305; Japan

ISLAM, Azad M.; Department of Physics; Suny College, Potsdam; Potsdam NY 13676; USA; 315-267-2282; 315-265-8531; E; 1.1; 4.1; 5.3

ISLER, Ralph C.; Fusion Energy Division; Oak Ridge National Laboratory; PO Box Y; Oak Ridge TN 37831; USA; 615-574-1174; FTS-624-1174; E; 5.4; Fusion plasmas--spectroscopy

ISLER, William E.; 12925 Valleywood; Silver Spring MD 20906; USA

ISOZUMI, Yasuhito; Radioisotope Research Center; Kyoto University; Kyoto; Japan; E.4; 2.2; 3.4; 5.8

ITANO, Wayne M.; Time and Frequency Division; National Bureau of Standards; 325 S. Broadway; Boulder CO 80303; USA; 303-497-5632; 303-499-4248; E; 1.1; 1.4; 3.2; 3.3; 3.6; Laser cooling of stored ions; High resolution spectroscopy of stored atomic ions; Frequency standards based on stored atomic ions; Atomic hyperfine interactions, g factors

ITO, Harumasa; 2-1-42-603 Sakuramachi; Koganei Tokyo 184; Japan

IVEY, Henry F.; Technology Assessment Department; Westinghouse Research Labs.; 1310 Beulah Road; Pittsburgh PA 15235; USA; 412-256-2683; 412-373-0162; E/T; 3.2; 3.6; 3.8; 3.1; Luminescence (photo-, electro-, cathodo-); Color centers in solids; Laser materials (solid state); Lamps and illumination

IVEY, Robert C.; Department of Physics; Abilene Christian University; PO Box 8127; ACU Station; Abilene TX 79699; USA; 915-677-1911 x2165; E; 1.1; 2.2; Gases and radicals--electron diffraction structure studies

IWINSKI, Zbigniew R.; Department of Physics; Brooklyn College of CUNY; Bedford Avenue & Avenue H; Brooklyn NY 11210; USA; 212-780-5578; T; 1.1; 2.2; Electron and positron scattering of atoms; Atomic processes--effects of screening

JACK JR., Hulan E.; 98-32 57th Avenue Apt. 8D Flushing NY 11368; USA

JACKSON, William M.; Department of Chemistry; University of California, Davis; 114 Chemistry Building; Davis CA 95616; USA; 916-752-0503; E/T; 3.8; 3.1; 4.1; 5.1; 5.7; Photodissociation dynamics using lasers; Reaction kinetics using lasers; Surface reactions using lasers and molecular beams; Cometary chemistry

JACOB, Elizabeth J.; Department of Chemistry; University of Toledo; Toledo OH 43606; USA; 419-537-2161; E; 1.1; 2.2; 3.3; Gas phase species via electron diffraction and microwave spectroscopy --molecular structure determination

JACOBS, Ralph R.; Lawrence Livermore National Lab.; PO Box 808 L-470; Livermore CA 94550; USA; 415-422-6153; E; 3.6; 3.8; 4.1; Advanced lasers for laser fusion and laser isotope separation applications; Gain, media gas, solid and liquid phases --kinetics studies; Linear and nonlinear spectroscopy

JACOBS, Stephen F.; Optical Sciences Center; University of Arizona; Tucson AZ 85721; USA; 602-626-2944; E; 4.1; 4.3; 5.5; IR detectors; Lasers; Interferometry

JACOBS, Stephen M.; Raychem Corporation; 300 Constitution Drive; Menlo Park CA 94025; USA;

JACOBS, Verne L.; Plasma Radiation Branch; Plasma Physics Division; US Naval Research Lab.; U.S. Navy; 4555 Overlook Avenue, SW; Washington DC 20375; USA; 202-767-3169; T; 1.4; 3.1; 3.4; 3.7; 3.9; Atomic collision theory and QED; Interaction of atomic systems with electric fields and EM radiation; Atomic collision and radiation processes in plasmas; Atomic spectral line shapes

JACOBSEN, E. H.; Department of Physics and Astronomy; University of Rochester; Rochester NY 14627; USA; 716-275-4374; E/T; 2.2; 4.3;

JACOBSON, Harry C.; Department of Physics and Astronomy; University of Tennessee; Knoxville TN 37916; USA; 615-974-4161; T; 3.2; Spectroscopy; Line shape problems

JACOX, Marilyn E.; Molecular Spectroscopy Division; National Bureau of Standards; Gaithersburg MD 20899; USA; 301-921-2754; E; 3.5; 3.1; Vibrational and electronic spectra of neutral and charged fragments of small molecules

JADUSZLIWER, Bernardo; M2/253; Chemistry and Physics Laboratory; The Aerospace Corporation; P.O. Box 92957; Los Angeles CA 90009; USA; 213-416-9217; E; 2.2; 3.6; 3.9; 4.3; 1.1; Electron-atoms collisions with and without state selection; Atomic and molecular polarizabilities

JAECKS, Duane H.; Department of Physics; Behlen Laboratory; University of Nebraska; Lincoln NE 68588; USA; 402-472-3274; E; 2.1; Atomic processes in energetic ion-atom and ion-molecule collisions

JAFFE, Charles; Department of Chemistry; Columbia University; New York NY 10027; USA

JAFFE, Hans H.; Department of Chemistry; University of Cincinnati; Cincinnati OH 45238; USA; 513-475-2262; T; 1.2; 4.1; Semi-empirical molecular quantum mechanics

JAFFE, Richard L.; Computational Chemistry Branch; Ames Research Center; NASA; 230-3; Moffett Field CA 95112; USA; 415-694-6458; T; 1.1; 3.5; 2.3; 1.3; 5.5; Ab initio quantum chemistry to determine molecular properties; Spectroscopic applications--gas phase; Polymer physical and mechanical properties; Classical trajectory calculations

JAHANI, Hamid R.; PO Box 688 NB11; Richardson TX 75080; USA

JAHODA, Franz C.; CTR Division; Los Alamos National Lab.; 1663 MSF639; Los Alamos NM 87545; USA; 505-667-4326; E/T; 5.4; Atomic and molecular science incidental to plasma diagnostics; Impurity radiation; Lithium atomic beam probing

JAIN, Manoj K.; AT&T Bell Labs.; Room 1L201 Crawfords Corner Road; Holmdel NJ 07733; USA

JAKACKY JR., John M.; 233 Burke Street; East Hartford CT 06118; USA

JAKAS, Mario M.; Department of Physics; Atomic Collision; Centro Atomico Bariloche; Argentine Atomic Energy Comm.; Bariloche RN 8400; Argentina; E/T; 2.1; 5.1; 4.2; Energy loss in ion-atom collisions

JAKLEVIC, R.; 31345 Old Cannon Road; Birmingham MI 48010; USA

JALUFKA, N. W.; 505 Brokenbridge Road; Yorktown VA 23692; USA

JAMERSON, F. E.; Department of Physics; GM Technical Center; General Motors Research Laboratory; Warren MI 48090; USA

JAMES, David R.; Biological and Radiation Physics; Health and Safety Research; Oak Ridge National Laboratory; 4500S H 150; Oak Ridge TN 37831; USA; 615-574-6205; E; 5.3; 5.1; 5.2; 3.2; Gaseous and solid dielectrics; High voltage physics

JAMISON, Keith A.; Applied Physics Branch; US Army Ballistics Research Lab.; Building 120; Aberdeen Proving Ground; Aberdeen MD 21005; USA; 301-278-4905; E; 2.1; 3.4; Soft X-ray production from 1-50 keV ion-atom collisions

JAMISON, Keith D.; Department of Physics; Rice University; PO Box 1892; Houston TX 77251; USA

JANOW, Richard H.; Department of Physics; City College of CUNY; 139th Street & Convent Ave.; New York NY 10031; USA; 914-592-4890; T; 5.1; Desorption processes driven by photons or charged molecules--modeling

JARNAGIN, Richard C.; Department of Chemistry 045A; University of North Carolina; Chapel Hill NC 27514; USA; 919-966-5433; E/T; 5.1; Small molecular adsorbates on solid surfaces--geometric and electronic structure

JARRETT, S. M.; 474 Los Ninos Way; Los Altos CA 94022; USA

JASON, Andrew J.; AT-3 MSH808; Accelerator Technology Division; Group AT-3; Los Alamos National Laboratory; PO Box 1663 MS808; Los Alamos NM 87545; USA; 505-667-2842; 505-662-4105; E; 3.9; 4.2; High energy beam diagnostics by interaction with atoms; Accelerator--related atomic physics; Beam detection by interaction with atomic beams or gases

JAVAN, Ali; Department of Physics; Massachusetts Institute of Technology; Building 6-208; 77 Massachusetts Avenue; Cambridge MA 02139; USA; 617-253-5088; E; 3.6; Laser physics

JAYAKUMAR, Raghavan; Plasma Physics Program; Physical Research Lab.; Navrangpura; Ahmedabad 38009; India

JAYARAM, Raman; Department of Physics and Astronomy; University of Pittsburgh; Allen Hall; Pittsburgh PA 15260; USA

JEFFERTS, K. B.; PO Box 363; Shaw Island WA 98286; USA

JEFFRIES, Jay B.; Molecular Physics Department; SRI International; 333 Ravenswood Avenue; Menlo Park CA 94025; USA; 415-859-6341; E; 2.3; 3.6; 3.8; 5.4; 5.5

JEN, Chih K.; Research Center; Applied Physics Laboratory; Johns Hopkins University; Johns Hopkins Road; Laurel MD 20707; USA; 301-953-6239; 301-953-6255; E/T; 3.3; 3.9; 1.1; 1.3; 3.11; Professor emeritus

JENKINS, David W.; Department of Physics; University of Notre Dame; Notre Dame IN 46556; USA

JENKINS, Leslie H.; 817 Whirlaway Circle; Knoxville TN 37923; USA

JENNINGS, Donauld A.; Time and Frequency Division 524.00; National Bureau of Standards; 325 S. Broadway; Boulder CO 80303; USA; 303-499-1000x3174; E; 4.1; 3.10; 3.5; 3.3; Laser frequency measurements; Far IR lasers to molecular spectroscopy--application

JENNINGS, William C.; Department of Electrical & Systems Engineering; Rensselaer Polytechnic Institute; Troy NY 12181; USA; 518-266-6087; 518-266-6488; E; 5.4; Particle beam probing diagnostic techniques for high temperature, magnetically confined plasmas--development

JENSEN, Barbara L.; Department of Physics; Boston University; 111 Cummington St.; Boston MA 02215; USA; 617-353-2610; 617-353-2600; 3.11; 3.10; 3.7; 3.3; 5.1

JENSEN, Craig C.; Department of Chemistry; Los Alamos National Laboratory; PO Box 1663; Los Alamos NM 87545; USA; 505-667-3519; E; 3.6; 3.9; Gas phase electronically excited atoms in electric field--trapping; Jet cooled molecular free radicals--laser induced fluorescence

JENSEN, Mark J.; 317-F Cheswick Place; Cary NC 27511; USA

JEON, Yoon H.; Department of Physics; Kum-Oh Institute of Technology; #180-1 Shinpyung Dong; Gumi-Si Kyungbu 641; Korea; E/T; 1.1; 3.11;

JEPSEN, Ove; Max-Planck Institute of FKF; Heisenbergstrasse 1; Stuttgart 80 70000; West Germany

JEPSEN, R. L.; 575 Stonegate; Eugene OR 97401; USA

JERJIAN, Khachig A.; Department of Physics; Louisiana State University; Baton Rouge LA 70803; USA

JETTE, A. N.; Applied Physics Laboratory; Johns Hopkins University; Johns Hopkins Road; Laurel MD 20707; USA; 301-953-6263; T; 1.1; 5.1; Atomic and molecular structure from electron spin resonance nuclear hyperfine data--determination; Surface studies

JEYS, Thomas H.; Department of Physics; Rice University; PO Box 1892; Houston TX 77251; USA; 713-527-8101x3571; E; 3.9; 1.2; 4.1; 3.6; Optical pumping of sodium atoms; Spin dependance of atomic collisions

JEZIORSKI, Bogumil S.; Department of Chemistry; University of Warsaw; Pasteura 1; Warsaw 02093; Poland

JIANG, Lin; 925 S. Waterview Drive, #201; Richardson TX 75080; USA

JIMENEZ-MIER, Jose; Department of Physics; Yale University; New Haven CT 06511; USA; 203-436-8783; E; 3.1; 3.2; 3.6; 3.5; 3.8

JOACHAIN, C. J.; Physique Theorique-Fac Science; University Libre De Bruxelles; Campus Plaine C P 227; - Bruxelles 1050; Belgium

JOHNSEN, Rainer; Department of Physics and Astronomy; University of Pittsburgh; Pittsburgh PA 15260; USA; 412-624-4352; E; 2.1; 2.2; Low-energy ion-neutral and electron-ion collisions

JOHNSEN, Russell H.; Department of Chemistry; Florida State University; Tallahassee FL 32306; USA; 904-644-3500; E; 4.2; 4.4; 5.2; Molecular ions by means of collisional activation mass spectrometry --characterization; Crystalline solids--kinetics of reactions

JOHNSON, A. W.; 12110 Princess Jeanne, NE; Albuquerque NM 87112; USA

JOHNSON, Brant M.; Department of Applied Science; Atomic and Applied Physics Division; Brookhaven National Laboratory; Building 901A Tandem; Upton NY 11973; USA; 516-282-4552-x4581; FTS-666-4552x4581; E; 2.1; 2.2; 3.2; 4.2; 3.4; Ion-atom collisions; Crossed electron-ion beams; Synchrotron-atomic physics

JOHNSON, Carol; Physics A251; Center for Radiation Research; National Bureau of Standards; Gaithersburg MD 20899; USA; 301-921-1000; E; 1.1; 2.1; 3.1; 3.6; 5.7

JOHNSON, Charles E.; Department of Physics; North Carolina State University; Raleigh NC 27695; USA; 919-737-2512; E; 3.3; RF spectroscopy of simple atoms and molecules

JOHNSON JR., Charles S.; Department of Chemistry; University of North Carolina; Chapel Hill NC 27514; USA; 919-966-5220; E/T; 2.3; 3.6; Laser light scattering; Magnetic resonance; Molecular motion and reaction rates --study

JOHNSON, David E.; Operations Group/Accelerated Division; Fermi National Lab.; PO Box 500; Mail Stop 306; Batavia IL 60510; USA

JOHNSON, Edward A.; Materials Character Division; Arm Materials and Mechanical Research Center; Watertown MA 02172; USA; 617-923-5042; E; 4.2; 2.1; 2.2; 5.1

JOHNSON, Fred M.; Department of Physics and Astronomy; California State University; Fullerton CA 92634; USA

JOHNSON, John L; 7709 Carlton Drive; Huntsville AL 35802; USA

JOHNSON, Lawrence W.; Department of Chemistry; York College of CUNY; Jamaica NY 11451; USA; 212-969-4097x4387; E; 1.3; 3.5; Molecular electronic spectroscopy; Porphyrins--excited states

JOHNSON, Lee K.; Department of Physics; Rice University; PO Box 1892; Houston TX 77251; USA; 713-527-4945; 713-527-8108x2471; E; 1.1; 2.1; 3.6; 4.1; Atomic polarization dependent associative ionization; Laser self frequency-locking to atomic transitions

JOHNSON, Norman J.; Block Engineering; 19 Blackstone Street; Cambridge MA 02139; USA; 617-868-6050; E; 4.3; 5.4; 5.6; Atmospheric characterization; IR and far IR; Spectroscopic instrumentation, FTS IR and far IR

JOHNSON, Peter D.; Independent Consultant; General Electric Co.; PO Box 8; 1100 Merlin Dr.; Schenectady NY 12309; USA; 518-785-5035; E; 3.2; 5.3; 5.4; Gas discharges--spectroscopy; Luminescence in solids--mechanism; Optical sources and measurements

JOHNSON, Philip M.; Department of Chemistry; State University of NY, Stony Brook; Stony Brook NY 11794; USA; 516-246-7663; E; 1.1; 1.2; 3.5; 3.7; Multiphoton processes in molecules

JOHNSON, Roy R.; KMS Fusion, Inc.; PO Box 1567; Ann Arbor MI 48106; USA; 313-769-8500; E; 1.1; 4.1; High-Z atoms with regard to radiation physics--excited states; Molecular vibration modes used for laser generators

JOHNSON, Stephen G.; Improved Fluorescent Laboratory; GTE-Sylvania; 100 Endicott Street; Danvers MA 01923; USA; 617-777-1900x2510; E; 5.4; Species found in low and high pressure plasmas --collision dynamics; Spectroscopy

JOHNSON, Walter R.; Department of Physics; University of Notre Dame; Notre Dame IN 46556; USA; 219-283-7463; T; 1.1; 1.4; 3.1; Heavy atoms--properties; Hartree Fock calculations of inner-electron binding including Lamb shift; Photoionization

JOHNSTON, Alan R.; 27-02 Fox Run Drive; Plainsboro NJ 08536; USA

JOHNSTON, David B.; Department of Chemistry; Radiation Laboratory; University of Minnesota; 137 Smith Hall; 207 Pleasant Street, SE; Minneapolis MN 55455; USA; 612-373-0151; E; 2.2; 3.2; Low energy-loss electron impact spectroscopy and fluorescence spectroscopy

JOHNSTON, Harold S.; Department of Chemistry; University of California; Berkeley CA 94720; USA; 415-642-3674; E; 3.1; Primary atomic and molecular products of photodissociation of simple molecules by visible and UV radiation

JOHNSTON, Lawrence H.; Department of Physics; University of Idaho; Moscow ID 83843; USA; 208-885-6380; 208-885-6745; E;

3.3; 3.6; 3.9; 4.1; Submillimeter-wave laser Stark spectroscopy of molecules

JOHNSTON JR., Milton D.; Department of Chemistry; Natural Sciences; Molecular Spectroscopy; University of South Florida; Tampa FL 33620; USA; 813-974-2535; E/T; 1.1; 3.5; NMR relaxation of solutions; Applications of paramagnetic ions; Statistical mechanics of NMR parameters

JOHNSTON, R. R.; 1694 Fallen Leaf; Los Altos CA 94022; USA

JOHNSTON JR., Thomas F.; Laser Products Division; Coherent, Inc., 10321; 3210 Porter Drive; Palo Alto CA 94304; USA; 415-858-7407; E; 4.1; 3.10; 5.3; Commercial lasers--design and development

JONA, F. P.; College of Engineering and Applied Science; State University of New York; Stony Brtook NY 11794; USA; 516-246-7649; 516-246-6759; E; 2.2; 3.2; Jones, Claude R.; Defense Research Programs; Group AP-2; Los Alamos National Lab.; PO Box 1663; MS F616; Los Alamos NM 87545; USA; 505-667-0402; E; 4.1; 5.3; 5.4; 2.3; 3.3; Lasers, high-energy, gas, optically prepared, chemical; Spectroscopy, infrared, visible

JONES, Dale R.; Department of Physics; University of Cincinnati; Cincinnati OH 45221; USA; 513-475-2352; E; 2.2; Low-energy electron-atom scattering

JONES, Douglas W.; Department of Physics; Montana State University; Bozeman MT 59717; USA; 406-994-3614x42; 406-587-7470; E/T; 1.1; Atomic line strengths

JONES, E. G.; Department of Chemistry; Wright State University; Dayton OH 43435; USA; 513-873-3122; E; 2.1; Luminescence arising from ion-atom collisions--analysis

JONES, H. W.; Department of Physics; Florida A&M University; Tallahassee FL 32307; USA

JONES, Keith W.; Department of Applied Science; Atomic-Applied Physics Division; Brookhaven National Laboratory; Building 901A; Upton NY 11973; USA; 516-282-4588x5125; FTS-666-4588x5125; E; 4.2; Accelerator based atomic physics

JONES, Larry A.; Physics Division; Group P-1; Los Alamos National Lab.; PO Box 1663 MS455; Los Alamos NM 87545; USA; 505-667-5364; E; 3.9; 5.4; Atomic physics as a diagnostic tool for high temperature high density plasma; Stark broadening in high density plasma

JONES, M. T.; Department of Chemistry; University of Missouri, St. Louis; 8001 Natural Bridge Road; St. Louis MO 63121; USA; 314-553-5311; E/T; 1.1; 5.2; 3.5; 5.2; ESR techniques used to study free radicals in solution and the solid state; Physical properties of organic metals

JONES, Patrick L.; Department of Chemistry; Ohio State University; 140 W. 18th Avenue; Columbus OH 43210; USA E; 2.1; 2.3; 3.1; 3.6; 4.1; Molecular energy transfer at surfaces

JONES, Roger C.; Electrical, Computer Engineering and Radiology; College of Engineering and Medicine; Quantum Electronics; University of Arizona; Building #20; Tucson AZ 85721; USA; 602-626-6102; 602-745-2795; T/E; 3.8; 5.3; Dense plasma molecular interactions; Laser beam-organic molecule interactions; Phonon-photon interactions in living molecules; Order-disorder phenomena in classical and quantum physical electronics

JONES, Scott; 320 Beaver Ave., Apt. #717; State College PA 16801; USA

JONES, Steven E.; Physics and Astronomy; Brigham Young University; Provo UT 84602; USA; 801-378-2749; 801-226-6873; E; 1.3; 2.3; 2.1; 1.1; 5.8; Muon-catalyzed fusion

JONES, Walter W.; U.S. Naval Research Laboratory; Code 4040; 4555 Overlook Avenue, SW; Washington DC 20375; USA; 202-767-3214; E; 5.5; Combustion chemistry

JONG, R. A.; Fusion Laboratory; United Technologies Research Center; Silver Lane; East Hartford CT 06232; USA; 203-727-7527; T; 5.4; RF interactions with mirror confined plasmas

JOPSON, Robert M.; 99 South Ward Avenue; Rumson NJ 07760; USA

JORDAN, Arthur K.; 9330 Boothe Street; Alexandria VA 22309; USA

JORDAN, Kenneth D.; Department of Chemistry; University of Pittsburgh; Pittsburgh PA 15260; USA; 412-624-5013; T; 1.1; 2.2; 2.3; Atoms and molecules--electron affinities; Positron-atom and positron-molecule interactions; Electronically excited atoms--reactions

JOSHI, Bhairav D.; Department of Chemistry; State University of NY, Geneseo; Geneseo NY 14454; 716-245-5305; 716-243-2903; T; 1.1; Molecular structure studies at or below the SCF level

JOSSEM, E. L.; Ohio State University; 174 W. 18th Avenue; Columbus OH 43210; USA

JOYCE, J. M.; Department of Physics; East Carolina University; Greenville NC 27834; USA; 919-757-6688; E; 2.1; 2.3; 3.4; 4.2; Ion-atom collisions: X-ray cross sections, electron capture, and chemical effects

JUDD, Brian R.; Department of Physics and Astronomy; Johns Hopkins University; Charles and 33rd Streets; Baltimore MD 21218; USA; 301-338-8693; T; 3.2; 5.2; Theoretical atomic spectroscopy; Group theory; Atoms in solids; Jahn-Teller effect

JUDD, O'Dean P.; Applied Photochemistry Division; Los Alamos National Laboratory; PO Box 1633 MS563; Los Alamos NM 87545; USA; 505-667-6250; T/E; 3.6; 3.7; 3.8; Polyatomic molecules--multiple-photon absorption; Laser chemistry; Laser isotope separation; Laser research

JUDGE, Darrell L.; Space Sciences Center; Letters, Arts & Sciences; University of Southern California; University Park MC 1341; Los Angeles CA 90089; USA; 213-743-6150; 213-743-2025; E; 1.1; 3.1; 5.1; Atomic and molecular structure; Vacuum UV absorption processes; Absolute cross sections for the production of all products

JULIAN, Maureen M.; Department of Chemistry; Hollins College; Hollins College VA 24020; USA; 703-362-6541; E; 5.1; Surface chemistry; Anthracene with UV--dimerization

JULIENNE, Paul S.; Molecular Spectroscopy Division; National Bureau of Standards; Gaithersburg MD 20899; USA; 301-921-2774; T; 2.1; 3.5; 3.8; Molecular spectroscopy; Scattering theory; Collision-induced radiative processes; Scattering in intense radiation fields

JUNKER, Bobby R.; Office of Naval Research; Code 421; 800 N. Quincy Street; Arlington VA 22217; USA; 703-696-4220; T; 1.1; 2.1; 2.2; Electron and heavy particle scattering; Atomic particles with fields--interaction

JURETSCHKE, H. J.; Polytech Institute; 333 Jay Street; Brooklyn NY 11201; USA

KACHRU, Ravinder; Molecular Physics Department; Chemical Physics Laboratory; SRI International; 333 Ravenswood Avenue; Menlo Park CA 94025; USA; 415-859-3727; E; 3.1; 3.10; 3.6; 5.2; 3.5; High-resolution spectroscopy of autoionizing Rydberg states coherent optical transient spectroscopy; Collisional and spectroscopic studies of atomic and molecular Rydberg states; Frequency modulation spectroscopy

KADLEC, Robert A.; Failure Analysis Associates; 11777 Mississippi Avenue; Los Angeles CA 90025; USA

KADLECEK, John A.; Department of Atmospheric Science; State University of NY, Albany; Building ES-324; 1400 Washington Avenue; Albany NY 12222; USA; 518-457-4930; E; 2.3; 5.6; Ions in air-like gas mixtures involving trace pollutants --identification of reaction channels

KAFLE, Surendra R.; 408 N. Fielder-Apt. 43; Arlington TX 76012; USA

KAHN, Luis; Battelle Memorial Laboratories; 505 King Avenue; Columbus OH 43201; USA; 614-424-4359; T; 1.1; Valence electron methods in molecular structure; Lattice structure on crack advance--simulation; Dislocation motion in solids

KALDOR, Andrew P.; Chemical Sciences Laboratory; Corporate Research; Exxon Research and Engineering Co.; Route 22 East; Clinton Township; Annandale NJ 08801; USA; 201-730-3004; E; 1.3; 3.8; 5.1; 5.2; 3.6; Chemical and physical properties of clusters; Laser chemistry of vibrationally excited molecules; Surface chemistry and physics; Laser chemistry in liquids

KALDOR, Uzi; School of Chemistry; Tel Aviv University; Tel Aviv 69978; Israel; 9723420590; T; 1.1; 2.2

KALLNE, Elisabeth; Jet Joint Undertaking; Abingdon Oxford; OX14 3EA; England; 442-352-8822; E; 3.4; 4.2; 5.2; 5.4; X-ray spectroscopy from plasma and solids

KALMAN, Gabor; Department of Physics; Boston College; Chestnut Hill MA 02167; USA

KALMAN, Otto F.; 2952 E. 5th Street; Tucson AZ 85716; USA

KAM, Steve S.; 1230 Bordeaux Drive; Sunnyvale CA 94086; USA

KAMARATOS, E.; Department of Chemistry; Tougaloo College; Tougaloo MS 39174; USA; 601-956-4941x262; T; 2.1; 2.3; 3.1; Chemical kinetics in a flow reactor to be applied to atomic and molecular energy transfer by collisions; Triatomic molecules--photolysis

KAMINSKY, Manfred S.; Physics Division; Argonne National Laboratory; 9700 S. Cass Avenue; Argonne IL 60439; USA; 312-972-4074; E; 4.2; 5.1; 5.2; Ion-solid interactions; Ions penetrating through foils--energy loss; Ions with solids--charge exchange; Helium injected into solids--depth distribution

KAMINSKY, Mark E.; 3228 Kimlee Drive; San Jose CA 95132; USA

KAMMASH, Terry; Department of Nuclear Engineering; University of Michigan; Cooley Bldg., North Campus; Ann Arbor MI 48109; USA; 313-764-0205; T; 5.4; Ionization, charge exchange and other atomic phenomena associated with plasma heating and fueling for fusion systems

KANDEL, Richard J.; Chemical Science Division; U.S. Department of Energy; Washington DC 20545; USA; 202-353-5820; E/T

KANE, P. P.; Department of Physics; Indian Institute of Technology; Powai Bombay 400076; India

KANEGSBERG, Edward; 16924 Livorno Drive; Pacific Palisades CA 90272; USA

KANG, Ik-Ju; Department of Physics; Southern Illinois University, Edwardsville; Edwardsville IL 62026; USA; 618-692-2985; 618-692-2472; T; 2.2; 3.1; 2.1; Electron-atom collisions; Atoms by electrons or photons--excitations and ionizations; Reactive scattering of atoms and molecules

KANIA, Don R.; Los Alamos National Lab.; PO Box 1663; MS D410; Los Alamos NM 87545; USA

KANTER, Elliot P.; Physics Division-203; Argonne National Lab.; 9700 S. Cass Avenue; Argonne IL 60439; USA; 312-972-4050; 312-972-3663; E; 4.2; 5.2; 2.1; 1.3; 1.1; Dissociation of fast molecular ions; Ion-solid interactions

KANTER, Helmut; Electronic Research Lab. M2 244; Aerospace Corporation; PO Box 92957; Los Angeles CA 90009; USA

KAPLAN, Alexander E.; School of Electrical Engineering; Purdue University; Room 313A; West Lafayette IN 47907; USA

KARO, Arnold M.; Chemistry and Materials Science Div.; Lawrence Livermore National Lab.; PO Box 808 L-325; Livermore CA 94550; USA; 415-422-7800; T; 1.1; Highly accurate interaction potentials--evaluation and use

KARPLUS, Martin; Department of Chemistry; Harvard University; 12 Oxford Street; Cambridge MA 02138; USA; 617-495-4018; T; 1.1; 2.3; 3.5; Molecules of biological interest--structure, spectra and dynamics

KARRAS, Thomas W.; Technical Support Dept.; Space Division; Electro-Optics; General Electric Co.; PO Box 8555; Philadelphia PA 19101; USA; 215-354-3928; 215-962-6737; E; 4.1; Metal vapor lasers--Electro-optics

KASNER, William H.; Westinghouse R&D Center; 1310 Beulah Road; Pittsburgh PA 15235; USA; 412-256-3848; E; 4.1; Carbon dioxide laser systems--research

KASPER, Jerome V.; Department of Chemistry; University of California, LA; 405 Hilgard Avenue; Los Angeles CA 90025; USA; 213-825-4225; E; 2.3; 3.3; 3.6; High resolution IR laser spectroscopy; Vibrationally excited hydrogen--kinetics

KASSAL, Thomas T.; Corporate Research Center; Chemical Physics; Grumman Aerospace Corporation; Mail Station A08-35; Bethpage NY 11714; USA; 516-575-7112; T; 5.3; IR radiation from hot gases; High resolution spectra of targets and backgrounds

KAST, John W.; Varian Associates; MS C-077; 611 Hansen Way; Palo Alto CA 94303; USA

KASTNER, Sidney O.; Laboratory for Astronomy and Solar Physics; Code 682; NASA Goddard Space Flight Center; Greenbelt MD 20771; USA; 301-344-8771; T; 5.7; Atomic physics in astrophysics---applications

KATASE, Akira; Department of Nuclear Engineering; Faculty of Engineering; Kyushu University; Hakozaki Fukuoka 812; Japan

KATAYAMA, Daniel H.; Ionospheric Physics Division; U.S. Air Force Geophysics Lab.; Hanscom AFB MA 01731; USA; 617-861-4042/3310; E; 3.6; 2.1; 3.2; Laser induced fluorescence of atmospheric ions and molecules; Molecular energy transfer

KATAYAMA, Ichiro; Research Center for Nuclear Physics; Osaka University; Mihogaoka; Ibaraki Osaka 567; Japan; 06-877-5111; E; 2.1; 4.2; 5.8

KATAYAMA, Mikio; Department of Pure and Applied Sciences; College of Arts and Sciences; University of Tokyo; 3-8-1 Komaba; Meguro-ku Tokyo 153; Japan; 03-467-1171; 0424-82-0097; E; 3.6; 3.1; 3.7; 3.8; 3.10

KATSANOS, Anastasios A.; Tandem Accelerator Laboratory; Physics; Nuclear Research Center DEMOKRITOS; Greek Atomic Energy Commission; 60225; Aghia Paraskevi Attiki 15310; Greece; (1) 65 18 770; (1) 65 13 111; E; 4.2; 2.1; 3.4; 4.3

KATSONIS, Konstantinos; Lab. Physique Des Gaz Et Plasmas; Atomic Data Centre Gaphyor; Bat. 212strasse 5;

Orsay 91405; France; 331-6941-6543; 331-69417250; E/T; 5.4; 5.1; 2.1; 2.2; Atomic and molecular data for fusion

KATZ, Ira; Plasma Physics Program; Systems, Science and Software; PO Box 1620; La Jolla CA 92069; USA; 714-455-0060; T; 5.3; 5.4; Moderate Z-atoms in laboratory and space plasmas--modeling ionization, recombination and radiation rates

KATZ, Joseph L.; Department of Chemical Engineering; Johns Hopkins University; Charles & 34th Streets; Baltimore MD 21218; USA; 301-338-8484; E/T; 3.8; Photoinduced condensation; Condensation with simultaneous chemical reaction; Homogeneous nucleation of gases and liquids

KAUFFMAN, Robert L.; Laser Fusion Program; Lawrence Livermore National Lab.; PO Box 808 L-479; Livermore CA 94550; USA; 415-422-0419; E; 5.1; 5.4; Laser fusion plasmas--highly ionized line spectra; Opacities and absorption in dense matter

KAUFMAN, Joyce J.; Department of Chemistry; Johns Hopkins University; Charles and 34th Streets; Baltimore MD 21218; USA; 301-338-7417; E/T; 1.1; 1.3; Quantum chemistry

KAUFMAN, Myron J.; Department of Chemistry; Emory University; Atlanta GA 30322; USA; 404-727-6619; E/T; 2.3; 5.6; 5.5; 4.4; Spectroscopy and chemical kinetics related to atmospheric science; Spectroscopy and chemical kinetics related to combustion

KAUFMAN, Stanley L.; School of Physics and Astronomy; University of Minnesota; 116 Church Street, SE; Minneapolis MN 55455; USA; 612-373-3124; E; 3.6; 4.2; Ultra-sensitive (single atom) laser spectroscopy used to determine isotopic shifts of rare isotopic species

KAUFMAN, Victor; Atomic and Plasma Radiation Div.; National Bureau of Standards; Gaithersburg MD 20899; USA; 301-921-2011; E; 3.4; 5.4; Spectra of highly ionized atoms of interest for magnetic confinement fusion--observation and interpretation; Inner-shell absorption from neutral and/or lightly ionized laser-excited vapors

KAUL, R. Dean; Department of Physics; Western Michigan University; Kalamazoo MI 49008; USA; 616-383-4057

KAUPPILA, Walter E.; Department of Physics and Astronomy; Wayne State University; Detroit MI 48202; USA; 313-577-2780; E; 2.2; 4.2; Low and intermediate energy (0.2 to 1000 eV) scattering of positrons and electrons by atoms and molecules

KAWAGUTI, Minato; Department of Information Science; Fukui University; 9-1-Bunkyo-3; Fukui 910 8202; Japan

KAY, Kenneth G.; Department of Chemistry; Kansas State University; Manhattan KS 66506; USA; 913-532-6697; T; 2.3; 3.7; Intramolecular vibrational energy transfer --dynamics; Molecular dissociation dynamics; Isolated molecules--statistical behavior

KAY, Richard B.; Department of Physics; American University; Massachusetts & Nebraska Aves., NW; Washington DC 20016; USA; E; 2.2; 3.6; 3.7

KAYE, Ronald J.; 1032 Lawrence Drive, NE; Albuquerque NM 87123; USA

KAZEK, Gregory J.; GTE Products Corporation; 60 Boston Street; Salem MA 01970; USA

KAZMERSKI, Lawrence L.; Photovoltaics Devices and Measurements Branch; Solar Energy Research Institute; 1617 Cole Boulevard; Golden CO 80401; USA; 303-231-1115; E; 5.1; SIMS, AES, UPS and XPS research on photovoltaic device surfaces and interfaces for improving solar cell performance and reliability

KEANE, Christopher J.; Plasma Physics Lab.; Princeton University; Princeton NJ 08540; USA; 609-683-2495; E/T; 3.2; 5.3; 5.4; Laser plasma spectroscopy; X-Ray lasers

KEBABIAN, Paul; Environmental Research and Tech., Inc.; 696 Virginia Road; Concord MA 01742; USA; 617-369-8910; E; 4.1; 4.3; 5.6; Neutral gas lasers for analysis and measurement of pollutants and other gases by IR absorption--development; NDIR gas analysis instruments--development

KEEFER, Dennis R.; Engineering Science and Mechanics; Center for Laser Applications; Univ. Tennessee Space Institute; Upper C-Wing; Tullahoma TN 37388; USA; 615-455-0631x256; E/T; 3.6; 3.2; 3.8; 5.3; 5.5; Laser/plasma interactions and plasma diagnostics; Advanced laser and electro-optic diagnostic instrumentation; Laser spectroscopy and laser fluorescence

KEEFFE, William M.; Sylvania Lighting Center; Research and Development Laboratory; GTE Products Corporation; 100 Endicott Street; Danvers MA 01923; USA; 617-777-1900; E; 5.3; 5.4; 5.1; 3.1; 3.2; Arc discharge studies in mercury-metal halide systems; Electrode effects in mercury-metal halide systems

KEENER, Bruce A.; 3080 Shaw Road; Marietta GA 30066; USA

KELIHER, Peter N.; Department of Chemistry; Villanova University; Villanova PA 19085; USA; 215-645-4871; E; 5.4; 5.6; Optical emission in high temperature direct current and microwave plasmas; Environmental analytical chemistry

KELLEHER, Daniel E.; Atomic and Plasma Radiation Div. 531; National Bureau of Standards; Room A167, Building 221; Gaithersburg MD 20899; USA; 301-921-2011; E/T; 3.9; 5.4; Plasma spectroscopy in dense plasmas for line broadening; Electric field effects on autoionizing levels; Laser scattering in plasmas

KELLER, Richard A.; Chemistry and Nuclear Chemistry; Group CNC-2; Los Alamos National Laboratory; PO Box 1663 MS732; Los Alamos NM 87545; USA; 505-667-3018; E; 3.6; 3.8; Laser interaction with atoms and molecules--analytical applications

KELLEY, J. Daniel; McDonnell Douglas Research Lab.; McDonnell Douglas Corporation; PO Box 516; St Louis MO 63166; USA; 314-233-2502; T; 2.1; 2.3; 3.8; Laser related kinetics and energy transfer processes; Molecular collision theory

KELLEY, Michael H.; Electron Physics Group; Center for Radiation Research; National Bureau of Standards; Building 220, Room B206; Gaithersburg MD 20899; USA; 301-921-2051; E; 2.1; 2.2; Spin polarized electrons from spin polarized atoms--spin dependent effects in scattering

KELLEY, Paul L.; Quantum Electronics Group; Solid State Division; Lincoln Laboratory; MIT; PO Box 73; Lexington MA 02173; USA; 617-863-5500 X7824; T; 3.10; 3.6; 4.1; 3.7; 3.3; IR molecular spectroscopy

KELLEY, Ralph E.; Directorate of Physics; U.S. Air Force; Office of Scientific Research; Building 410; Bolling AFB DC 20332; USA; 202-767-4908; Atomic and molecular physics --program manager

KELLMAN, Michael E.; Department of Chemistry; Columbia University; New York NY 10027; USA; 212-280-4342; T; 1.1; 3.1; Triatomic molecules--photodissociation; Two-electron atoms--collective motion; Nonrigid triatomic molecules

KELLY, Arnold J.; 8 Hathaway Drive; Princeton Junction NJ 08550; USA

KELLY, Hugh P.; Department of Physics; University of Virginia; Physics Building; Charlottesville VA 22901; USA; 804-924-3781; T; 1.1; Atomic and molecular properties using many-body methods--calculation

KELLY, John C.; Department of Physics; University of New South Wales; Kensington 2033; Australia; 6974574; 4491539; E/T; 4.2; 4.3; 5.1

KELLY, Raymond L.; Department

of Physics; Code 61; Naval Postgraduate School; Monterey CA 93943; USA; 408-646-2824; 408-646-2486; E/T; 3.2; 3.1; 3.3; Atomic spectra compilation

KELLY, Roger; T.J. Watson Research Center; IBM Corporation; PO Box 218; Yorktown Heights NY 10598; USA; 914-945-2309; E/T; 2.2; 3.6; 5.1; Sputtering due to ions, electrons and photons

KELSEY, Edward J.; AT&T Bell Laboratories; Crawford Corner Road; Holmdel NJ 07724; USA; 201-544-0610; T; 1.4; 1.2; 2.2; 1.1; 3.6

KEMPLE, Marvin D.; Department of Physics; Indiana University, Purdue; University at Indianapolis; PO Box 647; 1125 E. 38th Street; Indianapolis IN 46205; USA; 317-923-1321 x308; E; 1.1; 3.5; 5.2; Magnetic resonance, EPR, ENDOR, and NMR; Peptide interactions

KENAN, Richard P.; Physics, Electronics and Nuclear Technology Division; Battelle Memorial Laboratories; 505 King Avenue; Columbus OH 43201; USA; 614-424-5762; T; 4.1; Physical optics; Bistability in optics; Nonlinear optics

KENEFICK, Robert A.; Department of Physics; Texas A&M University; College Station TX 77843; USA; 409-845-3331; 409-845-7717; E; 2.1; 3.1; 3.4; 4.2; 4.3; Development and application of ion traps; Advanced ion source-E.G. EBIS; Ultracold antiproton physics and experimental technique

KENKRE, Vasudev M.; 56 Lynnwood Drive; Rochester NY 14618; USA

KENNEDY, Patrick W.; 8750 Georgia Avenue-Apt. 530A; Silver Spring MD 20910; USA

KENNEY, John W.; Department of Physical Sciences; Chemistry; Eastern New Mexico University; Station 33; Portales NM 88130; USA; 505-562-2152; 505-562-2174; E; 1.1; 1.3; 3.2; 3.6

KENNEY-WALLACE, Geraldine A.; Lash Miller Laboratory; University of Toronto; 80 St. George Street; Toronto Ontario M5S1A1; Canada

KEPPLE, Paul C.; Code 4720; Naval Research Labs.; Washington DC 20375; USA

KEPROS, John G.; Electro-Optics Technology Division; General Dynamics/Convair; PO Box 80847; MS 42-6210; San Diego CA 92138; USA; 714-571-7890; 714-277-8900x2279; E/T; 3.6; 3.8; Stimulated Raman scattering and laser physics; Inner-shell atomic theory; Plasma physics; Isotope separation

KEREMEDJIEV, George A.; Gary Mfg Division of Brand Rex; 1010 Jersey Avenue; N. Brunswick NJ 08902; USA

KERN, C. W.; Division of Chemistry; National Science Foundation; 1800 G Street, N.W.; Washington D.C. 20550; USA; 202-357-7947; T; 1.1; 1.2; 1.3; 3.9; Quantum chemistry; Quantum theory of molecular structure and spectra; Fine structure of molecules; Vibrational and rotational interactions

KERSHENSTEIN, John C.; Optical Sciences Division; Advanced Concepts Branch; U.S. Naval Research Laboratory; Code 6520; 4555 Overlook Avenue, S.W.; Washington DC 20375; USA; 202-767-3272; 202-767-3045; E/T; 3.3; 3.6; Chemical laser research

KESSEL, Quentin C.; Department of Physics; University of Connecticut; Storrs CT 06268; USA; 203-486-4113; E; 2.1; 3.4; 4.2; Dynamics of atomic collisions especially excitation of inner shells--accelerator based investigations

KESSLER JR., Ernest G.; Center for Basic Standards; Quantum Metrology Group; National Measurement Laboratory; National Bureau of Standards; A141 Physics Building; Gaithersburg MD 20899; USA; 301-921-2061; E; 3.4; 4.2; Precision X-ray wavelength measurements

KESSLER, Karl G.; Center for Absolute Physical Quantities; National Bureau of Standards; Gaithersburg MD 20899; USA; 301-921-2001; E; 1.4; 3.4; 4.2; Research programs in atomic and molecular physics--Administrator

KESTNER, Neil R.; Department of Chemistry; College of Basic Sciences; Louisiana State University; Baton Rouge LA 70803; USA; 504-388-1528; T; 1.1; 5.2; 2.2; 1.3; Intermolecular forces; Electron transfer processes in liquids; Trapped electrons in liquids

KETKAR, Suhas N.; Department of Physics; University of Texas; RLM 2 116; Austin TX 78712; USA

KETO, John W.; Department of Physics; University of Texas, Austin; Austin TX 78712; USA; 512-471-4151; E; 2.3; 3.6; 3.7; 3.8; 3.10; Energy transfer processes related to excimer lasers

KHADJAVI, Abbas; Department of Physics and Engineering; Fairfield University; Fairfield CT 06430; USA; 203-255-5411x564; E; 3.5; 3.6; 4.3; 5.6; IR upper atmospheric spectroscopy; Laser applications; Electronic instrumentation as applied to spectroscopy and thermometry

KHALID, Joseph M.; Research and Development Division; Square D Co.; 3700 6th Street, SW; Cedar Rapids IA 52406; USA; 319-365-4631; E; 5.4; Electric arcs; Research management

KHALIL, Omar S.; Abbott Diagnostics Division; Abbott Laboratories; 160-B Regal Row; Dallas TX 75247; USA; 214-630-8000x465; E; 1.3; Protein molecules--spectral properties; Metabolites in human sera

KHANDELWAL, Govind S.; Department of Physics; Old Dominion University; Hampton Boulevard; Norfolk VA 23508; USA; 804-440-3475; T; 3.4; Inner-shell ionization

KIELKOPF, John F.; Department of Physics; Moore Observatory; University of Louisville; 800 Old Zaring Road; Crestwood KY 40014; USA; 502-588-6787; 502-241-7841; E/T; 2.1; 3.2; 3.8; Atomic collisions --atomic spectroscopy and physics

KIJEWSKI, Louis J.; Department of Physics; Monmouth College; West Long Branch NJ 07764; USA

KIKUCHI, Chihiro; Department of Nuclear Engineering; Engineering Division; University of Michigan; Ann Arbor MI 48109; USA; 313-764-2369; E; 5.1; Chemical and radiation induced defects in solids; Nuclear power for nuclear peace

KILLINGER, Dennis K.; Quantum Electronics Department; Lincoln Lab.; MIT; PO Box 73; Lexington MA 02173; USA; 617-863-5500; E; 3.6; 4.1; Quantum electronics; Laser physics; Laser spectroscopy; Laser remote sensing

KIM, Eui-Hoon; Sung Dong-Ku; 147-62 Choong Kok-Dong; Seoul 133 2173; Korea

KIM, H. J.; Physics Division; Oak Ridge National Lab.; PO Box X; Building 6003; Oak Ridge TN 37830; USA; 615-574-4708; E; 2.1; 4.2; Atom-ion and ion-ion collisional physics for fusion purposes

KIM, Hyunwoo; Nagasaki Institute of Applied Science; 536 Abanachi; Nagasaki 851-01; Japan

KIM, Jick H.; Department of Physics; New York University; 4 Washington Place; New York NY 10003; USA

KIM, Jin J.; Dept. of Electrical and Computer Engineering; Institute for Modern Optics; University of New Mexico; Albuquerque NM 87131; USA; 505-277-4026; E/T; 4.1; 3.11; 3.10

KIM, Longhuan; Department of Physics and Astronomy; University of Pittsburgh; Pittsburgh PA 15260; USA; 412-624-4306; 412-624-4307; T; 2.2; 1.1; 3.4

KIM, Y. E.; Department of Physics; Purdue University; West Lafayette IN 47907; USA

KIM, Yong W.; Department of Physics; Building 16; Lehigh University; Bethlehem PA

18015; USA

KIM, Yong-Ki; Center for Radiation Research; National Bureau of Standards; Gaithersburg MD 20899; USA; 301-921-2071; T; 1.1; 2.1; 2.2; Atomic structure theory; Photon, electron and ions with atoms and molecules--interaction; Atomic and molecular physics to fusion technology--application

KIM, Young S.; 3242 Sawtelle Boulevard #2; Los Angeles CA 90066; USA

KIM, Yung S.; Department of Chemistry; College of Natural Science; Seoul National University; Seoul 151 0066; Korea

KIMBLE, Harry J.; Department of Physics; University of Texas, Austin; Austin TX 78712; USA; 512-471-1668; E/T; 3.11; 1.4; 3.6; 3.7; 4.1; Quantum optics--optical tests of Q.E.D. and of quantum mechanics; Nonlinear dynamics--quantitative investigations of dynamical instabilities in optical systems

KING, David B.; 425 Pine Shadows; Slidell LA 70458; USA

KING, John G.; Department of Physics; Massachusetts Institute of Technology; Room 26-457; Cambridge MA 02139; USA; 617-253-4180; E; 4.3; Molecular beams

KING, William T.; Department of Chemistry; Brown University; Providence RI 02912; USA; 401-863-2735; E/T; 1.1; 3.5; Molecular vibrational spectroscopy; Vibrational transition moments and line strengths

KINNEY, John H.; L-311; Lawrence Livermore Lab.; University of California; PO Box 808; Livermore CA 94550; USA

KINSEY, James L.; Department of Chemistry; Room 6-215; Massachusetts Institute of Technology; 77 Massachusetts Avenue-Rm 6-215; Cambridge MA 02139; USA; 617-253-1488; E/T; 1.1; 2.1; 2.3; 3.1; 3.5; Photodissociation dynamics of polyatomic species; Stimulated emission pumping; Rotationally and vibrationally inelastic collisions

KINSINGER, Richard E.; Corporate Research and Development; General Electric Co.; PO Box 8; Schenectady NY 12301; USA; 518-385-8442; T; 5.3; Gas and plasma properties from atomic and molecular data and use of macroscopic properties in models of electric arc behavior--calculation

KIRBY, Kate P; Harvard-Smithsonian Center for Astrophysics; Harvard College Observatory; 60 Garden Street; Cambridge MA 02138; USA; 617-495-7237; T; 1.1; 3.1; Molecules and radiation: photoionization, photodissociation, autoionization, photodetachment--interaction; Molecular structure and properties and transition properties--theoretical calculations

KIRCHHOFF, William H.; Center for Chemical Physics; Chemical Thermodynamics; National Measurement; National Bureau of Standards; Gaithersburg MD 20899; USA; 301-921-2133; 301-921-2131; E/T; 1.3; Protein thermodynamic properties, spectroscopy applied to membranes and proteins

KIRKPATRICK, Ronald C.; Group X-2; Los Alamos National Laboratory; PO Box 1663 MS220; Los Alamos NM 87544; USA; 505-667-5341/4812; T; 5.1; 5.7; Small fusion target design; Planetary nebulae--photoionization and recombination coefficients

KIRTMAN, Bernard; Department of Chemistry; Univ. of California, Santa Barbara; Santa Barbara CA 93106; USA; 805-961-2217; T; 1.1; 2.2; Electronic structure; Electron scattering

KISNER, Howard D.; 3217 Avenue Q; Wichita Falls TX 76309; USA

KIVELSON, Daniel; Department of Chemistry; Univ. of California, Los Angeles; Los Angeles CA 90024; USA; 213-825-1710; E/T; 5.2; Molecular motions in liquids and bilayers; Magnetic resonance and light scattering

KJELDAAS JR., T.; Polytech Institute of New York; 333 Jay Street; Brooklyn NY 11201; USA

KLAVAN, I. L.; 1308 Hudson Road; Teaneck NJ 07666; USA

KLEBAN, Peter H.; Department of Physics and Astronomy; University of Maine; Orono ME 04469; USA; 207-581-1033; T; 5.3; Electron swarms in molecular gases--calculations

KLEIBER, Paul D.; Department of Physics and Astronomy; Iowa Laser Facility; University of Iowa; Iowa City IA 52242; USA; 319-353-7081; E; 3.8; 2.3; 2.1; 3.6; 3.7

KLEIN, Lewis; Department of Physics and Astronomy; Howard University; Washington DC 20059; USA; 202-636-6251; T; 3.6; 3.7; 3.9; 3.10; 5.7; Laser spectroscopy

KLEINPOPPEN, H.; Institute of Atomic Physics; University of Stirling; Stirling UK FK34LA; Scotland

KLEMM, R. Bruce; Energy and Environment Division; Brookhaven National Laboratory; Building 527; Upton NY 11973; USA; 516-345-4022; E; 2.3; 5.5; Kinetics, gas phase, high temperature absolute rate measurements; Combustion research

KLEPPNER, Daniel; Department of Physics; Room 26-237; Massachusetts Institute of Technology; Cambridge MA 02139; USA; 617-253-6811; E; 1.2; 1.4; 3.3; 3.6; 3.9; Highly excited atoms --experimental physics

KLIEWER, K. L.; Building 221; Argonne National Lab.; 9700 South Cass Avenue; Argonne IL 60439; USA

KLINE, Laurence E.; Systems Sciences Division; Westinghouse R&D Center; 1310 Beulah Road; Pittsburgh PA 15235; USA; 412-256-7552; E/T; 5.3; 4.1; 2.2; 2.3; Plasma chemistry performance prediction/experiment; Gas discharge laser performance prediction/experiment; High voltage breakdown in bases

KLIWER, James K.; Department of Physics; University of Nevada; Reno NV 89557; USA; 702-784-6128; 702-784-6792; E; 4.2; 5.1; Channeling spectrometry--surface studies; Electronic stopping cross sections--measurement

KLOEPPING, William H.; PO Box 2691; Norman OK 73070; USA

KLOSE, Jules Z.; Atomic and Plasma Radiation Division; National Bureau of Standards; Gaithersburg MD 20899; USA; 301-921-2356; E; 1.1; 1.4; 3.1; 3.2; 5.4; UV radiometric standards; Atomic lifetimes

KLOSTERMAN, Elliot L.; Mathematical Sciences Northwest Incorporation; 2755 Northup Way; Bellevue WA 98004; USA; 206-827-0460; E/T; 2.3; 5.3; Chemical kinetics of XeCl laser system--theoretical modeling and experimentation; HgBr and HF/DF laser systems

KLOTS, Cornelius E.; Chemical Physics Section; Oak Ridge National Laboratory; P.O. Box X; Oak Ridge TN 37831; USA; 615-574-6234; E; 1.2; 1.3; 3.1; 3.8; Ion-molecule, photoionization, unimolecular electron attachment

KLUTH, Edward O.; 1107 W. 10th Street; Austin TX 78703; USA

KNABLE, Norman; DTS-54; Transportation Systems Center; Kendall Square; Cambridge MA 02142; USA

KNAUSS, Donald C.; Department of Chemistry; Oregon State University; Corvallis OR 97331; USA; 503-754-0123; T; 1.3; 5.2; Torsional motion in chain molecules in liquid state--Brownian dynamics

KNIGHT, Randall D.; Department of Physics; Ohio State University; Columbus OH 43210; USA; 614-422-8798; E; 3.9; 5.7; Atomic and molecular ions in ion traps--laboratory astrophysics studies

KNIPE, Richard H.; Research Division; US Naval Weapons Center; Code 3852; China Lake CA 93555; USA; 714-939-2051; E/T; 5.5; Combustion processes--chemical physics

KNIZE, Randall J.; Plasma Physics Laboratory; Princeton University; P.O. Box 451; Princeton NJ 08544; USA; 609-683-2121; 609-452-4388; E; 1.4; 2.1; 2.3; 4.1; 5.3; Excited state charge exchange; Impurity atom transport in plasmas

KNOCHENMUSS, Richard D.; Chemistry; University of Bern; Freiestrasse 3; 3012 Bern; Switzerland; 65-42-54; E/T; 1.1; 3.2; 2.2; 3.6; 3.1; Spectroscopy of multicenter transition metal complexes; Low-dimensional materials; Inelastic electron tunneling spectroscopy; Excited state properties

KNOWLES, David S.; 2049 Cambridge Ave.; Cardiff By The Sea CA 92007; USA

KNUTH, Eldon L.; Chemical Engineering; University of California, LA; 5531 Boelter Hall; Los Angeles CA 90024; USA; 213-825-8485; E/T; 2.3; 4.3; 4.4; 5.5; Molecular-beam mass-spectrometer systems used in sampling studies of chemically reacting gas mixtures; Relaxation phenomena in gases

KOBAYASHI, Hisao; Institute for Atomic Energy; Rikkyo (St. Pauls) University; 2-5-1 Nagasaka; Yokosuka Kanagawa 24001; Japan; 0468-56-3131; E; 1.1; 2.1; 3.2; 4.2; 5.2

KOBE, Donald H.; Department of Physics; North Texas State University; Box 5368; Denton TX 76203; USA; 817-565-3272; T; 3.6; Electromagnetic radiation with atomic and molecular theory --interaction

KOCH, H. W.; American Institute of Physics; 335 E. 45th Street; New York NY 10017; USA; 212-661-9404; 516-349-7800; E/T; 1.4

KOCH, Karl W.; Institute of Optics; University of Rochester; Rochester NY 14627; USA

KOCH, Mark E.; Research; Vought Millies and Advanced Programs; LTV Aerospace and Defence Co.; P. O. Box 225907; MS TH 85; Dallas TX 75265; USA; 214-266-1618; E

KOCH, Peter M.; Department of Physics; SUNY; Stony Brook NY 11794; USA; 516-246-6580; E; 3.6; 3.7; 3.8; 4.2; Highly-excited atoms; Laser spectroscopy and intense fields; Collision processes; Synchrotron radiation studies in atomic physics

KOCHER, Carl A.; Department of Physics; Oregon State University; Corvallis OR 97331; USA; 503-754-4631; E; 1.2; 2.1; 3.9; Highly excited states; Atomic and ionic collisions at very low energies

KOEL, Bruce E.; Chemistry-CIRES; University of Colorado; 215; Boulder CO 80309; USA; 303-492-7189; E; 5.1; 1.1; 1.3; 2.2; Electron spectroscopy of molecules on surfaces; Adsorption of atoms and molecules; Interactions of atoms and molecules with transition metals

KOENIG, Thomas W.; Department of Chemistry; University of Oregon; Eugene OR 97403; USA; 503-686-4601; E/T; 2.3; 3.1; Transient species --photoelectron spectroscopy

KOEPF, Gerhard; NASA Goddard Space Flight Center; Code 723; Greenbelt MD 20771; USA; 301-344-6745; E; 3.3; 3.6; High resolution submillimeter molecular line spectroscopy with new laser heterodyne receiver

KOEPPL, Gerald W.; Department of Chemistry; Queens College of CUNY; Flushing NY 11367; USA; 212-520-7437; T; 2.3; Chemical reaction rates--dynamical studies; Statistical theories of chemical reaction rates--theoretical studies theoretical studies

KOFFEND, Brooke; The Aerospace Corp.; Box 92957; Los Angeles CA 90009; USA; 213-648-5000; E; 2.3; 3.2; 3.5; Electronic spectroscopy and kinetics of small molecules

KOFFEND, John B.; Laser Kinetics and Spectroscopy; Aerophysics; Aerospace Corp.; 92957; Los Angeles CA 90009; USA; 213-648-7412; E; 2.3; 3.8; 3.6; Chemical laser kinetics and spectroscopy

KOHIN, Roger P.; Department of Physics; Clark University; Worcester MA 01610; USA; 617-793-7448; E/T; 3.5; Paramagnetic molecular ions--electron spin resonance

KOHL, John L.; Atomic and Molecular Physics Div.; Harvard-Smithsonian Center for Astrophysics; Harvard College Observatory; 60 Garden Street; Cambridge MA 02138; USA; 617-495-5436; E; 1.1; 2.2; 3.1; Multiple charged ions including measurements of dielectronic recombination, electron excitation cross sections and transition probabilities--studies; Neutral free radicals--measurements of photodissociation

KOHLER, Bryan E.; Department of Chemistry; Hall-Atwater Laboratory; Wesleyan University; Middletown CT 06457; USA; 203-347-9411x805; E; 1.3; 3.5; Linear polyenes --optical spectroscopy

KOLB, Alan C.; Maxwell Laboratories, Inc.; 8835 Balboa Avenue; San Diego CA 92123; USA; 714-279-5100; E; 4.1; 4.2; 5.4; Radiation process in high temperature dense matter; Atomic and molecular processes in gas lasers

KOLB JR., Charles E.; Center for Chemical and Environmental Physics; Research Group; Aerodyne Research, Inc.; 45 Manning Road; Billerica MA 01821; USA; 617-663-9500; E; 2.3; 5.6; 3.1; 5.5; 3.6; Chemical kinetics and molecular dynamics; Atmospheric and environmental science; Laser and spectral diagnostics; Gas-surface interactions

KOLENKOW, R. J.; 2435 Virginia Street; Berkeley CA 94709; USA

KOLLEN, Wendell; 1109 Winghaven Road; Maumee OH 43537; USA

KOMORNICKI, Andrew; Polyatomics Research Institute; Suite 420; 1101 San Antonio Road; Mountain View CA 94043; USA; 415-964-4013; T; 1.1; 2.3; 3.5; 5.6; Small molecules of interest in gas phase kinetics as applied to atmospheric research --quantum chemical calculations

KONOPNICKI, Marek J.; ARAO; US Air Force Weapons Lab.; Kirkland AFB NM 87117; USA; 505-844-0226; T; 4.1; High energy lasers--research and development; Propagation and coherent resonant interactions

KONOWALOW, Daniel D.; Department of Chemistry; State Univ. of NY, Binghamton; Binghamton NY 13901; USA; 607-777-6788; 607-722-7477; T; 1.1; 1.2; 1.3; 2.1; Atomic and molecular electronic structure and spectra; Excimer laser prospects; Long range interatomic interactions

KONRAD, Gerhard T.; SLAC; Stanford University; PO Box 4349; MB 30; Stanford CA 94305; USA

KOPELMAN, Raoul; Department of Chemistry; University of Michigan; Ann Arbor MI 48109; USA; 313-764-7541; 313-763-0288; E/T; 1.1; 1.3; 2.1; Energy transfer in molecules, molecular and biomolecular aggregates; Weak chemical forces, static and dynamic

KOREVAAR, Eric J.; Mechanical and Aerospace Engineering; Princeton University; D-414 Engineering Quad; Princeton NJ 08544; USA; 609-452-4738; 609-683-1224; E/T; 1.2; 3.1; 3.2; 3.6; 3.9; Semiclassical quantization; Crossed electric and magnetic fields

KORFF, Serge A.; Department of Physics; Meyer Hall of Physics; New York University; New York NY 10003; USA

KOSKI, Walter S.; Department of Chemistry; Johns Hopkins University; Charles and 34th Streets; Baltimore MD 21218; USA; 301-338-7418; E; 2.3; Ions--reactive scattering

KOSMAN, Warren M.; Department of Chemistry; Valparaiso University; Valparaiso IN 46383; USA; 219-464-5387; T; 1.1; 1.2; 3.1; 3.2; Ab initio calculations on atoms and small molecules

KOSTIUK, Theodor; Code 693;

NASA Goddard Space Flight Center; Greenbelt MD 20771; USA; 301-344-8431; E; 5.6; 5.7; 3.3; 3.10; 4.1; Atomic and molecular spectroscopy as applied to atmospheric and astrophysical species

KOSTROUN, Vaclav O.; Department of Applied and Engineering Physics; Nuclear Science and Engineering Division; Ward Reactor Laboratory; Cornell University; Ithaca NY 14853; USA; 607-256-4991; E/T; 2.1; 2.3; Charge transfer in low energy multiply charged ion collisions with atomic and molecular hydrogen; Impact parameter Born approximation calculation of atom-atom excitation cross sections

KOSTYK, Edmund; CAE Electronics Limited; PO Box 1800; St. Laurent PQ H4L4X4; Canada

KOSZYKOWSKI, Michael L.; Physical Research Division; Sandia National Laboratories; Livermore CA 94550; USA; 415-422-3257; T; 1.1; 2.1; 2.3; Chemical physics; Molecular collision theory; Semiclassical methods; Intramolecular relaxation

KOURI, Donald J.; Department of Chemistry and Physics; University of Houston Univ. Park; Cullen Boulevard; Houston TX 77004; USA; 713-749-2845; 713-749-4295; T; 2.1; 2.3; 5.1; Atom-molecule collisions-formal and computational studies/reactive collisions; Atom-molecular collisions-development of approximations; Molecule--surface collisions; Statistical mechanics of multiparticle systems

KOURLAS, James; Department of Science; Maritime College of SU, New York; Bronx NY 10465; USA; 212-892-3000x306; T; 3.6; Atomic resonance fluorescence

KOWALSKI, Frank V.; Department of Physics; Colorado School of Mines; Golden CO 80401; USA; 303-273-3000; E; 4.1; 4.3; Laser equipment for atomic and molecular research --development

KPONOU, Ahovi E.; Brookhaven National Lab.; Building 930; Upton NY 11973; USA

KRAMER, Peter B.; Department of Biophysics; Cambridge Research Laboratory; Johnson & Johnson; 195 Albany Street; Cambridge MA 02139; USA; 617-491-2300; E; 1.1; 3.6; 5.2; Immunoassays; NTIR Imaging; Medicine

KRAMER, Stephen L.; High Energy Physics Division; Argonne National Laboratory; Department of Energy; Building 362; 9700 S. Cass Avenue; Argonne Il 60439; USA; 312-972-6327; E; 3.4; 4.2; 5.1; Protons--stopping power; Charged particles--inner-shell ionization

KRAMER, Steven D.; Chemical Physics Section; Photophysics Group; Oak Ridge National Laboratory; P.O. Box X; 5500 Building; Oak Ridge TN 37830; USA; 615-574-5897; E; 3.10; 3.1; 2.3; 3.6; 4.4; Detection of single atoms and molecules; Laser ionization sources for mass spectrometers; Coherent vacuum ultraviolet light sources; Alkali atom reaction kinetics

KRANER, H. W.; Instrumentation Division; Brookhaven National Laboratory; Building 535B; Upton NY 11973; USA; 516-345-4238; E; 3.4; 4.2; X-ray spectroscopy; Trace element analysis; Ion beams for analysis

KRASE, Loren D.; 2504 W. Freeport; Broken Arrow OK 74012; USA

KRASINSKI, Jerzy S.; Laser Research and Development; Corporate Research and Development; Materials Laboratory; Allied Corporation; 7 Powderhorn Drive; Mt. Bethel NJ 07060; USA; 201-560-1750; 201-752-7987; E; 3.10; 4.1; 3.6; 3.7; 3.10; Laser physics; Laser development; Nonlinear optics; Raman effect; Quantum optics; Novel optical diagnostics; Moire techniques

KRAUS, Joseph S.; Radiation Physics Research Div.; Bell Telephone Laboratories; Room 1E457; 600 Mountain Avenue; Murray Hill NJ 07974; USA; 201-582-6892; E; 2.1; 4.2; Low energy atomic collisions; Optical radiation from excited particles after collision

KRAUSE, Herbert F.; Oak Ridge National Laboratory; 4500N, F-17; Oak Ridge TN 37830; USA; 615-574-5049; E; 2.3; 3.8; 4.2; 5.2; Thermal energy crossed molecular beam studies; Inelastic and reactive collisions; Energy transfer studies using lasers; Highly charged ions with matter in the solid state--interactions; Channeling phenomena

KRAUSE, Jeffrey L.; Department of Chemistry; University of Chicago; 5735 S. Ellis Avenue; Chicago IL 60637; USA; 312-962-7253; T; 1.1

KRAUSE, Lucjan; Department of Physics; University of Windsor; Windsor Ontario N9B3P4; Canada; 519-253-4232/2665; E; 2.1; 3.6; 1.1; 5.3

KRAUSE, Manfred O.; Chemistry Division; Oak Ridge National Laboratory; PO Box X; Building 4500N; Oak Ridge TN 37831; USA; 615-574-5019; E; 1.1; 3.1; 3.4; 4.2; Atoms including metal vapors using synchrotron radiation; Photoelectron spectrometry of atoms and molecules; Auger electron spectrometry

KRAUSS, Alan R.; Chemistry Division; Argonne National Laboratory; Building 200; 9700 S. Cass Avenue; Argonne IL 60439; USA; 312-972-3520; 312-972-3513; E; 5.1; 5.4; Particle-surface interactions, sputtering; Magnetic fusion

KRAUSS, Morris; Molecular Spectroscopy Division; National Bureau of Standards; Gaithersburg MD 20899; USA; 301-921-2165; T; 1.1; Quantum chemistry; Electronic structure, energies, properties of small molecules

KREMENS, Robert L.; 294 Wakely Terrace; Bel Air MD 21014; USA

KRENOS, John R.; Department of Chemistry; Wright-Rieman Laboratories; Rutgers University; New Brunswick NJ 08903; USA; 201-932-3048; E; 2.3; 1.3; 4.3; Chemiluminescence from reactions of metastable atoms and molecules--molecular beam studies

KRISHNA, N. R.; Department of Biochemistry; Comprehensive Cancer Center; University of Alabama, Birmingham; University Station; Birmingham AL 35294; USA; 205-934-5696; E; 1.1; Biomolecules --nuclear magnetic resonance studies

KRISHNAN, Mahadevan; Department of Engineering and Applied Science; Yale University; PO Box 2159; Yale Station; New Haven CT 06520; USA; 203-436-4565; E; 2.3; 5.4; Population inversions in laser produced plasmas and laser-initiated arcs; Physics of recombination and charge transfer pumped inversions in high-Z metal ions/scaling to short wavelength

KRIVANEK, Ondrej L.; Center for Solid State Science; Arizona State University; Tempe AZ 85281; USA

KROCEK, Ladislav; 3900 Linnean Avenue, N.W.; Washington DC 20008; USA

KROGSTAD, Robert T.; 58 Royal Street; Allston MA 02134; USA

KROHN, Burton J.; Theoretical Division; Los Alamos National Laboratory; Mail Stop J569; Los Alamos NM 87545; USA; 505-667-6458; 505-667-2097; T; 3.3; 3.5; 3.1; 1.1; 3.6

KROHN, Kenneth A.; Department of Radiology; School of Medicine; University of California, Davis; Sacramento CA 95817; USA; 916-453-3788; E; 2.3; 4.2; Polyatomics and halogens--hot atom chemistry

KROLL, Norman M.; Department of Physics; University of California, San Diego; La Jolla CA 92093; USA; 714-452-3647; T; 4.1; Nonlinear optics; Quantum electronics

KROMHOUT, Robert A.; Department of Physics; Florida State University; Tallahassee FL 32306; USA; 904-644-5567/2724; T; 1.1; 5.2; 5.1; Van der Waals forces; Nematics; Physisorption on metals

KRONENFELD, Jerrold E.; RFD 8; 4 Acropolis Avenue; Londonderry NH 03053; USA

KROTKOV, Robert V.; Department of Physics and Astronomy; Hasbrouck Laboratory; University of Massachusetts; Amherst MA 01003; USA; 413-545-0811; E/T; 2.1; 5.7; Coherent excitation in atomic/molecular collisions; Rotationally inelastic molecular collisions of astrophysically interesting species

KRUMHANSL, J. A.; Department of Physics; Cornell University; Ithaca NY 14853; USA

KRUPKE, William F.; Lawrence Livermore National Lab.; PO Box 808; Livermore CA 94550; USA; 414-422-5905

KRUSE, Theodore H.; Department of Physics; Serin Physics Laboratory; Rutgers University; Piscataway NJ 08854; USA; 201-932-2527; E; 3.2; 4.2; Beam foil spectroscopy with tandem Van de Graaff; Atomic research with synchrotron radiation

KU, Peh Sun; Environment Division; Consolidated Edison Co.; 4 Irving Place; New York NY 10003; USA; 212-460-6241; T; 2.3; 5.6; Atmospheric and aqueous pollutant dispersions--kinetics and thermodynamics

KU, Robert T.; Optics Division; Lincoln Laboratory; MIT; PO Box 73; Lexington MA 02173; USA; 617-862-5500-X5504; E; 3.6; 4.1; High power gas laser development; Discharge species and containment diagnostics; Tunable laser diagnostics of excited media

KUBIS, Joseph J.; KMS Fusion, Inc.; PO Box 1567; Ann Arbor MI 48106; USA; 313-769-8500; T; 1.1; Atomic energy levels for use in equation of state and radiative transfer calculations--models

KUCKUCK, Robert W.; L-Division; Lawrence Livermore Laboratory; PO Box 808; Livermore CA 94550; USA; 415-422-7374; E; 3.4; 4.2; Inner-shell ionization phenomena; X-ray scattering and diffraction; X-ray diagnostic techniques --development

KUGEL, Henry W.; Plasma Physics Lab.; Princeton University; PO Box 451; Princeton NJ 08544; USA

KUMAR, Prem; Research Laboratory of Electronics; 36-477; Massachusetts Institute of Technology; 50 Vasser Street; Cambridge MA 02139; USA; 617-253-8131; 617-489-4150; E/T; 3.11; 3.10; 3.7; 4.1; 3.6

KUNC, Joseph A.; Department of Physics SHSCI-274; University of Southern California; University Park MC-1341; Los Angeles CA 90089; USA

KUNG, Robert T.; Research and Development; Abiomed; 33 Cherry Holl Dr.; Danvers MA 01923; USA; 617-777-5410; E; 3.8; 4.1; 3.10; 3.7; 3.6; Laser-tissue interaction

KUNKEL, Wulf; Department of Physics; University of California; Berkeley CA 94720; USA

KUNZ, Christof; Hasylab; DESY; Notkestr 85; Hamburg 52 2000; West Germany; 040-8998-3706; E; 3.4

KUNZE, Hans-Joach D.; Fakultaet Fuer Physik and Astronomie; Institute Experimentalphysik V; Ruhr Universitat; 102148; Bochum 4630; West Germany; E; 5.4; 3.6; 3.2; 5.1

KUO, Chien-Yu; Rm 2C-205; AT&T Bell Laboratories; 555 Union Boulevard; Allentown PA 18103; USA; 215-439-7053; E; 4.1; 3.3; Optical fiber communications

KUPER, J. B.; Brookhaven National Lab.; Upton NY 11973; USA

KUPPERMAN, Aron; Department of Chemistry; California Institute of Technology; 1201 E. California Boulevard; Pasadena CA 91125; USA; 213-795-6811x2507; E/T; 2.2; 2.3; 3.1; 3.6; 3.8; Chemical dynamics; Chemical beam kinetics; Laser spectroscopy and photochemistry; Electron impact and photoelectron spectroscopy

KURBATOV, J. D.; PO Box 6; 5640 Batu Road; Victoria BC V8Z6K5; Canada

KURNIT, Norman A.; Group CHM-6; Los Alamos National Laboratory; PO Box 1663; MS J564; Los Alamos NM 87545; USA; 505-667-6002; E; 3.10; 4.1; 3.11; 3.6; 3.8

KURTZ, Henry A.; Department of Chemistry; Memphis State University; Memphis TN 38152; USA

KURTZ, Richard L.; Surface Science Division; Center for Chemical Physics; National Bureau of Standards; Gaithersburg MD 20899; USA; 301-921-2743; E; 5.1; Electronic and geometric structure of surfaces; Chemisorption and desorption from oxides

KURUCZ, Robert L.; Smithsonian Astrophysical Observatory; Smithsonian Institution; 60 Garden Street; Cambridge MA 02138; USA; 617-495-7429; FTS-830-7429; T; 1.1; 5.7; 3.1; Stellar atmospheres and spectra--atomic and molecular lines

KURYLO III, Michael J.; Center for Chemical Physics; Chemical Kinetics Division; National Bureau of Standards; Bldg. 222/Room A147; Gaithersburg MD 20899; USA; 301-921-2080; 301-921-2792; E; 2.3; 5.6; 5.5; Gas phase atomic and free radical reaction kinetics

KUSHAWAHA, Vikram S.; Department of Physics; Howard University; Washington DC 20059; USA

KUSHICK, Joseph N.; Department of Chemistry; Amherst College; Amherst MA 01002; USA; 413-542-2590/2342; T; 1.1; 1.3; Molecular motion in condensed phases--molecular dynamics simulation; Theoretical biophysical chemistry; Molecular dynamics computer simulation; Molecular mechanics

KUSHNER, Mark J.; Laser Analytic Spectroscopy Div.; Sandia National Laboratories; PO Box 5800; Albuquerque NM 87185; USA; 505-844-3233; E/T; 4.1; Gas phase chemical reactions in CVD reactors --laser Raman spectroscopy

KUSKO, Bruce H.; Crocker Nuclear Laboratory; Analytical Services; University of California, Davis; Davis CA 95616; USA; 916-752-1120; 916-752-1460; E; 5.6; 3.4; 4.2; Trace element analysis of art and archaeological objects using PIXE, XRF and NAA techniques.

KUYATT, Chris E.; Center for Radiation Research; National Bureau of Standards; Room C229 RADP; Gaithersburg MD 20899; USA; 301-921-2551; T; 2.2; Electron physics and electron optics

KVALE, Thomas J.; Physics and HASRD; Oak Ridge National Laboratory; Martin Marietta Energy Systems; Bldg. 5500; Oak Ridge TN 37831; USA; 615-574-4782; 615-482-2426; E; 1.1; 2.1; 4.2; Metastable atomic and molecular negative --ion spectroscopy

KWAN, Ching-Kwan CK; Department of Astronomy and Physics; Wayne State University; 666 W. Hancock; Detroit MI 48202; USA; 313-577-2726; E; 2.2; 4.2; 2.1; 3.1; 4.3

KWEI, George H.; Chemistry, Earth and Life Sciences Div.; Los Alamos National Laboratory; PO Box 1663; Los Alamos NM 87545; USA; 505-667-5893; E; 2.3; Chemical dynamics

KWIRAM, Alvin L.; Department of Chemistry; University of Washington; Seattle WA 98195; USA; 206-543-4020; E; 1.1; 1.3; 3.5; 5.2; Magnetic resonance; Lasers used to study triplet states in organic molecules and biological systems

KWOK, Munson A.; Chemical Kinetics Division; Aerophysic Laboratory; The Aerospace Corporation; PO Box 92957; MS 130-139; Los Angeles CA 90009; USA; 213-648-5441; E; 2.3; 3.6; 4.1; 5.3; 5.6; Chemical kinetics in chemical lasers

KWOK, Thomas Y.; Silicon Technology; Research Division; Thomas J. Watson Laboratory; IBM Research Center; PO Box 218; Yorktown Heights NY 10598; USA; 914-945-2495; 201-666-1509; E/T; 1.1; 5.1; Molecular dynamics studies in solids and interfaces; Monte Carlo simulation of interaction of particles and surfaces

KWONG, Hoi-Shuen V.; Department of Physics; University

of Nevada; 4505 Maryland Parkway; Las Vegas NV 89154; USA

KWONG, Robin B.; Hughes Aircraft Company; PO Box 9301; Albuquerque NM 87119; USA

KWONG, Victor H. S.; Physics; University of Nevada, Las Vegas; 4505 South Maryland Pkwy.; Las Vegas NV 89154; USA; 702-739-3539; 702-739-3563; E; 2.3; 5.4; 3.6; 5.1; 5.7

KYRALA, George A.; Physics Division, Group P-1; Los Alamos National Laboratory; PO Box 1663 MS455; Los Alamos NM 87545; USA; 505-667-7649; FTS-843-7649; E; VUV spectra and data used in characterization of plasma; Recombination/short wavelength lasers

LAANE, Jaan; Department of Chemistry; Texas A&M University; College Station TX 77843; USA; 409-845-3352; 409-693-5171; E/T; 3.3; 3.6; 1.1; 3.5; 3.10; Vibrational potential energy surfaces; Far-infrared spectroscopy; Raman difference spectroscopy; Matrix isolation spectroscopy

LACEY, Richard F.; Chemical Systems Department; Measurement Systems Lab.; Hewlett-Packard Laboratories; 1651 Page Mill Road; Palo Alto CA 94304; USA; 415-857-5253; E/T; 3.3; 4.3; 4.4; Atomic beam optics; IR spectroscopy

LACINA, William B.; Northrop Research and Technology Center; 1 Research Park; Palos Vds Peninsula CA 90274; USA; 213-377-4811x362; T; 4.1; 5.3; Molecular and electron kinetic processes applied to new laser systems--analytical modeling

LADANYI, Branka M.; Department of Chemistry; Colorado State University; Fort Collins CO 80523; USA; 303-491-5196; T; 1.3; 2.3; 5.2; Fluids--structure and optical response; Liquids--theory

LADISH, Joseph S.; Group L-4; Los Alamos National Laboratory; PO Box 1663 MS554; Los Alamos NM 87545; USA; 505-667-5789; E; 4.3; 5.4; Atomic physics and electron beam optics applied to the study of laser produced plasmas for fusion

LAFYATIS, Gregory P.; 242 Waverley Avenue; Watertown MA 02172; USA

LAGATTUTA, Kenneth J.; Department of Physics; University of Connecticut; Storrs CT 06268; USA; 203-486-4187; T; 2.2; Dielectronic recombination rates for heavy ions--calculation

LAGUNA, Glenn; Applied Physics Div., Group AP-4; Los Alamos National Laboratory; PO Box 1663; Los Alamos NM 87545; USA; 505-667-6838/7855; E/T; 2.3; 3.5; High resolution molecular spectroscopy; Photodissociation products --chemical kinetics

LAHIRI, Jayanti; Department of Physics; Southern Tech.; 1112 Clay Street; Marietta GA 30060; USA; 404-424-7215; 404-378-0315; T; 1.2; Photoabsorption characteristics of Rydberg states of atoms and ions

LAKE, Max L.; Universal Energy Systems, Inc.; 3195 Plainfield Road; Dayton OH 45432; USA; 513-253-3001; 513-254-9155; E; 2.2; Electron impact cross sections of selected atoms and molecules

LALAMA, Salvatore J.; 62 Burlington Road; Clifton NJ 07012; USA

LAM, Juan T.; Optical Physics Division; Hughes Research Laboratories; 3011 Malibu Canyon Road; Malibu CA 90265; USA; 213-456-6411; T; 3.8; 4.1; Nonlinear laser spectroscopy and collisional studies

LAM, Leo K.; Department of Physics; University of Southern California; Los Angeles CA 90007; USA; 213-743-6711; E; 4.1; Wavefront conjugation in sodium vapor; Nonlinear optics

LAMB JR., Willis E.; Department of Physics and Optical Sciences; University of Arizona; Tucson AZ 85721; USA; 602-626-4278; T; 1.1; 1.4; 3.6; Laser physics; Atomic structure; Quantum electrodynamics

LAMBERG, D. L.; Department of Astronomy; University of Texas; Austin TX 78712; USA; 512-471-4474; E; 5.7; Astronomical objects--quantitative spectroscopy

LAMBERT, David L.; Dept. of Astronomy; University of Texas; Austin TX 78712; USA; 512-471-7438; 512-471-4461; E; 5.7; 1.1; 3.1; 3.3; 2.3

LAMBERT, Robert H.; Department of Physics; University of New Hampshire; Durham NH 03824; USA

LAMBERTI, William A.; 825 Whittier Avenue; New Hyde Park NY 11040; USA

LAMBROPOULOS, Hector D.; Storage Systems Components; Advanced Development Division; Digital Equipment Corporation; ML 4-1/B 32; 146 Main Street; Maynard MA 01754; USA; 617-493-7450; E; 5.1; Thin solid films--vacuum deposition; Plasma interaction with solids; Ion beams for etching thin solid films

LAMBROPOULOS, P. P.; Department of Physics; University of Southern California; Los Angeles CA 90007; USA; 213-743-7123; T; 3.6; 3.7; 4.1; Lasers with atoms and molecules--interaction; Laser spectroscopy; Nonlinear optics

LAMONT JR., Lawrence T.; Vacuum Research and Development Div.; Varian Associates; 611 Hansen Way; Palo Alto CA 94303; USA; 415-969-1187; E; 5.3; Gas discharge physics

LAMPE, Frederick W.; Department of Chemistry; 152 Davey Laboratory; Pennsylvania State University; University Park PA 16802; USA; 814-865-1209; E; 2.3; 3.7; 3.8; Ion-molecule reactions; IR multiphoton decompositions; Vacuum UV photochemistry

LAND, David J.; Nuclear Branch; White Oak Laboratory; US Naval Surface Weapons Center; Silver Spring MD 20910; USA; 202-394-2256; T; 2.1; 3.4; 4.2; KeV-low MeV heavy ions--inelastic energy loss; Inner-shell ionization probability stopping powers

LANDMAN, Donald A.; Institute for Astronomy; University of Hawaii; 2680 Woodlawn Drive; Honolulu HI 96822; USA

LANE, Neal F.; Department of Physics; University of Colorado; Boulder CO 80309; USA; 303-492-6952; T; 2.1; Atomic and molecular physics with emphasis on collision phenomena

LANG JR., John C.; ARCO Exploration and Technology Co.; 2300 West Plano Parkway; Plano TX 75075; USA; 214-754-6907; 214-754-6401; E/T; 1.3; 2.3; 1.1; 1.3; 4.1; Chemical physics/physical chemistry; Multicomponent liquid mixtures especially containing surfactants/polymers; Physical measurements: light scattering, NMR, calorimetry, microscopy investigations of such system

LANG, Neil; Laser Program; Lawrence Livermore National Lab.; PO Box 808; Livermore CA 94550; USA; 415-422-6212; E; 1.1; 2.3; Gas phase elemental uranium--chemical and physical properties

LANGER, Lawrence M.; 1342 Southdows Drive; Bloomington IN 47401; USA

LANGHOFF, Peter W.; Department of Chemistry; Indiana University; Bloomington IN 47405; USA; 812-335-8621; 812-336-7295; T; 1.1; 3.1; 3.4; 3.10; 5.6; Molecular photoionization and electron scattering; Momentum distributions in atoms and molecules; Greens function methods for atoms and molecules; Molecular aeronomy and astrophysics

LANGHOFF, Stephen R.; Computational Chemistry Group; Physical Sciences Branch; NASA Ames Research Center; Moffett Field CA 94035; USA; 415-694-5236; T; 1.1; 3.5; 3.8; Molecular spectroscopy; Small molecules --calculation of potential surfaces; Photodissociation, predissociation, etc.--rates

LANGMUIR, Robert V.; Steel Lab. 116-81; Caltech; Pasadena CA 91125; USA

LANGSAM, Yedidyah; Chemistry, Computer, and Information Science; Brooklyn College of CUNY; Bedford Avenue and Avenue H; Brooklyn NY 11230; USA; 718-780-4161; E; 2.1; 3.6; 3.8; Energy transfer; Laser spectroscopy

LAPATOVICH, Walter P.; Physical Electronics; Materials Science; GTE Laboratories, Inc.; 40 Sylvan Road; Waltham MA 02254; USA; 617-466-2147; E; 1.1; 2.3; 3.6; 4.1; 5.4

LAPICKI, Gregory; Department of Physics; East Carolina University; Greenville NC 27834; USA; 919-757-6894; T; 1; 1

LAPP, Marshall; Combustion and Fuel Science Branch; Corporation Research and

Development; General Electric Co.; PO Box 8; Schenectady NY 12301; USA; 518-385-8383 /8082; E; 4.1; 5.5; Optical probes to study properties of flames--utilization

LARSON, Daniel J.; Department of Physics; University of Virginia; Charlottesville VA 22901; USA; 804-924-6782; 804-924-7591; E; 1.1; 3.3; 3.6; 4.1; Atoms and ions using lasers and/or microwaves--high resolution probing

LARSON, Everett G.; Department of Physics; Brigham Young University; Provo UT 84602; USA

LARTER, Raima; Department of Chemistry; Princeton University; Princeton NJ 08540; USA; 609-452-6453; T; 1.1; 2.1; 2.3; Inelastic molecular collisions; Collision dynamics--sensitivity analysis

LASKAR, Amulya L.; Department of Physics and Astronomy; Clemson University; Clemson SC 29631; USA; 803-656-3416; E; 5.2; Ionic defects; Ions with solids by measurement of conductivity diffusion and optical absorption--interaction

LASKOWSKI, Bernard C.; Analatom, Inc.; MS 230-3; 253 Humboldt Court; Sunnyvale CA 94089; USA; 408-734-4698; T; 1.1; 2.3; Alkali-rare gas and alkali-hydrides --structure studies; Metastable states of Cs- and Cs--determination of cesium anions and cations

LASSETTRE, Edwin N.; Department of Chemistry; Carnegie-Mellon University; 4400 Fifth Avenue; Pittsburgh PA 15213; USA; 412-578-3160; 412-781-0336; T; 1.1; 2.2; Momentum eigenfunctions and collision amplitudes in the complex momentum phase--mathematical behavior; Electron impact spectroscopy

LATTA, Bryan M.; Department of Physics; Acadia University; Box 256; Wolfville N.S. B0P 1X0; Canada; 902-542-2201 x228; 902-542-5615; T; 2.1; 5.1; 5.2; 5.4; Electronic and nuclear stopping powers; Numerical modeling; Nuclear lifetimes by Doppler-shift method

LATZ, Rudolf; Mainzerlandstrasse 326; Frankfurt 6000; West Germany

LAU, Albert M.; Exxon Research and Engineering Company; PO Box 45; Linden NJ 07036; USA; 201-474-2205; T; 3.6; 3.7; 3.8; Laser-radiation interactions with atoms and molecules in collision-free or collisional environments

LAUBERT, Roman; Department of Physics; East Carolina University; Greenville NC 27834; USA; 919-757-6483; E; 2.3; 3.4; 4.2; 5.2; Inner-shell vacancy creation; Electron capture to bound and continuum states; Energy loss of particles in matter

LAUDENSLAGER, James B.; Physics and Chemistry Division; Jet Propulsion Laboratory; 4800 Oak Grove Drive; Pasadena CA 91106; USA; 213-354-2259; E; 2.3; 3.5; 4.1; 4.4; Ion-molecule reactions using mass spectrometers; Laser spectroscopy of small molecules; Metastable rare gases--molecular beam studies; UV laser development and laser sensing of the atmosphere

LAUER, James L.; Department of Mechanical Engineering; Rensselaer Polytechnic Institute; Troy NY 12181; USA; 518-270-6260; E/T; 3.3; IR emission Fourier spectroscopy of thin layers

LAUFER, Allan H.; Chemical Kinetics Division; Center for Chemical Physics; National Bureau of Standards; Building 222, Room A145; Gaithersburg MD 20899; USA; 301-921-2151; E/T; 2.3; 2.8; Photochemistry and chemical kinetics

LAVILLA, Robert E.; Quantum Metrology Group; National Bureau of Standards; Building 221, Room A141; Gaithersburg MD 20899; USA; 301-921-2061; E; 3.1; 3.4; 4.2; Emission and absorption X-ray spectroscopy

LAW, H. David; Electro-Optics Division; Rockwell International; Electronics Research Center; 1049 Camino Dos Rios; Thousand Oaks CA 91360; USA; 805-498-4545x222; E; 5.2; Ion implantation in semiconductors

LAWLER, James E.; Department of Physics; University of Wisconsin; 1150 University Avenue; Madison WI 53706; USA; 608-262-2918; E; 1.1; 3.1; 3.2; 3.6; 5.3; Laser spectroscopy; Gas discharges --physics; Optogalvanic effects--laser diagnostics of gas discharges

LAWRENCE, George; Laboratory for Atmospheric and Space Physics; University of Colorado; Boulder CO 80309; USA; 303-492-6058/6803

LAWRENCE, Glen S.; PO Box 208; 182 Windsorville Road; Windsorville CT 06097; USA

LAWTON, Stan; Radiation Sciences; McDonnell Douglas Research Lab.; PO Box 516; St. Louis MO 63166; USA; 314-233-2547; E; 2.3; 3.6; 5.5; Laser diagnostic techniques applied to combustion

LAX, Benjamin; Department of Physics; Francis Bitter National Magnet Lab.; Massachusetts Institute of Technology; NW14-4104; 170 Albany Street; Cambridge MA 02139; USA; 617-253-5541; T; 3.9; Spectroscopy; Atoms in intense external fields

LAYER, Howard P.; Center for Absolute Physical Quantities; National Bureau of Standards; Building 221, Room B160; Gaithersburg MD 20899; USA; 301-921-3360; E; 1.4; Laser research applied to length standards and fundamental constant measurement

LAYNE, Clyde B.; Combustion Physics Div. 8351; Sandia National Laboratories; Livermore CA 94550; USA; 415-422-2246; E; 2.3; 5.5; Laser spectroscopy applied to combustion physics; Temperature and species diagnostics; Reaction kinetics

LAYZER, David; Harvard-Smithsonian Center for Astrophysics; 60 Garden Street; Cambridge MA 02138; USA; 617-495-2657; T; 1.1; 3.2; Complex atomic spectra in iso-electronic sequences

LAZAR, Norman H.; Plasma Physics Division; TRW, Inc.; 1 Space Park; Redondo Beach CA 90278; USA; 213-536-1370; E; 5.4; Atomic and molecular data from plasma production heating loss rates; and in diagnostic techniques--application

LEACH, Sydney; Lab. Photophys Moleculaire; Universite De Paris Sud-Bat 13; Batiment 213; Orsay 91405; France; 49417909; E/T; 3.1; 3.2; 1.2; 4.2; 4.4; Molecular photophysics; Synchrotron radiation uses in molecular physics; Molecular ions; Molecular astrophysics

LEARNER, Richard C.; Physics Department; Blackett Laboratory; Imperial College; Prince Consort Rd.; London SW71BZ; England; 01-589-5111; E; 1.1; 3.1; 3.2; Fourier transform spectroscopy of ultra-violet; Instrumentation for visible/U-V spectrometry

LEAVITT, John A.; Physics Department; 6MV Van De Graaff Lab.; University of Arizona; Tucson AZ 85721; USA; 602-621-6793; 602-621-2309; E; 5.1; 5.2; 3.2; 4.2; Ion beam analysis--backscattering and channeling; Beam-foil spectroscopy; Highly ionized systems--spectra and lifetimes

LEAVITT, Richard P.; 6112 Seminole Street; Berwyn Heights MD 20740; USA

LEBEDEFF, Sergej; 14 Jackson Drive; Norwalk CT 06851; USA

LEBOW, Paul; Institute for Physical Science and Technology; University of Maryland; College Park MD 20242; USA; 301-454-4369; E; 3.6; 5.6; Atmospheric gases--laser measurements

LEBRUN, Steven F.; Nuclear Physics Lab.; University of Illinois-Urbana; 23 Stadium Drive; Champaign IL 61820; USA

LEE, Anthony R.; Physics Department-Bundoora; La Trobe University; Melbourne; Australia

LEE, Chi-Hsiang; Department of Electrical Engineering; University of Maryland; College Park MD 20742; USA; 301-454-6852; E; 4.1; 3.10; 3.6; 1.1; 5.1; Picosecond technology; Picosecond optoelectronics

LEE, Ching-Tsung; Department of Physics; Alabama A&M University; PO Box 376; Normal AL 35762; USA; 205-859-7313; T; 3.11; 3.10; Free-electron lasers; X-ray lasers

LEE, Clarence E.; 3007 Brothers Boulevard; College Station TX 77840; USA

LEE, Edward K.; Department of Chemistry; University of California, Irvine; Irvine CA 92717; USA; 714-856-5021; 714-856-5998; E; 1.1; 2.3; 3.8; 4.1; 5.5; Matrix isolation, spectroscopy and photochemistry; Hydrogen bonding

LEE, Edward T.; Infrared Technology Division; U.S. Air Force Geophysics Lab.; LSI; Hanscom AFB MA 01731; USA; 617-861-4203; T; 2.1; 5.6; 1.2; Energy exchanges between upper atmospheric species

LEE, Edward Y.; 30675 Via La Cresta; Palos Verdes CA 90274; USA

LEE, Francis W.; Center for Quantum Electronics; University of Texas, Dallas; PO Box 688; Richardson TX 75080; USA; 214-690-2863; E; 2.3; 3.7; 4.2; Electron-beam pulsed high pressure reaction kinetics; Multiphoton spectroscopy and reaction kinetics

LEE, Ja H.; Space Technology Branch; Space Systems Division; Langley Research Center; NASA; MS 160; Hampton VA 23665; USA; 804-865-3781; E; 4.1; 5.3; 5.4; 3.1; 1.1; Optical pumping with dense plasma radiation; Solar pumped laser development; Atomic and molecular physics of dense plasmas; New applications of A&M and optical science

LEE, Jimmy; 4272 Warbler Loop; Fremont CA 94536; USA

LEE, Jonathan K.; Physics; Foster Radiation Laboratory; McGill University; 3610 University Street; Montreal Quebec H3A2B2; Canada; 514-392-4836; E; 5.8; 3.6; 3.1; Isotope shifts of radioactive atoms; Laser spectroscopy of atoms; Resonance ionization spectroscopy

LEE, Keum H.; Department of Physics; University of Missouri; Columbia MO 65211; USA

LEE, Long C.; Dept. of Electrical and Computer Engineering; San Diego State University; San Diego CA 92182; USA; 619-265-3701; E; 3.1; 3.2; 3.6; 4.1; 5.1

LEE, Min H.; Department of Physics; Inha University; Inchon; Korea

LEE, Min-Chang; 4 Fisher Road; Chelmsford MA 01824; USA

LEE, Paul L.; Department of Physics and Astronomy; California State Univ., Northridge; Northridge CA 91330; USA; 213-885-2775; E; 1.3; 3.4; 4.2; X-ray width measurement; Chemical shift; X-ray from pionic hydrogen--precision measurement

LEE, Sanboh; Department of Materials Sciences; National Tsing Hua University; Hsinchu Tiawan 300; Republic of China; (035) 715131/644; T; 5.2; 5.1; 1.3

LEE, Sang S.; Physics Department; Physics and Mathematics Division; Applied Optics Laboratory; Korea Advanced Institute of Science; PO Box 150; Chongyangni Seoul; Korea; 966-1931; 966-3836; E; 4.1; 3.10; 3.11; 3.6

LEE, Seongbok; Department of Physics; University of Notre Dame; Notre Dame IN 46556; USA

LEE, Siu-Au; Department of Physics; Colorado State University; Fort Collins CO 80523; USA; 303-491-6389; 303-491-6206; E; 1.4; 3.6; 3.10

LEE, Tong-Nyong; 3301 Stonesboro Road; Ft. Washington MD 20744; USA

LEE, Yim T.; Lawrence Livermore Lab. L-355; PO Box 808; Livermore CA 94550; USA

LEE, Yuan T.; Department of Chemistry; University of California; Berkeley CA 94720; USA; 415-486-6154; E; 3.8; 5.6; 3.1; Molecular collisions and interaction of light with single; Molecules--dynamics

LEEP, David A.; Electrosystems Division; National Bureau of Standards; Building 220, Room B344; Gaithersburg MD 20899; USA; 301-921-3121; E; 5.3; Electrical discharges in high-pressure gases of practical interest as dielectrics

LEGG, James C.; Department of Physics; James R. Macdonald Laboratory; Kansas State University; Manhattan KS 66502; USA; 913-532-6786; 913-532-6777; E; 2.1; 1.3; 2.3

LEHMANN, Kevin K.; Chemistry Lab.; Harvard University; PO Box 231; 12 Oxford Street; Cambridge MA 02138; USA

LEIBERICH, Andreas; 8C-C Phelps Avenue; New Brunswick NJ 08901; USA

LEIBY JR., Clare C.; RADC/ESO; Hanscom AFB MA 01731; USA; E; 3.6; 4.3; Laser interactions with atomic/molecular beams; Laser time and frequency standards

LELAND, Wallace T.; Laser Division; Los Alamos National Lab.; PO Box 1663 MS532; Los Alamos NM 87545; USA; 505-667-4075; E; 5.3; Electrical discharges in gas and laser gas kinetics

LEMPERT, Joseph; Westinghouse R&D Research Center; 1310 Beulah Rd 501-1W21; Pittsburgh PA 15235; USA; 412-256-7827; E; Magnetohydrodynamic generators--research

LENAMON, Larry L.; 2412 Colcord Avenue; Waco TX 76707; USA

LENGEL, Russell K.; Department of Chemistry; Purdue University; West Lafayette IN 47907; USA; 317-749-2551; E; 5.1; Interaction of gas atoms and molecules with solid surfaces-laser spectroscopic study

LENHARD, Joseph A.; Energy Research and Development; U.S. Department of Energy; PO Box E; Oak Ridge TN 37830; USA; 615-576-0723; E/T; 4.2; Research manager for large laboratory

LENIART, Daniel S.; Palo Alto Instrument Division; Varian Associates; 611 Hansen Way; Palo Alto CA 94022; USA; 415-493-4000 x3802; E/T; Electron paramagnetic resonance (EPR) instrumentation--manager of research and development

LEON, Melvin; MP-3, MS-H844; Medium Energy Physics; Los Alamos National Laboratory; 1663; Los Alamos NM 87545; USA; 505-667-5682; E/T; 1.3; 4.2; Muon-catalyzed fusion; Muonic atoms and molecules; Exotic atoms

LEONE, Stephen R.; Joint Institute for Laboratory Astrophysics; University of Colorado; Boulder CO 80309; USA; 303-492-5128; E; 2.1; 2.3; 3.1; Collisional excitation transfer; Product state distributions; Photodissociation and free radicals-studies

LEPAGE, G. P.; Department of Physics; Cornell University; Ithaca NY 14850; USA; 607-256-5151; T; 1.3; 4.1; Energy levels and decay rates in atoms, especially positronium and muonium--high precision QED calculations

LEPEK, Alexander; 9 Gordon Street; Kfar-Sava 44620; Israel

LEPPELMEIER, Gilbert W.; Institute for Technical Physics; Kernforschungszentrum Karlsruhe; 3640; D-7500 Karlsruhe 1; Fed. Rep. of Germany; (07247) 82 41 92; (0721) 68 48 26; E/T; 5.4; 5.1; 4.4; Atomic processes used for plasma diagnostics; Atomic processes in plasma-wall interactions--rate

LERMAN, Juan-Carlos; Department of Geosciences and Neurology; Colleges of Science and Medicine; Isotope Geochemistry/Sleep Disorder; University of Arizona; Tucson AZ 85724; USA; 602-626-7884; 602-626-6254; E/T; 4.4; 5.6; 3.10; Isotope geochemistry; Radiocarbon dating; Dynamical systems; Nonlinear mathematics

LEROI, George E.; Department

of Chemistry; Michigan State University; East Lansing MI 48824; USA; 517-353-9410/9100; E; 3.6; 3.10; 3.2; 1.1; 3.1; Optical molecular spectroscopy; Mass spectrometry of photoions and photofragments; Matrix isolation spectroscopy

LESK, Arthur M.; Department of Chemistry; Fairleigh Dickinson University; Teaneck NJ 07666; USA; 201-692-2333; T; 1.1; Chemical bonding by deriving general conditions --attempts to understand

LESLIE, Scott G.; Gas Laser Research and Development Division; Westinghouse R&D Center; 1310 Beulah Road; Pittsburgh PA 15235; USA; 412-256-7089; E/T; 4.1; 5.3; Laser discharge physics and discharge kinetics; Laser development

LESTER, Paul A.; Computer Engineering Center; Mellon Institute; 4616 Henry Street; Pittsburgh PA 15213; USA

LESTER JR., William A.; Department of Chemistry; Lawrence Berkeley Laboratory; University of California, Berkeley; Berkeley CA 94720; USA; 415-486-6722; T; 1.1; 2.1; 3.1; 2.3; Low energy heavy particle collisions-- computation of properties

LETAMENDIA, Louis; Physics; Molecular Optics; CPMOH; L.A. 283 CNRS; 351 Cours de la Liberation; Talence Gironde 33400; France; 56807006; 56802058; E/T; 2.3; 3.6; 1.1

LEUCHS, Gerhard; Sekiton Physik; University of Munchen; AM Coulombwall 1; D-8046 Garching; West Germany; 089-3209-5001; E; 3.1; 4.1; Photoelectron angular distribution in photoionization out of excited atomic states

LEUNG, Pui-Tak; Department of Physics; Tamkang University; Tamsui Taipei Hsien; Taiwan 251 0309; Republic of China

LEUNG, Yui-Chung; Department of Physics; University of Windsor; Windsor Ontario N9B3P4; Canada

LEVATTER, Jeffrey I.; Department of AMES; University of California, San Diego; Mail Code B-010; La Jolla CA 92093; USA; 714-452-2113; E; 5.3; Discharge pumped excimers--laser kinetics

LEVENSON, Marc D.; K32/281; Physical Science Division; IBM Research Laboratories; 5600 Cottle Road; San Jose CA 95193; USA; 408-256-6560; E; 3.10; 3.11; 3.6; 4.1; Coherent optical interactions; Nonlinear optics and applications; High resolution spectroscopy of solids

LEVENTHAL, Jacob J.; Department of Physics; University of Missouri, St. Louis; St. Louis MO 63121; USA; 314-553-5931; E; 1.1; 3.8; Ion and atomic beam studies of interactions; Laser initiated processes

LEVENTHAL, Marvin; Scattering and Low Energy Physics; Research Division 1115; Bell Telephone Laboratories; Room 1E-349; 600 Mountain Avenue; Murray Hill NJ 07974; USA; 201-582-3448; E; 1.4; 4.2; Shift measurements in highly ionized gases; Positron annihilation studies in very dilute gases

LEVI, Mark W.; 128 Arlington Road; Utica NY 13501; USA

LEVIN, Frank S.; Department of Physics; Brown University; Providence RI 02912; USA; 401-863-2291; T; 1.1; Bound state parameters for simple systems using the non-hermitian equations of many-body scattering theory--calculation

LEVIN, Jon C.; 2625 Riverview Street; Eugene OR 97403; USA

LEVINE, Alfred M.; Department of Applied Sciences; College of Staten Island; 130 Stuyvesant Place; Staten Island NY 10301; USA

LEVINE, Judah; JILA; University of Colorado; Boulder CO 80309; USA

LEVITT, Morris R.; Laser Focus and Electro-Optics; 119 Russell Street; Littleton MA 01460; USA

LEVY, Donald H.; James Franck Institute; University of Chicago; 5640 S. Ellis Avenue; Chicago IL 60637; USA; 312-962-7196; 312-962-7872; E; 3.5; 3.9; 4.1; Supersonic molecular beams spectroscopy

LEVY, Laurent P.; Solid State and Low Temperature; 11114; AT&T Bell Laboratories; 600 Mountain Avenue; Murray Hill NJ 07974; USA; 201-582-6590; E/T; 3.9; 3.10; 5.2; Level crossings, damping effects; Solitons, nonlinear spin dynamics

LEWIS, David A.; Department of Physics; Iowa State University; Ames IA 50011; USA; 515-294-8269; E; 1.1; 4.2; Atomic hyperfine structure

LEWIS, Lindon L.; Department 702; Efratom Division; Ball Corporation; 589; 9343 W. 108th Circle; Westminster CO 80020; USA; 303-460-5292; E; 1.4; 4.1; Time standards by optical pumping of cesium

LEWIS, Robert R.; Department of Physics; University of Michigan; Ann Arbor MI 48109; USA; 313-763-6729; T; 1.4; 3.9; 4.1; Atoms and molecules--weak interactions

LI, Eddie H.; Department of Mathematics; Hang Seng School of Commerce; Siu Lek Yuen; Sha Tin NT; Hong Kong

LI, Funming; 5747 Bates Street #151; San Diego CA 92115; USA

LI, Ming C.; Department of Physics; VA Polytechnic Institute and State Univ.; Blacksburg VA 24061; USA; 703-361-5430; T; 3.6; Laser light and atomic absorption-atomic scattering

LI, Tien K.; Los Alamos National Laboratory; PO Box 1663 MS539; Los Alamos NM 87545; USA; 505-667-6202; E; 4.2; Accelerator-related atomic physics

LI-SCHOLZ, Angela; Dept. of Science, Mathematics and Technology; Empire State College and State Univ., NY; 50 Wolf Road; Albany NY 12205; USA; 518-458-7150; E; 3.4; 4.2; Inner-shell ionization

LIAO, Paul F.; Physics and Optical Sciences; Bell Communications Research; Newman Springs Road; Red Bank NJ 07701; USA; 201-949-5640; E; 3.6; 4.1; Laser spectroscopy and nonlinear optics

LIBERMAN, Irving; Applied Physics; Applied Sciences; Research and Development Center; Westinghouse Electric Corporation; 1310 Beulah Road; Pittsburgh PA 15235; USA; 412-256-1571; E; 4.1; 3.10; 5.1; Infrared frequency generation; Visible radiation, generation and detection

LICHTEN, William L.; Department of Physics and Engineering and Applied Science; J. W. Gibbs Laboratory; Yale University; PO Box 6666; New Haven CT 06511; USA; 203-432-4114/4730; 203-436-3680; E; 3.6; Laser spectroscopy of atoms and molecules; Molecular beams

LIDE JR., David R.; Standard Reference Data; National Bureau of Standards; Gaithersburg MD 20899; USA; 301-921-2467; E/T; 1.1; 3.3; Reference data on atomic and molecular properties

LIEB, Elliot H.; Physics and Mathematics Departments; Jadwin Hall; Princeton University; PO Box 708; Princeton NJ 08544; USA; 609-452-4420; T; 1.1; Thomas-Fermi and related theories of atoms and molecules; Ionization of atoms; Interatomic potentials and molecular binding

LIEBENBERG, Donald H.; Physics Division; Los Alamos National Laboratory; PO Box 1663 MS764; Los Alamos NM 87545; USA; 505-667-5371; E; 3.6; 5.4; Plasma ion emission line studies; Line profile measurement; Brillouin scattering; Fluorescence spectroscopy

LIEBER, Michael; Department of Physics; University of Arkansas; Fayetteville AR 72701; USA; 501-575-2506; 501-575-5914; T; 2.1; 2.2; 2.3; 3.1; 1.4; Eikonal techniques to high and medium energy electron-atom scattering and electron capture in ion-atom and ion-ion collisions--application

LIEBERMAN, David H.; Department of Physics; SUNY, Stony Brook; Stony Brook NY 11794; USA

LIEBMAN, Joel F.; Department of Chemistry; University of Maryland, Baltimore County; 5401 Wilkens Avenue; Catonsville MD 21228; USA; 301-455-2549; T; 1.1; Organic molecules and ions--structure, energetics, regularities; Simple species-quantum chemical calculations; Ionization potentials; Proton affinities

LIEDHOLZ, Gerhard A.; Applied Research Div. 440; Coulter Electronics, Inc.; 440 W. 20th Street; Hialeah FL 33010; USA; 305-885-0131x251; E; 3.6; Improved spectroscopic methods

LIEFELD, Robert J.; Department of Physics; New Mexico State University; Las Cruces NM 88003; USA; 505-646-2206; E; 3.4; X-ray physics

LIGHT, Glenn C.; Chemical Physics Division; The Aerospace Corporation; 2350 E. El Segundo Boulevard; El Segundo CA 90245; USA; 213-648-7013; E; 2.3; Elementary gas phase reaction rates--experimental measurement

LIGHT, John C.; Department of Chemistry; James Franck Institute; University of Chicago; 5640 S. Ellis Avenue; Chicago IL 60637; USA; 312-962-7197; T; 2.1; 2.3; 1.1; Quantum atom-molecule scattering, inelastic and reactive; Vibrational levels and dynamics of polyatomic molecules; Numerical methods

LIGHTMAN, Allan J.; Research Institute; University of Dayton; 300 College Park; Dayton OH 45469; USA; 513-229-3938; E; 4.1; 3.6; 5.5; 3.10; 3.5; Atomic and molecular measurement probes applied to the study of reacting and nonreacting flowfields

LIM, Teck-Kah; Department of Physics and Atmospheric Science; Drexel University; 32nd and Chestnut Streets; Philadelphia PA 19104; USA; 215-895-2717; T; 1.3; 2.1; 2.3; Few-atom clusters; Atom-molecule scattering; Recombination rates

LIMBAUGH, Charles C.; Sverdrup Technology; Arnold AFS TN 37389; USA

LIN, Chii-Dong; Department of Physics; Kansas State University; Manhattan KS 66506; USA; 913-532-6823; T; 1.1; 2.3; 3.1; Atomic structure and electron correlation; Ion-atom collisions; Ionization and electron capture; Atoms and positrons ions--photoabsorption

LIN, Chun C.; Department of Physics; University of Wisconsin; 1150 University Avenue; Madison WI 53706; USA; 608-262-0697; E/T; 2.1; Atomic and molecular collision processes

LIN, J. T.; Code 6540; Naval Research Lab.; Washington DC 20375; USA

LIN, John; ITT Advanced Technology Center; 1 Research Drive; Shelton CT 06484; USA; 203-926-5270; T; 3.1; 3.6; 5.7; Quantum transport of electrons in solids

LIN, Jung; Department of Physics; Tennessee Technological University; Cookeville TN 38501; USA; 615-528-3477; E; 3.4; 3.6; 4.2; Nuclei far from stability--nuclear and laser spectroscopy

LINDAU, E. Ingolf; Stanford Electronics Lab.; Stanford University; Stanford CA 94305; USA

LINDER, Bruno; Department of Chemistry; Florida State University; Tallahassee FL 32306; USA; 904-644-5299; T; 1.1; 2.3; Interatomic and intermolecular interactions; Physisorption; Polarization phenomena; Intermolecular forces on NMR--effects; Electronic intensities; Fluids

LINDER, E. G.; RCA Laboratories (Retired); 16 Colonial Club Drive Apt. #205; Boynton Beach FL 33435; USA; 305-732-4049; E/T; 5.3; 5.2

LINDLE, Dennis W.; Lawrence Berkeley Lab.; MMRD Building 70A, Room 1115; Berkeley CA 94720; USA; 415-486-5666; E; 3.1; Resonance photoionization

LINDSAY, Mark D.; Physics Dept.; Lyman Lab.; Harvard University; Cambridge MA 02138; USA; 617-495-3386; 617-498-4250; E; 1.2; 3.5; 3.6

LINDSAY, Robert D.; 13619 Dronfield Avenue; Sylmar CA 91342; USA

LINEBERGER, William C.; Department of Chemistry; Joint Institute of Laboratory Astrophysics; University of Colorado; Boulder CO 80309; USA; 303-492-7834; E; 2.1; 3.1; 3.6; Photon interactions with ions; Molecular energy transfer; Free radicals--spectroscopy

LIPKIN, Daniel M.; 1717 Bantry Drive; Dresher PA 19025; USA

LIPPMANN, Bernard A.; Department of Physics and Mathematics; Stanford University; Stanford CA 94305; USA; 415-497-4508; T; 3.6; Atomic and molecular systems with electromagnetic radiation-interaction

LIPSCOMB, William N.; Chemistry; Gibbs Laboratory; Harvard University; 12 Oxford Street; Cambridge MA 02138; USA; 617-495-4098; 1.1

LIPSKY, Lester; Department of Computer Science; University of Nebraska; 115 Ferguson Hall; Lincoln NE 68588; USA; 402-472-6747; 402-472-2402; T; 1.1; 3.1; 2.2; 1.2; Calculation of 2 and 3 electron resonances; Computational physics

LIPSKY, Sanford; Department of Chemistry; Physical Chemistry; University of Minnesota; Minneapolis MN 55455; USA; 612-373-2373; 612-373-0151; E; 2.2; 3.2; 3.1; Electron energy loss spectroscopy; Emission spectroscopy; Photoabsorption and photoionization spectroscopy

LIPSON, Steven J.; 173 Hancock Street Apt. 7; Cambridge MA 02139; USA

LISBOA, Jorge A.; Universidade Federal DO RGS; Laboratо'Rio DO Laser; Instituto De Fisica; Bento Goncalves-G500; Porto Alegre RS 90000; Brazil; E; General structure and properties of atoms and molecules; Atomic and molecular collisions; Interaction of laser with atoms and molecules

LISY, James M.; Materials and Molecular Research Div.; Lawrence Berkeley Laboratory; University of California; Berkeley CA 94720; USA; 415-486-6447; E; 1.1; Intramolecular energy distribution in molecules

LITKE, John D.; Engineering Division; Photocircuits D. V.; 31 Sea Cliff Avenue; Glen Cove NY 11542; USA; 516-448-1352; E; 5.1; Adsorption, absorption and surface chemistry

LITOVCHENKO, Vladimir; Department of Physics; University of Oklahoma; 440 West Brooks; Norman OK 73019; USA

LITTLE, Roger G.; Spire Corporation; Patriots Park; Bedford MA 01730; USA; 617-275-6000

LITTMAN, Michael G.; Department of Mechanical and Aerospace Engineering; Princeton University; D418 Engineering Quadrangle; Princeton NJ 08544; USA; 609-452-5198; E; 3.6; 3.9; Laser spectroscopy; Laser diagnostics; Stark effect

LITVAK, Herbert E.; Central Research Division; Varian Associates; PO Box D-299; 611 Hansen Way; Palo Alto CA 94303; USA; 415-493-4000x4133; E; 4.4; 5.3; Atomic and molecular science to analytical chemistry-application; Gas discharges and mass spectrometry---spectroscopy

LITVAK, Marvin M.; Jet Propulsion Laboratory; California Institute of Technology; 4800 Oak Grove Drive; Pasadena CA 91103; USA; 213-354-7441; T; 3.5; 5.4; 5.6; Planetary plasma physics; Interstellar molecules

LIU, Cheng-Kuang; Electronic Engineering Department;

LIU, Jenn-Ying; Engineering; Research and Development; Quantronix Corp.; 225 Engineers Road; Smithtown NY 11788; USA; 516-273-6900x374; E; 4.1; 3.6; 3.9; 3.8

LIU, Yung S.; VLSI Laboratory; General Electric Research Center; PO Box 8; 1 River Rd.; Schenectady NY 12301; USA; 518-387-6436; 518-387-5085; E; 4.1; 3.8; 3.10; 3.6; Laser physics, chemistry, and engineering; Laser processing of semiconductors; Nonlinear optics; Laser modification of polymer surfaces

LIUZZI, Anthony; Department of Physics; University of Lowell; 1 University Avenue; Lowell MA 01854; USA

LIVINGSTON, A.E. Gene; Department of Physics; Notre Dame University; Notre Dame IN 46556; USA; 219-239-7554/7716; 219-239-6386; E; 1.1; 1.4; 3.2; 3.4; 4.2; Accelerator-based atomic physics; Highly ionized atoms spectroscopy; Highly ionized atoms: structure, spectra, excitation

LIVINGSTON, Ralph; Chemistry Division; Oak Ridge National Laboratory; PO Box X; Oak Ridge TN 37830; USA; 615-574-4983; E; 2.3; 3.3; Electron spin resonance spectroscopy; Short-lived free radicals--physical and chemical properties

LLAGUNO, Claro T.; Chemistry Department; University of the Philippines; Diliman; Quezon; Philippines

LO, Chaomei; 1653 Arthur Street; Eugene OR 97402; USA

LOCKWOOD, Grant J.; Division 1234; Sandia National Laboratories; PO Box 5800; Albuquerque NM 87185; USA; 505-846-1958; E; 2.1; 2.3; 4.2; Charge transfer and excitation cross section measurements

LODER, Rurik K.; U.S. Army Ballistic Research Lab.; Aberdeen Proving Ground MD 21005; USA; 301-278-3983; E/T; 2.3; 4.1; Physical optics; Chemical physics; Spectroscopy

LODGE, Elizabeth A.; Department of Metallurgy Science Math.; University of Oxford; Parks Road; Oxford OX13PH; United Kingdom

LODHI, Sattar K.; Physics; Physics and Chemistry; Sam Houston State University; 2012 SHSU; Huntsville TX 77341; USA; 409-294-1605; E/T; 1.1; 2.1; 3.4; 4.2; 5.1; Molecular stopping powers; Fine structure of x-ray lines of low Z elements; Interaction of radiation with matter; Charging of cosmic dust particles in space

LOGAN, Ralph A.; Research Division; Bell Telephone Laboratories; 600 Mountain Avenue; Murray Hill NJ 07974; USA; 201-582-2897; E; Solid state lasers

LOGOTHETIS, Eleftherio M.; Physics Division; Ford Motor Company Research; PO Box 2053; Dearborn MI 48121; USA; 313-323-4553; E; 5.1; Gas molecules with solids--interaction

LOHMANN, Birgit; Atomic and Molecular Physics Lab.; Australian National University; Canberra ACT 2601; Australia

LOHR JR., Lawrence L.; Department of Chemistry; University of Michigan; Ann Arbor MI 48109; USA; 313-764-3148; T; 1.1; 2.3; Molecular wavefunctions and potential energy surfaces--ab initio and semi-empirical calculation; Relativistic effects in chemical bonding--calculation; Reactive scattering cross-sections--calculation

LOIS, Lambros; 6104 Dunleer Court; Bethesda MD 20817; USA

LOMBARDI, Gabriel G.; MS R1-1196; TRW; 1 Space Park Drive; Redondo Beach CA 90278; USA

LONG JR., Edward R.; Applied Materials Branch; Materials Division; NASA Langley Research Center; MS 399; Hampton VA 23665; USA; 804-865-3892; E; 1.3; 2.2; 3.3; 4.2; IR and EPR spectroscopic analysis of effects of electron radiation on molecular structure of polymer systems

LONG, Keith A.; 177 Kriegsstrasse; Karlsruhe 1 75; West Germany

LONG JR., William H.; Laser Technology Laboratory; Northrop Research and Tech. Center; 1 Research Park; Palos Vrds Peninsula CA 90274; USA; 213-377-4811x233; T; 2.3; 5.3; 5.4; Gas discharge and laser modeling, including electron kinetics, vibrational kinetics, and plasma instabilities

LONGMIRE, Martin S.; PO Box 1417-D12; Alexandria VA 22313; USA

LONGWORTH, James W.; Biology Division; Oak Ridge National Laboratory; PO Box Y; Oak Ridge TN 37830; USA; 615-574-1214; FTS-624-1214; E; 1.3; Proteins and nucleic acids from their intrinsic components--luminescence

LOREE, Thomas R.; CHM-6; Los Alamos National Lab.; 1663; MS: J564; Los Alamos NM 87545; USA; 505-667-7705; E; 3.6; 4.1; 3.10; 3.2; 3.8; Atomic spectroscopy--detection

LORENTS, Donald C.; Chemical Physics Laboratory; SRI International; 333 Ravenwood Avenue; Menlo Park CA 94025; USA; 415-859-3167; E/T; 2.3; 3.6; 3.1; Kinetic processes in dense excited pure and doped rare gases; Laser physics

LOSONSKY, William; Department of Science; State Univ. of NY Maritime College; Bronx NY 10465; USA; 212-892-3000 x231

LOUIE, Doowah; 32-54 54th Street; Woodside NY 11377; USA

LOUISELL, William H.; Department of Physics; University of Southern California; University Park; Los Angeles CA 90007; USA; 213-741-2508/2226; T; 4.1; Quantum electronics; Quantum optics; Free electron lasers

LOVAS, Frank J.; Molecular Spectroscopy Division; National Bureau of Standards; Building 221 Room B265; Gaithersburg MD 20899; USA; 301-921-2023; E; 3.3; 3.5; 1.1; 1.3; High resolution microwave and IR molecular spectroscopy; Molecular radio astronomy

LOVOI, Paul; International Technical Associates; 1024 W. Maude, Suite 202; Sunnyvale CA 95070; USA; 408-746-3757; E; 1.1; 3.1; 4.2; H- photodissociation of 800 MeV beam with lasers to measure cross section, binding energy

LOWE, John P.; Department of Chemistry; 152 Davey Laboratory; Pennsylvania State University; University Park PA 16802; USA; 814-865-9661; T; 1.1; Molecular orbitals of properties of molecules of chemical interest

LOWITZ, David A.; Research and Development Center; Philip Morris Co.; PO Box 26583; Richmond VA 23261; USA; 804-271-3326; T; 1.1; Quantum mechanics to studies of molecular structure--applications

LOWRY, Jerald F.; Directed Energy Research; Applied Sciences; Westinghouse R&D Center; 1310 Beulah Road; Pittsburgh PA 15235; USA; 412-256-1633; E; 4.1; 5.3; 5.1; 4.3; Lasers, UV preionized electric discharge; Electron beam-sustained discharges

LOWRY, Ralph A.; University of Virginia; Thornton Hall; Charlottesville VA 22901; USA

LOXTON, Chris M.; Materials Research Laboratory; University of Illinois; 104 S. Goodwin; Urbana IL 61801; USA; 217-333-0386; E; 5.1; Sputtering and secondary ion emission processes--surface analysis

LOY, Michael T.; IBM Thomas J. Watson Research Center; Yorktown Heights NY 10598; USA; 914-945-2480; E; 3.7; 4.1; Multiphoton ionization of molecules; Coherent transients in molecules

LOYD JR., David H.; Department of Physics; Angelo State University; San Angelo TX 76901; USA; 915-942-2235;

E; 2.3; Charge exchange collisions between proton beams and gas targets
LU, Chun-Chian; 14140 S W 98 CT; Miami FL 33176; USA

LU, Jia-Jih; Department of Physics; University of California, Riverside; Riverside CA 92521; USA; 714-787-5331; E/T; 2.3; 4.2; Heavy-ion reactions

LUBELL, Michael S.; Department of Physics; City College of CUNY; Convent Avenue & 138th Street; New York NY 10031; USA; 212-690-8312; 212-690-8316; E; 2.2; 3.6; 1.1; 1.4; 3.1; Spin-tagged electron-atom collisions; Laser cooling and controlled atom research; Polarized electron sources and electron polarimeters; Positron-atom crossed beams collisions

LUCATORTO, Thomas B.; Radiation Physics Division; National Measurement Laboratory; National Bureau of Standards; A251, Physics Building; Gaithersburg MD 20899; USA; 301-921-2031; E; 3.6; 3.7; 1.1; Dense vapors--laser excitation and ionization; Atomic autoionizing states --spectroscopy; Multiphoton ionization

LUCCHESE, Robert R.; Department of Chemistry; Texas A&M University; College Station TX 77843; USA; 409-845-0187; T; 2.2; 3.1; 3.5; 5.1

LUCKEY, George W.; Special Research Laboratory; Eastman Kodak Co.; 240 Weymouth Drive; Rochester NY 14625; USA; 716-381-6956; E; 5.6; Fluorescence studies, environmental effects

LUDENA, Eduardo V.; QUIMICA; IVIC; Apartado 1827; Caracas 1010-A; Venezuela; (02)691-959x309; T; 1.1; 1.3; 5.1; Density matrix and density functional theory; Transition metal clusters and catalysis

LUKEN, William L.; Department of Chemistry; Duke University; Durham NC 27707; USA; 919-684-3130; T; 1.1; 3.2; Electronically excited states of small molecules; Highly ionized heavy atoms--calculations of spectra; Electron correlation effects in atoms and molecules

LUMSDEN III, Jesse B.; Rockwell International Science Center; PO Box 1085; Thousand Oaks CA 91360; USA

LUND, Clarence; Phoenix Corporate Research and Development Laboratories; Motorola Corporation; 2200 W. Broadway; Mesa AZ 85202; USA; 602-962-2759; E; 5.1; Ion implantation into silicon

LUNDEEN, Stephen R.; Department of Physics; Notre Dame University; Notre Dame IN 46556; USA; 219-239-5000; E; 1.2; 3.3; 4.2; High precision radio frequency spectroscopy of simple atoms using fast atomic beams

LUNDQUIST, Theodore R.; Product Division; Gatan, Inc.; 6678 Owens Drive; Pleasanton CA 94566; USA; 415-463-0200; E; 5.1; 4.4; 4.3

LUNNEY, James G.; Department of Physics; Trinity College; Dublin 2; Ireland; 772941 Ext. 1259; E/T; 5.1; 4.1; 3.10; 3.4; X-ray spectroscopy of laser produced plasmas; X-ray laser development; Optical bistability; Dielectronic satellite emission from plasmas

LUNTZ, Alan C.; Division K34; IBM Research Laboratory; 5600 Cottle Road; San Jose CA 95193; USA; 408-256-2675; E; 2.3; 3.8; 5.1; Chemical dynamics and chemical reactions--laser and molecular beam studies; Gas-solid surface interactions--laser and molecular beam studies

LUQUE, Antonio; Esti Telecomunicacion; Univ. Politecnica De Madrid; Madrid 3; Spain

LUTZ, Otto; Sandgrabenstr 14; Ammerbuch 1 D7403; West Germany

LYNDS, Lahmer; Chemical Physics Division; Mail Stop 76; United Technologies Research Center; Silver Lane; East Hartford CT 06108; USA; 203-727-7134; E/T; 1.1; 2.1; 3.6; 3.8; Laser excited reactive collisions between atoms and molecules; Electronic energy transfer; Laser spectroscopy and lifetimes of atomic levels (states)

LYNK, Edgar T.; GE CRD; PO Box 43; Building 37-Room 223; Schenectady NY 12345; USA

LYONS, Peter B.; Fusion Research Application; Physics Division; Los Alamos National Lab.; PO Box 1663 MS410; MS E527; Los Alamos NM 87545; USA; 505-667-6411; E; 5.4; 5.1; 4.1; Plasma diagnostics--atomic processes

LYOU, Jong-Hun; Department of Physics; University of Colorado; Campus Box 390; Boulder CO 80309; USA

LYSIAK, Richard J.; 1700 Ems Road West; Fort Worth TX 76116; USA

MACADAM, Keith B.; Department of Physics and Astronomy; University of Kentucky; Lexington KY 40506; USA; 606-257-3344; 606-257-6722; E; 1.2; 2.1; 3.9; 4.2; Atomic collisions; Rydberg atoms; Highly charged ions

MACARTHUR, Duncan W.; Q-Division; Los Alamos National Laboratory; MS-J562; Los Alamos NM 87545; USA; E; 3.6; 4.2; 1.4; 5.8

MACCALLUM, Crawford J.; Sandia Lab. 1231; Albuquerque NM 87185; USA

MACDONALD, J. Ross; Department of Physics and Astronomy; University of North Carolina; Chapel Hill NC 27514; USA; 919-962-3012; 919-967-5005; T; Ionic motion and effects at interfaces; Relaxation in condensed matter

MACDONALD, John; TRIUMF; University of British Columbia; Vancouver BC V6T1W5; Canada

MACDOWELL, Alastair A.; Science and Engineering Research Council; Daresbury Lab.; Warrington WA44AO; England; E; 3.4; 3.5; Instrumentation for soft x-ray beamlines on synchrotron radiation sources

MACEK, Joseph H.; Department of Physics and Astronomy; University of Nebraska; Lincoln NE 68588; USA; 402-472-2352; T; 1.1; 2.1; 2.2; Two electron atoms--atomic structure; Electron-ion and electron-atom collisions; Ion-atom collisions and the structure of small molecules

MACKELLAR, Alan D.; Department of Physics; University of Kentucky; Lexington KY 40506; USA

MACKNIGHT, Allen K.; Advanced Engineering Applications Div.; Garrett-AiResearch; 2525 W. 190th; Torrance CA 90509; USA; 213-323-9500x3049; E; 4.1; Chemical iodine laser development

MACLIN, Arlene P.; Board on Physics and Astronomy; National Academy of Sciences; 2101 Constitution Avenue; Washington, DC 20418; USA; 202-334-3520; T; Science policy and administration

MACPHERSON, Alistair K.; Department of Mechanical Engineering and Mechanics; Lehigh University #19; Bethlehem PA 18015; USA; 215-861-4105; T; 2.3; 5.1; Chemical reactions including ionization and decomposition of explosives--statistical mechanics

MACRAE, A. U.; Satellite Communications Lab.; AT&T Bell Laboratories; Crawfords Corner Road, Rm. 2F-335; Holmdel NJ 07733; USA; 201-949-6722; E; 5.1

MADAN, Rabinder N.; 20025-49 Community Street; Canoga Park CA 91306; USA

MADDEN, Keith P.; Department of Chemistry; Wayne State University; Detroit MI 48202; USA; 313-577-2597; E; Lipid bilayers using magnetic resonance techniques--molecular dynamics studies

MADDEN, Robert P.; Radiation Physics Division; National Bureau of Standards; A257, Physics Building; Gaithersburg MD 20899; USA; 301-921-2031; E; 3.1; 3.2; 4.2; 5.1; Atomic photoabsorption and photoionization; Thin films--optical properties

MADDOX, William E.; Department of Physics; Murray State University; Murray KY 42071; USA; 502-762-2993; E; 4.2; Accelerator-related atomic physics

MADEY, J. M.; High Energy Physics Laboratory; Stanford University; Stanford CA 94305; USA; 415-497-3034; E/T; 2.2; 4.1; 4.2; Free-electron lasers; Magnetic resonance experiments with low energy electron beams

MADEY, Theodore E.; Surface Science Division; National Bureau of Standards; Gaithersburg MD 20899; USA; 301-921-2188; E; 5.1; Molecules on surfaces--geometrical structure; Electronic excitations in surface molecules which result in decomposition and desorption

MADISON, Don H.; Department of Physics; Drake University; Des Moines IA 50311; USA; 515-271-3750; T; 2.1; 2.2; Atomic scattering

MADJID, A. Hamid; 326 Harris Drive; State College PA 16801; USA

MADSON, James M.; Department 223; McDonnell Douglas Research Lab.; McDonnell Douglas Corporation; PO Box 516; Building 110; St. Louis MO 63166; USA; 314-233-2544; E; 5.5; 5.4; 4.4; 5.3; Combustion-hydrocarbon-- combustion-soot generation; Plasma chemistry--electron attachment; Plasma generation--RF plasmas

MAEDA, Kaichi; Electrodynamics Branch; Laboratory of Extraterrestrial Physics; NASA Goddard Space Flight Center; Greenbelt MD 20771; USA; 301-344-8224; 301-345-7036; T; 4.2; 5.6; Energetic particles in the planetary magnetospheres and in the heliosphere; Aeronomy; Cosmic rays

MAGEE, John L.; Biology and Medicine Division; Lawrence Berkeley Laboratory; University of California; Building 29, Room 206; Berkeley CA 94720; USA; 415-486-6216; T; 2.3; Accelerated heavy particles--chemical effects

MAGEE JR., Norman H.; Theoretical Division, Group T-4; Los Alamos National Laboratory; PO Box 1663 MS212; Los Alamos NM 87545; USA; 505-667-5077; T; 2.2; 5.4; Collision excitation rates--calculations; Positive ions for use in plasma power losses--dielectronic recombination rates

MAGGIORA, Gerald M.; Computational Chemistry Support; Biotechnology; The Upjohn Company; 301 Henrietta Street; Kalamazoo MI 49001; USA; 616-384-9350; T; 1.1; 3.2; 2.3; 1.3; Quantum mechanics of large molecular systems; Empirical potential functions; Quantum mechanics of transition state structure and properties

MAGNUSON, Dale W.; Special Projects Department; Union Carbide Nuclear Co.; Oak Ridge Gaseous Diffusion Plant; PO Box P MS322; Oak Ridge TN 37830; USA; 615-576-0152; FTS-626-0152; E; 3.1; 3.3; High resolution IR spectroscopy using flow cooling; UV cross section measurements

MAGNUSON, Gustav D.; General Dynamics/Convair; San Diego CA 92138; USA; 619-224-7207; 619-547-5793; E/T; 2.1; 4.3; 4.2; 5.1; 5.3

MAHADEVAN, P.; Laser Chemistry; Aerophysics Laboratory; The Aerospace Corporation; M5747; PB92957; Los Angeles CA 90009; USA; 213-648-7615; E; 2.1; 2.2; 5.1; Electronic, atomic and ionic collision phenomena; Surface effects

MAHON, Rita; Department of Physics and Astronomy; Laboratory for Plasma Fusion; University of Maryland; College Park MD 20742; USA; 301-454-7076; E; 3.10; 3.6; 5.4; Collimated radiation in the VUV for use in measuring neutral hydrogen densities in Tokomak type plasmas--development of intense sources

MAHONEY, Francis W.; 1824 Fair Avenue; Simi Valley CA 93063; USA

MAHR, H.; Department of Physics; Cornell University; Ithaca NY 14853; USA; 607-256-3638; E; 4.1; Charge transfer laser in hydrogen plasma/alkali vapor systems --development

MAIER II, William B.; ESS-7; Los Alamos National Laboratory; Department of Energy; 1663; 430 Camino Cendantado; Los Alamos NM 87544; USA; 505-667-9657; 505-662-3843; E/T; 1.1; 2.1; 2.2; 2.3; 5.2; EMP from propagating beams; Spectra and photochemistry of solutions; Classical field theory; Optical emissions from propagating beams

MAINARDI, Raul T.; Department of Physics; Atomic Spectroscopy; FAMAF University NAC of Cordoba; Laprida 854; Cordoba Cordoba 5000; Argentina; 40669; E/T; 3.4; 3.1; X-ray spectrometry; Detection devices

MAJEWSKI, Walerian; Department of Physics and Astronomy; University of Maryland;

College Park MD 20742; USA

MAJMUDAR, Arun K.; Radar Science and Engineering Div.; Jet Propulsion Laboratory; MS 183/701; 4800 Oak Grove Drive; Pasadena CA 91103; USA; 213-354-2849; E/T; 4.1; Molecules by laser technique--remote sensing; Laser scattering, Fiber optics and electro-optical systems; Integrated optics; Optical communication

MAJUMDER, Protik K.; Department of Physics; Lyman Laboratory; Harvard University; Cambridge MA 02138; USA; 617-495-3386; E; 1.4; 3.7

MAKI, Arthur G.; Molecular Spectroscopy Division; National Bureau of Standards; Gaithersburg MD 20899; USA; 301-921-2755; E; 3.2; Molecules in the gas phase--high resolution molecular spectroscopy

MALIK, David J.; School of Chemical Sciences; University of Illinois; Urbana IL 61801; USA; 217-333-1728; T; Energy transfer; Schrodinger equation--exact and approximate solution

MALIK, F. Bary; Department of Physics and Astronomy; Southern Illinois University; Carbondale IL 62901; USA; 618-453-2643/2786; T; 2.2; Electron impact ionization and excitation of atoms and ions; Atomic processes--impact of relativistic effects

MALLARD, W. Gary; Exploratory Fire Research Group; National Bureau of Standards; Building 224, Room B260; Gaithersburg MD 20899; USA; 301-921-3771; E; 5.5; Molecular processes leading to soot formation in flames; Flames--electrical properties

MALLEY, Michael M.; Department of Chemistry; San Diego State University; San Diego CA 92182; USA; 619-265-6536; 619-265-5595; T; 3.6; 3.3; 3.10; Picosecond phenomena; Photophysics and photochemistry

MALLOW, Jeffry V.; Department of Physics; Loyola University of Chicago; Chicago IL 60626; USA; 312-508-3546; E/T; 1.1; 1.3; 3.6; Muonic atoms; Atomic and molecular self-consistent field theory; Laser spectroscopy

MALONE, Dennis P.; Department of Electrical Engineering; State University of NY, Buffalo; 223 Bell Hall; Amherst Campus; Amherst NY 14260; USA; 716-636-2422; E/T; 3.6; 5.4; Atomic interaction in plasmas; Molecular excitation using laser excitation; Plasma diagnostics

MANCA, Joseph J.; 627 Princeton Drive; Sunnyvale CA 94087; USA

MANDEL, Leonard; Department of Physics and Astronomy; University of Rochester; Rochester NY 14627; USA; 716-275-4361; E/T; 4.1; Atoms and coherent light--interaction between; Lasers --coherence properties; Quantum optics; Photon statistics and squeezing

MANDELBERG, Hirsch I.; Laboratory for Physical Sciences; 4928 College Avenue; College Park MD 20740; USA; 301-277-4190; E; 4.1; Laser development, electro-optics

MANDL, Alexander E.; Avco Everett Research Laboratory; 2385 Revere Beach Parkway; Everett MA 02149; USA; 617-389-3000; E; 2.3; Mercury-ammonia excimers

MANGELSON, Nolan F.; Department of Chemistry; Brigham Young University; Provo UT 84602; USA; 801-377-3669; E; 4.2; 3.4; 2.1; Proton induced X-ray emission analysis for element content of samples; Proton induced gamma-ray emission analysis for element content of samples

MANISTA, Eugene J.; M/S 142-2; Lewis Research Center; 21000 Brookpark Road; Cleveland OH 44135; USA

MANN, David M.; Plume Technology Office; US Air Force Rocket Propulsion Lab.; Edwards AFB CA 93523; USA; E; 2.3; Laser fluorescence spectroscopy; Oxidation reactions--kinetics

MANN, Joseph B.; Theoretical Division; Los Alamos National Laboratory; PO Box 1663 MS212; Los Alamos NM 87545; USA; 505-667-5962; FTS-843-5962; T; 2.2; 2.2; Electron impact excitation cross sections for ions; Electron impact ionization cross sections for ions

MANNING, Irwin; Condensed Matter/Radiation Science; US Naval Research Lab.; 4555 Overlook Avenue, SW; Washington DC 20375; USA; 202-767-6990; T; 2.1; 5.2; Penetration of matter by energetic atoms; Binary atomic collisions at high energies

MANOR, Robert E.; REM Agency; 6628 Texas Street; Whitehouse OH 43571; USA; 419-877-5502; T; 4.1; 4.3; Home laser fusion systems--development; Commercial plasma fusion systems--development

MANSIKKA, Kauko A.; Department of Physical Sciences; Theoretical Physics; Institute of Theoretical Physics; University of Turku; Turku 50 SF 20500; Finland; 921-645680; 921-362424; T; 1.1; 1.4; 3.1; 3.10

MANSKY II, Edmund J.; School of Physics; Georgia Institute of Technology; Atlanta GA 30332; USA; 404-894-5273; T; 2.2; 2.3; 5.3; 5.4; 1.2; Excitation in electron-metastable atom collisions; Recombination in dense plasmas; Heavy-particle collisions; Ion-atom charge-transfer collisions

MANSON, Joseph R.; Physics and Astronomy; Clemson University; Clemson SC 29631; USA; 803-656-3418; T; 5.1; 2.2; 1.1; Atom-surface interactions; Molecule-surface interactions

MANSON, Steven T.; Department of Physics and Astronomy; Georgia State University; Atlanta GA 30303; USA; 404-658-3221; 404-658-2271; T; 1.1; 2.1; 2.2; 3.1; 3.4; Atomic and molecular collisions and structure; Photoionization and photoelectron spectroscopy; Photoionization of excited states and ions; Electron spectroscopy in atomic collisions

MANZANARES, Elizabeth R.; Electrical and Computer Engineering; San Diego State University; San Diego CA 92182; USA; 619-265-3700; E; 1.1; 2.3; 3.6; Laser induced fluorescence from molecules adsorbed on aerosols

MAR, Jerry; 1495 Chukar Court; Sunnyvale CA 94087; USA

MARCHAND, Pierre D.; Department of Physics; Laval University; Quebec G1K7P4; Canada

MARCHETTI, Alfred P.; Emulsion Development Laboratory; Photographic Materials Research; Eastman Kodak Research Laboratories; Eastman Kodak Company; Building 59; 1999 Lake Avenue; Rochester NY 14650; USA; 716-477-4995; 716-385-1638; E; 3.9; 5.2; Light with silver halides--interaction; Molecular centers in silver halides--study; Light with dyes--interaction; Electron-hole recombination in semiconductors

MARCUS, Paul M.; IBM Research Center; Yorktown Heights NY 10598; USA

MARCUS, Rudolph A.; Department of Chemistry; California Institute of Technology; Building 127-72; Pasadena CA 91125; USA; 213-795-6811x2566; T; 2.3; 3.8; Electron transfer reactions; Semiclassical theory of bound states and of collisions; Intermolecular and intramolecular energy transfer; Energy localization in molecules

MARGENAU, Henry; Department of Physics and Philosophy; Yale University; 44 SPL; New Haven CT 06520; USA; 203-436-3314; T; Professor emeritus

MARIANI, David R.; Schlumberger-Doll Research; PO Box 307; Old Quarry Road; Ridgefield CT 06877; USA; 203-431-5342; 203-431-5345; E; 3.3

MARICQ, M. Matti; Chemistry; Brown University; Providence R.I. 02912; USA; 401-863-3767; E/T; 1.1; 3.1; 3.3; 2.1; 5.2; Autodetachment spectroscopy of weakly bound molecular anions; Vibrational relaxation by collisions of simple molecules in gases and liquids; Pulsed field effects in the NMR of solids

MARIELLA JR., Raymond P.; Electronic Optical Physics; Corporate Research; Corporate Technology; Allied Corporation; PO Box 1021R; Morristown NJ 07960; USA; 201455-2158; 201-455-2043; E; 5.1; 3.8; 2.3; Beam/surface growth kinetics--MBE, etc.

MARINO, Robert A.; Department of Physics and Astronomy; Hunter College of CUNY; 695 Park Avenue; New York NY 11791; USA; 212-570-5696; E; Quadruple coupling constants on molecular and ionic solids

MARKEVICH, Darlene J.; Office of Fusion Energy; ER-543; U.S. Department of Energy; MS G226 GTN; Washington DC 20545; USA

MARKISZ, John A.; Department of Nuclear Medicine; Harvard Medical School; 25 Shattuck Street; Cambridge MA 02115; USA; 617-732-6000; T; 4.2; Neutron scattering

MARKO, Kenneth A.; 2224 Highland Road; Ann Arbor MI 48104; USA

MARMAR, Earl S.; Plasma Fusion Center; MIT; 167 Albany Street; Cambridge MA 02139; USA; 617-253-5455; E; 5.4; 1.4; High temperature plasmas, including visible UV and X-ray regions--spectroscopy

MARMET, Paul; 1001 De Grenoble; STE-FCY PQ G1V2Z8; Canada

MARQUET, Louis C.; Optics Division; Lincoln Laboratory; MIT; 244 Wood Street; Lexington MA 02173; USA; 617-862-5500 x5427; E; 5.6; Laser applications: device physics and interactions with the atmosphere, including absorption, scattering, and kinetic cooling

MARRON, Michael T.; Molecular Biology Program; Office of Naval Research; University of Wisconsin, Parkside; 800 N. Quincy Street; Arlington VA 22217; USA; 202-696-4038; E/T; 1.3; 3.10; Mechanism of interaction between electromagnetic fields and organisms; Membrane effects of low frequency electromagnetic fields; Resonant absorption of microwaves by macromolecules

MARRUS, Richard; Department of Physics; University of California; Berkeley CA 94720; USA; 415-642-3686; E; 2.1; 3.2; 4.2; Spectroscopy and scattering with fast, highly ionized atoms

MARTIN, Eric; 3217 Hilldale Avenue; St. Anthony MN 55418; USA

MARTIN, Fred W.; Microscope Associates, Inc.; 50 Village Avenue; Dedham MA 02026; USA; 617-326-2288; E; 4.2; 5.1; 4.3; 3.4; Ion microbeams for atomic physics in solids --production; Production of ion microbeams for atomic collisions in solids

MARTIN, Georgia A.; Atomic and Plasma Radiation Div.; National Bureau of Standards; Building 221, Room A267; Gaithersburg MD 20899; USA; 301-921-2071; T; Atomic transition probability data from world scientific literature--critical evaluation and compilation

MARTIN, L. R.; Aerophysics Laboratory; The Aerospace Corporation; PO Box 92957; Los Angeles CA 90009; USA; 213-648-6920; E; 2.3; 5.6; 4.1; 4.3; 3.2; Vibrational quenching

MARTIN, Peter J.; Lecroy Research Systems Corporation; 700 S. Main Street; Spring Valley NY 10977; USA

MARTIN, Richard L.; Theoretical Division; Los Alamos National Lab.; Box 1663 MSJ569; Los Alamos NM 87545; USA; 505-667-7096; 805-961-3989; E; 1.3; 2.1; 3.1; 5.1; Electronic structure

MARTIN, Richard M.; Department of Chemistry; Univ. of California, Santa Barbara; Santa Barbara CA 93106; USA; 805-961-2628; 801-961-3989; E; 2.1; 5.1; 3.1; 3.8; 5.1; Crossed molecular beam studies of collisions of excited atoms; Surface reactions using excited atomic beams as a probe--studies

MARTIN, William C.; Atomic and Plasma Radiation Division; National Bureau of Standards; A167, Physics Building; Gaithersburg MD 20899; USA; 301-921-2011; E; 1.1; 3.2; Atomic spectra (optical) to determine energy levels--measurement and analysis; Theoretical characterization of levels; Critical compilation of such data

MARTINEZ, J. V.; Division of Chemical Sciences; U.S. Department of Energy; ER-141 Georgetown; Washington DC 20545; USA; 301-353-5820; Scientific Research Program Management

MARTINEZ, Robert M.; 4627 Merrill Street; Torrance CA 90503; USA

MARTINO, Anthony J.; 48 W. Eagle Road Apt. 109; Havertown PA 19083; USA

MASAKI, Kojima; Izumi-Shi; Kamo-4-29-9; Miyagi 981-31; Japan

MASE, Hiroshi; Department of Electronic Engineering; Ibaraki Univ-Ibaraki-Ken; 4-12-1 Naka-Narusawa; Hitachi-Shi 316; Japan

MASON, Arthur A.; Physics; Space Institute; University of Tennessee; Tullahoma TN 37388; USA; 615-455-0631; E/T; 3.3; 3.5; 3.6; 1.1; 5.6; IR Spectroscopy, intensity measurements and analyses

MASON, Edward A.; Department of Chemistry and Engineering; Brown University; Providence RI 02912; USA; 401-863-2447; 401-863-1533; T; 2.3; 5.3; 1.1; 2.1; Ion-molecule interactions; Swarm experiments--kinetic theory

MASSEY, Gail A.; Electrical and Computer Engineering; Engineering College; San Diego State University; San Diego CA 92182; USA; 619-265-5747; 619-265-5718; E; 5.5; 5.1; 3.10; 4.1; Turbulence monitoring by molecular fluorescence; Photoelectron generation at surfaces; Nonlinear generation of photoelectrons; Short wavelength laser development

MASSOUD, Hisham Z.; Department of Electrical Engineering; Duke University; Durham NC 27706; USA

MATCHA, Robert L.; Department of Chemistry; University of Houston; Houston TX 77004; USA; 713-749-4804; T; 1.1; Molecular properties and structure

MATESE, John J.; Department of Physics; University of Southwestern Louisiana; Lafayette LA 70504; USA; 318-264-6000x321

MATHEWS, C. Weldon; Department of Chemistry; Ohio State University; 140 West 18th Avenue; Columbus OH 43210; USA; 614-422-1574; E; 1.1; 3.2; 5.5; 3.1; 3.6; Gas-phase molecules with short lifetimes using conventional high-resolution spectrographs and laser-related techniques--molecular spectroscopy

MATHUR, Bhagwan P.; Department of Chemistry; Georgia Institute of Technology; Atlanta GA 30332; USA; 404-894-4070; E; 2.1; 2.2; 3.8; Electron-ion collisions; Atom-molecule collisions; Laser-matter interaction

MATHUR, Jagdish; Virginia Beach Campus; Tidewater Community College; 1700 College Crescent; Virginia Beach VA 23456; USA

MATSEN, F. Albert; Department of Chemistry; University of Texas; Austin TX 78712; USA; 512-471-4394; T; 1.1; Electronic properties of simple molecules; Electronic problems--applications of group theory

MATSUZAWA, Michio; Department of Engineering Physics; Univ. Electro Communications; 1-5-1 Chofugaok Chofu-Shi; Tokyo 182 8712; Japan; 0424(83)2161; 0426(36)6718; T; 1.2; 2.1

MATTAR, Farres P.; 73 Cranberry Street; Brooklyn NY 11201; USA

MATTHIAS, Eckart; Fachbereich Physik; Freie Universitat Berlin; Arnimallee 14; 1000 Berlin 33 1000; West Germany; 030-838-3340; 030-832-7825; E; 1.1; 1.2; 3.6; 5.2

MATTISON, Edward M.; Smithsonian Observatory; 60 Garden Street; Cambridge MA 02138; USA

MATTSON, Timothy G.; Chemistry; Natural Sciences II; Arthur Noyes Laboratory; California Institute of Technology; 127 72; Pasadena CA 91125; USA; 818-356-6513; T; 1.1; 2.3; 3.8; 4.1; Quantal reactive scattering calculations; Approximate potential methods in scattering theory; Parallel processor computer algorithms; Hypercube computer programming

MATULIC, Ljubomir; Department of Physics; St. John Fisher College; Rochester NY 14618; USA; 716-385-8168; T; 3.1; Short pulses through three-level atoms--propagation

MATYSIK, Kenneth J.; SES, Incorporation; Tralee Industrial Park; Newark DE 19711; USA; 302-731-0990; E; 5.1; Electron scattering from solid surfaces

MAUER IV, John L.; Geer Mountain Road; South Kent CT 06785; USA

MAVROYANNIS, Constantine; Division of Chemistry; Theoretical Chemistry; National Research Council; 100 Sussex Drive; Ottawa Ontario K1A0R6; Canada; 613-990-0951; T; 3.10; 3.11; 3.7; 3.6; 3.2

MAWARDI, O. K.; Department of Electrical Engineering; Case Western Reserve University; 517 Glennan Building; Cleveland OH 44106; USA; 216-368-4570; T; Dense plasma beams--acceleration; Shocks in highly ionized gases; Plasma beams--collective effects

MAXWELL, Louis R.; 5204 Moorland Lane; Bethesda MD 20814; USA

MAYO, Marguerite R.; CENG/SIG; 85X Grenoble CED; 38041; France

MAZGY, James D.; PO Box 230; Lake Hiawatha NJ 07034; USA

MAZUR, Eric; Department of Physics; Division of Applied Sciences; Gordon McKay Laboratory; Harvard University; 29 Oxford Street; Cambridge MA 02138; USA; 617-495-8729; 617-495-9616; E; 3.7; 3.6; 5.2; 3.8

MAZUR, Jacob; Center for Material Science; Polymer Division; National Bureau of Standards; Gaithersburg MD 20899; USA; 301-921-3344; T; 1.1; 2.3; Polymers--statistical mechanics; Polymer spectroscopy; Molecular dynamics and model chain calculations; Molecular conformation; Defect study

MAZUR, Jerzy H.; 2214 Haste Street #7; Berkeley CA 94704; USA

MAZUR, Ursula; Chemistry Department; Physical Division; Washington State University; Pullman Washington 99163; USA; 509-335-5822; E/T; 1.1; 2.2; 3.2; 3.3; 3.5; Vibrational spectroscopy of surfaces; Chemisorption and catalysis; Thin film synthetic method

MAZURE, Antonio J.; Department of Physics; University of Notre Dame; Notre Dame IN 46556; USA

MCABEE, Thomas L.; Department of Physics and Astronomy; University of North Carolina; Phillips Hall; Chapel Hill NC 27514; USA

MCAFEE JR., Kenneth B.; Physical and Inorganic Chemical; Research Laboratory; Bell Telephone Laboratories; 600 Mountain Avenue; Murray Hill NJ 07974; USA; 201-587-2887; E; 2.1; 2.3; Atomic and molecular collisions; Charge exchange

MCAFEE, Walter S.; 723 17th Avenue; Belmar NJ 07719; USA

MCARTHUR, David A.; Reactor Safety Research; Reactor-Safety Expts-6423; Sandia National Lab.; Department of Energy; PO Box 5800; Albuquerque NM 87185; USA; 505-844-3916; E/T; 4.1; 5.3; 2.1; Reactor-pumped lasers; Excitation of gases by nuclear reaction products

MCBRIERTY, Vincent J.; Department of Pure and Applied Physics; Trinity College; Dublin 2; Ireland

MCCALL, David W.; Chemical Research Laboratory; Bell Telephone Laboratories; 600 Mountain Avenue; Murray Hill NJ 07974; USA; 201-582-3467; E/T; 4.2; Research management

MCCALL, Samuel L.; Scattering and Low Energy; Physics Research Division; Bell Telephone Laboratories; 600 Mountain Avenue; Murray Hill NJ 07974; USA; 201-582-6305; T; 5.1; Surface enhanced Raman scattering

MCCAMMON, James A.; Department of Chemistry; University of Houston; Houston TX 77004; USA; 713-749-7351; T; 1.3; 5.2; Reaction rates, structure and dynamics of liquids and polymers, etc--statistical mechanics

MCCANN, Kevin J.; Department of Physics; Georgia Institute of Technology; Atlanta GA 30332; USA; 404-894-5262; T; 2.1; Collisional excitation and ionization

MCCLELLAN, Gene E.; Pacific-Sierra Research Corporation; 1401 Wilson Boulevard #1100; Arlington VA 22209; USA; 703-527-4975; E; 2.2

MCCLELLAND, Gary M.; Department of Chemistry; Harvard University; 12 Oxford Street; Cambridge MA 02138; USA; 617-495-1842; E/T; 2.3; 5.1; Rotational motion of highly vibrationally and rotationally excited molecules; Adsorption and desorption of molecules from surfaces

MCCLELLAND, Jabez J.; Electron Physics Group; Radiation Physics Division; National Measurement Laboratory; National Bureau of Standards; Bldg. 220 Rm. B206; Gaithersburg MD 20899; USA; 301-921-2051; E; 2.2; 3.6; 4.1

MCCLURE, Benjamin T.; Honeywell Corporate Research Center; 10701 Lyndale Avenue South; Bloomington MN 55420; USA; 612-887-4538; E; 5.2; 5.3; Ion implantation; Chemical agent detection by gaseous electronic techniques

MCCLURE, Donald S.; Department of Chemistry; Princeton University; Princeton NJ 08544; USA; 609-452-4980; E; 1.1; 3.5; 5.2; Interconfiguration electron transfer; Ions in crystals--photoionization transitions; Molecular spectroscopy; Multi-photon spectroscopy

MCCLURE, Gordon W.; Division 4244; Sandia National Laboratories; PO Box 5800; Albuquerque NM 87185; USA; 505-846-0172; E/T; 4.2; 4.3; Ion source development for particle beam fusion accelerators

MCCOLM, Douglas W.; Department of Physics; University of California, Davis; Davis CA 95616; USA; 916-752-2226; T; 1.1; Wave functions for light atoms--computation

MCCONKEY, J. W.; Department of Physics; University of Windsor; Windsor Ontario N9E3P4; Canada; 519-253-4232x467; E; Electron and photon collisions with atoms and molecules

MCCORKLE, R. A.; IBM Thomas J. Watson Research Center; PO Box 218; Yorktown Heights NY 10598; USA; 914-945-1731; E; 3.4; 4.2; 5.4; Atomic and ionic radiation/spectroscopy; Pulsed ion beams; Ion beam optics; X-ray (soft) emission from high charge states of ions

MCCORMICK, Larry D.; Bureau of Mines; Avondale Research Center; U.S. Department of Interior; 4900 LaSalle Road; Avondale MD 20782; USA; 301-436-7530; E; 4.2; 5.1; 3.4; Ion beams used to characterize surfaces and near surface areas of alloys

MCCOWN, Andrew W.; Gaseous Electronics Lab.; 607 East Healey; Champaign IL 61820; USA

MCCOY, Benjamin J.; Department of Chemical Engineering; University of California, Davis; Davis CA 95616; USA; 916-752-0400; T; 5.3; Transport phenomena

MCCUBBIN, T. K.; Department of Physics; Pennsylvania State University; 104 Davey Laboratory; University Park PA 16802; USA; 814-863-0016; E; 3.2; Molecular spectroscopy

MCCULLEN, John D.; Department of Physics; University of Arizona; Tucson AZ 85721; USA; 602-626-1625; T/E; 5.4; Plasma-neutral interactions, spectral excitation; Plasma-wall interactions

MCCULLOUGH JR., E. A.; Department of Chemistry and Biochemistry; Utah State University; Logan UT 84322; USA; 801-750-1630; E/T; 1.1; 2.3; 5.6; Molecules--numerical electronic structure calculations; Chemistry of small atmospheric molecules; Chlorofluorocarbon-ozone depletion problem; Catalysis

MCCURDY, C. W.; Department of Chemistry; Ohio State University; 140 West 18th Avenue; Columbus OH 43210; USA; 614-422-2278; 614-262-2324; T; 2.1; 2.2; 3.1; 2.3; Electron-atom and electron-molecules scattering and collisions

MCDANIEL, Earl W.; Department of Physics; Georgia Institute of Technology; Atlanta GA 30332; USA; 404-894-5214; E/T; Ionic transport and reactions in gases; Extraction of ion-neutral interaction potentials from transport data

MCDANIEL, Floyd D.; Department of Physics; North Texas State University; 5368; Denton TX 76203; USA; 817-565-3251; 817-565-2004; E; 2.1; 4.2; 5.2; Accelerator--based ion-atom collisions; Vacancy production by direct ionization and electron capture mechanisms

MCDERMOTT, Mark N.; Department of Physics FM-15; University of Washington; Seattle WA 98195; USA; 206-543-2770; 206-523-8808; E; 1.1; 3.3; 3.9; 4.1; Optical pumping

MCDERMOTT, William E.; ARAP; U.S. Air Force Weapons Laboratory; Kirkland AFB NM 87117; USA; 505-844-0196; E/T; Energy transfer processes involving electronically excited species

MCDIARMID, Ruth S.; Laboratory of Chemical Physics; National Institutes of Health; 2/BI-07; Bethesda MD 20205; USA; 301-496-6475; 301-496-1024; E; 3.2; Spectra of small organic molecules by vacuum UV--investigation; UV-visible, multiphoton ionization, and electron spectroscopy--investigation

MCDONALD, J. D.; Department of Chemistry; University of Illinois; 124L Noyes Laboratory; Urbana IL 61820; USA; 217-333-9700; E; 2.1; 3.2; Molecular beam scattering and spectroscopy

MCDOWELL, Charles A.; Department of Chemistry; University of British Columbia; 2036 Main Hall; Vancouver BC V6T1Y6; Canada

MCDOWELL, Harding K.; Department of Chemistry and Geology; Clemson University; Clemson SC 29631; USA; 803-656-3089; T; 1.1; 2.3; Atomic properties computation

MCDOWELL, Robin S.; CHM-4; Los Alamos National Laboratory; Box 1663 MSJ567; Los Alamos NM 87545; USA; 505-667-7071; E/T; 3.3; 3.5; 3.6; 3.8; High-resolution IR and Raman spectroscopy

MCFARLANE, Ross A.; Optical Physics Division; Hughes Research Laboratories; 3011 Malibu Canyon Road; Malibu CA 90265; USA; 213-456-6411; E; 4.1; Nonlinear optics in atomic and molecular systems

MCGEE, James F.; Department of Physics; St. Louis University; 221 N. Grand Boulevard; St. Louis MO 63103; USA; 314-658-2521; E/T; 5.4; X-ray focusing devices--physical and geometrical optics; X-ray microscopy; Laser fusion diagnostics

MCGEE, Thomas J.; Atmospheric Experiments Branch; NASA Goddard Space Flight Center; Code 615; Greenbelt MD 20742; USA; 301-344-5645; E; 3.6; 3.7; Laser spectroscopy of small molecules and radicals; Multiphoton effects in atoms

MCGLYNN, Sean P.; Department of Chemistry; Basic Sciences; Molecular Structure and Spectroscopy; Louisiana State University; Baton Rouge LA 70803; USA; 504-388-2945; 504-388-5833; E/T; 1.1; 1.2; 3.2; 3.8; 3.9; Molecular electronic structure and spectroscopy; Rydberg states in atoms and molecules; Field-effect spectroscopy; Optical techniques, lasers, optoacoustics, optogalvanics

MCGOWAN, J. W.; Department of Physics; University of Western Ontario; London Ontario N6A3U; Canada; 519-679-6332/2812; E; 2.2; 4.2; 5.1; Electron-ion recombination; X-ray absorption in hydrogenized cells; Ion-surface collisions

MCGREGOR, Douglas D.; 3108 Winchester Drive; Plano TX 75075; USA

MCGREGOR JR., Wheeler K.; Aeropropulsion Programs; Technology; Propulsion Diagnostics; Sverdrup Technology Inc.; MS 930; Arnold AFB TN 37389; USA; 615-454-7420/7663; E; 1.1; 2.3; 3.1; 5.5; Combustion diagnostic--temperature, species conclusion; Radiative properties of gases--UV to IR; Radiative properties of particulates--UV to IR; Raket plume phenomenology

MCGUIRE, Eugene J.; Sandia National Laboratories; Kirkland AFB, East; Albuquerque NM 87185; USA; 505-264-1178; T; 3.2; 3.7; Multiphoton ionization; Ionization cross sections and stopping powers for low Z ions; High Z ions--spectroscopy

MCGUIRE, James H.; Department of Physics; Kansas State University; Manhattan KS 66506; USA; 913-532-6786; 913-532-5629; T; 2.1; 2.2; 1.1; 4.2; 3.1; Theoretical scattering; Scattering theory; Single and multiple capture, ionization, excitation; Few body collisions

MCGUIRE, Michael D.; Physical Standards Research Lab.; Hewlett-Packard Laboratories; 3500 Deer Creek Road; Palo Alto CA 94304; USA; 415-857-5491; E; 1.1; Time and frequency standard based on hyperfine structure of positive mercury ions using ion storage techniques

MCGUIRE, Stephen C.; Physics; Alabama A&M University; P.O. Box 523; Huntsville AL 35762; USA; 205-859-7423; 205-852-4454; E; 5.4; 3.6; 2.2; Laser induced fluorescence diagnostics of weak plasmas; Electron collisional excitation cross section measurements

MCGUIRK, Michael; Optical Coating Department; Electro-Optical Division; Perkin-Elmer Corporation; M.S. 420; 761 Main Avenue; Norwalk CT 06859; USA; 203-834-4921; 203-834-1881; E/T; 2.3; 3.7; 5.2; Sensors for obtaining input data to a complete global model of the atmosphere, based on spectroscopic and physical optics principles--design; Chemical physics of film growth, directed to enhancing intense field performance

MCILRATH, Thomas J.; Institute for Physical Science and Technology; University of Maryland; College Park MD 20742; USA; 301-454-4843; E/T; 3.2; 3.4; 3.7; 3.10; 4.1; Spectroscopy of atoms and molecules in VUV; Multiphoton absorption in small molecules; Nonlinear optical mixing in gases and vapors; Lidar detection of small molecules

MCINTYRE, Adelbert; Optical Physics Division; U.S. Air Force Geophysics Lab.; Hanscom AFB MA 01731; USA; 617-861-3687; E; 2.1; 5.6; In situ measurements of H_2O (V3)-O and CO_2 (V3)-O collisional excitation

MCINTYRE JR., L. C.; Department of Physics; University of Arizona; Tucson AZ 85721; USA; 602-626-4275; E; 3.1; 4.2; 5.1; Time-of-flight molecular dissociation spectroscopy; Beam-foil spectroscopy

MCKAY, Kenneth G.; 200 E. 66th Street; New York NY 10021; USA

MCKENZIE, Robert L.; Photophysics Group; Experimental Fluid Dynamics Branch; Ames Research Center; NASA; Mail Stop 229-1; Moffett Field CA 94035; USA; 415-694-6158; E/T; 3.6; 3.10; Fluid mechanics

MCKIBBEN, Joseph L.; White Rock; 113 Aztec; Los Alamos NM 87544; USA

MCKINNEY, James T.; 3M Center; Building 236-3C; St. Paul MN 55101; USA

MCKNIGHT, Ronald H.; National Bureau of Standards; Building 220, Room B344; Gaithersburg MD 20899; USA; 301-921-3121;

E; 5.3; Ions produced by high voltage direct current transmission lines--study

MCKNIGHT, William B.; Department of Physics; University of Alabama, Huntsville; PO Box 1247; Huntsville AL 35807; USA; 205-895-6244; E/T; New molecular lasers and soft X-ray lasers--research

MCKOY, Vincent; Department of Chemistry; Noyes Laboratory of Chemical Physics; Pasadena CA 91125; USA; 818-356-6545; 818-356-6516; T; 3.1; 2.2; 3.7; Molecular photoionization processes; Electron-molecule collisions; Molecular multiphoton ionization

MCLAUGHLIN, Daniel J.; Department of Physics; University of Hartford; 200 Bloomfield Avenue; West Hartford CT 06117; USA; 203-243-4518; 203-486-3857; T; 1.2; 2.1; 2.2; 3.4; 3.9; Atomic and molecular processes in high temperature plasmas

MCLAUGHLIN, Ralph; Lawrence Berkeley Laboratory; University of California; Berkeley CA 94720; USA; 415-486-4641; E; 4.2; Atomic and molecular detection using Zeeman tuning of atomic emission lines--development of instrumentation

MCLEAN, Edgar A.; Plasma Physics Division; U.S. Naval Research Laboratory; Code 4732; 4555 Overlook Avenue, SW; Washington DC 20375; 202-767-2728; 202-767-2730; E; 5.4; 5.1; Laser-produced plasmas, including spectroscopic measurement of Te, Ne, Doppler velocity of ions and atoms--diagnostics

MCLENNAN, James A.; Department of Physics; Lehigh University; Bethlehem PA 18015; USA; 215-861-3917; T; 2.3; Non-equilibrium statistical mechanics; Kinetic theory for atoms and molecules

MCMAHON, Thomas R.; Fusion Division; General Atomic Co.; PO Box 81608; San Diego CA 92138; USA; 714-455-4375 /2270; E; 2.1; 4.2; Doppler spectroscopy of neutral hydrogen beams to determine beam species, energy distribution and divergence

MCMANAMON, Paul F.; 4161 Spruce Pine Court; Dayton OH 45424; USA

MCMENAMIN, Joseph C.; J.C. Schumacher Company; 580 Airport Road; Oceanside CA 92054; USA

MCMILLIAN, Gary B.; Physics; Rice University; 1892; 6100 South Main Street; Houston TX 77005; USA; 713-437-5945; E; 1.2; 2.1; 3.6; 3.9

MCNALL, John W.; 22 Salem Dr. Rd. #4; Greensburg PA 15601; USA

MCNALLY JR., J. Rand; Plasma Theory Section; Oak Ridge National Laboratory; PO Box Y; Building 9201-2, Room 301; Oak Ridge TN 37830; USA; 615-574-1351; T; Plasma theory

MCNEAL, Robert J.; Code EE-8; NASA Headquarters; Washington DC 20546; USA

MCNEIL, Laurie E.; Physics and Astronomy; University of North Carolina; Phillips Hall 039A; Chapel Hill NC 27514; USA; 919-962-1185; E; 5.2; Raman spectroscopy of molecular solids

MCPHERSON JR., Leroy A.; Department of Physics; Laser; University of Illinois, Chicago; 4348; Chicago IL 60680; USA; 312-996-5648; 312-996-5445; E; 4.1; 3.6; 3.4; 3.7; 3.2

MCRAE, E. G.; Surface Physics Research Department; Bell Telephone Laboratories; 600 Mountain Avenue; Murray Hill NJ 07974; USA; 201-582-4738; E/T; 5.1; Electron scattering at crystal surfaces

MCRAE, Thomas; Liquefied Gaseous Fuels Program; Lawrence Livermore National Lab.; Box 808 L-451; Livermore CA 94550; USA; 415-422-1576; E/T; 5.1; 5.5; IR absorption and Raman scattering diagnostic systems--design; Gas dynamic and combustion processes associated with large liquefied natural gas spills--modeling and analysis

MCTAGUE, John P.; Office of Science and Technology Policy; Executive Office Building; Room 5005; Washington DC 20506; USA; 202-456-7116; E; 3.4; 4.2; 5.8; Neutron, X-ray and light scattering spectroscopies

MCVEY, John B.; Rasor Associates, Inc.; 253 Humbolt Court; Sunnyvale CA 94089; USA; 408-734-1622; E/T; 5.1; 5.3; Thermonic energy conversion; Thermionic emissions devices; Plasma diodes

MCWILLIAMS, Bruce M.; Lawrence Livermore National Lab.; PO Box 808; L-372; Livermore CA 94550; USA

MEAD, C. Alden; Department of Chemistry; University of Minnesota; 207 Pleasant St, SE; Minneapolis MN 55455; USA; 612-373-7884; T; 1.1; Born-Oppenheimer approximation; Group theory to molecules--application

MEAD, Roy D.; Chemistry; Vacuum Ultraviolet; Mudd Building; Stanford University; Stanford CA 94305; USA; 415-497-0536; E; 3.1; 4.1; 3.10; 3.6; 1.1; Photodetachment from atomic and molecular negative ions with tumble lasers; Vacuum ultraviolet radiation generated by upconversion of ultraviolet laser

MEHLHORN, Herbert A.; Advanced Systems Center; Raytheon Company; Hartwell Road; Bedford MA 01730; USA; 617-274-7100x2960; E; 3.3; IR radiation--emission, absorption and detection/ measurement

MEISBURGER, William D.; 30 Beech Court; Fishkill NY 12524; USA

MEISELS, Gerhard G.; Department of Chemistry; University of Nebraska; Lincoln NE 68588; USA; 402-472-3501; E; 2.3; 3.1; 4.4; Photoionization mass spectrometry; Ionization processes and energetics; Collision dynamics of ion-molecules reactions; Ion structures and thermodynamics; Energy (state) selected ions and their reactions; Ion-molecule reactions--analytical applications; Ion drift properties

MEISENHEIMER, Robert G.; L-370; Lawrence Livermore National Lab.; University of California; PO Box 808; Livermore CA 94550; USA

MEITZLER, Charles R.; Serin Physics Lab; Frelinghuysen Road; Piscataway NJ 08854; USA

MELAMED, Nathan T.; Applied Physics Division; Westinghouse R&D Center; 1310 Beulah Road; Pittsburgh PA 15235; USA; 412-256-3647; E/T; 3.8; Molecular beams; Optical spectroscopy as applied to laser chemistry

MELDNER, Heiner W.; L-23; Lawrence Livermore National Lab.; PO Box 808; Livermore CA 94550; USA

MELIUS, Carl F.; Division 8341; Sandia National Laboratories; Livermore CA 94550; USA; 415-422-2651; T; Ion-atom collisions

MELLEN, Walter R.; Department of Physics and Applied Physics; University of Lowell; Lowell MA 01854; USA; 617-452-5000; 617-256-0170; T; 1.4; 3.1; De Broglie wave approach to Compton effect; Photon interactions with atoms in conservation of momentum system using an all wave approach

MELTON, Lynn A.; Department of Chemistry; University of Texas, Dallas; PO Box 688; Richardson TX 75080; USA; 214-690-2901; E; 5.5; 3.8; Fluorescence diagnostics for combustion

MELVEGER, Alvin J.; Analytical Chemistry; Research and Development; Ethicon, Inc.; Route 22; Somerville NJ 08876; USA; 201-524-3458; E; 1.1; 1.3; Structure/property relationships of polymers by IR, NMR spectroscopy and other techniques--determination; Analytical spectroscopy

MENASIAN, Stephen C.; PO Box 67; 22 S. Main Street; Allentown NJ 08501; USA

MENDELSOHN, Lawrence B.; Department of Physics; Brooklyn College of CUNY; Bedford Avenue and Avenue H; Brooklyn NY 11210; USA; 212-780-5805; T; 3.1; 5.1; X-ray and electron scattering; Compton effect in atoms, molecules and solids

MENDLOWITZ, Harold; Department of Physics and Astronomy; Howard University; Washington DC 20059; USA; 202-636-6252; T; 3.1; Charged particles with periodically varying fields; and radiation emitted --interaction

MENEGOZZI, Lionel H.; Research and Development Division; Research-Cottrell; PO Box 1500; Somerville NJ 08876; USA; 201-685-4903; E/T; 2.2; 4.2; 5.3; Gaseous electronics; Gaseous pollutants-electron beam treatment

MENENDEZ, Manuel G.; Department of Physics and Astronomy; University of Georgia; Athens GA 30602; USA; 404-542-2485; E; 2.1; 2.2; 4.2; Ion-atom scattering; Electron-atom scattering

MENES, Meir; Department of Physics; Polytechnic Institute; 333 Jay Street; Brooklyn NY 11201; USA

MENG, Hsien-Chun; Institute Fur Experimentalphysik V; Ruhr Univ. Bochum; Gebaude NB 05; Bochum 2 D-4360; West Germany

MENNE, Thomas J.; McDonnell Douglas Research Labs.; PO Box 516; St. Louis MO 63166; USA; 314-232-4687; E; 4.2; 5.4; 5.6; Chemical lasers; Particle beams; Magnetic confinement fusion; Atmospheric chemistry

MENOCAL, Serafin G.; Device Science and Technology Research; Solid State Science and Tech. Research; Bell Communications Research; Room 3D 201; 600 Mountain Avenue; Murray Hill NJ 07974; USA; 201-582-3574; E; 4.1; 3.6; 1.1; Semiconductor laser research; Single frequency sources; Laser measurements and characterization

MENSE, Allan T.; Department E-240; McDonnell Douglas Astronomy, Co.; PO Box 516; Building 81-Level 2-Room 216; St. Louis MO 63166; USA

MENYUK, Norman; Quantum Electronic Group; Solid State Division; Lincoln Laboratory; MIT; 244 Wood Street; Lexington MA 02173; USA; 617-863-5500/4741; E; 4.1; 3.6; 3.10; 5.6; Laser development; Remote sensing of molecules and impurities in atmosphere; Optical frequency conversion (SHG, THG summing)

MENZIES, Robert T.; Earth and Space Sciences Division; Jet Propulsion Laboratory; California Institute of Technology; 4800 Oak Grove Drive; Pasadena CA 91103; USA; 213-354-3787; E; 3.2; 5.6; Molecular and free radical trace species in Earth's atmosphere--spectroscopy and atmospheric measurements

MERIWETHER, J. R.; Department of Physics; University of Southwestern Louisiana; PO Box 44210; Lafayette LA 70504; USA; 318-264-6691; E; 3.4; 4.2; Applied X-ray spectroscopy using particle induced X-ray emission

MERKELO, Henri; Electrical and Computer; Physical Electronics; Quantum Electronics; University of Illinois; 1406 W. Green St.; Urbana IL 61801; USA; 217-333-2482; E/T; 2.3; 6,8; 1; Fluorescence, luminescence efficiency; Energy transfers, mechanism; Picosecond resolution of rapid fluorescence and transfers; Photochemical reactions at surfaces

MERRIFIELD, Richard E.; Central Research and Development; E. I. DuPont de Nemours and Co., Inc.; DuPont Experimental Station; Wilmington DE 19898; USA; 302-772-3603; T; 1.1; Molecular structure

MERRITT, Sally J.; 3206 South Fielder #101; Arlington TX 76015; USA

MERTS, A. L.; Theoretical Division, Group T-4; Los Alamos National Laboratory; PO Box 1663; Los Alamos NM 87545; USA; 505-667-5751; T; 1.2; 2.2; 3.1; 5.3; 5.4; Line broadening brought about through the many body interactions with dense plasmas (cluster effects); Electron and photon excitation cross sections with applications to radiation and electron transport; Astrophysics

MERZBACHER, Eugen; Department of Physics and Astronomy; Triangle Univ. Nuclear Laboratory; Univ. of North Carolina, Chapel Hill; 230 Phillips Hall; Chapel Hill NC 27514; USA; 919-962-3021; 919-942-5429; T; 2.1; 5.8; Ion-atom collisions at moderate and high velocities; Atomic collision theory, general atomic collisions and nuclear reactions

MESSMER, Richard P.; Corporate Research and Development; General Electric Co.; Schenectady NY 12301; USA; 518-375-8488; T; 1.1; Molecules and solids--electronic structure theory

METCALF, Harold J.; Department of Physics; State Univ. of NY, Stony Brook; Stony Brook NY 11790; USA; 516-246-6585; 516-246-6580; E; 3.6; 3.9; Atomic spectroscopy using pulsed lasers; Cooling and trapping neutral atoms

METIU, Horia I.; Department of Chemistry; Univ. of California, Santa Barbara; Santa Barbara CA 93106; USA; 805-961-2256; T; 2.3; Spectroscopy; Symmetry rules in chemical reactions

MEYER, Carl B.; Department of Chemistry BG-10; University of Washington; Seattle WA 98195; USA; 206-543-1647; E; 1.1; 5.6; Laser Raman spectrometry of sulfur molecule; Spectrometry of air impurities

MEYER, Fred W.; Physics Division; Oak Ridge National Laboratory; PO Box X; Oak Ridge TN 37831; USA; 615-574-4705; E; 1.2; 2.1; 4.2; 4.3; 5.4; Inelastic collisions involving multi-charged ions, Rydberg atoms, or hydrogen atoms- multi-charged ion source development crossed-beams, techniques for studying ion-atom and ion-electron collisions

MEYERAND JR., R. G.; United Technologies Research Center; 400 Main Street; East Hartford CT 06108; USA; 203-727-7300; E; Director of research

MEYERHOF, Walter E.; Department of Physics; Stanford University; Stanford CA 94305; USA; 415-497-4640; E; 2.1; 4.2; 5.8; Ion-atom collisions at energies between 1-1000 MeV and Z ranges from 1 to 92; Inner-shell vacancy production; Radiative processes; Nuclear time delays--effects

MIAN, Shaikh N.; Prastoli Hinoo; Ranchi Bihar 834002; India

MIAO, Cheng-Hsi; Department of Physics and Astronomy; University of New Mexico; Albuquerque NM 87131; USA

MICHA, David A.; Chemistry and Physics; University of Florida; 366 Williamson Hall; Gainesville FL 32611; USA; 904-392-1597; 904-392-7545; T; 1.1; 2.1; 2.3; Theoretical chemical physics and molecular collisions; Intermolecular forces and scattering theory; Molecular electronic structure; Computational methods

MICHAEL, Irving; Plasma, Particles and Fields Branch; Space Physics Division; US Air Force Geophysical Laboratory; Hanscom AFB MA 01731; USA; 617-861-2431; E/T; 4.4; Mass spectrometry; Plasmas

MICHAELS, Gordon E.; Advanced Isotope Separation Div.; Oak Ridge Gaseous Diffusion Plant; PO Box P; Oak Ridge TN 37830; USA; 615-574-8154; T; 2.1; 5.4; Ion sputtering; Uranium collisional cross sections

MICHELS, H. Harvey; Physics Division; United Technologies Research Center; 400 Main Street; East Hartford CT 06108; USA; 203-727-7489; T; 1.1; 2.1; 2.2; 2.3; Atoms and molecules--quantum mechanical studies of electronic structure; Collision cross sections and reaction rates

MICKISH, Roger A.; 564-207; Rocketdyne; Facilities and Industrial Engineering; Rockwell International; 6633 Canoga Avenue; Canoga Park CA 91303; USA; 818-700-4985; E; 5.3; 4.1; 3.1; 2.1; 5.5; High energy laser engineering development; Gas laser physics; Free electron laser; Particle beam

MIELCZAREK, Stanley R.; Radiation Physics Division; National Bureau of Standards; 7100 Beechwood Drive; Chevy Chase MD 20815; USA; E; 1; 2; 1; 1; 1

MIERS, Richard E.; Department of Physics; Indiana/Purdue University; 2101 Coliseum Boulevard East; Fort Wayne IN 46805; USA; 219-482-5593; 219-485-2568; E; 2.1; 3.6; 4.2; 4.3

MIES, Frederick H.; Molecular Spectroscopy Division; National Bureau of Standards; Gaithersburg MD 20899; USA; 301-921-2733; T; 2.1; 3.1; 3.6; 3.7; Atomic scattering theory; Molecular spectroscopy; Line shapes and collisions in intense laser fields

MILCHBERG, Howard M.; Research Division; AT&T Bell Laboratories; PO Box 451; 600 Mountain Avenue; Murray Hill NJ 07974; USA; 201-582-6344; E; 5.4; 5.1; 3.7; 5.3; 3.4

MILDE, Helmut I.; Ion Physics Division; High Voltage Engineering Co.; S. Bedford Street; Burlington MA 01803; USA; 617-272-2800; E; 2.2; 5.3; Electron beams used for crosslinking of cable; Ozone production; Pulsed corona discharge

MILDER, Frederic L.; 26 High Rock Way; Allston MA 02134; USA

MILES, R. O.; U.S. Naval Research Laboratory; Code 6570; 4555 Overlook Avenue, SW; Washington DC 20375; USA; 202-767-3298; E/T; 4.1; Solid state, semiconductor lasers and fibers

MILES, Richard B.; Department and Aerospace Engineering; School of Engineering and Applied Science; Princeton University; Room D-414; Olden St., Engineering Quadrangle; Princeton NJ 08544; USA; 609-452-5131; E; 1.1; 2.1; 3.6; 4.1; 5.3; Nonlinear Optics in gases and on surfaces; Laser diagnostics and molecular energy transfer

MILEY, George H.; Department of Nuclear Engineering; University of Illinois; 214 Nuclear Engineering Lab.; Urbana IL 61801; USA; 217-333-2294; E/T; 4.1; 5.4; Atomic physics related to gaseous lasers; Fusion plasmas

MILKMAN, Ingrid W.; Department of Physics; University of Oregon; Eugene OR 97403; USA

MILLER, Charles G.; Jet Propulsion Laboratory; California Institute of Technology; 4800 Oak Grove Drive; Pasadena CA 91103; USA; 213-577-9454; E; Surface migration at cryogenic temperatures

MILLER, David R.; Dept. of Applied Mechanics and Engineering Science; University of California, San Diego; La Jolla CA 92093; USA; 619-452-3182; E; 5.1; 2.3; 4.3; 1.3; Atomic and molecular beam scattering from solid surfaces; Gas phase kinetics and relaxation processes in free jet expansions

MILLER, Donald J.; Power Systems Technology Laboratory; TRW, Inc.; 1 Space Park; Redondo Beach CA 90278; USA; E; 2.3; Vibrational and electronic excitation of molecules by chemical reaction; Atomic and molecular energy transfer

MILLER, Donald L.; Research and Development Center; Westinghouse Electric Corporation; 1310 Beulah Road; Pittsburgh PA 15235; USA

MILLER, Glenn H.; Field Sciences Department; Test Planning Division 7112; Sandia National Laboratories; PO Box 5800; Albuquerque NM 87185; USA; 505-844-4376; E; 2.1; 2.3; 4.2; Atomic and molecular collisions using 10-100 KeV ion accelerator; Charge transfer processes in ion-molecule collisions; Accelerator-based studies of atomic and molecular collisions

MILLER, Hillard C.; Neutron Devices; General Electric Co.; 2908; Largo FL 34294; USA; 813-541-8639; E/T; 5.4; 5.3; Vacuum arcs; Electrical breakdown in vacuum

MILLER, James A.; Combustion Physics Division; Sandia National Laboratories; Livermore CA 94550; USA; 415-422-2759; T; 2.1; 2.3; Reaction rate theory; Molecular energy transfer; Molecular spectroscopy

MILLER, John C.; Chemical Physics Section; Health and Safety Research Division; Oak Ridge National Lab.; PO Box X; Building 4500S-Room S116; Oak Ridge TN 37831; USA; 615-574-6239; E; 3.6; 3.7; 3.10; 3.8; 1.2

MILLER, John H.; Radiological Sciences Division; Pacific Northwest Laboratories; Batelle Memorial Institute; PO Box 999; Richland WA 99352; USA; 509-376-5803; T; 3.1; 4.2; Energy transfer between ionizing radiation and molecules--basic mechanisms

MILLER, Kenneth J.; Department of Chemistry; Science; Rensselaer Polytechnic Institute; Troy NY 12181; USA; 518-266-8448; T; 1.1; 1.2; 1.3; Molecules with nucleic acids--interactions; Nucleic acid structures--conformational analysis; Carcinogens and DNA-quantum; Chemistry and chemical reactivity

MILLER, Phillip D.; EN-Tandem; Physics Division; Oak Ridge National Laboratory; PO Box X; Building 5500; Oak Ridge TN 37831; USA; 615-574-4781; FTS-624-4781; E; 4.2; 2.1; 5.2; 3.4; Multiply charged ions, charge transfer; Inner shell excitation of/by multiply charged ions; Channeling

MILLER, Roger E.; Chemistry; University of North Carolina; Chapel Hill NC 27514; USA; 919-966-5433; E; 1.3; 2.1; 3.1; 3.3; 3.5; Spectroscopy of Van der Waals dimers; Collisional energy transfer rates

MILLER, Steven M.; Infrared Dynamics Branch; Infrared Technology Division; Air Force Geophysics Laboratory; U.S. Government/Air Force; AFGL/LSI; 28 Garland Road; Hanscom AFB MA 01731; USA; 617-861-2810; E; 1.1; 2.1; 3.6; 4.1; 5.6; Chemical kinetics, dynamics and infrared fluorescent properties of atmospheric atoms and molecules

MILLER, Terry A.; Physical Chemistry Research Div.; Bell Telephone Laboratories; 600 Mountain Avenue; Murray Hill NJ 07974; USA; 201-582-4764; E/T; 2.3; 3.6; Anticrossing and laser magnetic resonance spectroscopy of simple atoms and molecules; Laser induced fluorescence studies of the spectroscopy and dynamics of molecular ions

MILLER, Thomas G.; DRSMI-RHST; U.S. Army Missile Command; Redstone Arsenal AL 35809; USA; 205-876-8272; E/T; 4.1; Laser research

MILLER, Thomas M.; Department of Physics and Astronomy; University of Oklahoma; 440 W. Brooks Street; Norman OK 73019; USA; 405-325-3961; E; 2.2; 3.1; 2.3; Electron scattering at low energies; Ions at thermal energies--interactions

MILLER, Walter B.; Department of Chemistry; University of Arizona; Tucson AZ 85721; USA; 602-626-2115; E/T; 2.1; Elastic, inelastic, and reactive collisions of neutral atoms and molecules --crossed molecular beam studies

MILLER, William H.; Department of Chemistry; University of California; Berkeley CA 94720; USA; 415-642-0653; T; 2.1; 2.3; Quantum and classical mechanics in molecular collision dynamics-role; Reactive and non-reactive molecular collisions

MILLER JR., William R.; Department of Physics; Pennsylvania State University; Capitol Campus; Middletown PA 17057; USA; 717-948-6098; E; 2.2; Low energy positrons

MILLETTE, Pierre A.; 6408 St. Louis Drive; Orleans Ontario K1C2Y2; Canada

MILLMAN, Sidney; 17 Fairview Avenue; Summit NJ 07901; USA

MILLS, Alfred P.; Department of Chemistry; University of Miami; Coral Gables FL 33124; USA; 305-284-2194; E/T; Liquid mixtures of nonelectrolytes--predictive systems for bulk physical properties

MILLS JR., Allen P.; Division 1.115; Bell Telephone Laboratories; 600 Mountain Avenue; Murray Hill NJ 07974; USA; 201-582-4162; E; 1.1; 1.3; 1.4; Positronium

MILLS, James W.; Department of Chemistry; Fort Lewis College; Durango CO 81301; USA; 303-247-0538; 303-247-7272; E/T; Molecular fluorescence; Laser excited coherence effects

MILONNI, Peter W.; Laser Optics Division; Perkin-Elmer Corporation; 50 Danbury Court; Wilton CT 06897; USA; 203-762-4763; T; Quantum electrodynamics and alternative theories; Laser theory

MIN, Hang-Gi; Department of Physics; West Virginia University; PO Box 6023; Morgantown WV 26506; USA

MIN, Kwang S.; Department of Physics; East Texas State University; Commerce TX 75428; USA; 214-886-5483/5488; T; 2.1; Charge exchange and atomic collisions

MINKOWSKI, Jan M.; Department of Electrical Engineering; Johns Hopkins University; 34th and Charles Streets; Baltimore MD 21210; USA; 301-338-7015; T; Quantum theory and information theory

MIRES, Raymond W.; Department of Physics; Texas Tech University; Lubbock TX 79409; USA; 806-742-3779; E/T; 1.1; 5.2; Atoms and ions in crystal fields--electronic structure; Electron correlation in atoms and ions

MISAKIAN, Martin; Applied Electrical Measurements; Electrosystems; National Bureau of Standards; Building 220-Room B344; Gaithersburg MD 20899; USA; 301-921-3121; E; 5.3; 5.6

MISEMER, David K.; Department of Physics; Iowa State University; Ames IA 50011; USA

MISHKIN, Eli A.; Professor of Applied Physics; Polytechnic Institute of New York; 333 Jay Street; Brooklyn NY 11201; USA; 718-643-4570; 718-858-5339; T; Thermonuclear fusion; Shock implosion analysis

MISKIMEN, Rory A.; Nuclear Physics Lab.; University of Illinois; 23 Stadium Drive; Champaign IL 61820; USA

MITCHELL, Robert R.; Systems Sciences Division; Westinghouse R&D Center; 1310 Beulah Road; Pittsburgh PA 15235; USA; 412-256-5187; T; 2.2; 5.3; Electron-heavy particle collisions as applied to electrical discharges in gases and gas laser kinetics --Boltzmann analysis

MITCHNER, M.; Department of Mechanical Engineering; Stanford University; Stanford CA 94305; USA; 415-497-1745; E/T; 1.1; 2.1; 2.2; 2.3; 3.1

MITROY, Jim; Institute for Atomic Studies; Flinders University; Bedford Park; SA 5042; Australia

MITTLEMAN, Marvin H.; Department of Physics; City College of New York; New York NY 10031; USA; 212-690-6841; T; 3.6; 3.8; Laser-atomic and molecular interactions

MIYAGAWA, Ichiro; PO Box 1921; University AL 35486; USA

MIZUSHIMA, Masataka; Department of Physics; University of Colorado; Boulder CO 80309; USA; 303-492-8707; T; 1.1; 3.8; Rotational energy levels of simple molecules; Spectral line shape (interaction between radiation field and molecules); Rotational relaxation of molecules by collisions

MJOLSNESS, Raymond C.; LASL-T3; PO Box 1663; MS 216; Los Alamos NM 87545; USA

MO, Charles T. C.; Nuclear Effects; Research and Development Associates; PO Box 9695; 4640 Admiralty Way; Marina del Rey CA 90291; USA; 213-822-1715; T; Classical electrodynamics and wave propagation, gas ionization/discharge, mathematical probability and statistics

MOAFI, Kambiz; West Roshanai; Shariatist Doulatst Baharave; Baharak Building No. 7; Tehran; Iran

MOAK, C. D.; Physics Division; Oak Ridge National Laboratory; PO Box X; Oak Ridge TN 37830; USA; 615-574-4790; E; 4.2; 5.2; Atomic collisions with accelerators; Stopping powers; Charge states of heavy ions; Crystal channeling

MOENY, William M.; Research and Development Division; TETRA Corporation; PO Box 4369; 4905 Hawkins NE; Albuquerque NM 87109; USA; 505-345-8623; E/T; 2.2; 4.1; 5.3; 3.1; High energy lasers, including electric discharge physics, high pressure glow discharges and fluid dynamics

MOHLER, Orren C.; Department of Astronomy; University of Michigan; 833 Dennison; Ann Arbor MI 48109; USA; 313-764-3454; E/T; 5.7; Atomic and molecular data to solar spectroscopy-- application

MOHNEN, Volker A.; Atmospheric Sciences Research Center; State University of New York, Albany; 1400 Washington Avenue; Albany NY 12222; USA; 518-442-3819; E; 2.3; 4.4; 5.6; Ion-molecule reactions of atmospheric importance

MOHR, Peter J.; Department of Physics; National Science Foundation; Washington DC 20550; USA; 202-357-7998; T; 1.4; 1.1; 1.3; 3.1; Quantum electrodynamics in simple atomic systems

MOLINA, Mario J.; Earth and Space Sciences; Jet Propulsion Laboratory; 4800 Oak Grove Dr.; Pasadena CA 91109; USA; 818-354-5752; E; 5.6; 3.2; 2.3; 3.1; Atmospheric chemistry

MOLITORIS, John D.; Department of Physics; Stanford University; Stanford CA 94305; USA; 415-497-3783; 415-497-4640; E; 2.1; 3.4; 4.2; 5.8; Relativistic ion-atom collisions; Molecular orbital X-ray studies; Interface of atomic and nuclear physics

MOLLOW, Benjamin R.; Department of Physics; University of Massachusetts, Boston; Boston MA 02125; USA; 617-287-1900x2368; E/T; 4.1; Quantum optics

MOLMUD, Paul; Advanced Technology Laboratory; Group R1/1086; TRW, Inc.; 1 Space Park; Redondo Beach CA 90278; USA; 213-535-2500; T; Brownian particles --resonant ion heating; Electromagnetic waves by inhomogeneous plasmas--scattering

MONAHAN, Kevin M.; Applied Physics Laboratory; Lockheed Palo Alto Research Lab.; 3251 Hanover Street; Palo Alto CA 94304; USA; 415-493-4411x5153; E; 4.2; 5.1; 5.4; Photon stimulated desorption; Matrix isolation; Soft X-ray and plasma physics; Spectroscopy

MONCE, Michael N.; Department of Physics and Astronomy; Connecticut College; New London CT 06320; USA; 203-447-1911/7346; E; 4.2; 2.1; 1.1; 5.6; 5.7; Collisions between ions and polyatomic molecules; Photon emission from collisional excitations; Spectra and photon emission cross sections

MONK, James L.; 4322 St. Patrick Drive; Oklahoma City OK 73120; USA

MONKHORST, Hendrik J.; Department of Physics WM361; University of Florida; Gainesville FL 32611; USA; 904-392-1597; 904-377-6277; T; 1.1; 1.3; 2.2; 5.2; 5.8; High-accuracy quantum chemistry-basis sets, beyond Born-Oppenheimer; Coupled-cluster method to electron correlation in molecules and solids--study and application; Muon-catalyzed fusion --atomic, molecular and nuclear aspects

MONTGOMERY, D. L.; 315 N. Associated Road, No. 2004; Brea CA 92621; USA

MONTGOMERY, Donald J.; Department of Metallurgy, Mechanics and Materials Science; Michigan State University; East Lansing MI 48824; USA; 517-355-5157/5141; T; 2.3; 5.2; Chemical physics; Stable isotopes used as a probe for solid state phenomena; Static electrification

MOODY, Elizabeth A.; Image /Signal Processing; Systems

Development; Electro-Optics Laboratory; Raytheon Company; PO Box 546; Beverly Farms MA 01915; USA; 617-927-6050; 617-274-7100x4244; E/T; 5.7; 5.6; 4.1

MOODY, Mitchell L.; Engineering Division; Marconi Avionics, Inc.; 4500 N. Shallowford Road; Atlanta GA 30338; USA; 404-394-7300; E; 3.6; Carbon dioxide TEA laser physics

MOODY, Stephen E.; Mathematical Science Northwest, Inc.; 2755 Northup Way; Bellevue WA 98004; USA; 206-827-0460; E; 4.1; Molecular physics for excimer laser development

MOONEY, Tim M.; Department of Physics and Astronomy; Univ. of North Carolina, Chapel Hill; Chapel Hill NC 27514; USA

MOORADIAN, Aram; Quantum Electronics Group; Lincoln Laboratory; MIT; PO Box 73; Lexington MA 02173; USA; 617-862-5500; E; Solid state laser--high resolution laser spectroscopy

MOORE, C. Bradley; Department of Chemistry; University of California; Berkeley CA 94720; USA; 415-642-3453; E; 3.8; 2.3; 3.6; 3.2; 3.3; Laser-induced chemical reactions; Energy transfer

MOORE, C. Fred; Department of Physics; University of Texas, Austin; RLM 5.208; Austin TX 78705; USA; 512-471-1236; 505-667-2925; E; 1.1; 2.1; 3.1; 4.2; 5.8; Atomic interactions in highly ionized atoms

MOORE, Edwin N.; Department of Physics; University of Nevada; Reno NV 89557; USA; 702-784-6789; 702-747-3451; T; 5.7; Astrophysical applications

MOORE JR., Frank L.; Department of Physics; Clarkson College; Potsdam NY 13676; USA; 315-268-2361; T/E; 3.2; 5.2; Atomic spectroscopy; Ion scattering in solids

MOORE, John H.; Department of Chemistry; University of Maryland; College Park MD 20742; USA; 301-454-4664; E; 2.2; 4.3; Molecular properties by electron scattering; Instrument development for spacecraft

MOORE-HEAD, M. E.; Department of Physics; University of New Orleans; Lakefront; New Orleans LA 70122; USA; 504-283-0341; E/T; 2.1; Excited electronic states--low-energy accelerator based studies

MOORHEAD, William D.; 3803 University Boulevard; Houston TX 77005; USA

MOOS, H. Warren; Department of Physics; Johns Hopkins University; Charles and 33rd Streets; Baltimore MD 21218; USA; 301-338-7337; E; 5.4; 5.7; Spectroscopic emissions from plasmas; Spectroscopic emissions from planetary atmospheres

MORAN, Thomas F.; Department of Chemistry; Georgia Institute of Technology; Atlanta GA 30332; USA; 404-894-4022; E; 2.1; 2.3; 4.4; Ions with molecules--reactions; Mass spectroscopy

MORAWITZ, Hans; Organic Solid State Division; Research Laboratory; IBM; 5600 Cottle Road; San Jose CA 95193; USA; 408-256-2174; T; 4.2; 5.1; 5.2; Superradiance on surfaces and Deed core excitations in cage molecules; Raman spectroscopy; Molecule-surface interactions; Organic molecules in solids--energy transfer and dephasing

MORE, Kenneth R.; 9901 E. Bexhill Drive; Kensington MD 20895; USA

MORGAN, Gerard P.; Department of Physics; University of Wisconsin; 1150 University Avenue; Madison WI 53706; USA

MORGAN, Thomas J.; Department of Physics; Wesleyan University; Middletown CT 06457; USA; 203-347-9411x2869; 203-347-9911x2758; E; 1.2; 2.1; 3.4; 3.6; 4.2; Heavy particle charge exchange; Negative ions; Rydberg states --laser preparation; Rydberg atoms--collisional properties

MORGAN, William L.; Department of Applied Science; Lawrence Livermore National Lab.; University of CA, Davis-Livermore; PO Bx 808 L-11; Livermore CA 94505; USA; 415-422-6289; T; 2.1; 2.3; 5.2; 5.3; 5.4; Atomic collision calculations; Ionic recombination rates calculations; X-ray laser research kinetics analysis

MOROI, David S.; Department of Physics; Kent State University; Kent OH 44242; USA; 216-672-2596; T; 3.1; Photoionization; Laser radiation with atomic and molecular systems--interaction

MORRIS, James H.; Heliotonics, Inc.; 5452 Oberlin Drive; San Diego CA 92121; USA

MORRIS, Roberta J.; Room 1E308; Bell Lab.; 600 Mountain Avenue; Murray Hill NJ 07974; USA; 201-582-7005; E; 3.4

MORRISON, Michael A.; Department of Physics and Astronomy; University of Oklahoma; 440 W. Brooks Street; Norman OK 73019; USA; 405-325-3961; T; 1.1; 2.1; 2.2; Atomic and molecular structure; Electron collisions; Heavy particle collisions; Electron-molecule collisions at low energies

MORSE, Philip M.; 126 Wildwood Street; Winchester MA 01890; USA

MORSE, Stephen S.; Department of Biological Sciences (Microbio); Nelson Lab.; Rutgers University; Room B308; New Brunswick NJ 08903; USA

MORTENSEN, Earl M.; Department of Chemistry; Cleveland State University; Cleveland OH 44115; USA; 216-687-2461; T; 2.3; Quantum mechanics to reactive scattering--application

MORTON, George A.; 1122 Skycrest Drive Apt. #6; Walnut Creek CA 94595; USA

MORTON, Richard G.; Electro-Optics Division; Research and Technology Center; Exxon Nuclear Company; 2955 George Washington Way; Richland WA 99352; USA; E; 4.1; Laser development for laser isotope separation

MORTON III, John R.; L Division; Lawrence Livermore National Lab.; PO Box 808; Livermore CA 94550; USA; 415-422-1215; E; 3.4; Nuclear weapons testing, mostly X-ray physics--aspects relating

MOSCATELLI, Frank A.; Department of Physics; Laser Spectroscopy; Swarthmore College; Swarthmore PA 19081; USA; 215-447-7256; 215-447-7570; E; 3.6; 5.4; 5.8; 3.9; 1.1; Optogalvanic detection laser spectroscopy; Rydberg atoms in high B field

MOSCOWITZ, Albert J.; Department of Chemistry; University of Minnesota; Minneapolis MN 55455; USA; 612-373-2349; T/E; 1.1; Electronic structure and stereochemistry of molecules; Molecular vibrations

MOSELEY, John T.; Department of Physics; University of Oregon; Eugene OR 97403; USA; 503-686-4753; 503-686-3186; E; 2.3; 3.1; 3.2; Photon interactions with molecular ions; Ion-molecule reactions; Molecular ion spectroscopy

MOSKOWITZ, Paul A.; Applied Research Division; IBM Thomas J. Watson Research Center; PO Box 218; Yorktown Heights NY 10598; USA; 914-945-1586; E; 3.6; Laser applications

MOSKOWITZ, Philip E.; General Engineering; GTE Products Corporation; 100 Endicott Street; Danvers MA 01923; USA; 617-777-1900; E; 3.6; 5.3; 5.4

MOSSBERG, Thomas W.; Department of Physics; Jefferson Physical Laboratory; Harvard University; Cambridge MA 02138; USA; 617-495-3768; E; 2.1; 3.6; 3.7; 3.11; 4.1; Coherent transient studies of neutral atom collisions; Transient strong-field resonance fluorescence; Homogeneous life widths of impurity atoms in solids; Transient optical phenomena

MOTZ, Joseph W.; X-Ray Division; National Bureau of Standards; Gaithersburg MD 20899; USA

MOWAT, J. R.; Department of Physics; North Carolina State University; Raleigh

NC 27607; USA; 919-737-2512/3775; E; 1.1; 4.2; 2.1; 5.4; Accelerator based atomic physics; Inner-shells of highly ionized atoms--structure; Ion-atom collisions; Electron and x-ray spectroscopy

MOWER, Lyman; Department of Physics; University of New Hampshire; Demeritt Hall; Durham NH 03824; USA; 603-862-1962; T; 1.1; 1.4; Quantum theory and quantum electrodynamics with application to atomic and molecular structure--formal investigation

MOZUMDER, A.; Department of Chemistry; Radiation Laboratory; University of Notre Dame; Notre Dame IN 46556; USA; 219-239-5363; 219-272-0997; T; 2.3; 3.8; 2.2; Electron interactions in gaseous and condensed media; theoretical radiation chemistry; Slow electron transport in liquids

MRUZIK, Michael R.; Research and Development Laboratory; Fairchild Camera and Instrument Corp.; 4001 Miranda Avenue; Palo Alto CA 94304; USA; 415-493-3100x2692; T

MSEZANE, Alfred Z.; Department of Physics; Atlanta University; Atlanta GA 30314; USA

MUELLER, Charles R.; Department of Chemistry; Purdue University; West Lafayette IN 47907; USA; 317-494-8117; T; 2.1; Scattering and stationary states of simple systems--convergent methods; Quantum mechanics

MUELLER, George P.; Radiation Matter Interactions Branch-Code 6652; Condensed Matter and Radiation Science; Naval Research Lab.; Washington DC 20375; USA; 202-767-2972; T; 5.1; 5.2

MUENTER, John S.; Department of Chemistry; University of Rochester; Rochester NY 14627; USA; 716-275-4223; E; 3.6; Molecular beam spectroscopy and laser spectroscopy

MUKHERJEE, Debasish; 28950 Burleson Street; Agoura Hills CA 91301; USA

MULLINS, Oliver C.; Department of Chemistry; Searle Lab.; University of Chicago; Chicago IL 60637; USA

MUMMA, Michael J.; Planetary Systems Branch; Lab. for Extraterrestrial Physics; NASA Goddard Space Flight Center; Greenbelt MD 20771; USA; 301-344-6994; E; 5.6; 5.7; 4.1; 3.3; 3.1; Atmospheric and astrophysical spectroscopy; Electronic, vibrational, and rotational spectroscopy of simple molecules; Advanced instrument development for quantitative spectroscopy

MURAD, Edmond; Space Physics Division; US Air Force Geophysics Laboratory; Hanscom AFB MA 01731; USA; 617-861-3176; 617-861-3046; E; 1.1; 2.3; 4.4; 5.6; Dissociation energies and ionization potentials; Reactive collision cross sections between ions and neutrals; Mass spectrometry; Planetary atmosphere

MURDAY, James S.; Chemistry Division; U.S. Naval Research Laboratory; Code 6170; 4555 Overlook Avenue, SW; Washington DC 20375; USA; 202-767-3550; T/E; 1.1; 2.2; 5.1; Auger electron spectroscopy; Line-shape theory

MURNICK, Daniel E.; Radiation Physics Research Department; Radiation Physics Division; AT&T Bell Laboratories; 600 Mountain Avenue; Murray Hill NJ 07974; USA; 201-582-4825; E; 1.4; 4.2; 1.4; 3.6; QED, PNC--fundamental studies; Laser spectroscopy; Hyperfine interactions; Accelerator-based A&M physics

MURPHY, John C.; Applied Physics Laboratory; Johns Hopkins University; Johns Hopkins Road; Laurel MD 20707; USA; 301-953-6214; E/T; 3.8; 4.1; 5.1; 3.10; 3.3; Thermal/thermoelastic imaging and spectroscopy; Photoprocesses, photoelectrolysis processes using metal oxide metal organic; Materials characterization/nondestructive evaluation using optical/thermal/microwave property determination

MURPHY, Randall E.; US Air Force Geophysics Laboratory; Hanscom AFB MA 01731; USA; 617-861-3630; E; 2.2; 2.3; 3.3; Atmospheric gases--IR spectroscopy and chemical kinetics via electron impact/excitation

MURRAY, Frank; High Power Microwave Group; Pomona; General Dynamics Corporation; PO Box 2507; 1675 Mission Blvd.; Pomona CA 91769; USA; 714-620-7511/1335; 714-626-7749; E/T; 3.11; 3.3; 2.2; 5.4; 5.3; Charged particle-field interactions

MURRAY, John R.; Laser Division; Lawrence Livermore National Lab.; PO Box 5508 L-470; Livermore CA 94550; USA; 415-422-6152; 415-820-5807; E/T; 4.1; 3.10; Laser development

MURRAY, Paul T.; Research Institute; University of Dayton; 300 College Park Avenue; Dayton OH 45469; USA; 513-255-5125; 513-229-3016; E; 5.1; 4.4; 1.3

MURRAY, R. B.; Department of Physics; University of Delaware; Newark DE 19711; USA; 302-738-2147; E; 5.1; Radiation damage and effects in solids

MUSCHLITZ JR., Earle E.; Department of Chemistry; University of Florida; Gainesville FL 32611; USA; 904-392-2006; E; 2.1; 2.2; Electronic energy transfer in molecular collisions using crossed molecular beams; Inelastic collisions of electrons with molecules

MUSTAFA, Adnan M.; Department of Physics; Damascus University; PO Box 4757; Damascus; Syria

MUTCH, George W.; Department of Chemistry; Andrews University; Berrien Springs MI 49104; USA; 616-471-3248; E; 2.3; Unimolecular decomposition of highly energized molecules; Collisional energy transfer between energized molecules and selected "cold" molecules

MUTTER, Walter E.; Department 084; Building 300-48A; IBM Corporation; Route 52; Hopewell Junction NY 12533; USA

MYERS, Edmund G.; Physics and Astronomy; Nuclear Physics; Rutgers University; PO Box 849; Frelinghuysen Rd.; Piscataway NJ 08852; USA; 201-932-2402; E; 3.6; 4.2; 4.1; 1.4; 5.8

MYERS, Gary D.; High Energy Laser Technology Div.; Bell Aerospace-Textron; PO Box 1 MS B49; Buffalo NY 14240; USA; 716-297-1000 x683; E/T; 4.1; Laser development

MYERS, Stephen A.; Self-employed consultant; 25 Nimitz Place; Old Greenwich CT 06870; USA; 203-637-2010; 203-637-3892; E; 4.1; 3.2; 3.6; 3.9; 3.3; Instrumentation development, analytical spectroscopy

NACHMAN, Paul; Joint Institute for Laboratory Astrophysics; University of Colorado; Boulder CO 80309; USA; 303-492-7855; E; 1.4; Energy levels--precision measurements; Atomic systems used for tests of basic principles, e.g. special relativity, isotropy of space

NADELMAN, Matthew S.; 8818 Hunting Lane #201; Laurel MD 20708; USA

NAGEL, David J.; Condensed Matter and Radiation Sciences Div.; Code 6680; US Naval Research Laboratory; U.S. Navy; 4555 Overlook Avenue, SW; Washington DC 20375; USA; 202-767-2931; E; 5.4; 3.4

NAGY, Paul J.; Department of Chemistry; Oxford College; Pierce Hall; Oxford GA 30267; USA; 404-786-7051; T; 2.3; Molecular dynamics; Energy flow within excited molecules

NAHAR, Sultana N.; Department of Physics and Astronomy; Wayne State University; Detroit MI 48202; USA; 313-577-2755; 313-831-5081; T; 2.2; Electron and positron collisions with atoms and molecules; Positronium formation from alkal; Atoms by scattering of positrons

NAKAYAMA, Yasuyuki; Institute for Chemical Research; Kyoto University; Yoshida Sakyo-Ku; Kyoto 606 8202; Japan

NALDI, Carlo; Ist Di Elettronica E Telecom; Politecnico; C So Duca Degli Abruzzi 24; Torini 10129; Italy

NAMIKI, Masatoshi; General Education; Physics Laboratory; Takachiho College; Ohmiya 2-19-1, Suginami-ku; Tokyo Japan 168; Japan; T; Low energy ion-atom collisions; Laser induced atomic collisions; Mathematical physics (sp: Stokes phenomena and non-adiabatic transitions)

NANES, Roger; Department of Physics; California State Univ., Fullerton; Fullerton CA 92634; USA; 714-773-2188; 714-773-3366; E; 3.1; 3.5; 3.6; 3.9; 3.2; Molecular spectroscopy and structure; Energy transfer and Laser-induced fluorescence; Spectroscopic intensity measurements; Zeeman effects and magnetic rotation spectroscopy

NARDUCCI, Lorenzo M.; Department of Physics; Drexel University; 32nd and Chestnut; Philadelphia PA 19104; USA; 215-895-2711; T; 3.6; 3.10; 3.11; Laser interaction with atoms; Nonlinear dynamical models of optical interactions; Amplified spontaneous emission from excited gases

NARUMI, Hajime; Department of Physics; Faculty of Science; Hiroshima University; Higashisenda-Machi Naka-Ku; Hiroshima 730 9104; Japan

NASH, Thomas J.; Lawrence Livermore Lab.; PO Box 808; L-401; Livermore CA 94550; USA

NASSER, Ibraheem M.; Department of Physics; University of Connecticut; Storrs CT 06268; USA

NATARAJAN, Marappan; Department of Physics; University of Nebraska; Lincoln NE 68588; USA

NATHEL, Howard; Building 70A-Room 4418; Lawrence Berkeley Lab.; 1 Cyclotron Road 70A-4418; Berkeley CA 94720; USA

NAUMAAN, Ahmed; Department of Electrical Engineering; University of Cincinnati; #30 Rhodes Hall; Cincinnati OH 45221; USA

NAUMANN, Robert A.; Departments of Chemistry and Physics; Princeton University; Princeton NJ 08545; USA; 609-452-4372 x4400; E; 1.3; 4.2; Exotic atoms; Negative muons--Coulomb capture

NAYFEH, A. H.; Department of Mechanical Engineering; Virginia Polytechnic Institute and State University; Blacksburg VA 24061; USA; 703-961-5453; E/T; 3.1; 3.6; 3.7; 3.10

NAYFEH, Munir H.; Department of Physics; Loomis Laboratory of Physics; University of Illinois; Urbana IL 61801; USA; 217-333-3774/0507; E; 1.3; 3.6; 3.8; Atomic collisions in the presence of electromagnetic fields; Colliding atoms--multiphoton ionization; Exotic atoms and chemical lasers--laser spectroscopy; Radiation with matter--coherent interaction

NAZAROFF, George V.; Department of Chemistry; Arts and Sciences; Indiana University, South Bend; 7111; 1700 Mishawaka Avenue; South Bend IN 46634; USA; 219-237-4254; T; 1.1; 2.2; Low-lying potential curves of H2; Electron-molecule dissociative attachment

NEE, Tsu-Jye A.; 11 Turnham Lane; Gaithersburg MD 20878; USA

NEHRING, Frederick W.; Computer Applications Division; Oak Ridge National Lab.; PO Box X; Oak Ridge TN 37830; USA; 615-576-2638; T; 5.4; Atomic physics pertaining to isotope separation

NEIL, George R.; Plasma Physics Division; TRW, Inc.; R1/1070; 1 Space Park; Redondo Beach CA 90278; USA; 213-536-1105; E/T; 3.2; 4.1; 4.2; U II, U III, U IV--spectroscopy; Free electron lasers

NELSON, Albert C.; Georgia Tech Research Institute; Georgia Institute of Technology; Atlanta GA 30332; USA; 404-424-9619; E; 3.3; 3.6; 5.6

NELSON, Leonard Y.; Laser Division; Mathemetical Sciences NW, Inc.; 2755 Northup Way; Bellevue WA 98004; USA; 206-827-0460; E; 3.3; 4.1; Optically pumped molecular IR lasers; IR and Raman spectroscopy; Laser development

NELSON, Robert N.; Department of Chemistry; Georgia Southern College; Landrum Box 8064; Statesboro GA 30460; USA; 912-681-5681; E/T; 2.1; 4.3; Cluster beam production of refractory molecules and clusters; Effusive flow-computer studies

NELSON, William H.; Department of Physics and Astronomy; Georgia State University; Atlanta GA 30303; USA; 404-658-2279; E; 3.1; Molecules damaged by ionizing radiation--magnetic resonance spectroscopy

NESBET, Robert K.; Physical Sciences K34/281; IBM Research Laboratory; 5600 Cottle Road; San Jose CA 95193; USA; 408-256-2673; T; 2.2; Atomic and molecular collision theory; Electron-atom and electron-molecule scattering

NESBITT, David J.; Joint Institute for Lab. Astrophysics; National Bureau Standards; Univ. of Colorado; Boulder CO 80309; USA; 303-492-8857; E; 1.1; 2.3; 3.6; 3.9; 3.3; IR spectroscopy of Van der Waals molecules; Stark/Zeeman saturation spectroscopy; Radical kinetics via time resolved laser spectroscopy; Gas-solid collision dynamics

NETZEL, Thomas L.; Photon Processes; Physical Technology; Amoco Research Center; Amoco Corporation; 400; Naperville IL 60566; USA; 312-420-5150; E; 2.3; 3.6; 4.1; 3.1; Picosecond spectroscopy-electron transfer reactions-organometallic photochemistry; Excited state spectra and relaxation mechanisms; Inorganic photochemical processes

NEUMANN, David B.; Chemical Thermodynamics Division; National Bureau of Standards; Gaithersburg MD 20899; USA; 301-921-3632; T; 1.1; 5.5; Molecular properties and interaction potentials of diatomic molecules--calculation of thermodynamic properties of organic and inorganic molecules

NEUMANN, Herschel; Department of Physics; University of Denver; Denver CO 80208; USA; 303-871-2238; T; 2.1; 5.1; 2.3; H/H+ collisions with rare gas atoms/molecules; Low energy ion scattering spectroscopy; Ultrasonic dispersion in gases--analysis

NEWBY, Neal D.; 228 E. Cordova Road; Santa Fe NM 87501; USA

NEWCOMB, Joal J.; Metz Trailer Park-Route 5/Lot 62; Picayune MS 39466; USA

NEWHALL, Herbert F.; Cornell University; Clark Hall;

Ithaca NY 14853; USA

NEWMAN, David E.; General Atomic Co.; PO Box 81608; San Diego CA 92138; USA

NEWMAN, James H.; Space Physics and Astronomy; Rice University; PO Box 1892; 6100 S. Main; Houston TX 77251; USA; 713-527-4932; E; 1.1; 2.1; 5.6

NEWMAN, John B.; Department of Physics; Towson State University; Baltimore MD 21204; USA

NEWMAN, Kathie E.; Department of Physics; University of Notre Dame; Notre Dame IN 46556; USA

NEWMAN, Leon A.; Electromagnetics and Physics Division; United Technologies Research Center; MS85; Silver Lane; East Hartford CT 06108; USA; 203-727-7262; E/T; 4.1; 5.3; Gas laser research

NEWSTEAD, C. M.; 1741 East Avenue; McLean VA 22101; USA

NEWTON, Marshall D.; Chemistry Division; Brookhaven National Laboratory; Upton NY 11973; USA; 516-345-4366; T; 1.1; Ab initio electronic structure calculations

NEYNABER, Roy H.; Department of Physics B-019; University of California, San Diego; La Jolla Institute; La Jolla CA 92093; USA; 714-452-3290; 714-454-3851; E; 2.1; Low-energy heavy particle collisions using molecular beams

NG, Ching-Yuen; Min Sheng Road East; 4/F #12-1 Alley 21 Lane 929; Taipei; Taiwan

NI, Wei-Tou; Department of Physics; JILA; National Tsing Hua University; Hsinchu Taiwan 300; Republic of China; (035)71531/469; E/T; 3.11; 5.2; Interferometers, squeezed states and quantum-mechanical noise; NMR of Li

NICHOLLS, Ralph W.; Department of Physics; CRFSS; York University; 4700 Keele Street; North York (Toronto) Ontario M3J1P3; Canada; 416-667-3833; 416-889-5093; E; 1.1; 3.1; 3.2; 3.5; 5.6; Experimental and theoretical molecular spectroscopy; Atmospheric extinction processes and remote sensing; Astrophysical molecular spectroscopy

NICOVICH, John M.; Molecular Sciences Branch; Solid State Science Division; Electro-Magnetics Laboratory; Georgia Institute of Technology; Engineering Experiment Station; Atlanta GA 30332; USA; 404-894-3424; E; 2.3; 5.6; Interaction of radicals and atoms with other species of interest in atmospheric studies

NIELSEN, Alvin H.; Department of Physics; College of Liberal Arts; Alvin H. Nielsen Physics Building; The University of Tennessee; Knoxville TN 37916; USA; 615-974-7817; 615-974-7850; E; 3.3; IR spectroscopy; High dispersion gratings; Vibration-rotation measurement and analysis

NIER, Alfred O.; Department of Physics and Astronomy; University of Minnesota; 116 Church Street, SE; Minneapolis MN 55455; USA; 612-373-3325; E; 5.6; 4.4; Terrestrial and planetary atmospheres--composition; Mass spectrometry-development

NIGHAN, William L.; Electronics and Electro-Optics Technology; Electro-Optics and Applied Physics; Theoretical Physics; United Technologies Research Center; Silver Lane; East Hartford CT 06108; USA; 203-727-7596; E/T; 2.3; 5.3; 4.1; Lasers; Gas discharge displays; Gaseous electronics; Atomic and molecular physics-chemical physics

NILES, Franklin E.; Electro-Optics Division; US Army Atmospheric Science Lab.; White Sands; Missile Range NM 88002; USA; 505-678-3721; E/T; Supervisory physicist in atmospheric sciences, research and development

NISHIZAWA, Jun-Ichi; 6-16 Komegafukuro 1 Chome; Sendai 980 8002; Japan

NITZ, David E.; Department of Physics; Saint Olaf College; Northfield MN 55057; USA; 507-663-3123; 507-663-3120; E; 1.1; 3.6; 2.1; Hyperfine interactions in molecules; Interactions of laser radiation with atoms and molecules; Atomic collisions-charge exchange

NIV, Tehuda; 120-C Cedar Lane; Highland Park NJ 08904; USA

NIXON, Eugene R.; Department of Chemistry; University of Pennsylvania; Philadelphia PA 19104; USA; 215-243-8313; E; 3.5; Molecular spectroscopy

NOBLE, Clifford J.; TCS Division; Daresbury Lab.; Daresbury Warr WA44AD; England

NOID, Donald W.; Chemistry Division; Oak Ridge National Laboratory; Oak Ridge TN 37830; USA; 615-574-4992; T; 2.3; 3.8; Kinetics of small molecules; Laser chemistry

NOLTING, Juergen; Siedlerweg 30; Minden D-4950; West Germany

NORCROSS, David W.; Quantum Physics Division (NBS); Joint Institute for Laboratory Astrophysics; University of Colorado; Boulder CO 80309; USA; 303-492-7858; T; 2.2; Electron collisions with atoms, ions and small molecules

NORRIS, Theodore B.; Department of Physics; University of Rochester; Rochester NY 14627; USA

NORTH, Dwight O.; Radio Corp. of America Laboratories; David Sarnoff Research Center; Princeton NJ 08540; USA; 609-734-2000; T; 4.1; Laser fundamentals; Quantum theory foundations

NORTHRUP, Scott H.; Department of Chemistry; Tennessee Technological University; PO Box 5055; Cookeville TN 38505; USA; 615-528-3748; T; 1.3; 2.3; Atomic motions in proteins and lipid bilayers--molecular dynamics computer simulation; Model biopolymers in solution--Brownian dynamics; Atom and molecules recombination dynamics in condensed media

NORTON, Robert N.; 17 Chamberlain Road; Claremont Cape Town 7700; South Africa

NOWAK, Edward J.; 51 White Birch Lane; Williston VT 05495; USA

NOYES, Richard M.; Department of Chemistry; University of Oregon; Eugene OR 97403; USA; 503-686-4611; 503-686-4601; E/T; 2.3; Chemical reactions--kinetics and mechanisms

NUSSBAUM, Gilbert H.; Mallinckrodt Institute of Radiology; Division of Radiation Oncology; Washington University Medical Ctr.; 510S Kings Highway; St. Louis MO 63110; USA

NYGAARD, Kaare J.; Department of Physics; University of Missouri, Rolla; Rolla MO 65401; USA; 314-341-4781/4703; E/T; 2.2; 3.1; 3.6; 5.3; Electron impact ionization and photoionization; Negative ions and laser kinetics; Electron transport coefficients; Electric discharges

O, Chun-Sing; Oak Ridge National Labs.; PO Box X-10; Building 5500; Oak Ridge TN 37831; USA; 615-574-4704; E/T; 1.1; 2.3; 3.1; 3.6; 4.2

O'BRIEN, John T.; 1620 S. E. Harris Drive; Bartlesville OK 74003; USA

O'BRIEN, Thomas J.; Department of Chemistry; Texas Tech University; Lubbock TX 79409; USA; 808-742-3087; T; 5.6; Chemical reactions occurring in the upper atmospheres; Coal combustion processes

O'CONNELL, Robert F.; Department of Physics and Astronomy; Louisiana State University; Baton Rouge LA 70803; USA; 504-388-6848; T; 3.6; 3.9; 3.11; 5.2; Atoms in blackbody radiation field; Atoms in magnetic fields; Atoms in laser fields

O'MALLEY, Thomas F.; M/C 611; General Electric Company; 175 Curtner Avenue; San Jose CA 95125; USA; 408-925-7513; T; 2.2; 5.3; 5.2; 2.3; Threshold behavior in elastic and inelastic electron collisions; Dissociative electron attachment and recombination; Molecular and atomic resonant states; Large scale computations

O'NEIL, Stephen V.; Joint Institute for Lab. Astrophysics; University of Colorado; Boulder CO 80309; USA; 303-492-7817; T; 1.1; Molecular electronic structure

O'REILLY, James M.; Materials Research Laboratory; Xerox Webster Research Laboratory; 800 Phillips Road; Webster NY 14580; USA; 716-422-3428; E; 1.3; 3.3; 5.8; Physical property studies of polymers using Fourier transform IR; Spectroscopy, differential scanning calorimetry, and small angle neutron scattering

OBLAS, Daniel W.; #312; GTE Laboratories, Inc.; 40 Sylvan Road; Waltham MA 02254; USA; 617-466-2456; 617-466-2816; E; 4.3; 4.4

OCHAB, John S.; Department of Physics; University of Maine; Orono ME 04469; USA

ODDERSHEDE, Jens; Kemisk Institut; Odense Universitet; Niels Bohrs Alle 25; DK-5230 Odense M; Denmark

ODOM, Robert W.; Charles Evans and Associates; 301 Chesapeake Drive; Redwood City CA 94063; USA; 415-369-4567; E; 4.4; 2.1; 5.1; Novel ionization methods--investigation; Adsorbed molecular species--surface ionization

OGRAM, Geoffrey L.; Research Division; Ontario Hydro; 800 Kipling Avenue; Toronto Ontario M8Z5S4; Canada

OHRN, Yngve; Department of Chemistry; University of Florida; Gainesville FL 32611; USA; 904-392-1597; T; 1.1; 2.3; 3.1; 2.1; 3.2; Quantum chemistry; Molecular electronic structure and spectra; Molecular reactive collisions; Propagator theory and methods

OHUCHI, Toshitomo; c/o Nippon Schlumberger K K; 34-12 Nishi-Rokugo/4 Chome; Ohta-Ku Tokyo 144; Japan

OKABE, Hideo; Chemical Kinetics Division; National Bureau of Standards; Gaithersburg MD 20899; USA; 301-921-2151; 202-636-6883; E; 1.1; 3.1; 3.8; 5.5; 5.6; Photochemistry of simple molecules; Dissociation processes; Electronically excited species; Chemical kinetics

OKANO, Masaharu; Rikagaku Kenkyusho; Wako-Shi Saitama 351; Japan

OKEKE, Cajetan E.; Department of Physics; University of Nigeria; Nsukka Anambra State; Nigeria

OKTAY, Erol; Office of Fusion Energy; U.S. Department of Energy; Washington DC 20545; USA; 202-353-4928; E/T; Fusion program--management

OLDENBORG, Richard C.; Photochemistry Group; Chemistry; Los Alamos National Laboratory; Los Alamos New Mexico 87545; USA; 505-667-2096; 505-667-3758; E; 3.6; 3.8; 2.3; 4.1; 5.5

OLEKSIK, John J.; Department of Chemistry; James Franck Institute; 5640 S. Ellis; Chicago Illinois 60637; USA; 312-962-7219; 312-962-7557; T; 1.1; 1.3; 1.4; 2.2; 3.1

OLESON, Norman L.; Department of Physics; University of South Florida; Tampa FL 33620; USA

OLLISON, William M.; Office of General Counsel; American Petroleum Institute; 1220 L Street, NW; Washington DC 20005; USA; 202-682-8262; 301-469-6345; E/T; 5.6; 2.3; 3.8; 4.4; 5.5; Pollutant formation, transformation, measurement, and health and material effects--analysis of literature

OLSEN, Gregory H.; Epitaxx, Inc.; 3490 Route 1; Princeton NJ 08540; USA

OLSEN, John; 1770 Dean Road; Paradise CA 95969; USA

OLSEN, Thomas; Department of Physics, Natural Science, and Mathematics; Lewis and Clark College; 15; 0615 SW Palatine Hill Road; Portland OR 97219; USA; 503-293-2751; 503-244-6161/6422; T; 3.7; 3.10; 3.1; 3.11; The study of the interaction of intense electromagnetic fields with matter, atoms in particular, including multiphoton ionization, the AC Stark Shift, and harmonic generation in particular

OLSGAARD, David A.; Physics Research East; Old Dominion University; Norfolk VA 23508; USA; 804-440-4619; T; 3.6; 3.5; 1.1; 3.10; Pulsed beam spectroscopy of sulfur dioxide, near UV

OLSON, David P.; Loomis Lab. of Physics; University of Illinois, U-C; 1110 W. Green Street; Urbana IL 61801; USA

OLSON, Gregory L.; Chemical Physics Division; Hughes Research Laboratories; 3011 Malibu Canyon Road; Malibu CA 90265; USA; 213-317-5457; E; 3.8; 5.1; Laser/solid interactions; Laser photochemistry; Laser annealing

OLSON, John R.; Department of Physics; University of Denver; Denver CO 80208; USA; 303-753-2170; E; 4.4; 5.6; Quadrupole mass spectrometry in the stratosphere

OLSON, Noble T.; Maxwell Lab., Inc.; 9244 Balboa Avenue; San Diego CA 92123; USA

OLSON, R. A.; International Laser Systems; 3404 N. Orange Blossom Trail; Orlando FL 32804; USA

OLSON, Robert A.; Research Applications Division; Systems Research Labs.; 2800 Indian Ripple Road; Dayton OH 45440; USA; 513-426-6000; E; 5.3; 5.4; 4.1; 3.9; 3.6

OLSON, Ronald E.; Molecular Spectroscopy Laboratory; SRI International; 333 Ravenswood Avenue; Menlo Park CA 94025; USA; 415-326-6200x2083; T; 2.1; Ion-atom and atom-atom collision processes--calculations

OMIDVAR, Kazem; Planetary Atmospheres Branch; Laboratory for Atmospheres; NASA Goddard Space Flight Center; Code 614; Greenbelt MD 20771; USA; 301-344-5766; 202-544-8338; T; 1.2; 2.2; 3.6; 5.6; Classical determination of oscillator strengths and liftimes of Rydberg states; Born and Bethe approximation in ionization collisions of electrons with atoms; Two-photon excitation cross sections in atoms; Dissociation of atmospheric oxygen by solar radiation and radiative transformation calculations

ONRUBIA, Isabel T.; Catedra Fisica Atomica Y Nucl; Fac De Ciencias Fisicas; U Complutense Ciudad Univ.; Madrid 3; Spain

OONA, Henn; 103 Beryl; Los Alamos NM 87544; USA

OOSTERHUIS, William T.; Division of Materials Research; National Science Foundation; 1800 G Street; Washington DC 20550; USA; 202-357-9791; E; 3.4; 4.2; X-ray scattering

OPPENHEIMER, Michael; Department of Astronomy; Harvard-Smithsonian Center for Astrophysics; 60 Garden Street; Cambridge MA 02138; USA; 617-495-3745; T; 2.1; 5.7; Scattering theory; Atomic and molecular process in astrophysics

OREL, Ann E.; L-472; Lawrence Livermore National Lab.; PO Box 808; Livermore CA 94550; USA

ORIENT, Otto J.; Electron Collision Processes; Earth and Space Science; Jet Propulsion Laboratory; California Institute of Technology; 4800 Oak Grove Drive; Pasadena CA 91109; USA; 818-354-7233; 818-354-6535; E; 1.1; 2.2; 4.3

ORLOFF, Jon H.; Applied Physics and Electrical Engineering; Oregon Graduate Center; 19600 NW Von Neumann Dr.; Beaverton OR 97006; USA; 503-690-1136; E; 4.3; 5.1; Electron optics; Electron and ion sources

ORMONDE, Stephen; Quantum Systems; PO Box 8575; Albuquerque NM 87108; USA; E/T; 1.1; Computer codes for atomic structure and interactions

ORVEK, Kevin J.; 6400 Independence Parkway #2703; Plano TX 75023; USA

OSBURN, Jean E.; Department of Chemistry; Tulane University; New Orleans LA 70118; USA

OSGOOD, Richard M.; Department of Electrical Engineering; Columbia University; New York NY 10025; USA; 212-280-4462; E; 2.3; 3.1; 5.1; Chemical physics: exciplex molecules; Light and atoms and molecules in gas and on surfaces--fundamental interactions

OSKAM, Hendrik J.; Department of Electrical Engineering; University of Minnesota; 123 Church Street, SW; Minneapolis MN 55455; USA; 612-373-5427; E/T; 5.3; 5.4; Collision processes in plasmas; Transport phenomena in plasmas; Plasma chemistry; Gaseous electronics

OSS, John P.; 7851 S. Carr Street; Littleton CO 80123; USA

OSTLUND, Neil S.; Department of Chemistry; University of Arkansas; Fayetteville AR 72701; USA; 501-575-4601; T; 2.3; Quantum chemistry; Computer science applications to physical science

OSTROVE, Neil; Bell Labs.; 6200 E. Broad Street; Columbus OH 43213; USA

OTT, William R.; Center for Radiation Research; Radiation Physics Division; National Bureau of Standards; Bldg. 220 Room B206; Gaithersburg MD 20899; USA; 301-921-3201; E; 5.4; 3.1; 2.2; 5.3; 5.1; Plasma spectroscopy in the vacuum UV; Gas discharge sources as intensity standards (VUV radiometry)--development; Synchrotron radiation applications; Transition probability and line-broadening measurements

OTTER, Fred A.; Physics of Solids Group-M.S. 75; Electronics and Electrooptics Tech.; United Technologies Research Center; Silver Lane; East Hartford CT 06108; USA; 203-727-7523; E; 2.1; 4.2; 4.4; 5.2; 5.3; Ion scattering spectrometry; Secondary ion mass spectrometry; Rutherford back scattering; Ion implantation and ion beam mixing

OTTLINGER, Michael E.; Department of Biophysics; The Johns Hopkins University; Baltimore MD 21218; USA

OURA, Kenjiro; Faculty of Engineering; Electron Beam Lab.; Yamadaoka Osaka University; Suita Osaka 565 1218; Japan

OVERBURY, Stephen H.; Chemistry Division; Oak Ridge National Laboratory; PO Box X; Oak Ridge TN 37830; USA; 615-574-5040; E; 4.2; 5.1; 5.2; Channeling of MeV ions in solids and studies of resultant electron emission; Ion scattering as a probe of solid surfaces

OVEREND, John; Department of Chemistry; University of Minnesota; Minneapolis MN 55455; USA; 612-373-2316; E/T; 3.5; 5.1; Surface species --vibrational spectroscopy; Overtone spectroscopy; Vibrational circular dichroism

OWENS, James C.; Engineering Physics Laboratory; Physics Division; Eastman Kodak Research Laboratories; 1999 Lake Avenue; Rochester NY 14650; USA; 716-458-1000; 716-477-7603; E; 4.1; 4.3; Optic and imaging systems materials; Electro-optics and acousto-optics; Spectroscopy of dyes; Novel imaging systems--chemical physics

OXMAN, John C.; 9 E. 10th Street #4R; New York NY 10003; USA

OZA, Dipak H.; Atomic and Plasma Radiation; National Bureau of Standards; Rm A167 Bldg. 221; Gaithersburg MD 20899; USA; 301-921-2356; T; 2.2; 3.1; 1.1; 5.4

PACE, Jerry H.; PO Box 587; Forest MS 39074; USA

PACK, John L.; Gas Laser Research and Development Division; Westinghouse R&D Center; 1310 Beulah Road; Pittsburgh PA 15235; USA; 412-256-5028; E; 4.1; Laser development

PACK, Russell T.; Theoretical Division T-12; Los Alamos National Laboratory; 1663 MS J569; Los Alamos NM 87545; USA; 505-667-5881; T; 1.1; 2.1; 3.1; Molecular scattering; Intermolecular potentials and photodissociation; Angular momentum decoupling approximations; Frame transformation theory

PADIAL, Nely T.; c/o Lee A. Collins; PO Box 476; Los Alamos NM 87544; USA

PAI, Robert Y.; Health and Safety Research Division; Oak Ridge National Lab.; PO Box X; Oak Ridge TN 37830; USA; 615-574-4662; E; 2.2; 5.3; Electron-molecule attachment relevant to gas dielectrics

PAIKEDAY, Joseph M.; Department of Physics; Southeast Missouri State University; Normal and Pacific; Cape Girardeau MO 63701; USA; 314-651-2172; T; 2.2; Differential scattering cross section of electrons scattered by atoms and molecules--calculation

PAISNER, Jeffrey A.; Advanced Isotope Separation Division; Y-Program; Lawrence Livermore National Lab.; PO Bx 808 L-467; Livermore CA 94583; USA; 415-422-6211; E; 3.6; 4.1; Atomic and molecular laser spectroscopy; Laser-atom interactions; Nonlinear optics; Coherent light sources

PALFREY, Stephen L.; IBM T. J. Watson Research Center; PO Box 218; Yorktown Heights NY 10598; USA

PALLMER, Paul; PO Box 3022; Pasco WA 99302; USA; 509-547-5091; E/T; 5.1; Radiation damage in solids; Surface-gas reactions

PALMER, Byron A.; Group CMB-1; Los Alamos National Laboratory; PO Box 1663 MS740; Los Alamos NM 87544; USA; 505-667-3528; E; 3.2; 3.5; Radioactive and actinide elements and molecules--high resolution optical spectroscopy

PALMER, Howard B.; Department of Materials Science and Engineering; Pennsylvania State University; 114 Kern Bldg.; University Park PA 16802; USA; 814-865-2516; E; 2.3; 5.5; Metal vapors with oxidants--spectroscopy and photon-yield chemiluminescent reactions

PALMER, Robert L.; Physical Sciences Division; IRT Corporation; PO Box 80817; San Diego CA 92138; USA; 714-565-7171; E; 5.1; Surface physics; Catalysis; Ion-surface interactions

PALUMBO, Louis J.; Laser Physics Branch; Code 6540; US Naval Research Laboratory; 4555 Overlook Avenue, SW; Washington DC 20375; USA; 202-767-2255; T; 5.3; High power gas lasers--computer modeling of kinetics

PAN, Liwen; Department of Physics and Astronomy; The Johns Hopkins University; Baltimore MD 21218; USA; 301-338-8455; 301-338-7360; T; 3.7

PAN, Y.; Department of Chemistry; Boston College; Chestnut Hill MA 02167; USA; 617-969-0100; T; 1.1; 2.3; Molecular spectra and structures; Quantum dynamics of chemical reactions

PANDINI, Davide; Via Polonia N. 12; Ferrara 44100; Italy

PANOCK, Richard; Bell Telephone Laboratories; Crawfords Corner Road; Holmdel NJ 07733; USA; 201-949-3421; E; 3.5; 4.1; 5.2; Deeply-bound species with metastable states--atomic and molecular spectroscopy; Nonlinear optics

PANTINAKIS, Apostolos; Physics Department; University Chemical Laboratory; University of Kent at Canterbury; Lensfield Road; Canterbury Kent CT27NR; England; E; 1.1; 3.4; 3.6; 3.10

PAPACOSTA, Pangratios; Department of Physics; Stetson University; PO Box 1342; Deland FL 32720; USA

PAPADIMITRIOU, Vaia D.; Room 508; International House of Chicago; 1414 E. 59th Street; Chicago IL 60637; USA

PAPAYOANOU, Aristotle; Autonetics Strategic Systems Division; Rockwell International Corp.; 4192; 3370 Miraloma Ave; Anaheim CA 92803; USA; 714-632-7044; E; 4.1; 5.3; 3.6; Laser physics and laser systems; Laser radars and communications systems; Interaction of optical radiation with materials

PARK, Changhwan; Department of Physics and Astronomy; Behlen Laboratory; University of Nebraska; Lincoln NE 68588; USA; 402-472-3688; 402-474-3578; T; 1.1; 2.1; 3.1

PARK, John T.; Department of Physics; University of Missouri, Rolla; Rolla MO 65401; USA; 314-341-4781; E; 2.1; 2.3; 4.2; Ion energy loss spectrometry; Differential cross sections for 15-200 KeV ion impact--measurement; Excitation-ionization elastic scattering; Charge transfer

PARKER, Gregory A.; Department of Physics and Astronomy; University of Oklahoma; 440 West Brooks Street; Norman OK 73019; USA; 405-325-3961; T; 2.1; Molecular scattering

PARKER, John W.; 5463 Mariners Cove Drive; Jacksonville FL 32210; USA

PARKER, Michael A.; 42858 Roberts Avenue; Fremont CA 94538; USA

PARKER, Paul M.; Department of Physics; Michigan State University; East Lansing MI 48824; USA; 517-353-9233; 517-355-9665; T; 3.5; 1.1; 3.3; Vibration-rotation theory of high-resolution molecular spectra

PARKINSON, W. H.; Atomic and Molecular Physics Division; Center for Astrophysics; Harvard College Observatory; 60 Garden Street; Cambridge MA 02138; USA; 617-495-4865; E; 3.1; 3.2; 3.5; Atomic and molecular spectroscopy and laboratory astrophysics

PARKS, Eric K.; Argonne National Laboratory; Building 200; 9700 S. Cass Avenue; Argonne IL 60439; USA; 312-972-3470; 312-972-3470; E; 1.3; 2.1; Molecular beam research; Metal cluster studies

PARKS, James E.; Department of Physics and Astronomy; Western Kentucky University; Bowling Green KY 42101; USA; 502-745-4357; E; 3.1; 3.6; Atoms and molecules using single and multiple photon processes with high power tunable dye lasers--ionization and excitation

PARKS, William F.; Department of Physics; University of Missouri, Rolla; Rolla MO 65401; USA

PARMENTER, Charles S.; Department of Chemistry; Indiana University; Bloomington IN 47405; USA; 812-337-3522; E; 2.1; 2.3; 3.1; 3.6; 3.8; Molecular energy transfer

PARMENTER, R. H.; Department of Physics; University of Arizona; Tucson AZ 85721; USA

PARPIA, Farid A.; 1444 South Bend Avenue; South Bend IN 46617; USA

PARR, Albert C.; Far Ultraviolet Physics Division; National Bureau of Standards; Building 221, Room A251; Gaithersburg MD 20899; USA; 301-977-7652; E; 3.1; 4.2; Synchrotron radiation in atomic and molecular photoionization

PARR, Christopher A.; Department of Chemistry; University of Texas, Dallas; PO Box 688; Richardson TX 75080; USA; 214-690-2187; T; 1.1; 2.3; Chemical reaction dynamics; Molecular vibration dynamics; Potential-energy hypersurfaces

PARR, Robert G.; Department of Chemistry; University of North Carolina; Chapel Hill NC 27514; USA; 919-933-1217; T; 1.1; Electronic structure of atoms and molecules

PARRY, Peter D.; Research Center; Western Electric Engineering; PO Box 900; Princeton NJ 08540; USA

PARSON, John M.; Department of Chemistry; Ohio State University; Columbus OH 43210; USA; 614-422-3267; 614-422-2088; E; 2.1; 2.3; Reactive and inelastic scattering using spectroscopic detection--molecular beam studies

PARSONS, Donald F.; CED Laboratory; Hewlett-Packard Corporation; PO Box 301; Loveland CO 80537; USA; 303-667-5000x3320; T; 4.1; 5.2; Atomic physics in solids

PARSONS, Michael L.; CHM-1; Chemistry; Los Alamos National Laboratory; MS-G740; Los Alamos NM 87545; USA; 505-667-9307; 505-667-4087; E/T; 3.2; 3.1; 1.1; 5.4; 5.5; Analytical atomic spectroscopy; Atomic emission excitation sources; Absorption spectra--fundamental constants; Atomic spectroscopy--analytical techniques

PARTLOW, William D.; Applied Plasma Research; Applied Sciences; Westinghouse R&D Center; 1310 Beulah Road; Pittsburgh PA 15235; USA; 412-256-7247; E; 5.3; 5.4; 2.3; 4.4; 3.2; Chemistry and physics of etching and deposition plasmas

PARTRIDGE III, Harry; NASA Ames Research Center; 230-3; Moffett Field CA 94035; USA; 415-965-6189; T; 1.1; 2.3; Ab initio quantum chemistry on small molecules; Multicenter Slater orbital integral package development

PASKE, William C.; Department of Physics and Astronomy; University of Oklahoma; 440 W. Brooks Street; Norman OK 73019; USA; 405-325-3961; E; 2.2; 2.3; Low energy electron impact on gases; Transfer rates and lifetime determination; Negative ion spectroscopy

PASSENHEIM, Burr C.; Radiation Effects; San Diego; Jaycor; PO Box 85154; 11011 Torreyana Rd.; San Diego CA 92138; USA; 619-453-6580; 619-278-7589; E; 3.2; 4.1; 5.1; Nuclear and space radiation effects on materials and electronics; Solid state physics and cryogenics and electro-optics; Electromagnetic pulse phenomena; Nuclear and directed energy effects research and testing

PATE, Bradford B.; SSRL Bin 69; Stanford University; PO Box 4349; 2575 Sand Hill Road; Stanford CA 94305; USA; 415-854-3300/3629; 415-854-3300/2874; E; 3.1; 3.4; 5.1; 2.2; Photoemission; Surface structure; Photostimulated-ion desorption; X-ray absorption near edge structure

PATEL, C. Kumar K.; Executive Director; Research, Physics and Academic Affair; AT&T Bell Laboratories; Bell Telephone Laboratories; 600 Mountain Avenue; Murray Hill NJ 07974; USA; 201-582-3425; E; 4.1; 1.4; 4.2; 5.6; 5.7; Molecular solids --opto-acoustic spectroscopy; Molecular lasers; Pollution detection; Molecular gases --spectroscopy

PATTERSON, Edward L.; Division 4212; Sandia National Laboratories; PO Box 5800; Albuquerque NM 87185; USA; 505-844-3886; E; 4.1; 4.2; Gas lasers processes

PATTERSON, Gary D.; Chemical Physics Division; Bell Telephone Laboratories; 600 Mountain Avenue; Murray Hill NJ 07974; USA; 201-582-2759; E; 5.1; Light scattering from liquids and amorphous solids

PATTERSON, Paul L.; Detector Engineering and Technology; 2212 Crampton Road; Walnut Creek CA 94598; USA

PATTERSON, Thomas A.; Extranuclear Labs., Inc.; PO Box 11512; Pittsburgh PA 15238; USA

PATTON, Robert J.; Improved Fluorescents; Lighting Products; Lighting Center; GTE Products Corportion; 100 Endicott St.; Danvers MA 01923; USA; 617-777-1900x2839; 617-692-3283; E; 5.2; 1.3; 1.2; Mossbauer spectroscopy of luminescent materials; Electric field gradients at fluorescing clusters; Bond lengths and angles and energy levels of fluorescing ions

PATTY, Richard R.; Department of Physics; North Carolina State University; Box 8202; Raleigh NC 27695; USA; 919-737-2521; E; 5.6; 1.1

PAU, Louis F.; Battelle Institute; C.P. 27; 5 Chemin Tressy-Cordy; Grand Lancy 2 Geneva CH1212; Switzerland; (022) 439831; E; 3.6; 5.1; X-ray and MMW induced currents in semiconductor junctions

PAUL, Derek A. L.; Physics; Arts and Science; University of Toronto; Toronto Ontario M5S1A7; Canada; 417-978-2971; 416-978-7135; 2.2; 1.3; 1.4; Electron and positron collisions; Positronium formation and anniliation; Left-handedness of the weak interaction

PAULSON, John F.; Ionospheric Physics; US Air Force Geophysics Lab.; Hanscom AFB MA 01731; USA; 617-861-3124; E; 2.3; 3.1; 5.6; Ion-neutral reactions; Ion photon reactions; Ionospheric chemistry

PAULSON, Ronald F.; US Air Force Avionics Laboratory; AFWAL/AADO; Wright Patterson AFB OH 45433; USA; 513-255-3086; E; 4.1; Gas and chemical lasers

PAVLOPOULOS, Theodore G.; U.S. Naval Ocean Systems Center; Code 8113; San Diego CA 92152; USA; 714-225-7329; E; 3.5; Laser dyes--spectroscopic characteristics

PAYNE, Marvin G.; Health Physics Division; Oak Ridge National Laboratory; PO Box X; Oak Ridge TN 37830; USA; 615-574-5898; E/T; 2.1; 3.1; 3.6; 3.7; Excited state populations in gases --kinetics; Laser fluorescence and spectral line broadening; Photoionization near resonance; Collision processes in strong laser fields and atom-atom collisions

PAYNE, Philip W.; Molecular Theory Program; Life Sciences Division; SRI International; 333 Ravenswood Locate 20541; Menlo Park CA 94025; USA; 415-859-3208; 415-859-2927; T; 1.1; 1.3; Computational quantum chemistry; Conformational analysis of macromolecules; Biophysics, electronic structure and conformation; Molecules and solids--electronic structure

PAYSEN, Robert A.; Chemistry; Bethany College; PO Box 477; Bethany WV 26032; USA; 304-829-7753; 304-829-4619; T; 1.1; 5.7; Modeling surfaces of solids with gas interactions, grain condensation processes in early solar system development

PEACHER, Jerry L.; Department of Physics; University of Missouri, Rolla; Rolla MO 65401; USA; 314-341-4786; T; 2.1; Atomic and molecular collision theory

PEARCE-PERCY, Henry T.; 1407 Marlboro; Richardson TX 75081; USA

PEARL, John C.; Laboratory for Extraterrestrial Physics; NASA Goddard Space Flight Center; Greenbelt MD 20771; USA; 301-344-5978; T; 3.3; 5.6; Gaseous and solid state spectral data to interpretation of planetary IR spectra (Voyager IR experiment)--application

PEARSON, E. Benson; 924 Cambridge Avenue; Sunnyvale CA 94087; USA

PEARSON, Edwin F.; Equipment Development Laboratory; Raytheon Corporation; MS C-35; Boston Post Road; Wayland MA 01778; USA; 617-358-2721; E; 3.3; 5.6; High resolution IR spectroscopy in environmental and industrial process monitoring --applications

PEATMAN, William B.; Bessy; Berlin Electron Storage Ring GMBH; Lentzeallee 100; D-1000 Berlin 33 33; Germany; 049-30-82004-152; E; 3.1; 4.1; 4.2

PECHUKAS, Philip; Department of Chemistry; Columbia University; New York NY 10027; USA; 212-280-4231; 212-280-2204; T; 1.1; 2.1; 2.3; Chemical kinetics; Semiclassical approximation; Nonlinear dynamics

PECORA, Robert; Department of Chemistry; Stanford University; Stanford CA 94305; USA; 415-497-0681; E/T; 1.3; 3.6; Light scattering from liquids and gases; Classical time correlation functions; Macromolecules

PEEK, James M.; Division 4211; Sandia National Laboratories; PO Box 5800; Albuquerque NM 87185; USA; 505-844-3442; T; 5.3; Theoretical models--construction; Atomic and molecular data used to extract predictions from such models; Primary atomic and molecular data--generation

PEGG, David J.; Department of Physics; University of Tennessee; Knoxville TN 37996; USA; 615-974-5478; E; 3.6; 4.2; Fast ion beams from accelerators--laser spectroscopy

PELLIN, Michael J.; Chemistry Division; Argonne National Laboratory; 9700 S. Cass Avenue; Argonne IL 60439; USA; 312-957-3510; E; 1.3; 5.1; Small metal clusters--fluorescence and structure; Sputtered atoms--doppler shift laser fluorescence

PENDLETON, Hugh N.; Department of Physics; Brandeis University; Waltham MA 02254; USA; 617-647-2839; 617-647-2835; T; 1.3; 1.4; 3.11; High precision quantum electrodynamic calculations on simple atoms; Positronium structure to two radiative loops

PENDLETON JR., William R.; Department of Physics; UMC 41; Utah State University; Logan UT 84322; USA

PENDYALA, Subra; Department of Physics; SUNY College, Fredonia; Fredonia NY 14063; USA

PENETRANTE, Bernie M.; Department of Physics; University of Pittsburgh; Pittsburgh PA 15260; USA

PENG, Jin-Sheng; Department of Physics; Research Group of Quantum Optics; Huazhong Normal University; Wuhan Hubei; Republic of China; 027-75601; E/T; 1.1; 2.2; 3.11; 5.1; Interaction potentials, hyperfine interactions; Electron and positron collisions with atoms and molecules; Quantum optics, resonance fluorescence, superradiance; Interaction of particles and radiation with surfaces

PENGRA, James G.; Department of Physics; Whitman College; Walla Walla WA 99362; USA; 509-527-5260; E; 2.3; 3.4; Electron capture ratios--measurement; Fluorescence yields; Inner-shell phenomena

PEPLINSKI, Daniel R.; Chemical Physics Division; The Aerospace Corporation; PO Box 92957; Los Angeles CA 90009; USA; 213-648-6928; E; 2.1; 4.3; 5.1; Molecular beam studies; Beam-surface studies; Beam source development

PEPMILLER, Philip L.; Physics; Oak Ridge National Laboratory; MMES/ORNL; PO Box X; Building 5500; Oak Ridge TN 37831; USA; 615-574-3104; E; 2.1; 4.2

PEPPER, David M.; Optical Physics Division; Hughes Research Laboratories; 3011 Malibu Canyon Road; Malibu CA 90265; USA; 213-456-6411; E/T; 1.1; 3.9; Stark spectroscopy; Hyperfine and Zeeman spectroscopy

PEPPER, George H.; Department of Physics and Astronomy; University of Kentucky; Lexington KY 40506; USA; 606-257-1995; 606-257-6721; E; 2.1; 3.4; 4.2; 5.7; PIXE trace element analysis using particle induced X-rays; Accelerator-based atomic physics; Heavy-ion collisions --spectroscopy and cross sections

PERCUS, Jerome K.; Department of Physics; Covrant Institute; New York University; 251 Mercer Street; New York NY 10012; USA; 212-460-7460; T; 1.1; Electron density and correlations--theoretical applications and numerical simulations

PEREL, Julius; Phrasor Scientific, Inc.; 1536 Highland Avenue; Duarte CA 91010; USA; 213-357-3201; E; 4.3; 4.4; 2.1; 2.3; 5.1; Ion source development; Charged liquid droplets and molecular clusters; Secondary ion mass spectrometry

PEREZ III, Joseph D.; Applied Physics Laboratory; Lockheed Palo Alto Research Lab.; 3251 Hanover Street; Palo Alto CA 94304; USA; 415-493-4411x5692; T; 3.4; 4.2; 5.4; X-ray line radiation from partially stripped atoms in hot plasmas--identification and use

PERIA, William T.; Department of Electrical Engineering; University of Minnesota; 123 Church Street, SE; Minneapolis MN 55455; USA

PERKOWITZ, Sidney; Department of Physics; Emory University; Atlanta GA 30322; USA; 404-329-4321; 404-329-6584; E; 3.3; 3.10; 4.1; 3.5; 1.1; Far infrared/submillimeter spectroscopy; Far infrared lasers; Infrared properties of water

PERRY, Robert A.; Combustion Physics Division 8351; Sandia National Laboratories; Livermore CA 94550; USA; 415-422-2298; E; 2.3; 5.5; Gas phase chemical kinetics including radical-radical, radical-molecules, ion-molecules, and radical particle reactions; Soot formation in flames

PERSON, James C.; Radiological and Environmental Research Division; Argonne National Laboratory; Building 203; 9700 S. Cass Avenue; Argonne IL 60439; USA; 312-972-4192; E; 3.1; Absorption cross sections and photoionization yields of molecules using monochromatic vacuum UV light--measurements

PETERSEN, Alan B.; Spectra Physics, Inc.; 1250 W. Middlefield Road; Mountain View CA 94042; USA; 415-961-2550; E; 3.6; 4.1; Ion lasers--investigation of processes; Potential new gas laser devices--speculative research; Laser radiation applications

PETERSON, Charles C.; 623 Greer Road; Palo Alto CA 94303; USA

PETERSON, Deane M.; Department of Earth and Space Sciences; State University of NY, Stony Brook; Stony Brook NY 11794; USA; 516-246-8250; T; 5.7; Collision rates used to characterize astrophysical plasmas

PETERSON, George A.; Department of Chemistry; Hall-Atwater Laboratories; Wesleyan University; Middletown CT 06457; USA; 203-347-9411/2379; T; 1.1; 2.3; 1.3; Many-body perturbation theory; Convergence to complete basis set limit; Orbital energies and chemical bonding; Potential energy surfaces and bulk rate constants

PETERSON, James R.; Molecular Physics Laboratory; SRI International; 333 Ravenswood Avenue; Menlo Park CA 94025; USA; 415-326-6200x3546; E; 2.1; 2.3; 3.1; Atomic and molecular collisions (charge transfer, scattering); Photodetachment; Photodissociation

PETERSON, Lennart R.; Department of Physics; University of Florida; Gainesville FL 32611; USA; 904-392-2145; T; 2.1; 2.2; 3.1; Electron impact cross section calculations and photon absorption processes; Scattering

PETERSON, Randolph S.; Department of Physics; University of Tennessee, Chattanooga; Chattanooga TN 37403; USA; 615-755-4711; E; 2.1; 3.4; 4.2; X-ray production in atom-ion collisions; Low energy ion-molecule and ion-atom scattering

PETERSON, William K.; Department 52/12; Space Science Laboratory; Lockheed Palo Alto Research Lab.; Building 205; 3521 Hanover Street; Palo Alto CA 94304; USA; 415-493-4411x45244; E; 5.7; Atomic and molecular data used in space research

PETICOLAS, Warner L.; Department of Chemistry; University of Oregon; Eugene OR 97403; USA; 503-686-4601; E; 1.1; 2.3; 3.5; Molecular structure and spectra; Chemical physics; Radiative processes, polarization and resonance

PETRASSO, Richard; American Science and Engineering; 955 Massachusetts Avenue; Cambridge MA 02139; USA; 617-868-1600; E; 3.4; 5.4; X-ray emissions from magnetically confined plasmas

PETRITZ, Richard L.; 2715 Springmede Court; Colorado Springs CO 80906; USA

PETUCHOWSKI, Samuel J.; Infrared Astrophysics Branch; Lab. for Extraterrestrial Physics; NASA/Goddard Space Flight Center; Code 697; Greenbelt MD 20771; USA; 301-344-5538; E; 5.7; 3.6; 4.1; Detection of molecules in interstellar media by means of submillimeter heterodyne radiometry

PEUSE, Bruce W.; Laser Products; Coherent, Inc.; 94303; 3210 Porter Dr.; Palo Alto CA 94303; USA; 415-858-7462; 408-245-6064; E/T; 4.1; 3.6; 3.10; 3.7; Tunable laser sources research and development

PEYTON, Bernard J.; Advanced Technology; AIL Division; Eaton Corporation Melville Plant; Walt Whitman Road; Melville NY 11747; USA; 516-595-4438; E/T; 3.3; 3.6; 5.4; Spectroscopy and atmospheric physics--analysis and experimental working; Infrared, RF and microwave spectrum; Interactions of laser radiation with atoms and molecules--laser spectroscopy; A&M physics in plasmas diagnostics of magnetic fusion

PFEIFFER, Gary V.; Department of Chemistry; Ohio University; Athens OH 45701; USA; 614-594-6951; T; 1.3; 5.1; Ab initio calculations (HF-SCF) of clusters of atoms as models for catalytic centers or as models of surfaces

PFEIFFER, Loren; Bell Lab., Inc.; Room 1-C-445; Murray Hill NJ 07974; USA

PFENDER, Emil; Department of Mechanical Engineering; Heat Transfer; High Temperature; University of Minnesota; 111 Church Street SE; Minneapolis MN 55455; USA; 612-373-3907; E/T; 5.4; 3.6; Plasma heat transfer; Arc technology; Plasma chemistry and plasmas processing

PHADKE, Laxman G.; Division of National Science and Mathematics; Northeastern State University; Tahlequah OK 74464 USA; 918-456-5511x241; 918-456-1072/Home; E; 4.1; 5.4; 5.5; Optical diagnostics; Relative intensity (F-value) measurement; Vision science-color contrast sensitivity

PHANEUF, Ronald A.; Physics Division; Oak Ridge National Laboratory; Building 6003; PO Box X; Oak Ridge TN 37831; USA; 615-574-4707; E; 2.1; 4.2; 2.2; 2.3; Multiply charged ions with atoms and electrons--inelastic collisions

PHELPS, Arthur V.; Joint Institute for Laboratory Astrophysics; Quantum Physics Division (NBS); University of Colorado; S. Broadway; Boulder CO 80309; USA; 303-492-7850; 303-497-3604; E; 2.2; 5.3; Electron collisions and transport in molecular gases

PHELPS III, Frederick M.; Department of Physics; Central Michigan University; 221 Brooks Hall; Mount Pleasant MI 48859; USA; 517-774-3249; E; High resolution atomic spectroscopy

PHELPS, James O.; Department of Physics; California State University; 6000 J Street; Sacramento CA 95819; USA

PHILLIPS III, Arthur C.; Products Development; Duracell; South Broadway; Tarrytown NY 10591; USA

PHILLIPS, Donald H.; General Instrumentation Research Board; Instrument Research Center; Langley Research Center; NASA; MS 234; Hampton VA 23665; USA; 804-865-2466; 804-898-8438; T; 5.6; 5.2; 3.5; 2.3; 1.3; Atmospheric molecules--spectroscopic and chemical properties; Molecular properties using computational techniques

PHILLIPS, John G.; Department of Astronomy; University of California; 601 Campbell Hall; Berkeley CA 94720; USA; 415-642-1952; E; 3.5; Diatomic molecules--analysis of high dispersion spectra

PHILLIPS, William D.; Electrical Measurements and Standards Division; National Bureau of Standards; Building 220 Room B258; Gaithersburg MD 20899; USA; 301-921-3806 x2007; 301-921-2250; E; 3.6; 3.7; 3.9; 4.1; 1.4; Laser spectroscopy; Atoms with strong laser fields--interactions; Laser cooling and trapping of atoms; Stark effect

PHILPOTT, Michael R.; Surface Science Division K33/281; IBM Research Laboratory; 5600 Cottle Road; San Jose CA 95193; USA; 408-256-1600; T/E; 5.1; Molecules at interfaces--properties

PICARD, Richard H.; Infrared Dynamics Branch (LSI); Infrared Tech Division; AF Geophysics Lab.; Hanscom Air Force Base; Bedford MA 01731; USA; 617-861-2222; T; 5.6; 3.11; 3.6; Atmospheric physics/optics, especially airglow and aurora; statistical methods in optics/physics; Laser and molecular spectroscopy, including lineshape

PICHANICK, Francis M.; Department of Physics and Astronomy; University of Massachusetts; Amherst MA 01003; USA; 413-545-0977; E; 1.1; 1.4; 2.2; Free atoms involving magnetic resonance--structure; Low-energy electron-atom and electron-molecule collisions with high-energy resolution

PICKETT, Herbert M.; Jet Propulsion Laboratory; California Institute of Technology; 4800 Oak Grove Drive; Pasadena CA 91103; USA; 213-354-6861; E/T; 3.3; Microwave and submillimeter wavelength molecular spectroscopy

PIEBOLD, Gerald J.; Chemistry; Brown University; Providence Rhode Island 02912; USA; 401-863-3586; E; 1.1; 2.3; 3.8; 4.1; 5.1; Molecular photodissociation; Photoacoustic effect

PIERCE, Daniel T.; Electron Physics Group; Radiation Physics Division; National Bureau of Standards; Bldg. 220, Room B206; Gaithersburg MD 20899; USA; 301-921-2051; 301-921-2052; E; 5.1; 2.2; Spin polarized electron scattering; Sources and detectors of electron spin polarization

PIERCE, Thomas M.; PO Box 726; Berkeley CA 94701; USA; 415-540-5788; 3.10; 3.11; 1.3

PILLOFF, H. S.; Office of Naval Research; 800 N. Quincy Street; Arlington VA 22217; USA; 703-696-4221; T; 4.1; Superradiance calculations

PILTCH, Martin S.; Tunable Laser CHM-6; Chemistry 2; Los Alamos National Laboratory; MS J-564; P.O. Box 1663; Los Alamos NM 87545; USA; 505-667-7102; E; 3.6; 4.1; 3.1; 3.10; 3.8; IR and UV laser physics

PIMENTEL, George C.; Department of Chemistry; Laboratory for Chemical Biodynamics; University of California; Berkeley CA 94720; USA; 415-486-4355; E; 3.8; Absorption and chemiluminescent spectroscopy in cryogenic samples

PINDZOLA, Michael S.; Department of Physics; Auburn University; Auburn AL 36849; USA; 205-826-4127/4264; T; 2.2; 2.3; 3.1; 4.2; Electron excitation and ionization of atomic ions; Photoionization and photoexcitation of atoms; Rare gas-halogen kinetic studies

PINE, Alan S.; Lincoln Laboratory; MIT; 244 Wood Street; Lexington MA 02173; USA; 617-862-5500x5330; E; 3.3; 3.5; 3.6; IR molecular spectroscopy; Tunable lasers

PINGS, Cornelius J.; Chemistry Engineering Laboratory; California Institute of Technology; Pasadena CA 91125; USA; 213-795-6811x2388; E/T; 3.4; X-ray diffraction

PINKERTON, John H.; PO Box 6447; College Station TX 77844; USA

PINSON, James W.; PO Box 8415; Southern Station; Hattiesburg MS 39406; USA

PIPKIN, Francis M.; Department of Physics; Lyman Laboratory; Harvard University; Cambridge MA 02138; USA; 617-495-2910; 617-495-3386; E; 1.4; 2.3; 3.6; 1.2; 3.1; Fine structure of hydrogenic atoms--measurements; Charge exchange collisions--measurements; Molecular Rydberg states;

Molecular autoionization--measurements

PISTORESI, Denis J.; Physics Division; The Boeing Company; PO Box 3999; Seattle WA 98124; USA; 206-773-1035; T; 4.1; Electrical carbon monoxide and carbon dioxide lasers--development; Chemical hydrogen fluoride and deuterium fluoride lasers--development

PITCHFORD, Leanne C.; Manager, Discharge Physics Department; GTE Laboratories, Inc.; 40 Sylvan Road; Waltham MA 02254; USA; 617-466-2704; T; 5.3; Electron transport --Boltzmann calculation

PITTMAN, Timothy L.; National Bureau of Standards; Building 221 Room A167; Gaithersburg MD 20899; USA; 301-921-2356; E; 5.4; Plasma spectroscopy; Line broadening in plasmas

PITTS JR., James N.; Statewide Pollution Research; Fawcett Laboratory; University of California, Riverside; Riverside CA 92521; USA; 714-787-4584; E/T; 2.3; 3.7; 5.6; Photochemistry; Chemical kinetics; Air pollution

PITZER, Kenneth S.; Department of Chemistry; University of California; Berkeley CA 94720; USA; 415-642-6000; T/E; 1.1; Relativistic quantum chemistry

PITZER, Russell M.; Department of Chemistry; Ohio State University; 140 W. 18th Avenue; Columbus OH 43210; USA; 614-422-7063; T; 1.1; Electronic structure and properties of molecules

PLATT, Paul E.; 42765 Caldas Court; Fremont CA 94538; USA

PLEASANCE, Lyn D.; Lawrence Livermore National Lab.; PO Box 808; Livermore CA 94550; USA; 415-422-6155; E/T

PLEASANT, Melvin; 11310 Leesburg Place; Louisville KY 40222; USA

PLECIOUS, Robert C.; Radiation Physics Division; National Bureau of Standards; Building 245 Room C216; Gaithersburg MD 20899; USA; 301-921-2201; E; 3.4; 4.2; X-ray spectroscopy; Radiography; X-ray image processing

PLUMMER, Patricia L. M.; Department of Physics; Lab. for Atomic and Molecular Physics; University of Missouri, Rolla; 116 Physics; Rolla MO 65401; USA; 314-341-4790; T; 1.3; 2.3; 5.6; 1.6; Small polyatomic molecules and aggregates of molecules --characteristics and properties; Theoretical studies of physics and chemistry of small clusters

POE, Robert T.; Department of Physics; University of California, Riverside; Riverside CA 92521; USA; 714-787-5334/5340; T; 2.1; 2.2; Electron-atom, electron-molecule and atom-molecule scattering

POIRIER, John A.; Department of Physics; University of Notre Dame; Notre Dame IN 46556; USA; 219-283-6386; E; 3.2; 4.2; Beam foil UV spectroscopy

POL, Victor; 6 Woodsbluff Run; Fogelsville PA 18051; USA

POLAK-DINGELS, Penelope M.; Department of Chemistry; University of Maryland; College Park MD 20742; USA; 301-454-6090; E; 3.8; Laser-induced reactions in crossed beam experiments with alkali atoms

POLIAKOFF, Erwin D.; Radiological and Environmental; Research Division; Argonne National Laboratory; Building 203; 9700 S. Cass Avenue; Argonne Il 60439; USA; 312-972-7720; E; 3.1; 4.2; Photoelectron angular distributions and fluorescence polarization --investigation of photoionization and photodissociation

POLITZER, Peter; Department of Chemistry; University of New Orleans; New Orleans LA 70148; USA; 504-286-6850; 504-286-7216; T; 1.1; 5.1; Atomic and molecular energies and the electrostatic potentials at the nuclei of atoms and molecules--development of relationship

POLLACK, Edward; Department of Physics; University of Connecticut; Storrs CT 06268; USA; 203-486-3670; E; 2.1; 2.3; 4.2; Low KeV energy atomic collisions

POLO, Santiago R.; Department of Physics; Pennsylvania State University; 327 Davey Laboratory; University Park PA 16802; USA; 814-865-4972; T; 3.5; Polyatomic molecules--molecular spectroscopy (IR, Raman, resonance; Raman and fluorescence)

POMILLA, Frank R.; Department of Physics; York College of CU of New York; Jamaica NY 11451; USA; 212-969-4415; T; 2.2; 2.1; Atom-atom scattering

POMRANING, Gerald C.; Nuclear Engineering; School of Engineering; UCLA; 6291 Boelter Hall; Los Angeles CA 90024; USA; 213-825-1744; 213-476-8895; T; 5.7; 5.3; 5.4; 5.5; 5.6

POPLE, John A.; Department of Chemistry; Carnegie-Mellon University; 4400 5th Avenue; Pittsburgh PA 15213; USA; 412-578-3132; T; 1.1; Molecular structures and energies

PORTER, John H.; Computing Center; Boston University; 111 Cummington Street; Boston MA 02215; USA

PORTER, Leonard E.; Department of Physics and Astronomy; University of Montana; Missoula MT 59812; USA; 406-243-2073; 406-243-6223; E/T; 2.1; 2.2; 5.2; Stopping power of matter for charged projectiles; Energy loss spectra; Straggling of charged projectiles

PORTER, Richard N.; Department of Chemistry; State University of New York, Stony Brook; Stony Brook NY 11794; USA; 516-246-6067; T; 1.1; 3.2; 3.5; Atoms and molecules--structure, spectra and dynamics

POSHUSTA, Ronald D.; Department of Chemistry; Washington State University; Pullman WA 99164; USA; 509-335-3362; T; 1.1; Small polyatomic molecules--structure; Atoms and molecules--correlation effects; Non-adiabatic effects; Applied quantum mechanics--methods; Group theory

POST JR., Douglass E.; Plasma Physics Laboratory; Princeton University; PO Box 451; Princeton NJ 08544; USA; 609-683-2619; E/T; 5.4; 5.1; Fusion plasmas and development of diagnostic techniques for plasmas--modeling

POST, Richard S.; Department of Nuclear Engineering; University of Wisconsin; Madison WI 53706; USA; 608-263-4970; E; 5.1; Sputtered particles--characteristics

POULSEN, Peter; M-Division; Lawrence Livermore National Lab.; PO Box 808; Livermore CA 94550; USA; 415-422-6692; E; 5.1; 5.4; Neutral beams for fusion applications; Plasma/gas surface interactions

POULTNEY, Sherman K.; Sensor Systems; Military System Division; Optical Group; Perkin-Elmer Corporation; MS 950; 100 Wooster Heights Road; Danbury CT 06897; USA; 203-797-5032; E/T; 5.6; 3.1; 4.1; 3.3; 5.5

POWE, Ralph E.; Engineering and Industrial Research Station; Mississippi State University; PO Drawer DE; Miss State MS 39762; USA; 601-325-2266

POWELL, Cedric J.; Surface Science Division; National Bureau of Standards; Chemistry B-248; Gaithersburg MD 20899; USA; 301-921-2188; E; 2.2; 3.4; 3.1; 5.1; Electronic excitations in solids; Electronic excitations in surface spectroscopies (e.g., AES, XPS); Electronic de-excitation processes in atoms and solids

POWELL, Edward; Plasma Theory Branch; Plasma Physics; Naval Research Laboratory; Office of Naval Research (ONR); NRL Code 4790; 4555 Overlook Ave. SW; Washington DC 20375; USA; T; 4.1; 4.2; 5.1; 5.4

POWELL, Howard T.; L-490; Y-Division; Lawrence Livermore National Lab.; PO Bx 808; Livermore CA 94550; USA; 415-422-6149; E; 4.1; 5.3; 5.4; Atomic and molecular processes relevant to lasers for fusion purposes; High power laser development for laser fusion

POWER, Edwin A.; Department of Mathematics; University College; Gower Street; London WC1 4550; England

POWERS, E. L.; Department of Zoology; Center for Fast Kinetics Research; University of Texas; 131 Patterson; Austin TX 78712; USA; 512-471-4615; E; 3.8; 4.2; 5.2; 3.6; 4.1; Fast kinetics initiated in solution by pulsed lasers and electron beams; Fast kinetics to biological research--application

PRADHAN, Anil K.; Joint Institute for Laboratory Astrophysics; University of Colorado; P.O. Box 440; Boulder CO 80309; USA; 303-497-5257; T; 2.2; 3.2; 3.3; 1.1; Atomic processes in laboratory and astrophysical plasmas

PRATT, David W.; Department of Chemistry; University of Pittsburgh; Pittsburgh PA 15260; USA; 412-624-5074; E/T; 1.1; 1.4; 3.3; 3.6; 3.8; Structure and dynamics of molecular excited states; Laser spectroscopy; Magnetic resonance; Gas phase and condensed phase

PRATT, Richard H.; Department of Physics and Astronomy; University of Pittsburgh; 3941 O'Hara Street; Pittsburgh PA 15260; USA; 412-624-4304; T; 3.1; 2.2; 5.4; 1.1; Atomic processes at relativistic energies; Inner shell radiation processes; e-y processes at high temperature and density

PRATT, Stephen T.; Building 203 C-141; Argonne National Lab.; 9700 S. Cass Avenue; Argonne IL 60439; USA; 312-972-4199; E; 3.1; 3.6; 3.8; 3.7; Multiphoton ionization mass spectrometry and photoelectron spectrometry; Photoionization of small molecules and atoms; Electronic spectroscopy of Van der Waals molecules

PRELAS, Mark A.; Department of Nuclear Engineering; University of Missouri; Columbia MO 65211; USA; 314-882-3568; 314-882-8201; E/T; 5.4; 5.3; 3.6; 4.1; 5.2; Fusion; Direct energy conversion; Gaseous electronic;

Laser physics

PRENTICE, John K.; Orlando Technology Incorporated; 2137 Francella, NW; Albuquerque NM 87104; USA; 505-842-6606; 505-243-7694; T; 2.1; 2.2; 5.1; 3.4

PRESENT, Richard D.; Department of Physics; University of Tennessee; Knoxville TN 37916; USA; 615-974-6655; T; 1.1; Intermolecular (van der Waals) forces

PRESES, Jack M.; Chemistry Department; Brookhaven National Laboratory; Upton NY 11973; USA; 516-282-4374; 515-282-5509; E; 2.1; 3.8; 4.2; Vibrational energy transfer; IR laser chemistry; Electronic energy transfer using synchrotron radiation

PRESSMAN, Jerome; 4 Fessenden Way; Lexington MA 02173; USA

PRESTON, Steven C.; 4026 Cheena; Houston TX 77025; USA

PRICE, Jack L.; Department of Physics; North Texas State University; PO Box 5368; North Texas Station; Denton TX 76203; USA; 817-565-2004; 817-566-3938; E; 2.1; 4.2; 5.4

PRICE, Stephan D.; Infrared Branch (OPI); Optical Physics (OP); US Air Force Geophysics Laboratory; Hanscom AFB MA 01731; USA; 617-861-4552; E; 3.3; 5.6; 5.7; Rocket surveillance of sky in IR; IR astronomical emission

PRIESTLEY, Eldon B.; Corporate Research Laboratories; Exxon Research and Engineering Co.; PO Box 45; Linden NJ 07036; USA; 201-474-2544; E/T; 2.3; 3.6; 3.8; Chemical physics; IR laser chemistry; Energy deposition, flow and utilization in molecules; Gas phase laser spectroscopy

PRIOR, Michael H.; Department of Physics; Materials and Molecular Research; Lawrence Berkeley Laboratory; University of California; Building 88; Berkeley CA 94720; USA; 415-486-5088; 415-642-3686; E; 1.1; 2.3; 3.6; 4.2; Metastable ionic states--radiative lifetimes; Trapped ions--laser fluorescence; Collision and spectroscopic studies with low energy multiply charged ions; Few electron ions--fine and hyperfine structure

PRITCHARD, David E.; Department of Physics; MIT; Room 26-231; Cambridge MA 02139; USA; 617-253-6812/6816; E; 1.1; 2.1; 3.5; 3.8; Atomic and molecular collisions; Line broadening; Molecular spectroscopy; Atom-field interactions

PRITCHARD, Richard H.; Applied Technology Division; BDM Corporation; 1801 Randolph Road, SE; Albuquerque NM 87106; USA; 505-848-5568; T; 1.1; 2.3; Molecular structure-biological activity relationships

PRITT JR., Alfred T.; Chemistry and Physics Division; Rockwell International Science Center; PO Box 1085; Thousand Oaks CA 91360; USA; 805-373-4274; 213-398-7002; E; 2.1; 2.3; 3.6; 3.8; Real time and fast flow techniques used to study dynamics of chemical reactions and energy transfer among atoms and molecules and spectroscopy of new states

PRONKO, Peter P.; Materials Science Division; Argonne National Laboratory; 9700 S. Cass Avenue; Argonne IL 60439; USA; 312-972-5050; E; 4.2; 5.2; Atomic collisions and scattering in solids; Ion implantation; Rutherford scattering; Channeling in crystals

PROSNITZ, Donald; L-470; Lawrence Livermore National Lab.; PO Box 808; Livermore CA 94550; USA

PRUETT, James G.; Department of Chemistry; University of Pennsylvania; 250 S. 34th Street; Philadelphia PA 19104; USA; 214-243-4640; E; 2.3; 3.5; 5.5; Molecular beam reactions of vibrationally excited molecules; Diatomic molecular spectroscopy; Metastable state monitoring in flame combustion

PRYOR, Roger W.; 102 Silver Ridge Road; New Canaan CT 06840; USA

PUERTA, Julio J.; Dept. De Fisica; Univ. Simon Bolivar; APDO 80659; Caracas 108 6840; Venezuela

PUGH JR., Henry L.; Department of Physics; U.S. Air Force Academy; USAF Academy CO 80840; USA; 303-472-2487; E/T; 2.3; 3.1; 3.8; Nitroaromatics and nitrainines thermochemical/photochemical decomposition; EPR analysis and molecular orbital calculations to determine reaction kinetics and mechanisms

PULLEN, Bailey P.; Chemistry and Physics; Science; ORNL; Southeastern Louisiana University; PO Box 842; University Station; Hammond LA 70402; USA; 504-549-2319; 504-549-2159; E

PUTLITZ, zu Gisbert; Physics Institute; University of Heidelberg; Philosophenweg 12; Heidelberg D-6900; West Germany; 49-6221-569211; E; 1.1; 1.3; 1.4; 3.3; 3.6; Atomic spectroscopy; Elementary particles, exotic atoms; Atom-nucleus interactions

PYLE, Robert V.; Magnetic Fusion Energy Division; Lawrence Berkeley Laboratory; University of California; Berkeley CA 94720; USA; 415-486-5011; E; 2.1; 4.2; 5.4; Fusion-oriented collision physics

PYYKKO, V. P.; Department of Chemistry; University of Helsinki; Et Hesperiankatu 4; Helsinki 00100; Finland; 358-0-410566x448; 358-0-410612; T; 1.1

QUARLES, C. A.; Department of Physics; Texas Christian University; Fort Worth TX 76129; USA; 817-921-7375; E; 2.2; 3.4; 4.2; Electron scattering; Inner-shell ionization; Bremsstrahlung

QUICK JR., Charles R.; Chemistry Division; Brookhaven National Laboratory; Building 555; Upton NY 11973; USA; 516-345-4372; E; 3.2; 3.3; 3.8; Spectroscopy; IR and UV photochemistry of small molecules

QUIGLEY, Gerard P.; Group AP-4; Applied Photochemistry Division; Los Alamos National Laboratory; PO Box 1663 MS567; Los Alamos NM 87545; USA; 505-667-7753; E; 3.8; Laser-induced molecular fluorescence; Laser isotope separation

QUINN, Jarus W.; Optical Society of America; 1816 Jefferson Place, N.W.; Washington DC 20036; USA; 202-223-8130; E; 2.1; 3.5

QUIVERS, William W.; 287 Harvard Street #19; Cambridge MA 02139; USA

RAAB, Frederick J.; Department of Physics-FM 15; University of Washington; Seattle WA 98195; USA; 206-543-9196; 206-543-2770; E; 1.4; 4.1; 3.9; Time reversal non-invariance in atoms; Gravitational tests in atoms; Lorenta invariance tests in atoms

RABIN, Sheldon G.; 10 Cambridge Road; East Hanover NJ 07936; USA

RABINOVITCH, B. Seymour; Department of Chemistry; University of Washington BG10; Seattle WA 98195; USA; 206-543-1636; E; 2.1; 2.3; Intramolecular and intermolecular energy relaxation

RABITZ, Herschel; Department of Chemistry; Princeton University; Princeton NJ 08540; USA; 609-452-3917; T; 2.1; 2.3; Molecular collisions; Relaxation phenomena; Chemical reactions

RADFORD, Harrison E.; Atomic and Molecular Physics Division; Harvard-Smithsonian Center for Astrophysics; 60 Garden Street; Cambridge MA 02138; USA; 617-495-7474; E; 3.3; Millimeter and far-IR spectroscopy of free radicals

RADICATI, Filippo; Charles Evans and Associates; 301 Chesapeake Drive; Redwood City CA 94063; USA; 415-369-4567; E; 4.2; 4.4

RADOJEVIC, Vojislav; Theoretical Physics; Lab. 020; Boris Kidric Inst Vinca; PO Box 522; Beograd 11001; Yugoslavia

RADZIEMSKI, Leon; Group AP-4; Los Alamos National Laboratory; PO Box 1663; Los Alamos NM 87545; USA; 505-667-7484; E/T; 3.8; Uranium-plasma spectroscopy; Organic laser photochemistry

RAFTOPOULOS, Vassilis; Ritter OBS; University of Toledo; Toledo OH 43606; USA

RAGENT, Boris; Ames Research Center; NASA; MS 245-1; Moffett Field CA 94035; USA

RAGHAVAN, Pramila; Department of Physics; Busch Campus; Rutgers University; Freylinghuysen Road; Piscataway NJ 08854; USA

RAGLE, John L.; Department of Chemistry; University of Massachusetts; Graduate Research Center Tower A; Amherst MA 01003; USA; 413-545-2375; 413-545-2054; E/T; 1.1; 5.2; 3.3; Nuclear quadrupole resonance studies of molecular and complex structure in solids; SCF calculations of molecular properties

RAHBEE, Alfred; Optical Physics Lab.; Air Force Cambridge Research Lab.; Hanscom Field; Bedford MA 01730; USA

RAHIMZADEH, Ebrahim; 1301 15th Street, N.W., Apt. 516; Washington DC 20005; USA

RAHMAN, Aneesur; Materials Science and Technology; Argonne National Laboratory; DOE; 9700 S. Cass Avenue; Argonne IL 60439; USA; 312-972-5528; T; 5.2; Condensed systems: solids, fluids and superionic conductors; Atomic, ionic and molecular interactions and resulting condensed matter properties

RAITH, Burkhard; Inst F Experimental Physik III; Ruhr-Univ Bochum; Postfach 102148; Bochum 1 D4630; West Germany; 0234-700-3597; E; 2.1; 5.6

RAITH, Wilhelm; Physics; Fakultaet F Physik; University of Bielefeld; 8640; Universitaetsstrasse; Bielefeld D-4800; West Germany; 521-106-5394; E; 2.2; 4.2; 4.3; Spin-dependent electron scattering; Time-of-flight spectroscopy

RAJAGOPAL, A. K.; Department of Physics and Astronomy; Louisiana State University; Baton Rouge LA 70803; USA; 504-388-6835; T; 5.4; Density-functional methods: relativistic systems, finite temperature effects

RAJNAK, Katheryn; Department of Physics; Kalamazoo College; Kalamazoo MI 49007; USA; 616-383-8450; T; 1.1; 3.2; Atomic energy levels of U I and II--analysis; Crystal fields of lanthanides and actinides

RAJU, Datla V.; National Bureau of Standards; A167 Building 221; Gaithersburg MD 20899; USA

RAMAKER, David E.; Department of Chemistry; George Washington University; Washington DC 20052; USA; 202-676-6934; 202-767-3250; T; 3.1; 3.4; 5.1; Atoms and molecules in solids and at surfaces --physics and chemistry

RAMBAUSKE, Werner; 170 Acton Street; Carlisle MA 01741; USA

RAMBOW, Frederick H.; 5526 Darnell Street; Houston TX 77096; USA

RAMLER, Warren J.; RPC Industries; PO Box 327; Plainfield IL 60544; USA; 815-436-2304; E; 1.3; 2.3; 4.2; Complex molecular chains--accelerator-related chemistry pertaining to physical properties

RAMSEY, John M.; Analytical Chemistry Division; Oak Ridge National Laboratory; PO Box X; Oak Ridge TN 37830; USA; 615-574-4921; E; 3.6; Lasers united for chemical and physical analysis

RAMSEY, Norman F.; Department of Physics; Harvard University; Cambridge MA 02138; USA; 617-495-2864; E; 1.4; 3.3; 3.5; 4.2; 5.8; Parity and time reversal symmetry with atomic beams and neutron beams

RAND, Stephen C.; Physical Science Division; IBM Research Laboratory; 5600 Cottle Road; San Jose CA 95193; USA; 408-256-1169; E/T; 5.2; Dynamic nuclear spin interactions on electronic decay of metastable states of ions in solids--influence

RANDALL, Russ R.; Dresser Atlas DC-14; PO Box 1407; Houston TX 77251; USA

RANGANATHAN, Dilip; B-6/19 Safdarjung Enclave; New Delhi 029 7251; India

RANK, David H.; Department of Physics; Pennsylvania State University; University Park PA 16802; USA; 814-865-7533; E/T; 3.6; 4.1; Optics and spectroscopy; Laser physics

RAO, Pemmaraju V.; Department of Physics; Emory University; Atlanta GA 30322; USA; 404-727-4297; Atomic and molecular collisions; A&M physics and nuclear physics

RAPP, Donald; Department of Environmental Sciences; Mechanical and Chemical Systems; Jet Propulsion Laboratory; California Institute of Technology; MS 157-316; 4800 Oak Grove Drive; Pasadena CA 91109; USA; 818-354-4931; E/T; 2.1; 1.1; Use of concurrent processing in computation

RASOR, Ned S.; Rasor Associates, Inc.; 253 Humboldt Court; Sunnyvale CA 94086; USA; 408-734-1622; E/T; 5.1; 5.3; Cesium vapor thermionic energy converter --emission and surface physics, and gaseous electronics

RAST JR., Howard E.; Marine Sciences and Technology; Electronic Material Sciences; Optical Electronics Branch; U.S. Naval Ocean Systems Center; 271 Catalina Boulevard; San Diego CA 92152; USA; 619-225-7976; 619-225-6591; E; 1.1; 3.2; 3.3; 3.6; 3.10; Rare earth ions in condensed matter, optical properties; Optical spectroscopy of solids, liquids, gases; Electro-optical properties of materials; Lattice and acoustical vibrations of solids

RATKOWSKI, Anthony J.; Department of Defense; Infrared Technology Division; Air Force Geophysics Lab.; Atmospheric Backgrounds Branch(LSP); Hanscom AFB MA 01731; USA; 617-861-4910; E/T; 2.1; 2.2; 2.3; 3.1; 3.2; Infrared atmospheric phenomenology

RATNER, Mark A.; Department of Chemistry; Northwestern University; Evanston IL 60201; USA; 312-491-5652; T; 1.1; 1.3; Molecular electronic structure and response phenomena

RAU, A. R.; Department of Physics and Astronomy; Louisiana State University; Baton Rouge LA 70803; USA;

504-388-6841; 504-767-0153; T; 1.2; 2.2; 3.9; 3.1; 5.4; Highly excited states in external fields--strong perturbations; Doubly-excited states and connection to threshold ionization of atoms; electrons and yield of ion pairs in a medium due to an incident energetic charged particle--variational calculations

RAUSCHENBACH, Kurt; Ithaca Commons; Center Ithaca Apt 335; Ithaca NY 14850; USA

RAVISHANKARA, Akkihebbal R.; Molecular Sciences Branch; Electromagnetics Laboratory; Aeronomy Laboratory; National Oceanic and Atmospheric Admin.; R/E/AL2; 325 Broadway; Boulder CO 80303; USA; 303-497-5821; E; 2.3; 5.6; 5.5; 3.6; 3.8; Atmospheric chemistry; Chemical kinetics; Combustion chemistry; Photochemistry

RAY, Gary W.; Molecular Physics and Chemistry Section; Jet Propulsion Laboratory; 4800 Oak Grove Drive; Pasadena CA 91103; USA; 213-354-3800; E; 2.3; 3.8; 4.4; 5.6; Gas phase chemical kinetics and photochemistry of atmospheric interest; Photoionization mass spectrometry

RAY, J. A.; Instrument and Controls Division; Oak Ridge National Laboratory; PO Box X; Building 3042; Oak Ridge TN 37830; USA; 615-574-5535; E; 4.3; 5.4; 4.2; Atomic and molecular physics experiments--instrumentation and diagnostic development

RAY, Pinaki S.; Fakultaet Fuer Physik; Univ Bielefeld; Universitaetsstrasse 25; Bielefeld 1 4800; West Germany

RAYBORN, Grayson H.; Department of Physics and Astronomy; University of Southern Mississippi; Hattiesburg MS 39401; USA; 601-266-7206; E; 3.1; 4.4; Neutral species of carbon--photoionization mass spectrometric study; Ionization and appearance potentials--measurement

RAYBURN, Louis A.; Department of Physics; University of Texas, Arlington; PO Box 19059; Arlington TX 76019; USA; 817-273-2972; E; 3.4; 4.2; L-shell X-ray production cross sections by protons--measurement; Human hair--trace element analysis

RAYMAN, Marc D.; Joint Institute for Laboratory Astrophysics; University of Colorado; Boulder CO 80309; USA; 303-492-7387; 303-492-7789; E; 4.1

RAYMER, Michael G.; Institute of Optics; University of Rochester; Rochester NY 14627; USA; 716-275-4846; E/T; 3.6; 3.7; 3.10; 3.11; 4.1

RAYMONDA, John W.; High Energy Laser Technology Div.; Bell Aerospace-Textron; PO Box 1; Buffalo NY 14240; USA; 716-297-1000x7453; E; 4.1; 5.3; Chemical laser research and development; New atomic and molecular laser candidates--identification and testing; Laser medium--diagnostics and spectroscopy

RAYNOR, Susanne; Department of Chemistry; Harvard University; 12 Oxford Street; Cambridge MA 02138; USA; T; 2.1; 3.2; Molecular collisions; Hn species--spectra; Hydrogen ions--complex

READER, Joseph; Atomic Spectroscopy Group; Atomic and Plasma Radiation Division; National Measurement Laboratory; National Bureau of Standards; Gaithersburg MD 20899; USA; 301-921-2011; E; 1.1; 3.2; Highly ionized atoms--spectroscopy

READING, John F.; Department of Physics; Texas A&M University; College Station TX 77843; USA; 713-845-7717 x5073; T; 2.1; Ion-atom collisions in the MeV/amu energy range--calculations

READING, Melissa M.; 1240 Asti Court; Livermore CA 94550; USA; 415-467-5032; E; 3.7

RECK, Gene P.; Department of Chemistry; Wayne State University; Detroit MI 48202; USA; 313-577-2602; E; 2.3; 3.8; 3.6; Laser driven chemistry; Negative ions; Atomic and molecular beams

REDD, Emmett R.; Department of Physics; University of Missouri, Rolla; Rolla MO 65401; USA

REDDING, Rogers W.; Department of Physics; North Texas State University; Denton TX 76203; USA; 817-565-2630; T; 1.1; 2.3; 3.3; Vibration-rotation in small molecules; Curve crossing phenomena

REDDY, K. V.; Department 223, Building 110; McDonnell Douglas Research Lab.; PO Box 516; St. Louis MO 63166; USA; 314-233-2542; E; 2.3; 3.6; State selected species and laser spectroscopy--dynamics

REDDY, Satti P.; Department of Physics; Molecular Physics; Memorial University of Newfoundland; Elizabeth Avenue; St. John's NF A1B3X7; Canada; 1-709-737-8838; 1-709-737-8879; E; 1.1; 2.1; 3.2; 3.3; 3.6; Collision-induced infrared spectra of simple molecules and simple molecules; Electronic spectra of simple molecules; Laser-induced fluorescence in simple molecules; Infrared spectra of molecules hydrogen, nitrogen, oxygen etc.

REDI, Martha H.; Department of Physics; Princeton University; Princeton NJ 08544; USA; 609-452-4367/4400; T; 1.1; 1.3; 2.3; Quantum theory to structure-function of biomolecules--application; Tunneling phenomena in molecular physics

REDI, Olav; Department of Physics; New York University; 4 Washington Place; New York NY 10003; USA; 212-598-3422; E; 1.1; 3.2; 4.2; Isotope shift and hyperfine structure in atomic spectra by high resolution spectroscopy--measurements

REDINGTON, Rowland W.; GE Research and Development Center; PO Box 8; Schenectady NY 12301; USA

REDISH, Edward F.; Department of Physics; University of Maryland; College Park MD 20742; USA

REDMON, Michael J.; Chemical Dynamics Corporation; Suite N140; 1550 W. Henderson Road; Columbus OH 43220; USA; 614-459-2145; T; 2.1; Energy transfer in atomic and molecular collisions

REE, Francis H.; Physics H-Division; Lawrence Livermore National Lab.; PO Box 808 L355; Livermore CA 94550; USA; 415-422-7234; T; 1.1; Intermolecular potentials by ab initio methods--computations

REED, David A.; Charles Evans and Associates; 301 Chesapeake Drive; Redwood City CA 94063; USA; 415-369-4567; E; 5.1

REED, Kennedy; Department of Physics; V-Divisions; Lawrence Livermore National Lab.; Box 808 L-296; Livermore CA 94550; USA; 415-423-1112; T; 2.2; 2.1; 1.1; 5.4; Atomic and molecular collisions; Structure and properties of atoms and molecules; Interface between A&M and other areas of science and technology

REED JR., Sidney G.; 11104 Waycroft Way; Rockville MD 20852; USA

REES, M. H.; Geophysical Institute; University of Alaska-Fairbanks; Fairbanks AK 97775; USA; 907-479-7564; E/T; 5.6; Atomic and molecular processes to auroral and airglow; spectroscopy to aeronomic processes--application

REEVES, Richard L.; Eastman Kodak Research Laboratories; 1999 Lake Avenue; Rochester NY 14650; USA; 716-477-7811; E; 2.3; 1.3; 2.3; 5.2; Large organic ions with metal ions, polymers and colloidal aggregates in solution--kinetics and thermodynamics of interaction

REEVES, Tricia M.; Department of Physics; Kansas State University; Manhattan KS 66506; USA; T; 2.1

REGAN, Thomas H.; Research; Administration; Eastman Kodak Co.; 1899 Lake Avenue; Rochester NY 14650; USA; 716-477-7594

REGISTER, David F.; Physics and Chemistry Division; Code 183-601; Jet Propulsion Laboratory; 4800 Oak Grove

Drive; Pasadena CA 91103; USA; 213-354-7233; E; 2.2; 3.8; Electron-atom/molecule differential cross sections and spectroscopy; Laser-excited electron scattering; Electron-photon coincidence measurements

REICHARDT, John W.; Kaman Science; PO Box 7463; Colorado Springs CO 80933; USA; 303-599-1883; E; 5.3; Ions--production and transport

REID, Charles E.; Quantum Theory Project; Department of Chemistry; University of Florida; Gainesville FL 32611; USA; 904-392-1597; T; 1.1; Lower bounds to energy eigenvalues of atoms and molecules--calculations

REID, Ian; Faculty of Engineering; University of Western Ontario; London Ontario N6A5B9; Canada

REID JR., Roderick V.; Department of Physics; University of California, Davis; Davis CA 95616; USA; 916-752-1139/1500; T; 1.1; Relativistic g-factors in simple molecules

REIDY, James J.; Department of Physics; University of Mississippi; University MS 38677; USA; 601-232-5322; E; 1.3; 1.4; 1.1; 5.8

REIHL, Bruno; IBM Zurich Research Lab.; Saumerstrasse 4; Ruschlikon 8803; Switzerland

REILLY, James P.; Department of Chemistry; Indiana University; Bloomington IN 47405; USA; 812-335-1980; E; 2.3; 3.5; 3.1; 3.6; 3.8; Transients--molecular spectroscopy

REINER, Robert H.; Materials and Chemistry; Technical Division; Oak Ridge Gaseous Diffusion Plant; Martin Marietta Energy Systems; PO Box P MS271; Oak Ridge TN 37830; USA; 615-576-1501; 615-574-9966; E; 3.6; 3.8; 2.3; 5.4; 5.5; Uranium isotope separation; Fluoride chemistry; Micon temperature materials

REINHARDT, William P.; Joint Institute for Laboratory Astrophysics; University of Colorado; Boulder CO 80309; USA; 303-492-8857; T; 2.3; 3.7; 3.8; Atoms in intense fields; Multiphoton processes; Chaotic/classical quantum dynamics of molecules; Mathematical physics and molecular photabsorption

REINOVSKY, Robert E.; Plasma Physics Branch; Advanced Technology Division; AF Weapons Laboratory; Kirkland AFB NM 87117; USA; 505-844-3672; E; 3.4; 5.4; High energy density plasma physics

REISS, Howard R.; Department of Physics; American University; Washington DC 20016; USA; 202-686-2548; T; Multiphoton and intense field electromagnetic interactions with atoms, ions and molecules--calculational techniques

REISS, Howard; Department of Chemistry; University of California, LA; 405 Hilgard Avenue; Los Angeles CA 90024; USA; 213-825-3029; E/T; 1.3; 3.7; 3.8; Nucleation and growth to detection and amplification in photochemistry, spectroscopy and thermodynamics--application

REMPFER, Gertrude F.; PO Box 268-B; Route 1; Forest Grove OR 97116; USA

REN, Shangfen; Department of Physics; Texas A&M University; College Station TX 77843; USA; 409-845-4853

RENDINA, Rodney A.; 15 Barisano Way; Nashua NH 03063; USA

RENNER, Darwin S.; 1314 Cedar Hill Avenue; Dallas TX 75208; USA

RENTZEPIS, P. M.; Physical and Inorganic; Chemistry Division; Bell Telephone Laboratories; 600 Mountain Avenue; Murray Hill NJ 07974; USA; 201-582-4785; E; 2.3; 4.1; 3.6; 3.8; Picosecond spectroscopy

REPKO, Wayne W.; Department of Physics; Michigan State University; East Lansing MI 48824; USA; 517-353-0833; T; 1.3; 1.4; QED and weak interactions in atomic systems and exotic systems such as positronium and muonium--tests

RESCIGNO, Thomas N.; Physics; V-Division; Lawrence Livermore National Lab.; University of California; 808; Livermore CA 94550; USA; 415-422-6210; T; 2.2; 3.1; 1.1; Electron-molecule scattering; Photoionization; Resonance phenomena

RETZLOFF, David G.; Department of Chemical Engineering; University of Missouri; 1030 Engineering; Columbia MO 65211; USA; 314-882-4036; T; 2.3; 3.9; 4.2; 5.2; Molecules in the presence of external fields or through interaction with crystal catalysts and ultimate rearrangement to other molecular species--destabilization

REUTHER, James J.; Manufacturing and Advanced Materials; Mechanical Engineering; Battelle Columbus Laboratories; Pennsylvania State University; 505 King Avenue; Columbus OH 43201; USA; 614-424-7916; E; 5.5; 2.3; 3.2

REZ, Peter; Charlwoods Road; VG Microscopes, Ltd.; East Grinstead; Sussex RH19 2JQ; England

RHEE, Moon-Jhong; Electrical Engineering Department; University of Maryland; College Park MD 20742; USA

RHIM, Won-Kyu; Information Science Division; Jet Propulsion Laboratory; Code 183-401; 4800 Oak Grove Drive; Pasadena CA 91103; USA; 213-354-5547; E/T; 1.1; 5.2; Molecular structure and motions in solids using solid state NMR--characterization

RHODES, Charles K.; Department of Physics; University of Illinois, Chicago; PO Box 4348; Chicago IL 60680; USA; E/T; 2.1; 3.1; 3.4; 3.7; 3.10

RHODES II, Richard A.; 205 NW Monroe Circle N.; St. Petersburg FL 33702; USA

RHODIN, Thor N.; Department of Applied Physics; College of Engineering; Cornell University; Ithaca NY 14853; USA; 607-256-4068; 607-273-7062; E/T; 4.4; 5.1; Synchrotron radiation physics; Electronic structure of metals; Semiconductor interfaces; Surface/interfaces solid state physics

RICE, James K.; Laser Physical Chemistry Div 4218; Sandia National Laboratories; PO Box 5800; Albuquerque NM 87185; USA; 505-844-4135; E; 3.1; 3.8; 4.2; Laser-induced chemical reactions in small molecules; Photochemistry; Photoionization

RICE, Stuart A.; Chemistry and James Franck Institute; Physical Sciences; University of Chicago; 5640 Ellis Avenue; Chicago IL 60637; USA; 312-962-7199; 3.8; 5.2; Elementary process in photochemistry; Intramolecular energy transfer in isolated molecules and induced by collisions

RICE, T. Maurice; Institute Theoretische Physique; ETH-Honggerberg; Zurich CH8093; Switzerland

RICH, Arthur; Department of Physics; University of Michigan; Ann Arbor MI 48109; USA; 313-764-2408; E; 1.3; 1.4; QED and weak interactions using positrons and positronium--fundamental tests

RICH, Joseph W.; Physical Sciences Department; Experimental Research Division; Advanced Technology Center; Calspan Corporation; PO Box 400; Buffalo NY 14225; USA; 716-631-6728; E; 2.1; 2.2; 2.3; 3.8; 3.9; Gas laser development, gas laser discharges; Laser induced chemistry and isotope separation; Plasma chemistry; Energy transfer in molecular collisions

RICH, Nathan H.; Department of Physics; Memorial University of Newfoundland; St. John's Newfoundland A1B 3X7; Canada; 709-737-8886; E; 3.10; 3.3; 3.5

RICHARD, Patrick; Department of Physics; J.R. Macdonald Laboratory; Kansas State University; Cardwell Hall; Manhattan KS 66506; USA; 913-532-6783; 913-532-6777; E; 4.2; 2.1; 2.3; Ionization, excitation and charge exchange in ion-atom collisions; X-ray spectroscopy of few electron ions

RICHARDS, Thomas J.; 1011 W. Stratford Drive; Peoria

IL 61614; USA

RICHARDSON, James W.; Department of Chemistry; Purdue University; West Lafayette IN 47907; USA; 317-494-5258; T; 1.1; 3.2; 5.2; Bonding theory; Transition-metal atoms, ions, complex ions and crystalline compounds--spectroscopy

RICHARDSON, Jeffery H.; Chemistry and Materials Science Div.; Lawrence Livermore National Lab.; PO Box 808; Livermore CA 94550; USA; 415-422-6350; E; 3.6; 3.8; 4.1; Laser spectroscopy; Photoelectrochemistry; Laser instrumentation and techniques

RICHARDSON, Martin C.; College of Engineering and Applied Science; Laboratory for Laser Energetics; University of Rochester; Rochester NY 14627; USA

RICHARDSON, Ralph J.; SRI International; 333 Ravenswood Avenue; Menlo Park CA 94025; USA; 415-326-6200x3948; E; 3.6; 4.1; Excitation techniques for molecular oxygen and kinetics of the chemical oxygen-iodine transfer laser

RICKS, Douglas W.; Optical Sciences Center; University of Arizona; Tucson AZ 85721; USA

RIES, Richard R.; Scientific, Technological and International Affairs; Office of Assistant Director; National Science Foundation; 1800 G Street NW; Washington DC 20550; USA; 202-357-7477; 703-360-6122; E; 3.9; 4.3; 4.4

RIGROD, William W.; Laser Technology Division L-9; Los Alamos National Laboratory; PO Box 1663; Los Alamos NM 87545; USA; T; 3.6; 4.1; 5.2; Solids and gases to high-intensity laser radiation--response; Nonlinear optics

RILEY, Merle E.; Laser Theory Division 4211; Sandia National Laboratories; PO Box 5800; Albuquerque NM 87185; USA; 505-844-3141; T; 2.2; 2.3; 3.7; Electron scattering activity; Resonant multiphoton excitation; Chemical kinetics

RIMBEY, Peter R.; Boeing Aerospace Co.; POB 3999 MS4240; Seattle WA 97124; USA; 206-655-9130; T; 5.2; Electromigration; Hydrogen in metals; Atomic diffusion in solids

RING, James W.; Department of Physics; Hamilton College; Clinton NY 13323; USA; 315-859-7510; E/T; 1.1; Hydrogen-bonded chains of molecules--vibrational modes

RINK, John P.; Group AP-3; Los Alamos National Laboratory; PO Box 1663 MS565; Los Alamos NM 87545; USA; 505-667-6686; E; 3.6; 3.8; 4.1; Laser research, IR and UV; Laser development; Laser isotope separation of uranium

RINKER JR., George A.; T-1 MSB212; Theoretical; Los Alamos National Laboratory; 1663; Los Alamos NM 87545; USA; 505-667-2516; 505-667-5061; T; 1.3; 1.4; 1.1; 5.2; 5.4

RISLEY, John S.; Department of Physics; North Carolina State University; Raleigh NC 27695; USA; 919-737-2524; E; 1.1; 2.1; 2.2; Atomic collisions with protons and negative hydrogen ions; Autodetaching states; Electric field mixing on hydrogen atom states produced in collisions--effect; Electron impact VUV photoemission; Absolute cross sections

RITCHIE, R. H.; Health and Safety Research Division; Oak Ridge National Laboratory; Oak Ridge TN 37830; USA; 615-574-6208; T; 5.1; Particle-solid state physics; Collective electron states in matter; Charged particles in matter--energy loss

RITTNER, Edmund S.; 8800 Fallen Oak Drive; Bethesda MD 20817; USA

RIZZO, Joseph; Laboratory for Laser Energetics; University of Rochester; 250 E. River Road; Rochester NY 14627; USA; 716-275-5101; E; 3.6; 5.1; Lasers to inertial confinement fusion studies--application

ROBERDS, Richard M.; Clemson University; 126 Freeman Hall; Clemson SC 29631; USA

ROBERTS, James R.; Atomic and Plasma Radiation Division; National Bureau of Standards; Building 221, Room A167; Gaithersburg MD 20899; USA; 301-921-2356; E; 2.2; 5.4; Transition rates in highly ionized atomic species--determination

ROBERTS, Thomas G.; U.S. Army Missile Command; DRSMI-RHB; Redstone Arsenal AL 35898; USA; 205-876-5176; E; 3.6; 4.2; 5.4; Laser saturated absorption spectroscopy; F-values and development of spectroscopic diagnostic techniques for plasmas and lasers--measurements

ROBERTSON, H. S.; Department of Physics; University of Miami; PO Box 248046; Coral Gables FL 33124; USA; 305-284-2323; T/E; 5.3; 5.4; Direct current glow discharges--moving striations; Plasma physics

ROBERTSON, M. M.; Division 2152; Sandia National Laboratories; PO Box 5800; Albuquerque NM 87185; USA; 505-844-2472; E; 4.1; 4.3; Electro-optics including optical fibers; Optical sources including atomic transition devices

ROBERTSON, Munro V.; 4320 E. Pan American NE Apt. #270; Albuquerque NM 87107; USA

ROBERTSON, William W.; 3505 Cherry Lane; Austin TX 78703; USA

ROBIN, Melvin B.; Chemical Physics Division; AT&T Bell Laboratories; 600 Mountain Avenue; Murray Hill NJ 07974; USA; 201-582-6372/6702; E; 1.1; 1.2; 3.1; 3.2; 3.5; Nonlinear laser spectroscopy of large molecules

ROBINSON, David B.; PO Box 530342; 10540 N.E. 4th Avenue; Miami FL 33138; USA; 305-754-0027; T; 2.3; Cesium phenomena--physics and chemistry

ROBINSON, Edward J.; Department of Physics; New York University; 4 Washington Place; New York NY 10003; USA; 212-598-3695/3661; T; 3.6; 3.7; 3.10; 2.1; 2.2; Dynamics of systems with finite numbers of states; Fundamental quantum mechanical problems

ROBINSON, Edward L.; Department of Physics; University of Alabama, Birmingham; University Station; Birmingham AL 35294; USA; 205-934-5249; 4333/4736; E; 4.2; 3.4; 5.7; 5.7; Inner shell effects by collision and PIXE

ROBINSON, G. W.; Department of Chemistry; Picosecond/Quantum Radiation Lab.; Texas Tech University; PO Box 4260; Lubbock TX 79409; USA; 806-742-3099; E/T; 3.8; 5.2; 3.8; Properties of liquid water and chemical reactions in liquid water; Ultrafast time-resolved spectroscopy and chemical mechanistics using lasers

ROBINSON, Hugh G.; Department of Physics; Duke University; Durham NC 27706; USA; 919-684-8226; E; 1.4; 2.3; Atomic constants--precision determinations; Alkalis and noble gases--Penning ionization cross sections

ROBISCOE, R. T.; Department of Physics; Montana State University; Bozeman MT 59717; USA; 406-994-3614; E/T; 1.4; Parity violation in atomic hydrogen

ROCKWOOD, Stephen D.; Associate Director; Los Alamos National Laboratory; PO Box 1663; Los Alamos NM 87545; USA; 505-667-8682; E; 4.1; 5.3; 3.8; 3.6; 3.7

RODGERS, James E.; Radiation Med/Bles Building; Georgetown University Hospital; 3800 Reservoir Road, NW; Washington DC 20007; USA

RODRIGUEZ-TRELLES, Felix L.; Dept. De Fisica; Fac CS Exactas Y Naturales; Pabellon 1 Ciudad Univ.; Buenos Aires 1428; Argentina

RODRIGUEZ-VIDAL, Maximino; Fac Ciencias Fisicas; Univ. Complutense; Ciudad Universitaria; Madrid 3; Spain

ROELLIG, Leonard O.; 167 East 67th Street; New York NY 10021; USA

ROESLER, Fred L.; Department

of Physics; University of Wisconsin; 1150 University Avenue; Madison WI 53706; USA; 608-262-1495; E; 2.1; 2.3; 3.1; Atoms with photons, electrons and ions--interaction

ROGERS, James W.; Office of Environment and Energy (AEE-300); Federal Aviation Administration; 800 Independence Avenue, SW; Washington DC 20591; USA; 202-755-8933; E/T; 5.6; High altitude pollution

ROGERS, Max T.; Department of Chemistry; Michigan State University; East Lansing MI 48824; USA; 517-353-9410; E/T; 1.1; 5.2; Molecular structure and valence theory; NMR, e-spin resonance and nuclear quadropole resonance spectroscopy; Electric dipole moments and transition metal complexes; Fluorine compounds

ROGERS, Wade T.; Electron Physics Group; Radiation Physics Division; National Bureau of Standards; Building 220, Room B206; Gaithersburg MD 20899; USA; 301-921-2051; E; 2.2; 3.8; Spin-polarized electron-atom scattering; Optical pumping

ROHATGI, Vijay K.; Plasma Physics Division; Bhabha Atomic Research Center; Trombay; Bombay Maharashtra 400085; India; 5510558; 8120521; E; 5.1; 5.4; 5.5; 4.3

ROL, Pieter K.; Dept. of Chemical and Mechanical Engineering; College of Engineering; Wayne State University; 1191 MacKenzie Hall; Detroit MI 48202; USA; 313-577-3879; 313-577-3800; E; 5.1; 2.3; Ion implantation in metals; Ion molecule reactions

ROLLEFSON, R.; Department of Physics; University of Wisconsin; 1150 University Avenue; Madison WI 53706; USA; 608-263-6289; E; 1.1; 3.3; Charge distribution in small molecules and IR intensities

ROLSTON, Steven L.; Department of Physics; SUNY, Stony Brook; Stony Brook NY 11794; USA

ROMAN, Ward C.; Plasma/Fluid Dynamics Laboratory; United Technologies Research Center; MS 16; Silver Lane; East Hartford CT 06108; USA; 203-727-7590; E/T; 3.9; 4.4; 5.4; High temperature UF6 plasma research; Uranium reactors; Plasma systems--mass spectrometric diagnosis; Plasma systems with magnetic fields--interaction

ROMESSER, Thomas; Plasma Physics Division; TRW Defense and Space Systems Group; R1-2044; 1 Space Park; Redondo Beach CA 90278; USA; 213-536-3266; E; 5.4; Isotope separation; Atomic, ionic and molecular research data

RON, Akiva; Racah Institute Physics; Hebrew University Jerusalem; Jerusalem; Israel

RONN, Avigdor M.; Department of Chemistry; Brooklyn College of CUNY; Bedford Avenue and Avenue H; Brooklyn NY 11210; USA; 212-780-5108; 212-790-4225; E; 2.1; 3.8; 3.7; 4.1; 5.5; Laser induced photochemistry; Energy transfer; Kinetics; Isotope separation

ROOT, John W.; Department of Chemistry; University of California, Davis; Davis CA 95616; USA; 916-752-0942/0838; E/T; 2.3; 4.2; Chemical kinetics with atomic fluorine and other unstable radical species; Very high energy neutral polyatomic molecules--behavior

ROOT, Robert G.; 8 Forrester Road; Wakefield MA 01880; USA

ROOTHAN, Clemens C.; Department of Physics; University of Chicago; Chicago IL 60637; USA; 312-753-8315; Theoretical atomic and molecular physics

ROSADO, John A.; 10519 Edgemont Drive; Adelphi MD 20783; USA

ROSARIO-GARCIA, Efrain; Department of Physics; University of Puerto Rico; Mayaguez 00708; Puerto Rico

ROSE, David J.; Massachusetts Institute of Technology; Room 24-109; Cambridge MA 02139; USA

ROSE, Timothy L.; Research Division; EIC Laboratories, Inc.; 111 Downey Street; Norwood MA 02062; USA; 617-769-9450; E; 2.3; 3.8; 5.1

ROSEN, Arne; Department of Physics; Chalmers University of Technology; Gothenburg S41296; Sweden; 00946-31-810100

ROSEN, David I.; Physical Sciences, Inc.; 30 Commerce Way; Woburn MA 01801; USA; 617-933-8500; E; 3.6; 5.4; 5.5; Laser produced plasmas

ROSEN, Gerald; Physics and Atmospheric Sciences; Drexel University; 415 Charles Lane; Wynnewood PA 19096; USA; 215-896-8727; 215-895-2721; T; 1.2; 2.3; 5.5; Theoretical quantum mechanics--general theory and analytical relations pertaining to atomic bound states

ROSENBERG, Leonard; Department of Physics; New York University; 4 Washington Place; New York NY 10003; USA; 212-598-7635; T; 2.2; 3.8; Bound state amd scattering problems--development of approximation methods; Electron-atom scattering in a laser field

ROSENBERG, Richard A.; Synebrotran Radiation Center; Research; University of Wisconsin; 3725 Sehneider Drive; Stoughton WI 53589; USA; 608-873-6651; E; 1.1; 2.3; 3.1; 3.4; 4.2; Photon-stimulated description; Surface chemistry; Molecular solids

ROSENBERGER, Albert T.; Department of Physics and Atmospheric Sciences; Drexel University; Philadelphia PA 19104; USA; 215-895-2781; E/T; 3.8; 4.1; 5.4; Superradiance; Optical bistability; Collision broadening rates--calculation

ROSENBERGER, Franz E.; Department of Physics; University of Utah; Salt Lake City UT 84112; USA; 801-581-8372; E/T; 5.2; High-temperature vapors via matrix isolation and mass spectroscopy--characterization; Effusion vapor pressure studies

ROSENBLATT, Gerhard M.; Department of Chemistry; Pennsylvania State University; 152 Davey Laboratory; University Park PA 16802; USA; 814-865-7242; E; 1.1; 4.1; 5.1; Solid-state vapor kinetics and equilibria; Raman spectroscopy; High-temperature molecules--properties

ROSENBLUH, Michael; Department of Physics; Francis Bitterman Division; National Magnet Laboratory; Massachusetts Institute of Technology; NW14-2225; 150 Albany Street; Cambridge MA 02139; USA; 617-253-5536/7790; E; 1.2; 3.6; 3.9; Helium Rydberg states--laser magnetic resonance spectroscopy; Atomic spectroscopy in intense magnetic fields

ROSENDORFF, Simcha; Department of Physics; Technion; Haifa; Israel

ROSENTHAL, Myron M.; 48 Tall Oaks Drive; Wayne NJ 07470; USA

ROSNER, Baruch; Department of Physics; Technion; Haifa; Israel

ROSNER, S. David; Department of Physics; University of Western Ontario; London Ontario N6A3 K7; Canada; 519-679-2996; 519-679-2568; E; 3.6; 1.4; 3.3; 1.1

ROSOCHA, Louis A.; National Research Group, Inc.; PO Box 5321; Madison WI 53705; USA; 608-231-2242; T/E; 3.6; 5.3; Plasma chemistry--laser and oxygen discharges; Energy deposition into various species in gases subjected to electrical discharges

ROSS, John; Department of Chemistry; Stanford University; Stanford CA 94305; USA; 415-497-9203; T; 2.3; Chemical reactions--dynamical theory

ROSS, William R.; Department of Physics; University of New Brunswick; PO Box 4400; Fredericton NB E3B5A3; Canada

ROSSI, Angelo R.; Department of Chemistry; University of Connecticut; Storrs CT 06268; USA; 203-486-3460; T; 1.2; Molecular Rydberg states; Positive ions (radical cations); Far UV photochemistry

ROSTAS, M. Francois; Reside Voltaire 1; Chatenay Malabry 92290; France

ROSZMAN, Larry; Atomic and Plasma Radiation Division; National Bureau of Standards; Building 221, Room 2071; Gaithersburg MD 20899; USA; 301-921-2071; T; 5.4; Atomic processes and plasmas--interaction between

ROTENBERG, Manuel; Dept. of Electrical Engineering and Computer Science; University of California, San Diego; Q-003; La Jolla CA 92093; USA; 714-452-3557; T; 1.1; 2.2; Atomic structure; Electron collisions

ROTHBERG, Gerald M.; Department of Mathematics and Metallurgy Engineering; Stevens Institute of Technology; Hoboken NJ 07030; USA

ROTHE, Erhard W.; RIES; Engineering College; Wayne State University; Detroit MI 48202; USA; 313-577-3865; 313-577-3800; E; 2.1; 3.6; 3.1; 3.8; Molecular beam studies of atomic collisions; Laser interactions with molecules; Steric effects in AMC

ROTHENBERG, Joshua E.; IBM Watson Research Center; PO Box 218; Yorktown Heights NY 10598; USA; 914-945-3749; 914-945-1493; E; 1.1; 3.10; 3.11; 4.1; 5.1; Ultrafast (picosecond and sub-picosecond) measurements of atoms, molecules, and solid state systems

ROTHMAN, Laurence S.; Infrared Physics Branch; Optical Physics Division; Air Force Geophysics Laboratory; US Air Force; Hanscom AFB MA 01731; USA; 617-861-2336; T; 3.3; 1.1; Data base of basic spectroscopic constants for atmospheric gases active in IR and microwave--maintenance; Transmission and emission through atmospheric paths--calculation

ROTHSTEIN, Jerome; Department of Computer and Information Science; Ohio State University; 2036 Neil Avenue Mall; Columbus OH 43210; USA

ROTHSTEIN, Stuart M.; Department of Chemistry; Brock University; St. Catharines Ontario L2S 3A1; Canada; 416-688-5550; 416-945-1338; T; 1.1

ROUNTREE, Steven P.; 2727 Kaliste Saloom Road, Ste. 600; Lafayette LA 70508; USA

ROUSSEL, Robert L.; 208 W. 23rd Street Apt. #912; New York NY 10011; USA

ROWAN, William L.; Fusion Research Center; University of Texas; Austin TX 78712; USA; 512-471-4559; E; 5.4; Tokamak plasmas--effects of impurities

ROWE, John E.; Department of Physics; University of Florida; Gainesville FL 32611; USA

ROY, Amitava; Department of Materials Science Engineering; Cornell University; Bard Hall; Ithaca NY 14853; USA

ROY, Denis; Physics Department; CRAM; Universite Laval; Pav Vachon; Quebec QC G1K 7P4; Canada; 418-656-5365; 418-656-3120; E; 2.2; 3.1; 5.1; 4.3; 1.1; Electron spectroscopy for study of absorbed and gaseous species; Electron optics applied to electron spectroscopy

ROY, Ronald A.; PO Box 790; Yale Station; New Haven CT 06520; USA

ROY, Swati; Department of Physics; Texas Christian University; Fort Worth TX 76129; USA

ROYCE, Barrie S.; Dept. of Mechanical and Aerospace Engineering; Applied Physics and Materials; Princeton University; D416 Duffield Hall (Engineering Quadrangle); Princeton NJ 08544; USA; 609-452-4681; E/T; 3.3; 4.1; Photoacoustic spectroscopy; Fourier transform IR-PAS; Catalytic materials

ROZSNYAI, Balazs F.; Physics T-Division; Lawrence Livermore National Labs.; PO Box 808; Livermore CA 94550; USA; 415-422-4085; T; 2.2; 5.4; 5.7; Atomic physics applied to high temperature astrophysical plasmas; State and radiative properties--equation; Electron scattering in hot/dense plasmas

RUAN, Ken Y.; 366 E. Mosholu Parkway S., Apt. 30; Bronx NY 10458; USA

RUBERTICCHIO, Donna M.; 4240 N. Marmora Drive; Chicago IL 60634; USA

RUBIN, Kenneth; Department of Physics; City College of New York; 138th Street and Convent Avenue; New York NY 10031; USA; 212-690-6895; E/T; 2.2; 3.6; Atomic beam techniques used in electron atom cross section measurements; Atomic and laser beams--interaction

RUDD, M. E.; Department of Physics and Astronomy; University of Nebraska; Lincoln NE 68588; USA; 402-472-2792; E; 2.1; 2.2; 3.4; Electron ejection by ion, neutral, and electron impact, autoionization and Auger effect

RUEDENBERG, Klaus; Departments of Chemistry and Physics; Iowa State University and Ames Lab. USDOE; Ames IA 50011; USA; 515-294-5253; T; 1.1; 2.3; Quantum chemistry and theory of valency; Ab initio approach to molecular structure/properties and chemical reactions; Specific systems--calculation; Many-electron quantum mechanics --fundamental theoretical developments

RUFF, George A.; Department of Physics; Bates College; Lewiston ME 04240; USA; 207-786-6322; E; 1.2; 2.1; 3.1; 4.1; Optical and electron spectroscopy of atomic Rydberg states; Laser spectroscopy of atoms and molecules; Stabilization of tunable lasers

RUGGE, Hugo R.; M2-254; Aerospace Corporation; PO Box 92957; Los Angeles CA 90009; USA

RULE, Donald W.; R-41 Nuclear Branch; White Oak Laboratory; US Naval Surface Weapons Center; Silver Spring MD 20910; USA; 301-394-2270; T; 5.1; 5.4; Beam density effects in energy loss by particles passing through solids

RUMBLE, John R.; OSRD; National Bureau of Standards; A323 Physics Bldg.; Gaithersburg MD 20899; USA; 301-921-3441; T; 1.1; 2.2; 3.1; Electron scattering by molecules and atoms; Small ions and atoms--properties; Photoionization of molecules

RUSH JR., John E.; Department of Physics; University of Alabama, Huntsville; PO Box 1247; Huntsville AL 35807; USA; 205-895-6569; T; 2.1; Molecular collision cross sections--calculations

RUSKAI, Mary B.; Mathematics; College of Science; University of Lowell; Lowell MA 01854; USA; 617-646-9377; 617-452-5000x2520; T; 1.1; 1.4; Schrodinger operators; Mathematical properties of density matrices; Density functionals

RUSS, Lawrence E.; 7C-501; Bell Laboratories; 600 Mountain Avenue; Murray Hill NJ 07974; USA

RUSSEK, Arnold; Department of Physics; University of Connecticut; Storrs CT 06268; USA; 203-486-4978; T; 1.1; 2.1; Atomic and molecular collisions; Molecular structure; Scattering theory

RUSSELL, Gary R.; Lasers Group; Jet Propulsion Laboratory; 4800 Oak Grove Drive; Pasadena CA 91103; USA; 213-354-3547; E/T; 3.6; 4.1; Laser research and development; Laser spectroscopy; Laser kinetics

RUSSELL, Geoffrey A.; Analytical Sciences Division; Eastman Kodak Research Labs; Kodak Park; Building 82, 1999 Lake Avenue; Rochester NY 14650; USA; 716-458-1000; Ext.76301; E; 1.1; 1.3; Polymers and low molecular weight species--interactions between; Chemical structure on polymer-polymer interactions--influence

RUSSELL, James E.; Department of Physics; University of Cincinnati; Cincinnati OH 45220; USA; 513-475-3271; T; 1.3; 2.3

RUSSO, Onofrio L.; 2 Fredericks Court; Elmwood Park NJ 07407; USA

RUSTGI, Om P.; Department of Physics; SUNY College, Buffalo; 1300 Elmwood Avenue; Buffalo NY 14222; USA; 716-878-5201; 716-632-5768; E/T; 3.1; 3.4; 5.1

RUTHERFORD, John A.; Chemistry; "S" Cubed; "S" Cubed Div. of Maxwell Corp.; 92038-1620; 3398 Carmel Mountain Road; San Diego CA 92121; USA; 619-587-8397; E; 1.3; 2.3; 3.1; 4.4; 5.3; Charge exchange and ion molecule reactions; Crossed beam techniques; Chemical physics; Electron and atomic physics

RUTLEDGE, Wyman C.; Central Research Lab.; Mead Corporation; 8th and Hickory; Chillicothe OH 45601; USA

RYAN, Dave; Geochemistry; Stable Isotope; Spectrion Anal of Trace Elements Inc.; 1234 Woodroffe Avenue; Ottawa Ontario K2C 2T5; Canada; 613-225-4123; 613-224-0734; E; 4.4; 4.3; 1.1; Thermal ionization, mass scattering, stable isotope ratio analysis; Dispersion (magnetic/electrostatic) UHV techniques; Thermal conductivity, electron dispersive X-rays (EDX)

RYAN, Frederick M.; Applied Physics; Applied Sciences; Optical Physics; Westinghouse R&D Center; 1310 Beulah Road; Pittsburgh PA 15235; USA; 412-256-1656; E; 3.1; 5.5; 4.1; 3.2; 3.3; Tunable laser spectroscopy

RYAN, Laurence J.; 4130 Willow Grove; Dallas TX 75220; USA

RYAN, Stewart R.; Department of Physics; University of Oklahoma; 440 W. Brooks Street; Norman OK 73019; USA; 405-325-3961; E; 2.1; 2.2; 4.2; Electron-molecule scattering near threshhold; Time-of-flight studies of molecular dissociation processes; Metastable atom-molecular collision processes

RYBKA, Theodore W.; RF Systems Engineering Optics Group; Electronics; Microelectronics and Optics Labs.; General Dynamics; 85310 MZ7203R; Convair Dr. and Missile Rd.; San Diego CA 92138; USA; 573-7220; 444-0670; E/T; 4.1; 5.1; 3.10; 2.1; 4.2; Electrooptics--novel device for use in communications; Interaction of radiation with surfaces--holes burned in thin films by laser; Wave mixing--RF modulation of laser diodes for fiber optic applications; Ion-ion collisions--production of directional beams

RYDING, Geoffrey; PO Box 485; Beverly MA 01915; USA

RYERSON, Peter S.; 2971 Bainbridge Avenue; New York NY 10458; USA

SABBAS, Albert M.; P-3; Los Alamos National Lab. MS D456; Los Alamos NM 87545; USA

SABELLI, Nora H.; Computer Center; University of Illinois, Chicago Circle; PO Box 4348; Chicago IL 60680; USA; 312-279-3386; 312-996-2473; T; 1.1; Resonances--electronic structure, wavefunction and lifetimes; Molecules as a function of their environment--properties

SABIN, John R.; Department of Physics; Quantum Theory Project; University of Florida; Gainesville FL 32611; USA; 904-392-1597; T; 1.1; 2.1; 5.1; 5.7; Stopping properties of atoms, molecules, and solids; Excitation spectra of molecules with astrophysical interest

SABY, John S.; 8 Tamarac Terrace; Hendersonville NC 28739; USA

SACHDEV, Subir; 11 Madison #3; Cambridge MA 02138; USA

SACHS, Judith G.; Department of Chemistry; University of Illinois at Chicago Circle; PO Box 4348; Chicago IL 60680; USA; 312-996-0774; T; 2.1; Energy exchange in collisions of atoms or small molecules; at low energy--calculation

SACHS, Lester M.; 8823 Stonehaven Road; Randallstown MD 21133; USA

SADLEJ, Andrzej J.; Chemical Center; University of Lund; PO Box 740; Lund 22007; Sweden

SAFINYA, Kambiz A.; Schlumberger-Doll Research; PO Box 307; Old Quarry Road; Ridgefield CT 06877; USA

SAFRON, Sanford A.; Department of Chemistry; Florida State University; Tallahasee FL 32306; USA; 904-644-5239; E; 2.3; Ion-molecule crossed beam reactive scattering

SAGAN, David C.; Department of Physics; Cornell University; Clark Hall; Ithaca NY 14853; USA

SAGE, Martin L.; Department of Chemistry; Syracuse University; Syracuse NY 13210; USA; 315-423-2713; T; 1.1; 1.3; High energy vibrational states of molecules; Intramolecular energy transfer; Unimolecular reactions; Van der Waals molecules

SAHA, Bidhan C.; Department of Physics and Astronomy; University of Oklahoma; 440 West Brooks-Room 131; Norman OK 73019; USA; 405-325-3961; T; Wave functions, polarizability and interaction potentials; Electron and positron collisions with atoms and molecules; Rydberg states studies; Rearrangement collisions

SAHA, Haripada; Department of Chemistry; Indiana University; Bloomington IN 47405; USA; 812-335-8864; T; 1.1; 2.2; 2.3; 3.1; 3.6; Structure and properties of atoms and molecules; Atomic and molecular collisions; Interaction of radiation with atoms and molecules

SAHNI, Omesh; IBM Thomas J. Watson Research Center; PO Box 218; Yorktown Heights NY 10598; USA; 914-681-5592; E/T; 5.3; Luminescence in gases and solids for display devices; High field transport in solids of electrons

SAIGUSA, Toshifumi; Ryukoku University; Fukakusa Fushimiku; Kyoto 612 0598; Japan

SAINT-CLAIR, Jonathan M.; Department of Physics FM-15; University of Washington; Seattle WA 98195; USA

SALESKY, E. T.; Los Alamos National Lab.; PO Box 1663; MS E543; Los Alamos NM 87545; USA

SALK, Sung-Ho S.; Physics and Graduate Center for Cloud Physics; University of Missouri at Rolla; Rolla MO 65401; USA; 314-341-4340; 314-364-5117; T; 1.1; 2.1; Atom-diatomic molecule collisions (reactive scattering); Molecular structure

SALMAN, Salman M.; PO Box 586; Upton NY 11973; USA

SALOMAN, Edward B.; Far Ultraviolet Physics; Radiation Physics Division; National Bureau of Standards; A251 Physics Building; Gaithersburg MD 20899; USA; 301-921-2031; E; 3.1; 3.9; 3.4; 3.6; 3.2; Synchrotron radiation to study dynamics of excited A&M states; Radiometry using synchrotron radiation; Electric field effects on autoionizing states; Photoabsorption cross section data center

SALOP, Arthur; Molecular Physics Laboratory; SRI International; 333 Ravenswood Avenue; Menlo Park CA 94025; USA; 415-326-6200

SALOUR, Michael M.; Tacan Corporation; 7910 Ivanhoe Avenue Suite 417; La Jolla CA 92037; USA

SALTSBURG, Howard M.; Department of Chemical Engineering; University of Rochester; Rochester NY 14627; USA; 716-275-4582; E; 5.1; Molecular beam scattering from solid surfaces

SAMPSON, Douglas H.; Department of Astronomy; 525 Davey Laboratory; Pennsylvania State University; University Park PA 16802; USA; 814-865-0261; T; 2.2; 3.1; 1.1; Electron inelastic collisions with highly charged ions; Radiative transition rates; Autoionization transition rates; Wave functions and energy levels for highly charged ions

SAMSON, James A.; Department of Physics; University of Nebraska; Lincoln NE 68588; USA; 402-472-2791; E; 3.1; 3.4; 4.4; 5.6; Photon interaction in the vacuum UV; Photoelectron and ion spectroscopy; Fluorescence of gases

SAMUEL, Mark A.; Department of Physics; Oklahoma State University; Stillwater OK 74074; USA

SANCHE, Leon; Faculty of Medicine; University of Sherbrooke; Stoke Road; Sherbrooke PQ J1H5N4; Canada

SANDERS JR., Frank C.; Department of Physics and Astronomy; Southern Illinois University; Carbondale IL 62901; USA; 618-536-2117; 618-453-2643; T; 1.1; 3.1; 1.4; Calculations of high precision for atomic and molecular energies and properties via Z-dependent perturbation theory

SANDERS, Steven G.; Northern Arkansas Telephone Co.; 301 Main Street; Flippin AR 72634; USA

SANDLIN, Glenn D.; U.S. Naval Research Laboratory; Code 4163S; 4555 Overlook Avenue, SW; Washington DC 20375; USA; 202-767-2649; T; 1.1; 5.4; 5.7; Energy levels and constitution of highly-ionized ions in astrophysical plasmas--analysis and determination

SANDNER, Wolfgang; Fak F Physik; University of Freiburg; 3 Hermann Herder; Freiburg D7801; W. Germany; 011-49-761-203-371; E; 1.1; 1.2; 2.2; 3.4; 3.6; Electron spectroscopy of autoionizing Rydberg states; Doubly excited Rydberg atoms; Electron-electron coincidences

SANDO, Kenneth M.; Department of Chemistry; University of Iowa; Iowa City IA 52242; USA; 319-353-3788; T; 1.1; 2.1; 3.6; Atomic spectral lines--shapes; Atomic low-energy collision theory; Molecular electronic structure

SANDS, Richard H.; Physics and Astronomy Building; University of Michigan; Room 745; Ann Arbor MI 48109; USA

SANSONETTI, Craig J.; Atomic and Plasma Radiation Division; National Bureau of Standards; A167 Physics Building; Gaithersburg MD 20899; USA; 301-921-2011; E; 3.6; 3.2; 1.1; 1.2; Laser spectroscopy; Analysis of spectra

SANZONE, George; Department of Chemistry; Virginia Polytechnic Institute and State University; Blacksburg VA 24061; USA; 703-961-5394; E/T; 2.3; 4.4; 5.3; Excimer and exciplex formation and charge exchange kinetics --molecular; Beam studies; Shock-tube chemistry; Mass spectrometry

SAPOROSCHENKO, Mykola; Department of Physics; Southern Illinois University; Carbondale IL 62901; USA; 618-453-3732; E; 2.3; 4.4; 5.3; Ions in gases--mobility;

Ion-molecule reactions; Mass spectrometry

SARACHMAN, Theodore N.; Department of Physics; Whittier College; Whittier CA 90608; USA; 213-693-0771x290; T; 1.1; Rotation-vibration interaction in polyatomic molecules

SARGENT III, Murray; Optical Sciences Center; University of Arizona; Tucson AZ 85721; USA; 602-626-1145; T; 3.6; Laser physics; Laser spectroscopy

SARKA, Benjamin N. M.; Research Applications Division; Systems Research Laboratories, Inc.; 2800 Indian Ripple Road; Dayton OH 45440; USA; 513-252-2706; 513-252-4264; E; 4.1; 3.6; 5.3; 3.10; 5.5

SARRAF, Sanwal P.; Engineering Physics Lab.; Physics Division; Kodak Research Lab.; Eastman Kodak Company; Building 82A; Rochester NY 14650; USA; 716-477-3081; E/T; 4.1; 5.4; 5.1; Development and use of special techniques and instruments; Interface between A&M and other areas of science

SARTWELL, Bruce D.; Code 6675; Naval Research Laboratory; Overlook Avenue; Washington DC 20375; USA; 202-767-4800; E; 5.1; 3.4; 4.2; 4.3; Ion implantation of materials to alter mechanical and chemical properties; Ion-induced x-ray emission and Auger electron spectroscopy of surfaces

SAUDER, William C.; Department of Physics; Virginia Military Institute; Lexington VA 24450; USA; 703-463-6225; 703-463-0506; E; 3.4; High resolution X-ray spectroscopy

SAUERS, Isidor; Health and Safety Research Division; Oak Ridge National Laboratory; PO Box X; Oak Ridge TN 37830; USA; 615-574-6203; E; 5.3; Applied physics related to decomposition processes; in dielectric states

SAVAGE, William R.; Department of Physics and Astronomy; University of Iowa; Iowa City IA 52242; USA

SAWADA, Fred H.; 1022 Tomahawk Trail; Scotia NY 12302; USA

SAXENA, Krishan M.; Department of Chemistry; University of Alberta; Edmonton AB T6G2E1; Canada

SAXON, Roberta P.; Molecular Physics Department; SRI International; 333 Ravenswood Avenue; Menlo Park CA 94025; USA; 415-859-2663; 415-859-3643; T; 1.1; 3.1; 2.1; Molecular structure of excited states; Photodissociation; Atomic and molecular collisions

SAYLOR III, Tillman K.; Department of Physics and Astronomy; University of Pittsburgh; Pittsburgh PA 15260; USA; 412-624-4347; E; 4.2; Accelerator-based atomic physics

SCALETTAR, Richard; Lake Forest; 24782 Winterwood Drive; El Toro CA 92630; USA

SCARL, Donald B.; Department of Physics; Polytechnic Institute of New York; Farmingdale NY 11735; USA; 516-454-5144; E; 3.11; 3.6; 4.1; Quantum optics; Atomic and optical coherence

SCHADT, Randall J.; Department of Physics; University of Illinois at Urbana; Physics Building; Urbana IL 65401; USA; E; 4.1; 5.3; 5.6

SCHAEFER III, Henry F.; Department of Chemistry; University of California; Berkeley CA 94720; USA; 415-642-1957; T; 1.1; Electronic structure of atoms and molecules

SCHAEFER, Juergen A.; Department of Physics; Montana State University; Bozeman MT 59717; USA

SCHAEFFER, Norman M.; Radiation Research Associates, Inc.; 3550 Hulen Street; Fort Worth TX 76107; USA; 817-731-2711; T; 5.2; 5.6; Fission reactor shielding analysis/design; Radiation environments from fission sources--air/ground transport

SCHAFER, Trudy P.; 820 San Ramon Way; Sacramento CA 95825; USA; 916-482-8108

SCHAPPERT, Gottfried T.; P-1; Physics Division; Los Alamos National Lab.; PO Box 1663; Los Alamos NM 87545; USA; 505-667-1294; E/T; 3.1; 3.4; 4.1; 1.1; Atomic physics; Laser physics and kinetics

SCHATZ, George C.; Department of Chemistry; Northwestern University; Evanston IL 60201; USA; 312-491-5657; T; 2.1; 2.3; Molecular collisions; Reactive scattering; Collisional energy transfer

SCHATZ, Paul N.; Department of Chemistry; University of Virginia; McCormick Road; Charlottesville VA 22901; USA; 804-924-3249; 804-924-3714; E/T; 3.1; 3.2; 3.5; 3.9; Magnetic circular dichroism spectroscopy

SCHAWLOW, Arthur L.; Department of Physics; Stanford University; Stanford CA 94305; USA; 415-497-4356; 415-497-4357; E; 3.5; 3.6; Laser spectroscopy of atoms and small molecules

SCHEARER, Laird D.; Department of Physics; Laboratory for Atomic Physics; University of Missouri, Rolla; Rolla MO 65401; USA; 314-341-4792; 314-341-4702; E; 2.1; 3.6; 3.9; 4.1; 5.3; Metastable interactions; Afterglows; Dissociative excitation; Penning ionization

SCHECTMAN, Richard M.; Department of Physics and Astronomy; University of Toledo; Toledo OH 43606; USA; 419-537-2341; E; 1.1; 2.1; 4.2; 5.1; Ion-solid interactions, ion-atom interactions, and atomic properties; Accelerator-based studies

SCHEER, Milton D.; 811 N. Belgrade Road; Silver Spring MD 20902; USA

SCHEFF, Victor A.; 1799 Euclid; Berkeley CA 94709; USA

SCHEIBE, Murray; Mission Research Corporation; PO Drawer 719; Santa Barbara CA 93102; USA

SCHENCK, Peter; Division 561; National Bureau of Standards; Gaithersburg MD 20899; USA; 301-921-2859; E; 3.1; 5.5; Laser enhanced atomic and molecular ionization in flames--measurement of; Combustion diagnostics --spectroscopic applications

SCHEPS, Richard; Western Research Corporation; 8616 Commerce Avenue; San Diego CA 92121; USA; 714-785-5885; T/E; 2.2; 2.3; 3.5; Excimer state kinetics and spectroscopy; Electron beam dynamics in rare gases and rare gas halides

SCHER, Gary M.; Desktop Computer Division; Hewlett-Packard Company; 3404 E. Harmony Road; Fort Collins CO 80525; USA

SCHERER, James R.; Plant Development-Quality Research Unit; Agricultural Research Service; Western Regional Research Center; U.S. Department of Agriculture; 800 Buchanon; Albany CA 94710; USA; 415-486-3631; 415-486-3348; E/T; 3.6; 3.2; Raman and IR spectroscopy

SCHERRER, Victor E.; 1300 Swan Creek Road; Fort Washington MD 20744; USA

SCHIAVONE, James A.; Kinetic Chemistry Research Division; Bell Telephone Laboratories; 600 Mountain Avenue; Murray Hill NJ 07974; USA; 201-582-3856; E; 1.2; 2.2; 3.5; Small molecules and rare gas atoms--translational spectroscopy; Highly excited Rydberg states--electron impact excitation

SCHIEVE, W. C.; Department of Physics; University of Texas; Austin TX 78712; USA; 512-471-7253; T; 4.1; Quantum optics

SCHIMA, Francis J.; Radioactivity Group 536; National Bureau of Standards; C114 B245; Gaithersburg MD 20899; USA; 301-921-2396; E; 5.4; 5.8; Isotopic separation, using resonant ionization; Nuclear orientation and alignment using optical perm

SCHLACHTER, Alfred S.; Magnetic Fusion Energy Group; Acceleration and Fusion Research Division; Lawrence Berkeley Laboratory; University of California; MS4-230; Berkeley CA 94720; USA; 415-486-5011; FTS-451-5011; E; 2.1; 2.3; 4.2;

3.4; 3.6; Atomic charge-transfer and ionization cross sections

SCHLESSINGER, Leonard; Pacific Sierra Research Corporation; 12340 Santa Monica Boulevard; Los Angeles CA 90025; USA; 213-820-2200; T; 2.1; 2.2; 3.6; 3.7; 3.9

SCHLIE, LaVerne A.; U.S. Air Force Weapons Laboratory; ARAP; Kirkland AFB NM 87117; USA; 505-844-9536; E/T; 4.1; 4.2; Lasers

SCHLUETER, Warren A.; Mission Research Corporation; PO Drawer 719; Santa Barbara CA 93102; USA; 805-963-8761

SCHMALZ, Thomas G.; Department of Marine Sciences; Theoretical Chemical Physics Group; Texas A&M University at Galveston; PO Box 1675; Galveston TX 77553; USA; 409-740-4497; T; 1.1; 2.1; 2.3; Atomic and molecular collisions; Electron transfer collisions; Vibrational-electronic interactions

SCHMELLING, Stephen G.; 222 S. Country Club Road; Ada OK 74820; USA

SCHMELTEKOPF, A.; NOAA/ERL; Building 24; Boulder CO 80302; USA

SCHMIEDEKAMP, Ann M.; Department of Physics; Pennsylvania State University, Ogontz; 1600 Woodland Road; Abington PA 19001; USA; 215-886-9400 x317; T; 1.1; Ab initio molecular orbital calculations

SCHMIEDER, Robert W.; 8348; Sandia National Laboratories; Livermore CA 94550; USA; 415-422-2821; 415-422-2057; E; 4.2; 2.1; Highly ionized atoms molecules--spectroscopy; Electron beam ion sources; Low energy high charge state ion collisions

SCHMUCKER, Ronald F.; The Village Green Apt. #68R; Budd Lake NJ 07828; USA

SCHNEIDER, Barry I.; Theoretical (T-12); Los Alamos National Laboratory; PO Box 1663; Los Alamos NM 87545; USA; 505-667-5488; T; 2.2; 1.1; Electron scattering

SCHNEIDER, Richard T.; Department of Nuclear Engineering Science; 202 Nuclear Science Center; University of Florida; Gainesville FL 32611; USA; 904-392-1407; E; 3.3; 4.1; 3.1; 5.1; Nuclear pumped lasers; IR technology; Rocket plume spectroscopy

SCHNEPP, Otto; Department of Chemistry; University of Southern California; University Park; Los Angeles CA 90089; USA; 213-741-2961; E; 1.1; 3.5; 3.9; Molecular spectroscopy; Excited states of molecules; Raman effect in polyatomic molecules

SCHNOPPER, Herbert W.; Harvard-Smithsonian Center for Astrophysics; 60 Garden Street; Cambridge MA 02138; USA; 617-495-7145; E; 3.4; X-ray spectroscopy

SCHOEN, Richard; National Science Foundation; Washington DC 20550; USA

SCHOFIELD, Keith; Quantum Institute; University of California, Santa Barbara; Santa Barbara CA 93106; USA; 805-961-3526; 805-961-2582; E; 2.3; 3.6; 5.5; Laser fluorescence spectroscopy; Combustion chemistry; Atomic and molecular excited states chemistry

SCHOLZ, W. W.; Department of Physics; State University of New York; Albany NY 12222; USA

SCHOOLEY, Willard A.; 228 Ferndale Drive; Collinsville VA 24078; USA

SCHOWENGERDT, Franklin D.; Department of Physics; Colorado School of Mines; Golden CO 80401; USA; 303-273-3830; E; 1.1; 2.2; 3.3; Auger electron spectroscopy; Low energy electron scattering; Elastic resonances

SCHRADER, David M.; Department of Chemistry; Marquette University; Milwaukee WI 53233; USA; 414-224-3332; 414-224-7065; E/T; 1.3; 2.2; 3.9; 5.2; Quantum mechanical calculations on atoms; and molecules containing a positron

SCHREIBER, Paul W.; Aero-Propulsion Laboratory; AFWAL/POOC-3; Wright-Patterson AFB OH 45433; USA; 513-255-2923; E; 3.5; 5.4; 5.5; Molecules and their application to plasma and combustion diagnostics; Nonlinear spectroscopy

SCHRODT, Ariel G.; Dover Industrial Chrome, Inc.; 2929 N. Campbell Avenue; Chicago IL 60618; USA

SCHROEER, Juergen M.; Department of Physics; Illinois State University; Normal IL 61761; USA; 309-438-5193/6656; 309-438-8756; E/T; 4.4; 5.1; Secondary ion mass spectrometry

SCHUBERT, David C.; Aerospace Division; Westinghouse Electric; PO Box 1521; Mail Stop 550; Baltimore MD 21230; USA

SCHUELER, Bruno W.; Charles Evans and Associates; 301 Chesapeake Drive; Redwood City CA 94063; USA; 415-369-4567; E; 4.4; 3.6

SCHUESSLER, Hans A.; Department of Physics; Laser and Stored Ion Physics; Texas A&M University; College Station TX 77843; USA; 409-845-5455; E/T; 1.1; 1.2; 1.4; 2.1; 3.6; Laser physics and stored ion physics; On-line laser spectroscopy of short lived isotopes; On-line spectroscopy of lightly charged ions; Collision studies involving excited atoms

SCHULER, Robert H.; Department of Chemistry; Radiation Laboratory; University of Notre Dame; Notre Dame IN 46556; USA; 219-283-7502; E; 3.8; 4.2; 2.3; 3.10; Radiation chemistry and photochemistry

SCHULMAN, M. Bruce; 633 Langdon Apt. #101; Madison WI 53703; USA

SCHULTZ, John W.; Department of Chemistry; U.S. Naval Academy; Annapolis MD 21402; USA; 301-267-3403; E; 3.3; 3.5; 3.6; 3.10; IR spectroscopy; IR absorption intensities

SCHULZ, Peter A.; Joint Institute for Laboratory Astrophysics; University of Colorado; Boulder CO 80309; USA; 303-492-7751/7764; E; 3.1; Molecular negative ions--photodetachment

SCHUMER, Douglas B.; Ohaus Scale Corporation; 29 Hanover Road; Florham Park NJ 07932; USA; 201-377-9000x306; E; 4.1; 3.10

SCHWARZSCHILD, Bertram M.; AIP Physics Today; 335 E. 45th Street; New York NY 10017; USA

SCHWEITZER JR., Walter G.; Center for Absolute Physical Quantities; National Bureau of Standards; B164, Physics Building; Gaithersburg MD 20899; USA; 301-921-2001; E; 4.1; Stabilized lasers for application to wavelength and frequency standards

SCHWENDEMAN, Richard H.; Department of Chemistry; Michigan State University; East Lansing MI 48824; USA; 517-353-9412; E/T; 3.3; 3.6; IR laser and microwave spectroscopy

SCHWIER, Hartwig; Fakultaet Physics; University of Bielefeld; Universittaets-Strasse 25; Bielefeld 1 D-4800; West Germany

SCHWIRZKE, Fred R.; Dept. of Physics; Code 61 SW; Naval Postgraduate School; Monterey CA 93940; USA; 408-646-2806; E/T; 5.1; 5.4; 5.3; 4.3; 3.8

SCOFIELD, James H.; L 296 Theoretical Atomic Physics; Lawrence Livermore National Lab.; PO Box 808; Livermore CA 94550; USA; 415-422-4098; T; 1.1; 2.2; 3.1; 3.4; 5.4; Energy levels and transition rates in highly ionized atoms; Photo and electron ionization cross sections

SCOTT JR., John E.; Department of Mechanical and Aerospace Engineering; University of Virginia; Charlottesville VA 22901; USA; 804-924-7421/3796; E; 3.6; 4.3; Isotope separation; Uranium hexafluoride--spectroscopy; Instrumentation development

SCULLY, Marlan O.; Department of Physics and Astronomy; Institute for Modern Optics; University of New Mexico;

800 Yale Boulevard, NE; Albuquerque NM 87131; USA; 505-277-2726; T; 3.6; 4.1; 4.2; Quantum optics; Laser spectroscopy

SEABAUGH, Alan C.; National Bureau of Standards; Room A331; Gaithersburg MD 20899; USA

SEAGRAVE, John D.; P-1/MS E526; Los Alamos National Lab.; PO Box 1663; Los Alamos NM 87545; USA

SECREST, Don; Department of Chemistry; University of Illinois; 155 Noyes Laboratory; Urbana IL 61801; USA; 217-333-1728; T; 1.1; 2.1; Mathematical approach to quantum mechanical collision--application; Small molecule quantum mechanics

SEELY, David G.; Department of Physics; University of Missouri at Rolla; Room G10; Rolla MO 65401; USA; 314-341-4781; 314-364-8475; E; 2.1; 4.2; Intermediate energy ion-atom energy loss spectroscopy

SEELY, John F.; Space Sciences Division; Code 4174; US Naval Research Laboratory; 4555 Overlook Avenue, SW; Washington DC 20375; USA; 202-767-3529; E/T; 5.4; 1.4; 3.4; Atomic processes in laser-produced plasmas

SEEM, Dennis A.; Ocean Acoustics Division; NOAA/AOML; 4301 Rickenbacker Csy; Miami FL 33149; USA

SEGAL, Gerald A.; Department of Chemistry; University of Southern California; University Park; Los Angeles CA 90089; USA; 213-743-6201; T; 1.1; Electronic structure of molecules and their ions including scattering states --calculations

SEGALL, Herbert; Department of Physics; Occidental College; 1600 Campus Road; Los Angeles CA 90041; USA

SEGREDO, Eugenio; Applied Research Division; IBM Thomas J. Watson Research Center; PO Box 218; Yorktown Heights NY 10598; USA; 914-945-2479 x2922; E; 5.1; Electro-optics; Thermionic emission; In-situ chemical reactions--surface studies; Chemical contaminants --processing control

SEIDL, Milos; Department of Physics; Stevens Institute of Technology; Hoboken NJ 07030; USA

SEITZ, Frederick; Rockefeller University; 1230 York Avenue; New York NY 10021; USA

SEKA, Wolf; Laboratory for Laser Energetics; University of Rochester; 250 E. River Road; Rochester NY 14618; USA; 716-275-3815; E; 3.4; 5.4; Highly charged plasmas produced by laser-plasma interaction; X-ray spectroscopy

SEKI, Hajimes; Division K33/82; IBM Almaden Research Center; 650 Harry Road; San Jose CA 95120; USA; 408-256-4084; 408-927-2345; E; 3.3; 3.10; Tribology related to head/disk and head/tape interaction; IR and Raman spectroscopy of electrode/ electrolyte

SEKI, Ryoichi; Department of Physics and Astronomy; California State University, Northridge; Northridge CA 91325; USA; 213-885-2775; T; 1.3; Mesonic and exotic atoms

SELIGER, Howard H.; Department of Biology; Johns Hopkins University; Baltimore MD 21218; USA; 301-338-7307; 301-338-7334; E; 1.3; 2.3; 3.5; Chemical production of excited states

SELL, Jeffrey A.; Physics Division; General Motors Research Lab.; General Motors Technical Center; 12 Mile and Mound Roads; Warren MI 48090; USA; 313-575-3431; 313-575-2836; E; 3.6; 4.1; 5.5; 3.1; Photoacoustic and photothermal spectroscopy; Laser induced chemical reactions; ignition; IR diode laser spectroscopy

SELLIN, Ivan A.; Physics Department; Physics Division; Oak Ridge National Laboratory; University of Tennessee at ORNL; PO Box X; Oak Ridge TN 37830; USA; 615-574-4793; 615-483-6909; E; 2.1; 4.2; 5.2; 1.4; 2.3; Highly ionized matter--physics

SELTZER, Stephen M.; Radiation Theory Group; Ionizing Radiation Division; Center for Radiation Research; National Bureau of Standards; Gaithersburg MD 20899; USA; 301-921-2685; 301-340-1710; T; 5.2; 2.2; 3.1; 3.4; Transport of electrons and photons through bulk matter; Cross-section development for transport calculations

SEMAN, Michael L.; Department of Physics; Northeastern Illinois University; 550 N. St. Louis Avenue; Chicago IL 60625; USA

SEMON, Mark D.; Department of Physics and Astronomy; Bates College; Lewiston ME 04240; USA; 207-786-6324; 207-786-3467; T; 2.1; Coulomb scattering; Nonrelativistic scattering theory

SENBA, Masayoshi; TRIUMF; University of British Columbia; 4004 Wesbrook Mall; Vancouver BC V6T2A3; Canada; 604-222-1047; E; 2.3; 2.1; 4.2; 5.1; 5.2; Behavior of muon and murium in gas phase; Muonium molecular ion formation and its reactions; spin exchange of muonium with atoms and molecules; chemical reactions of muonium

SENITZKY, Benjamin; Department of Electrical Engineering; Polytechnic Institute of New York; Route 110; Farmingdale NY 11735; USA; 516-454-5085; E/T; 3.3; Radiation and matter--interaction

SETSER, D. W.; Department of Chemistry; Kansas State University; Manhattan KS 66506; USA; 913-532-6692; E; 2.3; 3.6; 5.3; Atoms and molecules studied by spectroscopic techniques--chemical dynamics of; Electronically excited atoms and molecules

SEYBOLD, Paul G.; Department of Chemistry and Biological Chemistry; Wright State University; Dayton OH 45435; USA; 513-254-0308; T/E; 1.1; 3.5; Luminescence spectroscopy of molecules; Molecular orbital calculations

SHAFFNER, Richard O.; 655 South Fairoaks Avenue Apt. P-308; Sunnyvale CA 94086; USA

SHAFROTH, Stephen M.; Department of Physics and Astronomy; University of North Carolina; Chapel Hill NC 27514; USA; 919-933-3016; E; 2.2; 3.4; 4.2; 5.1; K,L,M-shell ionization cross sections; Radiative electron capture and target thickness effects; Double-K vacancy production and dielectronic recombination; Resonant Raman X-ray scattering and Auger electrons

SHAH, Akshay V.; Department of Physics; Rochester Institute of Technology; 1 Lomb Memorial Drive; Rochester NY 14623; USA

SHAH, Rajiv R.; 2116 Newcombe; Plano TX 75075; USA

SHAHIN, Issa S.; Department of Physics; University of Jordan; Amman; Jordan

SHAHIN, Michael M.; 12 Widewaters Lane; Pittsford NY 14534; USA

SHAKESHAFT, Robin; Department of Physics; Arts and Sciences; University of Southern California; Los Angeles CA 90089; USA; 213-743-8055; 213-743-6458; T; 3.7; 2.1; Multiphoton processes; Atomic collisions in the presence of radiation

SHAKIBI, Farhad; Mojahedin Ave.; Mahdavipor Fagihalmolk Alley; Iran St. Ali Akbar No. 31 1; Tehran 11586; Iran

SHALIMOFF, George V.; Materials and Molecular Research Division; Lawrence Berkeley Laboratory; University of California; 1 Cyclotron Road; Berkeley CA 94720; USA; 415-486-5141; E; 1.1; 3.2; Rare earth and actinide elements--emission and absorption spectroscopy; Magnetochemistry

SHALVOY, Richard B.; Eastern Research Center; Stauffer Chemical Company; Dobbs Ferry NY 10522; USA

SHAM, Tsun K.; Department of Chemistry; Brookhaven National Laboratory; Upton NY 11973; USA; 516-282-4345; E; 3.4; 4.2; 5.1; 5.2; Photoemission spectroscopy of solids and surfaces; X-ray absorption spectroscopy;

Synchrotron radiation spectroscopy

SHAND, Michael L.; Optical Physics Division; Allied Chemical Co.; PO Box 1021R; Morristown NJ 07960; USA; 201-455-4077; E; 3.6; 5.2; Rare earth ions in solids--laser-related research

SHANK, Charles V.; Electronics Research Division; AT&T Bell Laboratories; Holmdel NJ 07733; USA; 201-949-4532; E; 3.6; Laser spectroscopy and VUV generation

SHAPERO, Donald C.; Board on Physics and Astronomy; National Academy of Sciences; 2101 Constitution Ave.; Washington DC 20418; USA; 202-334-3520; E; Science policy

SHARMA, Suresh C.; Department of Physics; University of Texas at Arlington; PO Box 19059; Arlington TX 76019; USA

SHARPTON, Francis A.; Department of Physics; Mathematics and Natural Sciences; Northwest Nazarene College; Nampa ID 83651; USA; 208-467-8881; 208-467-8011; E; 1.1; 2.3; 3.6; 4.3; 5.3; Electron excitation of oxygen, nitrogen, and hydrogen from 0 to 500 eV; Electron excitation of neon, krypton, and xenon from 0 to 200 eV

SHATAS, Romas A.; Southern Technologies, Inc; 110 Wynn Drive; Huntsville AL 35802; USA; 205-837-2461; T; 3.1; 5.6; Collisionless, radiative transitions in atomic and molecular neutrals, and ions under natural and artificial excitations in the exosphere

SHAVITT, Isaiah; Department of Chemistry; Ohio State University; 140 West 18th Avenue; Columbus OH 43210; USA; 614-422-1668; T; 1.1; Ab initio calculations of molecular electronic structure; Electron correlation effects in molecules; Configuration interaction and perturbation theory methods; Computational methods development

SHAW, David T.; Department of Electrical Engineering; State University of New York, Buffalo; 4232 Ridge Lea Road; Buffalo NY 14226; USA; 716-831-3059; E; 4.4; Ion-cyclotron resonance mass spectroscopic investigation; of heteromolecular nucleation

SHAW, Gilbert B.; Department of Computer Science; University of Oregon; Eugene OR 97403; USA

SHAY, Thomas M.; Chemistry; Los Alamos National Laboratory; P.O. Box 1663; MS/E535; Los Alamos NM 87545; USA; 505-667-8390; E/T; 4.1; 3.6; 3.7; 2.3; 5.1; Ultrashort pulse laser development and theory excimer laser development; Narrow bandwidth wide field of view optical filter; Frequency stable laser development

SHE, Chiao Y.; Department of Physics; Colorado State University; Fort Collins CO 80523; USA; 303-491-6261; E; 3.10; 3.10; Applications of nonlinear optics for practical measurements; High resolution laser spectroscopies on simple atoms and molecules; Resonance fluorescence

SHEATS, James R.; Department of Chemistry; Stanford University; Stanford CA 94305; USA; 415-497-1097; E; 3.8; Chemical and hydrodynamic instability phenomena; Non-equilibrium statistical mechanics; Lasers to chemical dynamics--application of

SHEERS, William S.; 4200 Lockfield #901; Houston TX 77092; USA

SHELDON, John W.; Department of Physics; Florida International University; Miami FL 33199; USA; 305-554-2608; 305-554-2605; E; 2.1; 2.3; Metastable atomic and molecular beam scattering

SHELTON, W. Neil; Department of Physics; Florida State University; Tallahassee FL 32306; USA; 904-644-6552; 904-385-3989; T; 2.2; 3.3

SHEN, Guangfu; PO Box 5; 4 Washington Place; New York NY 10003; USA

SHEN, Y. R.; Department of Physics; University of California; Berkeley CA 94720; USA; 415-642-4856; E; 3.6; 5.1; 3.10; Laser interaction with molecular beams and adsorbed molecules

SHENG, King C.; 21 Lee Avenue; Troy NY 12180; USA

SHER, Mark H.; Department of Applied Physics; Ginzton Laboratory; Stanford University; P.O. Box 27; Applied Physics Department; Stanford CA 94305; USA; 415-497-0246; E; 3.1; 3.4; 3.6; XUV laser research

SHEREFF, Sidney L.; 136-30 Sanford Avenue; Flushing NY 11355; USA

SHERIDAN, John R.; Department of Physics; College of Natural Sciences; University of Alaska, Fairbanks; 306 Tanana Drive; Fairbanks AK 99775; USA; 907-474-6107; 907-455-6851; E; 2.1; 3.1; 2.3; 3.2; 3.5; Collisional excitation and quenching cross sections; Radiative lifetimes and excited-state correlations; Beam growth and decay equations; Atmospheric implications of atomic and molecular parameters

SHIBIB, M. Ayman; Wyomissing Hills; 7 Tewkesburry Drive; Reading PA 19610; USA

SHIMADA, Katsunori; Electrical Power and Propulsion Section; Jet Propulsion Laboratory; 4800 Oak Grove Drive; Pasadena CA 91103; USA; 213-354-4147; E; 5.5; Thermionic and thermoelectric energy conversion

SHIMAKURA, Noriyuki; General Education Department; Niigata University; Nino-Cho 8050 Ikarashi; Niigata 950-21; Japan

SHIMAMORI, Hiroshi; Fukui Inst of Technology; Gakuen 3-618; Fukui 910; Japan

SHIMIZU, Sakae; Institute of Chemical Research; University of Kyoto Sakyo; Kyoto; Japan; 075-711-1380; E; 2.2; 3.4; 5.8

SHIMKAVEG, Gregory M.; Department of Physics; Spectroscopy Laboratory; Massachusetts Institute of Technology; Room 6A-220, 77 Mass. Avenue; Cambridge MA 02139; USA; 617-253-5077; E; 3.6; 4.1; 5.8; 1.1

SHIN, Hyung K.; Department of Chemistry; University of Nevada; Reno NV 89557; USA; 702-784-6041; T; 2.1; Molecular collisions

SHIN, Seung H.; Department of Physics; College of Natural Sciences; Kangweon National University; Chuncheon Kangweondo 200; Korea

SHINADA, Masaki; Laboratory of Physics; The University of Electro-Comm; 1-5-1 Chofugaoka Chofu-Shi; Tokyo 182; Japan

SHIPSEY, Edward J.; Department of Physics; University of Texas, Austin; Austin TX 78712; USA; 512-471-7467; T; 2.1; 2.3; Computer computation of collisions in the molecular representation; Theory and computation of chemical reactions

SHIRK, James S.; Department of Chemistry; Illinois Institute of Technology; 3300 S. Federal; Chicago IL 60616; USA; 312-567-3446; 202-767-4871; E; 3.8; 3.5; 5.2; 3.6; IR photochemistry and photophysics in solids; Laser instrumentation development

SHIRLEY, David A.; Materials and Molecular; Research Division; Lawrence Berkeley Laboratory; University of California; Berkeley CA 94720; USA; 415-486-4000 x5111; E/T; 3.4; X-ray photoelectron spectroscopy; Director of research

SHIRLEY, John A.; Chemical Physics Division; United Technologies Research Center; Silver Lane; East Hartford CT 06108; USA; 203-727-7227; E; Applicability of coherent anti-Stokes Raman spectroscopy to combustion; systems research

SHIRLEY, Jon H.; Consultant for Time and Frequency; Boulder Laboratories; National Bureau of Standards; Salina Star Route; Boulder CO 80302; USA; 303-497-3125; T; 3.7; 3.10; 3.11; Doppler-free spectroscopy; Atomic frequency standards

SHIRTS, Randall B.; Department of Chemistry; University of Utah; Salt Lake City UT 84112; USA; 801-581-5736; 801-582-1012; T; 1.1; 2.3; 3.3; 3.6; 3.7; Semiclassical mechanics of small molecules; Intramolecular energy transfer

SHIVELY, John E.; Private Consultant; 404 Plymouth Court; Benicia CA 94510; USA; 707-745-8274; T; 2.2; 5.2; Electron scattering by single metal atoms; Metals --calculation of bulk properties

SHKAROFSKY, Issie P.; 1959 Cinton Avenue; Montreal PQ H3S1L2; Canada

SHKEDI, Zvi; 601 N. Alta Vista Boulevard; Los Angeles CA 90036; USA

SHNIDMAN, Robert; 3712 Bancroft Road; Baltimore MD 21215; USA

SHOBATAKE, Kosuke; Institute for Molecular Science; Myodaiji; Okazaki 444 1215; Japan

SHOEMAKER, Richard L.; Optical Sciences Center; University of Arizona; Tucson AZ 85721; USA; 602-626-3030; E; 2.1; 3.6; Coherent transient effects and relaxation processes in atomic, molecular, and solid state systems

SHORE, Bruce W.; Lawrence Livermore National Lab.; PO Box 808; Livermore CA 94550; USA; 415-422-6204; T; 3.7; 4.1; 3.6; Photon dynamics

SHORER, Philip; Atomic and Molecular Physics Division; Harvard-Smithsonian Center for Astrophysics; 60 Garden Street; Cambridge MA 02138; USA; 617-495-5873; T; 2.1; 3.1; 5.4; 5.7; Atomic and molecular processes relevant to fusion and astrophysics; Relativistic effects; Photon-atom interactions; Collision processes

SHUGART, Howard A.; Department of Physics; University of California; Berkeley CA 94720; USA; 415-642-0596; 415-642-3686; E; 1.1; 3.3; 3.6; 3.9; Metastable atomic states--lifetimes

SHULL, Michael; Department of Astrophysics; Joint Institute for Lab. Astrophysics; University of Colorado; Boulder CO 80309; USA; 303-492-7827; T; 3.2; 3.4; 5.7; Astrophysical plasmas--modeling; X-ray and UV emission spectra of supernovae remnants; O-star winds

SHYN, Tong-Wha; Space Physics Research Laboratory; University of Michigan; 2455 Hayward Street; Ann Arbor MI 48105; USA; 313-763-6214; 313-764-2125; E; 2.2; 1.1; 1.4; Electron impact scattering cross section measurements of gases; Elastic and ionization cross sections; Excitation cross sections

SIBATA, Claudio H.; Inst de Radioprotecao E; Dosimetria; Caixa Postal 37025 22602; B Tijuca-Rio De Janeiro; Brazil

SICKAFUS, Ed N.; Sensors and Integrated Circuit Processes; Research Staff; Scientific Laboratory; Ford Motor Company; P.O. Box 2053; Dearborn MI 48121; USA; 313-322-3912; E; 5.1; 1.1

SIEBERT, Donald R.; Department of Chemistry; Drew University; Madison NJ 07940; USA; 201-377-3000x367; E; 3.6; Lasers via new spectroscopic techniques--analytical applications

SIECKMANN, Everett F.; 509 So. Polk; Moscow ID 83843; USA

SIEGEL, Jon; Radiological and Environmental; Research Division; Argonne National Laboratory; 9700 S. Cass Avenue; Argonne IL 60439; USA; 312-972-4255; T; 2.2; 3.1; Electron-molecule collisions and molecular photoionization

SIEGEL, Melvin W.; Research, Development and Application; Extranuclear Lab., Inc.; PO Box 11512; Pittsburgh PA 15238; USA; 412-782-3884; E; 2.3; 4.4; 5.4; Ionization processes; Ion-molecule reactions; Weak plasmas; Analytical mass spectrometer systems development

SIEGMAN, Anthony E.; Department of Electrical Engineering; Edward F. Ginzton Laboratory; Stanford University; Stanford CA 94305; USA; 415-497-0222; E; 3.6; Picosecond spectroscopy using lasers

SIERRA, Rafael A.; Group L-10; Los Alamos National Laboratory; PO Box 1663 MS533; Los Alamos NM 87545; USA; 505-667-6888; E; 4.1; Antares carbon dioxide laser--research in support of construction

SILBEY, Robert J.; Department of Chemistry; MIT; Cambridge MA 02139; USA; 617-253-1470; T; 5.2; Spectroscopy and relaxation of molecules and atoms in condensed phases

SILCOX, John; School of Applied and Engineering Physics; Cornell University; Clark Hall; Ithaca NY 14853; USA; 607-256-3332; E; 5.1; Electron loss spectroscopy in solids; Microspectroscopy

SILFVAST, William T.; 11311; Electronics Research Laboratory; AT&T Bell Laboratories; Crawfords Corner Road; Holmdel NJ 07733; USA; 201-949-6331; E; 3.1; 3.4; 1.1; 4.1; 3.6; Recombination in atoms and ions; Photoionization of atoms with high fluxes from laser-produced plasmas; Long-lived autoionizing levels in ions

SILVER, David M.; Applied Physics Laboratory; Johns Hopkins University; 11100 Johns Hopkins Road; Laurel MD 20707; USA; 301-953-6265; T; 1.1; 2.1; 2.3; 5.6; Electronic structure of atoms and molecules; Many-body perturbation theory; Inelastic and reactive collision phenomena; Chemical reactions

SILVER, Joel A.; Center for Chemical and Environmental Physics; Research Group; Aerodyne Research, Inc.; 45 Manning Road; Billerica MA 01821; USA; 617-663-9500; E; 2.1; 2.3; 3.1; 3.6; 5.5; T-V excitation cross sections --measurements; Kinetic rate mechanisms--measurements and modeling

SILVER, Joshua D.; Department of Physics; Clarendon Laboratory; University of Oxford; Parks Road; Oxford OX13PU; England; 0865-59291; 0865-59911; E; 1.4; 3.2; 3.6; 4.2; Relativistic and QED effects in highly ionized atoms

SILVERMAN, J. N.; Theor Chem-Fakultat Chemie; University of Bielefeld; Bielefeld 1 D-4800; Fed. Rep. of Germany

SILVERMAN, Mark P.; Department of Physics; MC Cook Science Center; Trinity College; Hartford CT 06106; USA; 203-527-3151; E/T; 3.6; 3.10; 3.7; 1.2; 1.3

SILVERS, Stuart J.; Department of Chemistry; Virginia Commonwealth University; Richmond VA 23284; USA; 804-257-1298; E; 1.1; 4.1; Molecular structure and spectroscopy; Level crossing and double resonance studies of excited states of; gas phase molecules

SILVERSTONE, Harris J.; Department of Chemistry; Johns Hopkins University; Charles and 34th Street; Baltimore MD 21218; USA; 301-338-7431; T; 1.1; Quantum theory to determine atomic and molecular properties

SIM, Douglas W.; 44-409 Kaneohe Bay Drive; Kaneohe HI 96744; USA

SIMON, Barry; Department of Mathematics and Physics; Princeton University; Princeton NJ 08544; USA; 609-452-4325x5650; T; 1.1; 3.9; Complex scaling theory of resonances; Atoms and molecules in constant electric magnetic field; Hamiltonian operators and wavefunctions --mathematical properties

SIMONS, Adrian L.; 267 Country Club Lane; Scotch Plains NJ 07076; USA

SIMONS, Donald G.; White Oak Laboratory; U.S. Naval Surface Weapons Center; Silver Spring MD 20910; USA; 301-394-2272; E/T; 3.4; 4.2; 5.1; Heavy ion stopping powers; X-ray production cross sections; Ion beam materials analysis

SIMONY, Paul R.; Department of Physics; Jacksonville University; Jacksonville FL 32211; USA; 904-744-3950x6377; 904-744-8028; T; 2.1; 2.2

SIMPSON, Charles G.; Code 6604; Naval Research Laboratory; 4555 Overlook Ave., SW; Washington DC 20375; USA; 202-767-1168; E; 2.2; 3.1; 5.2

SIMPSON, J. A.; 312 Riley Street; Falls Church VA 22046; USA

SIMPSON, W. T.; Department of Chemistry; University of Oregon; Eugene OR 94703; USA; 503-686-4601; T; 1.1; 3.5; Molecular electronic spectra of molecules; Quantum mechanics of molecular electronic states

SIMS, James S.; Chemical Metallurgy Group; Metallurgy Division; National Bureau of Standards; Gaithersburg MD 20899; USA; 301-921-2913; T; 1.1; First row atoms--wavefunctions and properties

SINANOGLU, Oktay; Department of Chemistry; Yale University; 225 Prospect Street; New Haven CT 06520; USA; 203-436-2444; T; 1.1; 2.3; Many-electron theory of atoms and molecules; Atomic structure and atomic transition probabilities; Electron correlations; Chemical kinetics

SINCLAIR, Rolf M.; Physics Division; National Science Foundation; Washington DC 20550; USA; 202-357-7997; Administration

SINGER, Sidney; Group P-1; Los Alamos National Laboratory; PO Box 1663; Los Alamos NM 87545; USA; 505-667-3104; E; Highly excited matter--physics; Hot dense plasmas --physics

SINGH, Gurbax; Natural Sciences; School of Arts and Sciences; University of Maryland, Eastern Shore; Princess Anne MD 21853; USA; 301-651-2200x319; 301-921-3621

SINGH, Jag J.; Electronics Directorate; Instrument Research Division; Langley Research Center; NASA; Mail Stop 235; Hampton VA 23665; USA; 804-865-3907; 804-865-2449; E/T; 4.1; 4.2; 4.4; 5.6; 5.8; Reconstruction from computed interferograms; Ion-induced nucleation in gaseous mixtures

SINGH, Surendra P.; Department of Physics; University of Arkansas; Fayetteville AR 72701; USA; 501-575-5930; 501-575-2506; E/T; 3.11; 3.10; Quantum statistical properties of radiation, laser theory; Phase transitions in nonlinear light-matter interaction

SINGLETARY, Lillian; 32759 Seagate Drive #106; Rancho Palos Verdes CA 90274; USA

SINHA, Ashok K.; Communications Satellite Corp.; 20008 Wanegarden Court; Germantown MD 20767; USA; 301-554-6891; T; 2.1; Atomic and molecular collisions

SIOMOS, Konstadinos; Health and Safety Research Division; Oak Ridge National Laboratory; PO Box X; 4500 S. H-158; Oak Ridge TN 38931; USA; 615-574-6203; E; 3.6; 5.2; 3.8; 3.1; 3.10; Laser nonlinear molecular and negative ion spectroscopy in liquid media; Laser-induced excitation and ionization spectroscopy of atoms and molecules; Photophysics in high-density and liquid environment spectroscopy; Laser spectroscopy in analytical applications

SIPLER, Dwight; AFGL/OPA; Hanscom AFB MA 01731; USA

SISK, Raymond W.; A310 Villager Apartments; 316 Highland Road; Ithaca NY 14850; USA

SISKA, Peter E.; Department of Chemistry; University of Pittsburgh; Pittsburgh PA 15260; USA; 412-624-5064; E/T; 1.1; 2.3; Intermolecular forces and reaction dynamics; Crossed molecular beam and theoretical studies; Metastable rare gases

SITTERLY, Charlotte M.; 3711 Brandywine Street, NW; Washington DC 20016; USA; 202-966-9044; T; 3.2; 5.7; Atomic spectra--critical compilation of data on analyses of; XUV solar spectra as to chemical origin--identification of lines

SKIBOWSKI, Michael; Inst Exp Physik; Universitaet Kiel; Olshausenstr 40-60; Kiel 1 Schleswig/Holstein 2300; FR Germany; 0431-880-3853; 0431-880-3850; E;
5.1; 5.2

SKIFSTAD, James G.; Department of Mechanical Engineering; Purdue University; West Lafayette IN 47907; USA; 317-749-2634; T; 4.1; High power gasodynamic and chemical lasers; Nonequilibrium thermodynamics; Atomization physics

SKINNER, Charles H.; Plasma Physics Laboratory; Princeton University; PO Box 451; James Forrestal Campus; Princeton NJ 08544; USA; 609-683-2214; E; 2.3; 3.4; 3.6; 3.7; 3.10

SKINNER, Gordon B.; Department of Chemistry; Wright State University; Dayton OH 45435; USA; 513-873-2028; 513-873-2855; E; 2.3; 3.2; 3.6; 5.4; Resonance absorption spectroscopy of atoms; Atom concentrations in reacting gas mixtures; Mechanisms of combustion reactions-role of atoms

SKOFRONICK, James G.; Department of Physics; Florida State University; Tallahassee FL 32306; USA; 904-644-5497; E; 5.1; 4.3; Collisional behavior of He, D2 and HD with crystal surfaces; Nozzle beam development

SKOWRONEK, Maurice; Plasma Denses Tour 12 E5; University P. et M. Curie; 4 Place Jussieu; Paris Cedex 05 75230; France; 4314-4336 2525; 1.3; 3.1; 3.6; 3.10; 3.11

SKUPSKY, Stanley; Laboratory for Laser Energetics; University of Rochester; Rochester NY 14623; USA; 716-275-3951; T; 3.4; 5.4; X-ray spectra from laser compressed plasmas--computer simulation

SKUTLARTZ, Alexander E.; Department of Physics; James R. Mcdonald Laboratory; Kansas State University; 117 Cardwell Hall; Manhattan KS 66506; USA; 913-532-6786; 913-532-6777; E; 2.1; 4.2; 1.1; Dynamics of collision processes investigated via electron spectroscopy; Recoil-ion TOF in coincidence with scattered projectile (impact parameter dependence), electron capture and loss; Processes in heavy ion-atom collisions

SKUTNIK, Bolesh S.; E. B. Industries; 660 Hopmeadow Street; Simsbury CT 06070; USA

SLANGER, Tom; Physical Sciences; Chemical Physics Laboratory; SRI International; 333 Ravenswood Avenue; Menlo Park CA 94025; USA; 415-859-2764; E; 2.1; 2.3; 3.1; 3.2; 3.6; Metastable atoms and molecules--spectroscopy and kinetics; Chemistry of terrestrial and planetary atmospheres; Airglow phenomena

SLATER, Richard; Atomic and Molecular Physics; Directed Energy Technology; Avco Everett Research Laboratory; 2385 Revere Beach Parkway; Everett MA 02149; USA; 617-381-4470; E; 4.1; 3.10; 5.3; 5.6; Laser induced chemistry; Plasma chemistry; Atomic and molecular metastable production methods; Particle beam propagation

SLIMEY JR., James G.; Department 8453 Building 160; Optics and Laser Technology Div.; Aerojet ElectroSystems Co; PO Box 296; 1100 W. Hollyvale; Azusa CA 91702; USA; 213-334-6211x7112; T; 3.3; IR spectroscopy

SLINEY JR., James G.; MS 35A; Allied/EO Products; 31717 La Tienda Drive; West Lake Village CA 91362; USA

SLOANE, Christine S.; Environmental Sciences; General Motors Research Laboratory; General Motors Technical Center; 12 Mile and Mound Roads; Warren MI 48051; USA; 313-575-3490; T; 5.6; Aerosol optics

SLOANE, Thompson M.; Physical Chemistry Division; General Motors Research Lab.; General Motors Technical Center; 12 Mile and Mound Roads; Warren MI 48090; USA; 313-575-6553; E; 4.4; 5.5; Elementary reactions important in combustion; in crossed molecular beams; Mass spectrometer sampling of flames

SLOCUM, Robert E.; New Product Development Division; Texas Instruments, Inc.; PO Box 6015; Dallas TX 75222; USA; 214-867-9804; E; 3.8; 4.3; Optical pumping in helium to devise highly sensitive magnetometers

SMALLEY, R. E.; Department of Chemistry; Rice University; PO Box 1892; Houston TX 77001; USA; 713-527-4014; E; 1.1; 3.6; Molecular structure and dynamics via laser spectroscopic measurements --in superionic beams

SMARS, Erik; AGA AB Innovation; Lidingo S18181; Sweden

SMIO, Tani; Department of Physics; Marquette University; Milwaukee WI 53233; USA; 414-224-7069; T; 1.1; 2.2; Separable form-factor expansion of the Coulomb potential; and its application to electron-hydrogen atom scattering

SMITH, Allan L.; Department of Chemistry; Drexel University; 32nd and Chestnut Streets; Philadelphia PA 19104; USA; 215-895-2667; E/T; 3.6; Diatomic molecules --laser spectroscopy

SMITH, Arthur A.; Department of Research and Development; Research Applications; Laser/Electro-Optics; Systems Research Laboratories; 2800 Indian Ripple Road; Dayton OH 45440; USA; 513-426-6000 x33; E; 3.6; 3.2; 5.5; 4.1; Laser/electro-optical diagnostics of hostile environments i.e., solid and liquid combustion systems

SMITH, Barry T.; PO Box 1042; Route 3; Hayes VA 23072; USA

SMITH, Darwin; Department of Chemistry; University of Georgia; Athens GA 30602; USA; 404-542-2626x33; T; Electronic systems--wavefunctions

SMITH, David A.; Precision Products Division; Northrop Corporation; 100 Morse Street; Norwood MA 02062; USA; E/T; 4.3; Ring laser gyroscope-- development of

SMITH, David B.; Department of Physics and Astronomy; University of Kentucky; Lexington KY 40506; USA; 257-2761; 257-6722; E; 1.2; 2.1; Collisions of ions with laser excited atoms

SMITH, David C.; Physics Division; United Technologies Research Center; Silver Lane; East Hartford CT 06108; USA; 203-727-7281; E/T; 3.6; 5.6; Laser development and atmospheric physics related to propagation of laser radiation

SMITH, David; Department of Chemistry; PA State University, Hazelton; Hazelton PA 18201; USA; 717-454-8731; T; Hindered rotation of polyatomic ions in solids

SMITH, DeForest F.; Special Projects Branch; Enrichment Technologies Division; Oak Ridge Gaseous Diffusion Plant; Union Carbide Corporation; PO Box P MS322; Oak Ridge TN 37830; USA; 615-572-0150; E; 3.6; Molecular laser isotope separation process; for uranium hexafluoride--development; Nonlinear absorptions in uranium hexafluoride

SMITH, Earl W.; Time and Frequency Division; National Bureau of Standards; 325 S. Broadway; Boulder CO 80302; USA; 303-499-1000x3357; T; 5.4; Hydrogen atoms in a plasma--spectral line broadening

SMITH, Felix T.; Molecular Physics Laboratory; SRI International; 333 Ravenswood Avenue; Menlo Park CA 94025; USA; 415-326-6200x2398; T; 2.1; 2.2; Collision theory; New dynamical approaches to few-body systems--formulation

SMITH, Gary A.; Building K-1; Research and Development; General Electric Company; PO Box 8; Room 2C25; Schenectady NY 12301; USA

SMITH, George F.; Research Laboratories; Hughes Aircraft Co.; 3011 Malibu Canyon Rd.; Malibu CA 90265; USA; 213-317-5200; E; 4.1; Laser applications; Management of research

SMITH, George W.; Physics; Physics Division; General Motors Research Lab.; General Motors Technical Center; 12 Mile and Mound Roads; Warren MI 48090; USA; 313-575-2907; 313-575-2836; E/T; 5.2; 5.3; 5.5; 5.6; Molecular and chemical physics; Phase behavior; Calorimetry and kinetic and mechanical properties; Liquid crystals: Properties and applications; Nucleation: theory and experiment

SMITH, Gregory P.; Chemical Kinetics; Chemical Kinetics Division; Chemical Physics; SRI International; 333 Ravenswood Avenue; Menlo Park CA 94025; USA; 415-859-3496; E; 3.5; 5.5; 3.6; Spectroscopy of small molecules; Quenching and energy transfer of small molecules in flames

SMITH, Henry J.; Visidyne, Inc.; 5 Corporate Place, S. Bedford St.; Burlington MA 01813; USA

SMITH, Jerel A.; 1223 E. Arques Avenue; Sunnyvale CA 94086; USA

SMITH, Joseph A.; Transportation; U.S. Coast Guard; USCG Research and Development Center; Marine Systems; Box 261; Suzane Lane; Brooklyn CT 06234; USA; 203-779-0635; 203-441-2656; E; 1.1; 4.1; Computer models

SMITH, Leslie N.; Department of Chemistry; Princeton University; Princeton NJ 08540; USA; 609-452-3905; T; 2.1; 3.6; Atom-molecule scattering; Molecules and electromagnetic fields-- interaction

SMITH, Lloyd P.; 565 Woodside Drive; Woodside CA 94062; USA

SMITH, Neville V.; Bell Telephone Laboratory; Murray Hill NJ 07974; USA

SMITH, Peter L.; Harvard College Observatory; Atomic and Molecular Physics Division; Harvard University; 60 Garden Street, MS-50; Cambridge MA 02138; USA; 617-495-4984; 617-495-5475; T; 1.1; 3.1; 3.2; 3.6; 5.7; Transition mobilities; UV wavelength; Atomic and molecular measurement; Instrumentation; Spectroscopic high-resolution; UV satellite-borne

SMITH, Richard R.; Magnetic Fusion Energy Division; Lawrence Berkeley Laboratory; 1 Cyclotron Road; Berkeley CA 94720; USA; 415-486-5534; E; 2.1; 4.2; 5.1; 5.4; Neutral beam diagnostics: light emission, surface analysis; Charge exchange, ionizing collisions

SMITH, Sidney T.; 4514 Bee Street; Alexandria VA 22310; USA

SMITH, Stephen J.; Quantum Physics Div. (NBS); Joint Institute for Laboratory Astrophysics; University of Colorado; Campus Box 440; Boulder CO 80309; USA; 303-492-7788; 303-472-7783; E; 3.7; 3.11; 3.1; 3.10; Photoelectron angular distributions from excited states

SMITH, Winthrop W.; Department of Physics U-46; University of Connecticut; Storrs CT 06268; USA; 203-486-3573; 203-486-5112; E; 2.1; 3.6; 3.1; 3.4; 4.2; Photon and electron emission resulting from ion-atom collisions; spectroscopic studies; Accelerator-based atomic physics; Laser spectroscopy

SMITHER, Robert K.; Fusion; Chemical Technology; Argonne National Lab.; Department of Energy; 9700 S. Carr; Argonne IL 60439; USA; E/T; 5.1; 5.4; 5.5; 5.7; Properties of other special atoms and molecules; e.g. exotic atoms, macromolecules, polymers, clusters; Atomic and molecular collisions excluding electron collisions; Inner shell transitions including X-ray absorption and emission

SMYTH, Kermit C.; Exploratory Fire Research Group; National Bureau of Standards; Building 224, Room B260; Gaithersburg MD 20899; USA; 301-921-3771; E; 3.6; 5.5; 2.3; Premixed and diffusion flames--optical studies; Laser-enhanced ionization and laser-induced fluorescence experiments; Mass spectrometric sampling studies

SNAVELY, Benjamin B.; Special Optical Systems; Research and Engineering; Eastman Kodak Co.; 121 Lincoln Avenue; Rochester NY 14650; USA; 716-253-2394; E; 4.1; 5.4; Physical optics; Atomic and molecular physics in solids; Laser applications

SNOW, William R.; Pacific Western Systems, Inc.; 505 E. Evelyn Avenue; Mountain View CA 94041; USA

SNYDER, James J.; Quantum Metrology Group; National Bureau of Standards; Gaithersburg MD 20899; USA; 301-921-2061; E; 3.6; 3.10; 4.1; 4.4; Laser spectroscopy

SNYDER, Lewis E.; Department of Astronomy; 341 Astronomy Building; University of Illinois; 1011 W. Springfield Avenue; Urbana IL 61801; USA; 217-333-5530; 217-333-3090; E; 3.5; 5.7; Interstellar molecules--observation and detection

SNYDER, Ralph B.; Department of Physics; University of Connecticut; Storrs CT 06268; USA; 203-486-4187; 203-486-4915; T; 1.1; 2.1; Atomic structure; Heavy particle collisions

SO, Sung-Leung I.; Department of Physics; SUNY, Stony Brook; Stony Brook NY 11794; USA

SOBEL, Alan; Lucitron, Inc.; 1918 Raymond Drive; Northbrook IL 60062; USA; 312-564-8383; E; 5.3; Gas-discharge display devices

SOKOLOFF, Jack; Department of Physics; York University; 4700 Keele Street; Toronto Ontario M3J1P3; Canada

SOLARZ, Richard W.; Y-Division; Lawrence Livermore National Lab.; PO Bx 808 L-468; Livermore CA 94550; USA; 415-422-6218; E; 3.6; 3.8; Laser spectroscopy and photochemistry of atoms and small molecules

SOLIMENE, Nicholas; 87-27 94th Street; Woodhaven NY 11421; USA

SOLLID, Jon E.; Antares Project; Group P-5; Los Alamos National Laboratory; MS 533; Los Alamos NM 87545; USA

SOLLNER, Gerhard; Department of Electrical and Computer Engineering; University of Massachusetts; Amherst MA 01003; USA; 413-545-2894/0638; T; 5.7; Interstellar molecules --low-noise instrumentation for observation

SOLOMON, Jerry E.; Department of Physics; San Diego State University; San Diego CA 92182; USA; 714-265-6157; E; 4.1; Saturation and coherent transient spectroscopy in simple molecules; Photoacoustic spectroscopy

SOLTANOLKOTABI, Mahmood; Khiabane Nazareh Gharbi Kouyeh; Ghainan #89; Isfahan; Iran

SOLTIS, Paul J.; Department of Engineering; Engineering Division (SSIED); Engineering Materials; US Naval Air Engineering Center; Building 562; Lakehurst NJ 08733; USA; 201-323-2285; 215-674-2314; T; 1.3; 3.2; 5.7; Excitons, muonium, positronium nature; Valence electron orbits to line spectra relations

SOMERVILLE, Lawrence P.; Physics Division; Argonne National Lab.; Building 203, Room H113; Argonne IL 60439; USA

SOMORJAI, Gabor A.; Department of Chemistry; University of California; Berkeley CA 94720; USA; 415-642-4053; E; 5.1; Surfaces by low-energy electron diffraction --structure; Surface reactions and decomposition by electron spectroscopy

SONG, Ru-Wang; 83-14 Britton Avenue; Elmhurst NY 11373; USA

SOREM, Michael S.; Tunable Lasers and Applications Group; Chemistry Division; Los Alamos National Laboratory; P.O. Box 1663 MS564; Los Alamos NM 87545; USA; 505-667-9739; E; 4.1; 3.6; 5.4; Optically pumped laser systems; Laser development, optically pumped lasers; Laser spectroscopy; Isotope separation

SOSNIAK, Jacob; 7464 Springmill Road; Indianapolis IN 46260; USA

SOUDER, Paul A.; Department of Physics; Yale University; 217 Prospect Street; New Haven CT 06520; USA; 203-436-8785; E; 1.3; 4.2; Muonium and muonic helium atoms

SOUZA, Steven P.; Department of Physics and Astronomy; Williams College; Williams MA 01267; USA; 413-597-2247; E; 5.1; Atomic hydrogen on frozen surfaces relating to atomic hydrogen frequency standards and production of; Bose condensation in spin-polarized hydrogen gas--behavior

SPALEK, George; 1311 Lejano Lane; Santa Fe NM 87501; USA

SPARKS, Randal K.; Department of Chemistry and Chemical Engineering; California Institute of Technology; 1201 E. California Boulevard; Pasadena CA 91125; USA; 213-795-6811; E; 2.3; 3.1; Crossed molecular beam studies of chemical reactions; and photofragment spectroscopy

SPARROW, Julian H.; X-Ray Physics Division; National Bureau of Standards; Building 245, Room C216; Gaithersburg MD 20899; USA; 301-921-2201; E; 2.2; 3.4; Low-Z target to produce standard X-ray line calibration sources; electron excitation of K and L lines

SPEAR, Arthur W.; 201 East Mimosa Circle; San Marcos TX 78666; USA

SPENCE, David; Radiological and Environmental; Research Division; Argonne National Laboratory; 9700 S. Cass Avenue; Argonne IL 60439; USA; 312-972-4191; E; 2.2; 3.2; 3.5; Atomic and molecular spectroscopy and processes; by low energy electron scattering

SPENCER, William P.; MIT; Room 26-237; Cambridge MA 02139; USA

SPEZESKI, Joseph J.; Department of Physics; University of Arizona; Tucson AZ 85721; USA; 602-626-4271/4276; E; 3.1; Time-of-flight studies of dissociation fragments; from simple molecules

SPICER, Leonard D.; Department of Biochemistry and Radiology; Medical Center; Duke University; Durham NC 27710; USA; 919-684-4327; E; 2.1; 2.3; 3.6; 3.8; Hot atom reaction dynamics; Molecular collisions energy transfer and reaction of electronically and rotationally-vibrationally excited small molecules; Photo-assisted transition metal catalysis; Laser-induced fluorescence

SPICER, William M.; Electrical Engineering, Materials Science, and Applied Physics; School of Engineering; Stanford Electric Lab.; Stanford University; McCullough Bldg., Room 228; Stanford CA 94303; USA; 415-497-4643; E/T; 5.1; Surfaces and interfaces; Photoemission in solids

SPIERENBURG, G. Pieter; Bohturmweg 10 CH8437; Switzerland

SPOONER, David W.; Department of Physics; Stanford University; Stanford CA 94305; USA; 415-497-4640; 415-497-0655; E; 2.1; 4.2; 5.8

SPOSITO, Garrison; Department of Soil and Environmental Sciences; Environmental Sciences; University of California, Riverside; Riverside CA 92521; USA; 714-787-3757x5103; 714-787-3843; E; 5.6; 5.2; 3.3; 3.2; 3.6; Clay mineral structure (aluminosilicates); Absorbed water structure on inorganic surfaces; Organic functional group-metal interactions

SPRINGER, Robert H.; LR&TSO #1310; General Electric Lighting Business Group; Nela Park; East Cleveland OH 44112; USA; 216-266-3147; E; 5.3; 5.4; Physical electronics and spectroscopy related to lighting; Discharges and electric phenomena; Electrode science

SPROUSE, Gene D.; Department of Physics; State University of New York at Stony Brook; Stony Brook NY 11794; USA; 516-246-5051/7110; E; 3.6; 4.2; Hyperfine structure of nuclei far from stability; Highly ionized atoms

SPRUCH, Larry; Department of Physics; New York University; 4 Washington Place; New York NY 10003; USA; 212-598-7636; T; 2.1; 2.3; 3.8; Scattering theory: charge transfer, variational principles, effects of; Long-range potentials, radiative corrections

SPYROU, Spyros M.; 711 W. Vanderbilt; Oak Ridge TN 37830; USA

SRINIVASAN, Varadarajan; 10400 Pineville Avenue; Cupertino CA 95014; USA

SRIVASTAVA, B. N.; Aerophysics Research Committee; Avco Everett Research Laboratory; 2385 Revere Beach Parkway; Everett MA 02149; USA; 617-389-3000x559; T; 3.6; 5.3; Kinetic modelings; Discharge modelings; Laser physics

SRIVASTAVA, Santosh K.; Molecular Physics and Chemistry Section; Jet Propulsion Laboratory; 4800 Oak Grove Drive; Pasadena CA 91103; USA; 213-354-3246x2301; E; 2.2; Collisions involving electrons, atoms and molecules; Spectroscopy of atoms and molecules; Collision cross sections for electron-atom /molecule interactions

ST. JOHN, Robert M.; Department of Physics and Astronomy; University of Oklahoma; 440 W. Brooks Street; Norman OK 73019; USA; 405-325-3961; E; 1.1; 2.2; 4.2

ST. JOHN III, Willard M.; Department of Chemistry; Oakland University; Rochester MI 48053; USA; 313-377-2100; E/T; 1.1; 2.3; Experimental molecular electron affinities; Quantum chemistry

ST. PETERS, Richard L.; General Electric Research and Development Center; PO Box 8; Schenectady NY 12301; USA; 518-385-8470; E/T; 3.5; 3.6; 4.2; Applied molecular spectroscopy; Optical spectroscopy with lasers; EXAFS; Nonlinear optical spectroscopic techniques

STADELMANN, James L.; 7233 South Campbell Avenue; Chicago IL 60629; USA

STAHL JR., Edward A.; 12777 Gearhart Road; Greencastle PA 17225; USA

STALDER, Kenneth R.; 515 King Street; Redwood City CA 94062; USA; 415-367-1359; 415-327-8548; E; 2.1; 3.6; 4.2; 5.1; 5.3

STALEY, Frederick J.; PO Box 2287; Fort Collins CO 80521; USA

STALEY, Ralph H.; Department of Chemistry; MIT; Cambridge MA 02139; USA; 617-253-4537; E; 2.3; 3.8; Gas phase metal ion chemistry and photochemistry using ion cyclotron resonance spectroscopy

STALLCOP, James R.; Physical Sciences Branch; Ames Research Center; Moffett Field CA 94035; USA; 415-965-6140; T; 1.1; 3.6; Atomic and molecular structure calculations; Photon-atom-molecule scattering

STAMPS, George M.; 268 Hillspoint Road; Westport CT 06880; USA

STANTON, Alan C.; Applied Sciences Division; Aerodyne Research, Incorporated; Crosby Drive; Bedford MA 01730; USA; 617-275-9400; E/T; 2.1; 3.6; Diode laser spectroscopy for detection of radical species; and measurement of fundamental spectroscopic parameters; Molecular beam studies of collisional energy transfer cross sections

STANTON, Joyce L.; 55 Austin Place; New York NY 10304; USA

STANTON, Richard E.; Department of Chemistry; Canisius College; 2001 Main Street; Buffalo NY 14208; USA; 716-883-7000; 716-835-8612; T; 1.1; Dirac-Coulomb computational methods; SCF convergence theory; Applications of group theory; Correlation energies

STAPOR, William J.; 12921 Lyme Bay Drive; Herndon VA 22071; USA

STARACE, Anthony F.; Department of Physics and Astronomy; University of Nebraska; Lincoln NE 68588; USA; 402-472-2795/2770; T; 3.1; 3.9; 2.1; 2.2; 3.7; Atomic photoionization processes; Atoms in high magnetic fields; Hyperspherical coordinate methods

STARR, Walter L.; Atmospheric Experiments Branch; Space Science Division; NASA Ames Research Center; M/S 245-5; Moffett Field CA 94035; USA; 415-694-5503; 415-941-9175; E; 5.6; 3.2; 3.1

STAUBER, Michael C.; 79 Long Drive; Hempstead NY 11550; USA

STAUFFER, Allan D.; Department of Physics; York University; 4700 Keele Street; Downsview Ontario M3J1P3; Canada; T; 2.2; 1.1; 5.3; 5.4

STEARNS, John W.; Magnetic Fusion Energy; Accelerator Division; Lawrence Berkeley Laboratory; University of California; Mail STOP 5-119; Berkeley CA 94720; USA; 415-486-5011; FTS 451-5011; E; 2.1; 4.2; 3.4; 3.6; Intense neutral beam development--analysis; Cross sections for atomic and molecular, ion-target interactions; Surface plasma interactions

STEARNS, Mary B.; Physics Division; Ford Research Laboratory; PO Box 2053; Room S3039; Dearborn MI 48121; USA; 313-323-1714; E/T; 5.2; Nuclear magnetic resonance and hyperfine fields in alloys; EXAFS studies of dilute impurities in iron and nickel

STEBBINGS, Ronald F.; Department of Space Physics and Astronomy; Rice University; PO Box 1892; Houston TX 77251; USA; 713-527-8101 x3531; 713-527-4996; E; 2.3; 5.6; 2.1; 1.2; 5.7; Differential ion-atom and atom-atom scattering; Spectroscopy, photodissociation; Rydberg atom collisions

STECHEL, Ellen B.; Department of Chemistry; University of California, LA; Los Angeles CA 90024; USA; 213-825-9035; T; 1.1; 2.3; Low energy reactive collisions; Intramolecular energy transfer in small anharmonic molecules

STEEL, Duncan G.; Physics and Electrical Engineering; Randall Laboratories; University of Michigan; Ann Arbor MI 48109; USA; 313-764-2386; 313-764-4437; E; 1.1; 3.6; 3.10; 2.1; 3.11; Frequency domain nonlinear laser spectroscopy; Application of A&M spectroscopy to condensed matter physics

STEELE, William A.; Department of Chemistry; 152 Davey Laboratory; Pennsylvania State University; University Park PA 16802; USA; 814-865-3711; T/E; 1.1; 5.1; 3.10; Intermolecular interactions; Statistical thermodynamics of molecular systems; Gas-solid interactions; Nonlinear interaction of radiation with atoms and molecules

STEGEMAN, G. I.; Optical Sciences Center; University of Arizona; Tucson AZ 85721; USA

STEHLE, Philip; Department of Physics and Astronomy; University of Pittsburgh; Pittsburgh PA 15260; USA; 412-624-4294; T; 3.6; 3.7; 3.11; Intense light with atoms--interactions

STEHMAN, Robert M.; Department of Physics; Northeastern Illinois University; 5500 N. St. Louis Avenue; Chicago IL 60625; USA; 312-583-4050; x683/746; T; 3.1; Atomic and molecular negative ions--photodetachment

STEIN, Samuel R.; Efratom; Ball Corporation; PO Box 589; Broomfield CO 80020; USA; 303-460-5227; E; 3.3; 3.6; 4.1; Atomic frequency standards

STEIN, Talbert S.; Department of Physics and Astronomy; Wayne State University; Detroit MI 48202; USA; 313-577-2713; 313-577-2721; E; 2.2; 4.2; Collisions of low and intermediate energy positrons and electrons with atoms and molecules

STEINBERG, Martin; Quantum Institute; Univ. of California, Santa Barbara; Santa Barbara CA 93106; USA; 805-961-3278; E; 2.1; 3.8; 5.5; Atomic and molecular fluorescence in flames; Quenching; Vibrational and rotational relaxation; Multiplet mixing; Laser induced chemistry

STEINER, Bruce W.; Materials B-308; National Bureau of Standards; Gaithersburg MD 20899; USA

STEINFELD, Jeffrey I.; Department of Chemistry; Massachusetts Institute of Technology; Room 2-221; Cambridge MA 02139; USA; 617-253-4525; E/T; 2.3; 3.5; 3.8; Molecular spectroscopy; Kinetics: Laser-induced processes in molecules and at surfaces

STEINHAUS, David W.; Department of Physics; Southwestern College, Memphis; 2000 North Parkway; Memphis TN 38112; USA; 901-274-1800x378; E; 3.2; Atomic spectrum of uranium--measurement

STELSON, Paul H.; Physics Division; Oak Ridge National Laboratory; PO Box X; Oak Ridge TN 37830; USA; 615-574-4773; E; 4.2; Research management

STEPH, Nick C.; Department of Physics; Lake Forest College; Sheridan and College Roads; Lake Forest IL 60045; USA

STERN, Richard C.; Advanced Isotope Separation Program; Lawrence Livermore National Lab.; PO Box 808 L-46; Livermore CA 94550; USA; 415-422-6213; E/T; 2.3; 3.8; Chemical kinetics using crossed molecular beams; Laser excitation

STERNHEIMER, Rudolph M.; Physics Division; Brookhaven National Laboratory; Building 510A; Upton NY 11973; USA; 516-282-3759; T; 1.1; Sternheimer quadrupole antishielding factors--calculations and comparison with experiment; Calculations of the electronic polarizabilities for several ions and neutral atoms (particularly the alkali metal atoms)

STERNLIEB, Abraham; Applied Physics Department; Soreq Nuclear Research Center; Israel Atomic Energy Commission; Yavne 70600; Israel

STETTLER, John D.; Research Directorate; U.S. Army Missile Command; DRSMI-RRO; Building 7770; Redstone Arsenal AL 35809; USA; 205-876-3820; T; 2.1; 3.3; 5.6; Energy transfer between molecules (vibrational and rotational); Submillimeter radiation--atmospheric absorption

STEUER, Malcolm F.; Department of Physics and Astronomy; University of Georgia; Athens GA 30602; USA

STEVENS, Walter J.; Quantum Chemistry Group; Molecular Spectroscopy Division; Center For Chemical Physics; National Bureau of Standards; Bldg. 221, Room B268; Gaithersburg MD 20899; USA; 301-921-2774; 301-921-2021; T; 1.1; 1.3; 3.1; Molecular spectroscopy

STEVERDING, Bernard; U.S. Army Research; PO Box 65; FPO New York NY 09510; USA

STEWART, Howard A.; 540 Treetop Lane; Hixson TN 37343; USA

STEWART, Richard E.; Lawrence Livermore Lab.; PO Box 808; L-401; Livermore CA 94550; USA

STIDHAM, Howard D.; Department of Chemistry; University of Massachusetts; Amherst MA 01003; USA; 413-545-2583; E; 3.5; 3.6; Vibrational, spin resonance (NMR) molecular spectroscopy; Laser Raman spectroscopy

STINESPRING, Charter D.; Center for Chemical and Environmental Physics; Research Group; Aerodyne Research, Inc.; 45 Manning Road; Billerica MA 01821; USA; 617-663-9500; E; 5.1; 3.1; XPS of absorbed atoms and molecules

STITCH, Malcolm L.; Laser Enrichment Division; Exxon Nuclear Research and Technology Center; 2955 George Washington Way; Richland WA 99352; USA; 509-375-7262; E; Uranium isotope separation process using selective absorption; of laser radiation--development

STOCKBAUER, Roger; Surface Science Division; National Bureau of Standards; Building 222, Room B248; Gaithersburg MD 20899; USA; 301-921-2096; E; 3.1; 4.2; 4.4; 5.1; Photoionization mass spectrometry; Variable wavelength photoelectron spectroscopy; Photosimulated desorption of atoms and molecules from surface adsorbed species

STOCKDALE, John A.; Chemical Physics Section; Chemical Physics Division; Oak Ridge National Laboratory; PO Box X; Oak Ridge TN 37830; USA; 615-574-6238; 615-487-0075; E/T; 1.1; 3.6; 3.7; 2.1; 2.2; Negative ions--properties of; Atomic and molecular polarizabilities; Multiphoton ionization; Nonlinear interactions of radiation with atoms and molecules

STOCKLI, Martin P.; Department of Physics; J.R. Macdonald Laboratory; Kansas State University; Cardwell Hall; Manhattan KS 66506; USA; 913-532-6784; 913-532-6786; E; 4.2; 2.1

STOCKMAL, Alexander J.; 29 Katrina Circle; Bethel CT 06801; USA

STOCKTON, Marilyn; 1031 Scott Boulevard Apt. #B2; Decatur GA 30030; USA

STOFFELS, James J.; 1930 George Washington Way #203; Richland WA 99352; USA

STOGRYN, Daniel E.; Department of Physical Sciences and Mathematics; Mount St. Mary's College; 12001 Chalon Road; Los Angeles CA 90049; USA; 213-476-2237x239; T; 1.1; Intermolecular forces; Statistical mechanics

STOICHEFF, Boris P.; Department of Physics; University of Toronto; 60 St. George Street; Toronto Ontario M5S1A7; Canada; 416-978-2948; E; 3.6; 3.10; 1.1; 1.2; 4.1

STOJANOFF, Christo G.; Desert Research Institute; 1500 Buchanan Boulevard; Boulder City NV 89005; USA

STOLARSKI, Richard S.; NASA Code 964; Goddard Space Flight Center; Greenbelt MD 20771; USA

STONE, Craig D.; Department of Physics; North Carolina State University; Cox Hall; Raleigh NC 27650; USA

STONE, Edward J.; Space Physics Research Laboratory; University of Michigan; 2455 Hayward; Ann Arbor MI 48109; USA; 313-763-9940; E; 4.2; Rocket probe sampling in the middle atmosphere

STONE, Jack A.; 113 Cross Street; Middletown CT 06457; USA

STONE, Philip M.; Development and Technology Division; Office of Fusion Energy; U.S. Department of Energy; Washington DC 20545; USA; 202-353-3734; T; 1.1; 2.2; 5.4; 2.3; Fusion energy--research contracting and management of projects related to data evaluation for plasma radiation and energy transport

STONE, Sam; Physics Department, L-295; Lawrence Livermore National Lab.; PO Box 808; Livermore CA 94550; USA; 415-422-4302; T; 5.3; 5.4; 5.7

STONEMAN, Robert C.; Department of Physics; University of Virginia; Charlottesville VA 22901; USA; 804-924-6781; E; 1.2; 3.9; 2.1; 3.3

STONER JR., John O.; Department of Physics; University of Arizona; Tucson AZ 85721; USA; 602-621-6814; 602-624-1881; E; 1.1; 2.2; 4.2

STORM, David A.; 35 Valley View Terrace; Montvale NJ 07645; USA

STORM, Ellery; HSEI; MS F692; University of California; PO Box 1663; Los Alamos NM 87545; USA

STORY, Troy L.; Department of Chemistry; Morehouse College; Atlanta GA 30314; USA; 404-681-2800; T; 1.1; Large vibration-rotation interactions in molecules

STOTLAR, Suzanne C.; Materials Technology; Metallurgy and Ceramics; Materials Science and Technology; Los Alamos National Laboratory; University of California; P.O. Box 1663 MS430; Los Alamos NM 87545; USA; 505-662-7729; 505-667-6914; E; 4.1; 4.3; 3.3; 3.10; 5.1; Development and characterization of high speed diagnostics for broadband, applications and pyroelectric detectors; Development and characterization of semiconductor, photocathodes development, and characterization of solid state materials and devices

STOVALL JR., Emory J.; 363 Andanada; Los Alamos NM 87544; USA

STOVALL, Gregory T.; PO Box

6179; College Station TX 77844; USA

STOWELL, David Y.; 7730 Hanover Parkway #102; Greenbelt MD 20770; USA

STRAND, Oliver T.; X-ray Measurements Department; L Division; Lawrence Livermore National Lab.; Box 808, L-379; Livermore CA 94550; USA; 415-423-0528; E; 3.4; 5.4; X-ray spectroscopy; Plasma spectroscopy using VUV radiation

STRATT, Richard M.; Department of Chemistry; Brown University; Providence RI 02912; USA; 401-863-3418; T; 5.2; 1.1; Intramolecular statistical mechanics; Highly quantum atomic/molecular degrees of freedom with a classical environment-- interactions

STRICKLER, Stewart J.; Department of Chemistry; University of Colorado; P.O. Box 215; Boulder CO 80309; USA; 303-492-7367; E; 1.1; 3.5; 3.7; 3.8; Electronic spectroscopy of polyatomic molecules; Vibrational-electronic interactions; Lifetimes; Two-photon spectroscopy; Photochemistry

STRICKLETT, Kenneth L.; Department of Physics and Astronomy; University of Nebraska, Lincoln; Lincoln NE 68588; USA; 402-472-2786; 402-472-2770; E; 2.3; 2.2; 3.8

STROH III, William; 9 Sinclair Terrace; Short Hills NJ 07078; USA

STROKE, H. Henry; Department of Physics; New York University; 4 Washington Place; New York NY 10003; USA; 212-598-3204/3661; E; 3.6; 3.8; 4.2; 5.4; 4.1; Hfs and isotope shifts for nuclear EM and atomic structure studies; Collisionally aided radiative excitations and radiatively aided collisions; Laser probing of fluorescent lamp plasmas; Calorimeter development for eV spectroscopy of charged particles and X-rays

STRONG, David K.; 17360 NW West Union Road; Portland OR 97229; USA

STROUD JR., Carlos R.; Institute of Optics; University of Rochester; Rochester NY 14611; USA; 716-275-2598; 716-275-2471; E/T; 1.2; 1.4; 3.6; 3.11; 4.1; Rydberg atomic state dynamics; Instabilities of lasers

STROZIER JR., John A.; 51 Dyke Road; Setauket NY 11733; USA

STRUENSEE, Michael C.; 1810 Kenwood Avenue; Austin TX 78704; USA

STUBB, Tor H.; Semiconductor Lab.; State Inst. Tech. Research; Otaniemi; Finland

STUMPF, Bernhard J.; Department of Physics; New York University; 4 Washington Place; New York NY 10003; USA; 212-598-3332; E/T; 1.1; 2.2; 3.6; 3.8; 3.9

STWALLEY, William C.; Department of Chemistry; Iowa Laser Facility; University of Iowa; Iowa City IA 52242; USA; 319-353-7081; E/T; 1.1; 2.1; 3.6; 3.8; 4.1; Atomic and molecular interactions; Laser spectroscopy; Spin-polarized atoms; Molecular beams

STYRIS, D. L.; Department of Chemical Technology; Research Division; Pacific Northwest Laboratories; Battelle Memorial Institute; PO Box 999; Building 320, Area 300; Richland WA 99352; USA; 509-376-1907; E; 4.4; 5.1; 4.2; Mass spectroscopy and atomic absorption spectroscopy applied to atomic and molecular emissions from high temperature surfaces; Ion impact radiation

SUCHANNEK, Rudolf G.; Department of Physics; University of California, LA; 405 Hilgard Avenue; Los Angeles CA 90024; USA; 213-825-9531; E; 5.4; Spectral line profiles of atoms and ions in high temperature plasmas--measurement

SUCHER, Joseph; Department of Physics and Astronomy; University of Maryland; College Park MD 20742; USA; 301-454-4894; T; 1.4; Parity violation in atomic physics

SUCK, Sung Ho; Department of Physics; Graduate Center for Cloud Physics; University of Missouri, Rolla; Rolla MO 65401; USA; 314-341-4341; T; 1.1; 2.1; 2.3; Atom-diatomic molecule collisions (reactive scattering); Molecular orbital calculations

SUCKEWER, S.; Plasma Physics Laboratory; Princeton University; Princeton NJ 08544; USA; 609-683-3149; E/T; 3.6; 5.4; Plasma physics; Plasma spectroscopy in Tokomak; Lasers

SUDWORTH, Keith; Electronics Tech.; Raychem R&D; Faraday Road Dorcan; Swindon Wiltshire; England

SUGAR, Jack; Atomic and Plasma Radiation Division; National Bureau of Standards; A163 Physics Building; Gaithersburg MD 20899; USA; 301-921-2011; E; 1.1; 3.2; Absorption and emission spectra of atoms and ions; Atomic structure--calculation of

SUHRE, Dennis R.; Directed Energy Research Department; Applied Sciences; Westinghouse R&D Center; Westinghouse Electric Corporation; 1310 Beulah Road; Pittsburgh PA 15235; USA; 412-250-1649; E/T; 4.1; 5.3; Development of laser optical systems; Modeling and diagnostics of laser systems

SULLIVAN, Gregory W.; 327B S. Lake Heights; Carbondale IL 62901; USA

SUN, James C.; Department of Physics; University of California, Riverside; Riverside CA 92521; USA; 714-787-5331; T; 2.2; 2.3; Electron-molecule scattering; Atom-molecule reactive collision

SUNDARAM, Bala; Department of Physics and Astronomy; University of Pittsburgh; Allen Hall; Pittsburgh PA 15260; USA

SUNDARAM, S.; Department of Physics; Univ. of Illinois, Chicago Circle; PO Box 4348; Chicago IL 60680; USA; 312-996-5346; E/T; 3.2; 3.3; Diatomic molecules--electronic spectra; Raman and IR spectra of polyatomic molecules

SURKO, Clifford M.; Physics Division; AT&T Bell Laboratories; 600 Mountain Avenue; Murray Hill NJ 07974; USA; 201-582-2873; E; 5.4; 5.2; Study of waves and turbulence in Tokomak fusion plasmas using light scattering techniques; Study of the onset of time dependence and chaos in Rayleigh-Benard; Convection (fluid dynamics)

SUTCLIFFE JR., Victor C.; Semiconductor Research Laboratory; Texas Instruments, Inc.; PO Box 225012; MS 82; Dallas TX 75265; USA; 214-995-2479; E/T; 5.1; Electron beam interactions with solids

SUTO, Masako; Electrical and Computer Engineering; San Diego State University; San Diego CA 92182; USA; 619-265-4550; E; 2.3; 3.1; 5.1; 5.6; 5.7; Photoabsorption, photodissociation; Kinetics of radicals; Atmospheric chemistry; Astrophysical applications of atomic and molecular physics

SUTTON, David G.; Aerophysics Laboratory; The Aerospace Corporation; El Segundo CA 90245; USA; 213-648-5049; E; 2.1; 2.3; Electronic energy transfer between small molecules and atoms; Chemiluminescence

SUZUKI, Shigeru; Department of Physics and Astronomy; Univ. of North Carolina, Chapel Hill; Chapel Hill NC 27514; USA

SWAMY, Nyayapathi V.; Department of Physics; Oklahoma State University; Stillwater OK 74074; USA; 405-624-5814; T; 3.6; 3.9; One-particle Hamiltonian with magnetic fields--symmetry; Relativistic one-particle Hamiltonian with magnetic fields--symmetry

SWANN, Charles P.; Bartol Research Foundation; University of Delaware; Newark DE 19711; USA; 302-738-8113; E; 4.2; 5.1; Angular distribution and film thicknesses of sputtered gas

SWANSON, Nils; Far UV Physics; A251 Physics; National Bureau of Standards; Gaithersburg MD 20899; USA; 301-921-2031; E; 4.2; 5.1; 3.1; UV and soft x-ray detector calibrations; Photoemission from surfaces; Photoabsorption and photoionization

SWEETMAN, Eric; 208 White Pine Circle; Lawrenceville NJ 08648; USA

SWENSON, David R.; Department of Physics; 1150 University Avenue; Madison WI 53706; USA

SWIDER, William; United States Air Force; Ionospheric Physics Division; US Air Force Geophysics Laboratory; Hanscom AFB MA 01731; USA; 617-861-3891; 617-861-4832; T; 5.6; Stratospheric and low ionospheric modeling

SWITZER, Gary L.; Research Applications; Systems Research Laboratories; 2800 Indian Ripple Road; Dayton OH 45440; USA; 513-252-2706; 513-426-6000; E; 4.1; 3.10; 3.6; 3.5; 3.2

SYED, Bashir A.; MD 277; c/o AESD General Electric Company; French Road; Utica NY 13503; USA

SYNEK, M.; Div. of Earth and Physical Science; University of Texas, San Antonio; San Antonio TX 78285; USA; 512-691-5457; 512-684-3525; E/T; 1.1; 4.1; 5.1; 5.7; Laser-crystal energy efficiency

SZALEWICZ, Krzysztof; Department of Physics; University of Florida; Gainesville FL 32611; USA; 904-392-6616; 904-392-1597; T; 1.1; 1.30; 2.2; 3.1; 5.8; High-accuracy quantum chemistry; Coupled-cluster and perturbation methods for atoms and molecules; Muon-catalyzed fusion --atomic, molecular and nuclear aspects; Intermolecular interactions-symmetry adapted perturbation approach

SZE, Robert C.; CHM-5; Chemistry Division; Los Alamos National Laboratory; University of California; 1663 MS J566; Los Alamos NM 87545; USA; 505-667-4300; E; 4.1; 3.10; 3.6; High energy gas laser research; Nonlinear interactions of radiation with atoms and molecule Raman scattering

SZILAGYI, Mike N.; Department of Electrical Engineering; University of Arizona; Tucson AZ 85721; USA

SZOKE, Abraham; T-Division; Lawrence Livermore National Lab.; PO Box 808; Livermore CA 94550; USA; 415-422-4099; T; 3.6; 4.1; 3.7; Free electron lasers; Resonant Raman effect; Radiation from high temperature and density matter

TABATABAIE, Nader; Semiconductor Research Group; University of Illinois; 155 EEB; 1406 W. Green; Urbana IL 61801; USA

TAI, Chen-Yu; Department of Physics and Astronomy; University of Toledo; 2801 W. Bancroft; Toledo OH 43606; USA

TAKEISHI, Yoshiyuki; 24-6 Inokashira 3-Chome; Mitaka Tokyo 181; Japan

TAKEO, Makoto; Department of Physics; Portland State University; PO Box 751; Portland OR 97297; USA; 503-229-4230; E/T; 5.4; Atomic line shapes for hot dense gases; Surface effect on line shapes

TALMAN, James D.; Applied Mathematics Department; University of Western Ontario; London Ontario N6A5B9; Canada; 519-679-3666; T; 1.1; 3.1; 3.4; Effective atomic potentials calculated variationally; Multiconfiguration optimized potential models; Multicenter integral calculations for molecular theory; Satellite structure in photoelectron and Auger spectra

TAM, Andrew C.; K06; Research; IBM Research Lab.; IBM Corp.; 5600 Cottle Road; San Jose CA 95193; USA; 408-256-5599; E; 2.3; 1.1; 3.1; Laser interaction with matter; Laser metrology

TAMBE, Balkrishna R.; Department of Physics; Southern Technical Institute; Marietta GA 30060; USA; 404-424-7215; T; 3.1; Photoionization of atoms

TANENHOLTZ, Stanley; Middle Road; Southboro MA 01772; USA

TANG, Fu-Ching; Department of Physics; City College of CUNY; Convent Avenue and 138th Street; New York NY 10031; USA; 212-690-8312; 212-568-3108; E; 1.1; 2.2; 4.3; 5.6; Polarized electron beam and polarized hydrogen atom beam; Spin-dependence, coherence and correlation in atomic collisions

TANG, K. T.; Department of Physics; Pacific Lutheran University; Tacoma WA 98447; USA; 206-531-6900x318; T; 1.1; 2.1; 2.3; Atomic and molecular collisions; Elastic, inelastic, reactive scatterings; Interatomic interactions; Potential models

TANG, Kenneth Y.; Research and Development; Western Research Corporation; 955 Distribution Avenue; San Diego CA 92121; USA; 619-578-5885; E; 2.3; 3.10; 4.1; 5.1; Reaction kinetics of electronically excited rare gases; Raman conversion in rare gas halide lasers; Narrowband rare gas halide laser development; Photoionization of thin film lithium

TANG, Sheng Y.; IPAPS B-029; University of California, San Diego; La Jolla CA 92093; USA; 619-452-4149; E; 2.1; 3.8; 4.3; Atomic and molecular collisions; Beam technology; Laser chemistry

TANI, Smio; Department of Physics; Marquette University; Milwaukee WI 53233; USA

TANIS, John A.; Accelerator and Fusion Division; Lawrence Berkeley Laboratory; University of California; Berkeley CA 94720; USA; 415-486-5011; E; 2.1; 4.2; Atomic collision experiments related to fusion research; Neutral beam injectors

TANNEN, Peter D.; 8920 Matthew, NE; Albuquerque NM 87112; USA

TAPE, James W.; Group Q-4; Los Alamos National Lab.; MS E-541; Los Alamos NM 87545; USA; 505-667-7777; E; 3.4; 5.2; XRF and photon transmission near K and L edges applied to analysis of special nuclear materials in solids and solutions

TARTER, C. Bruce; Theoretical Division; Physics Department; Lawrence Livermore National Lab.; PO Box 808; Livermore CA 94550; USA; 415-422-4169; T; 1.1; 2.1; 3.1; 5.4; 5.6; Astrophysics

TATEWAKI, Hiroshi; Research Institute for Catalysis; Hokkaido University; 060; Kita 10, Nish: 10; Sapporo; Japan; 011-716-2111; T; 1.1; 1.2; 1.3

TAWIL, Maxime M.; 4133 Wadsworth Court; Annandale VA 22003; USA

TAYAL, Swaraj S.; Physics and Astronomy; Atomic Physics; Louisiana State University; Baton Rouge LA 70802; USA; 504-388-8311; T; 1.1; 2.1; 3.1

TAYLOR, Barry N.; Electricity Division; National Bureau of Standards; Building 220, Room B258; Gaithersburg MD 20899; USA; 301-921-2701; T; 1.4; Fundamental constants data analysis

TAYLOR, David J.; Chemistry Division; Los Alamos National Laboratory; PO Box 1663 MS567; Los Alamos NM 87545; USA; 505-667-5309; 505-667-7886; E; 5.5; 4.1; 3.10; Optics for combustion research

TAYLOR, Gary; Plasma Physics Lab.; Princeton University; PO Box 451; James Forrestal Campus; Princeton NJ 08544; USA

TAYLOR, Howard S.; Department of Chemistry; LAS; University of Southern California; Los Angeles CA 90089; USA; 213-743-2590; T; 2.1; 2.2; 2.3; 3.7; 3.9; Quantum chemistry; Electron scattering; Atom-molecule scattering; Electronic, atomic and molecular reactions

TAYLOR, James W.; Department of Chemistry; University of Wisconsin; 1101 University Avenue; Madison WI 53706; USA; 608-262-4561; E; 3.1; 4.4; 5.1; Photoionization mass spectrometry; Supersonic molecular beams; VUV absorption/photoionization cross sections; Photoelectron spectroscopy; without ESCA; Molecular surface desorption

TAYLOR, John R.; Department of Physics; University of Colorado; Boulder CO 80309; USA; 303-492-6714/6952; T; 2.1; Foundations of quantum scattering theory

TAYLOR, Kenneth T.; Theory Group; Daresbury Lab.; Science Research Council; Daresbury Warrington WA44AD; England

TAYLOR, Lyle H.; Directed Energy Research; Applied Sciences; Westinghouse Corporate R&D Center; 1310 Beulah Road; Pittsburgh PA 15235; USA; 412-256-1650; T; 4.1; 3.1; 5.1; 1.1; 3.8; Laser kinetics; Molecules in intense laser fields; Vibrational energy levels of large molecules; Laser light propagation

TAYLOR, Raymond L.; CVD Incorporated; 185 New Boston Street; Woburn MA 01801; USA; 617-933-9243; E/T; 5.3; 5.6; 2.3; 3.6; 3.8; Atomic and molecular cross section data for laser modeling; Atomic and molecular cross section data for high temperature chemistry

TAYLOR, Robert C.; Department of Chemistry; University of Michigan; Ann Arbor MI 48109; USA; 313-764-7362; E/T; 3.3; 3.5; Vibrational spectra and molecular structure

TAYLOR, Ronald D.; Department of Physics; Berkeley Research Associates, Inc.; 852; Springfield VA 22150; USA; 703-750-3434; T; 2.1; 2.3; 3.6; 3.8; 5.4

TEAGUE, Michael R.; L-626; Lawrence Livermore National Lab.; PO Box 808; Livermore CA 94550; USA; 615-423-6269; 615-423-4387; T; 4.2; 4.1; 3.7; 5.4; 1.4

TEAL, G. K.; 5222 Park Lane; Dallas TX 75220; USA

TEEGARDEN, Kenneth J.; Institute of Optics; University of Rochester; Rochester NY 14627; USA; 716-275-2320; E; 4.1; 5.2; Color center lasers; Point defects and impurities in rare gas crystals; Spectroscopy

TEETS, Richard E.; Physics Division; General Motors Research Lab.; General Motors Technical Center; 12 Mile and Mound Roads; Warren MI 48090; USA; 313-575-7936; E; 3.5; 3.6; 5.5; 5.3; Flame chemistry; Laser spectroscopy for analytical applications

TEHRANI, Mohammad M.; Systems and Research Center; Avionics and Defense; Honeywell Inc.; 312; 2600 Ridgeway Parkway; Minneapolis MN 55440; USA; 612-378-4731; E; 4.1; 3.10; 3.11; Ring laser gyros, optical sensors; Nonlinear optical materials; Coherence properties of optical fields

TEICH, Malvin C.; Electrical Engineering; SEAS; Columbia Radiation Laboratory; Columbia University; 520 W. 120th Street; New York NY 10027; USA; 212-280-3117; E/T; 3.11; 3.7; 3.1; 4.1

TELLINGHUISEN, Joel B.; Department of Chemistry; Vanderbilt University; Nashville TN 37235; USA; 615-322-4873; E/T; 3.2; 3.5; 3.7; UV-visible spectroscopy of diatomic molecules; Excimer lasers; Multiphoton processes

TEMKIN, Aaron; Atomic Physics Office; Laboratory for Astronomy; and Solar Physics; NASA Goddard Space Flight Center; Greenbelt MD 20771; USA; 301-344-8091; T; 2.2; 1.1; 1.2; 5.7; 5.6; Administration; Electron-atom/molecule interaction

TEMKIN, Richard J.; Plasma Fusion Center; N.W. 16-138; MIT; Cambridge MA 02139; USA; 617-253-5528

TEMMER, Georges M.; Nuclear Physics Laboratory; Rutgers University; New Brunswick NJ 08903; USA; 201-932-2400; E; 3.3; 4.2; 5.2; Nuclear physics; Beam foil spectroscopy; Electron channeling in crystals

TERHUNE, Robert W.; Ford Research Laboratory; PO Box 2053; Room 52076; Dearborn MI 48121; USA; 313-322-6785; E; 2.3; 3.5; 4.1; Molecular spectroscopy; Chemical physics; Physical optics

TETU, Michel; Department of Electrical Engineering; Laval Universite; Cite Universitaire; Quebec Que 10; Canada

TEUBNER, Peter J.; School of Physical Science; Finders University; Stuart Road; Bedford Park Australia 5042; South Australia; (08) 275-2232; (08) 275-2328; E; 2.2; 4.1

THADDEUS, Patrick; Center for Astrophysics; 60 Garden Street; Cambridge MA 02138; USA; 617-495-7312; E/T; 3.3; 5.7; Microwave spectroscopy to support radio and far IR astronomy

THALER, William J.; Department of Physics; Georgetown University; 37th and O Streets, NW; Washington DC 20057; USA; 202-625-4144; E/T; 4.1; 5.2; Atomic and molecular interaction in liquid, solid, and gas lasers

THATCHER, Everett W.; 3803 Liggett Drive; San Diego CA 92106; USA

THEODOSIOU, Constantine E.; Department of Physics and Astronomy; University of Toledo; 2801 W. Bancroft Street; Toledo OH 43606; USA

THIRUMALAI, Devarajan V.; Institute for Physical Sciences and Technology; University of Maryland; College Park MD 20742; USA; 301-454-2636; T; 1.2; 1.3; 2.3; 5.1; 5.2

THOE, Robert S.; Department of Physics; University of Tennessee; Knoxville TN 37920; USA; 615-974-6760; E; 4.2; 5.8; Charge exchange for nuclei on gaseous atomic targets

THOMAS, Brian K.; Department of Physics; Brooklyn College of CUNY; Bedford Avenue and Avenue H; Brooklyn NY 11210; USA; 212-780-5812; T; 2.1; 2.2; Intermediate and high energy scattering; Eikonal and Glauber theory; Classical binary encounter theory; Formal scattering theory

THOMAS, Bruce R.; Department of Physics and Astronomy; Carleton College; Northfield MN 55057; USA; 507-663-4389; E; 1.1; 2.2

THOMAS, Dan M.; Experimental Science; G. A. Technologies, Inc.; 85608/L-350; San Diego CA 92138; USA; 619-455-2403; E; 5.4; 4.2; 3.6; 3.9; 4.1; Plasma diagnostics using merged neutral atomic and laser beams; Neutralization of charged particle beams; Ion source physics

THOMAS, Edward W.; Department of Physics; Georgia Institute of Technology; Atlanta GA 30332; USA; 404-894-5200; E; 5.2; 2.1; Atomic collisions in solids

THOMAS, H. Ronald; Molecular and Organic Materials Area; Xerox Webster Research Center; Xerox Square W114; Rochester NY 14644; USA; 716-422-2314; E; 3.2; 3.3; 3.8; 4.2; X-ray and UV photoelectron spectroscopy; Laser induced IR and UV chemistry

THOMAS, John E.; Spectroscopy Laboratory; G-014; MIT; Cambridge, MA 02139; USA; 617-253-6791

THOMAS, Richard G.; Department of Physics; Natural Sciences; Prairie View A&M College; AA; Prairie View TX 77446; USA; 409-857-3518; 409-857-4513; 3.4

THOMAS, Robert; Brookhaven National Laboratory; Upton NY 11973; USA; E; Neutron scattering; Phase transitions, phase diagrams and critical phenomena in systems exhibiting magnetic order

THOMAS, T. D.; Department of Chemistry; Oregon State University; Corvallis OR 97331; USA; 503-754-2081; E; 3.1; Photoelectron and Auger spectroscopy of small molecules and atoms

THOMASON, Mike; L-Division, Group L-9; Los Alamos National Laboratory; PO Box 1663 MS535; Los Alamos NM 87545; USA; 505-667-5867; E; 2.3; 3.6; Energy transfer in carbon dioxide collisions by double resonance spectroscopy

THOMPSON, Al C.; Building 70A; Lawrence Berkeley Lab.; 1 Cyclotron Road; Berkeley CA 94720; USA

THOMPSON, Donald L.; Group CNC-2; Los Alamos National Laboratory; PO Box 1663; Los Alamos NM 87545; USA; 505-667-4686; T; 2.3; Chemical dynamics; Energy transfer and nucleation reactions

THOMPSON, Eric; Department of Electrical Engineering and Applied Physics; Case Western Reserve University; Cleveland OH 44106; USA; 216-368-4080; T; 5.1; Atomic and molecular beam scattering from solids

THOMPSON JR., H. Bradford; Chemistry; University of Toledo; Toledo OH 43606; USA; 419-537-4579; 419-536-0668; E/T; 1.1; Theory of the covalent bond; Geometry calculations in large molecules and structures; Conformational equilibria and steric exclusions

THOMPSON, Richard T.; EG&G Energy Measurements Group; PO Box 1912; M/S G-01; Las Vegas NV 89101; USA

THOMSEN, John S.; Department of Physics; Johns Hopkins University; Charles and 34th Streets; Baltimore MD 21218; USA; 301-467-8849; 301-338-7347; T; 1.1; 3.4; X-ray spectroscopy--wavelengths, line widths, asymmetry and energy levels

THOMSON, David B.; Physics Division, Group P-1; Los Alamos National Laboratory; PO Box 1663; Los Alamos NM 87545; USA; 505-667-4897/4834; E; 5.4; Radiation and opacity studies in plasmas; Atomic processes in high temperature plasmas; High stages of ionization of high-Z elements

THOMSON, George M.; Engineering Technology and Physics Branch; System Engineering and Concepts Analysis Division; US Army Ballistics Research Lab.; U.S. Army Material Command; AMXBR-SECAD; Building 120; Aberdeen Proving Ground MD 21005; USA; 301-278-4905; 301-278-3614; E; 2.1; 3.4; 3.6; 4.2; 5.6; Atomic collision induced X-ray production; Auger effects; Laser induced fluorescence for remote molecular detection; X-ray fluorescence for environmental monitoring; X-ray radiography for dynamic measurements

THONNARD, Norbert; 114 Ridge City Center; Oak Ridge TN 37830; USA

THORSON, Walter R.; Department of Chemistry; University of Alberta; Edmonton AB T6G2G2; Canada; 403-432-3687; T; 2.1; Vibrational dynamics in hydrogen-bonded systems

THOURET, Wolfgang E.; Duro-Test Corporation; 2321 Kennedy Boulevard; North

Bergen NJ 07047; USA

TIERNAN, Thomas O.; Department of Chemistry; Brehm Laboratory; Wright State University; Dayton OH 45435; USA; 513-873-2202; E; 2.2; 2.3; 4.4; Low energy ion-neutral collisions; State-to-state ion chemistry; Low energy electron-neutral reaction kinetics; Chemical ionization mass spectrometry

TILAK, Anup S.; 55 Garretson Road; Bridgewater NJ 08807; USA

TIMP, Gregory L.; 81 Poplar Road; Briarcliff NY 10510; USA

TIMSIT, Roland S.; Kingston Laboratory; Alcan International Limited; PO Box 8400; Kingston Ontario K7L4Z4; Canada; 613-549-4500; 613-544-6784; E/T; 3.5; Scattering of light by rough surfaces; Laser treatment of surfaces

TIO, T. K.; R&D Associates; PO Box 9695; Marina Del Rey CA 90291; USA; 213-877-1715 x477; T; 2.2; 2.3; Low-energy electron-molecule scattering; Predissociation; Chemical reactions

TIPPING, Richard H.; Department of Physics; University of Nebraska, Omaha; Omaha NE 68182; USA; 402-554-2510; T; 1.1; 3.5; Molecular spectroscopy, lineshape and intensity calculations of diatomic molecule

TISONE, Gary C.; Laser Research and Development Division; Sandia National Laboratories; PO Box 5800; Albuquerque NM 87185; USA; 505-844-8506; E; 4.1; 4.2; Laser development for laser fusion; Rare gas-halogen lasers and HF lasers; Electron beam excitation

TITTEL, F. K.; Department of Electrical Engineering; Rice University; PO Box 1892; Houston TX 77001; USA; 713-527-4833

TOBIAS, Sigmond; 128 Versailles Road; Rochester NY 14621; USA

TOBUREN, Larry H.; Radiological Physics Division; Battelle Memorial Institute; Pacific Northwest Laboratory; PO Box 999; Richland WA 99352; USA; 509-376-3348; E; 2.1; 4.2; Differential ionization cross sections in fast ion-atom collisions

TOENNIES, Jan Peter; Max Planck Inst Fur Stromungsf; Boettingerstrasse 10; Goettingen 3400; Fed. Rep. Germany; 551-709-2600; 0551-57172; E/T; 2.1; 2.3; 5.1; 3.6

TOLEDO, James; PIO Valdivieso 268; Quito; Ecuador

TOLK, Norman H.; Radiation Physics Research Division; Bell Telephone Laboratories; Central Service Organization; 600 Mountain Avenue; Murray Hill NJ 07974; USA; 201-582-6096; E/T; 4.2; 5.1; Ion-surface/electron-surface inelastic collisions; Anisotropic atomic state preparation

TOLLIVER, David E.; 5211 Vallecito Drive; Westminster CA 92683; USA

TOMKINS, Frank S.; Chemistry Division; Argonne National Laboratory; 9700 S. Cass Avenue; Argonne IL 60439; USA; 312-972-3635; E; 3.1; 3.2; 4.2; Atomic spectroscopy; Laser spectroscopy; High resolution spectroscopy

TOMLINSON, W. J.; Lightwave Component Technology Research; Network Technology Research; Bell Communications Research; Box 7020; 331 Newman Springs Road; Red Bank NJ 07701; USA; 201-758-2976; 201-671-1180; E/T; 3.10

TON-THAI, Dinh; 15681 Williams Street #85; Tustin CA 92680; USA

TOPP, Michael R.; Department of Chemistry; University of Pennsylvania; 34th and Spruce Streets; Philadelphia PA 19174; USA; 215-243-4859; E; 3.8; 1.3; 1.1; 3.1; 3.6; Photochemistry and chemical physics; Laser spectroscopy; Fast reactions; Molecular relaxation phenomena; Transient molecular interactions

TORR, Douglas G.; Center for Atmospheric and Space Science; UMC 41; Utah State University; Logan UT 84322; USA; 801-750-2779; E/T; 2.3; 5.6; Chemical processes in the upper atmosphere

TORRES, Barbara W.; PO Box 478; Tijeras NM 87059; USA

TORRES, Ignacio A.; Inst De Fisica; AP 20-364; Mexico 20 DF; Mexico City; Mexico

TOTEN JR., Arvel D.; PO Box 164; Route 6; Denton TX 76201; USA

TRABERT, Elmar C.; Physics and Astronomy; Experimentalphysik III; Dynamitron-Tanden Laboratoriun; Ruhr Universitat; Postfach 102148; Universitatsstrasse 150; D-4630 Bochun 1; F.R. Germany; 0234-700-7310; 0234-700-6205; E; 3.2; 4.2; 1.1; 1.4; 2.1; Lifetimes of long-lived states of highly ionized atoms; Recoil ion spectroscopy

TRACY, David H.; Optical Group Research Division; Perkin-Elmer Corporation; Main Avenue, MS 283; Norwalk CT 06856; USA; 203-762-6049; E; 3.2; Atomic spectroscopy for quantitative elemental analysis

TRACY, J. Charles; Physics Division; General Motors Research Lab.; General Motors Technical Center; 12 Mile and Mound Roads; Warren MI 48090; USA; 313-575-2838; E; Research management

TRAINOR, Daniel W.; Directed Energy Technology Organization; Avco Everett Research Laboratory; 2385 Revere Beach Parkway; Everett MA 02149; USA; 617-389-3000; 617-381-4467; E; 2.3; 3.10; 4.1; 2.1; 3.8; UV/visible laser research and development; Electron quenching of excited exciplex molecules; Electron dissociative attachment; Stimulated Raman scattering

TRAJMAR, Sandor; Molecular Physics and Chemistry Division; Jet Propulsion Laboratory; California Institute of Technology; 4800 Oak Grove Drive; Pasadena CA 91103; USA; 213-354-2145; E; 2.2; 3.2; Electron-atom (molecule) collision phenomena; Spectroscopy

TREACY, Peter B.; Department of Nuclear Physics; Res. School of Physical Sciences; Australian National University; 4 GPO; Canberra Act 2601; Australia; 062-492089; 062-492083; E/T; 1.1; 1.2; 2.1

TREGAY, George W.; High Energy Laser Technology Division; Bell Aerospace-Textron; PO Box 1; Buffalo NY 14240; USA; 716-297-1000; E; 4.1; 5.3; Chemical laser development; Chemical laser diagnostic measurements

TRUE, Nancy S.; Department of Chemistry; University of California, Davis; Davis CA 95616; USA; 916-752-0874; E; 1.1; 2.3; 3.3; 3.5; Microwave spectroscopy; Gas phase NMR spectroscopy

TRUHLAR, Donald G.; Department of Chemistry and Chemical Physics; Smith Hall; University of Minnesota; 207 Pleasant Street SE; Minneapolis MN 55455; USA; 612-373-5018; T; 2.2; 2.3; 1.1; 5.1; Chemical dynamics and electron scattering calculations; Supercomputer calculations

TRUJILLO, Stephen M.; IRT Corporation; PO Box 80817; San Diego CA 92138; USA; 714-565-7171x365; E; Applied research

TRUMP, Darryl D.; Research Applications Division; Systems Research Laboratories, Inc.; 2800 Indian Ripple Road; Dayton OH 454406; USA; 513-252-2706; E; 3.10; 4.1; 3.6; 3.2; 5.5

TSAI, Bilin P.; Department of Chemistry; University of Minnesota, Duluth; Duluth MN 55812; USA; 218-726-7220; E; 3.1; 3.2; Spectroscopy on gas phase ions; Photoionization

TSAI, Chin-Chi; Oak Ridge National Lab.; PO Box Y; Building 9201-2; Oak Ridge TN 37830; USA

TSANG, Won-Tien; Electrophotonics Research; Communication Science; Electronics; AT&T Bell Lab.; Room 4F-437; Holmdel NJ 07733; USA; 201-949-5164; E; 4.1

TSENG, Hsiang-Kuang; Department of Physics; National Central University; Chung-

Li Taiwan 32054; Republic of China; 034-427151x5460; T; 1.1; 2.2; 3.1; 3.4; 5.4

TSENG, Tien-Jiunn; Department of Physics; Chung Yuan Christian University; Chung Li Taiwan 320; Republic of China; 034-563-171x259; T; 1.1

TSERRUYA, Itshak; Department of Nuclear Physics; Weizmann Institute of Science; Rehovot 76100n 320; Israel; 972-8-482030; 972-8-482651; E; 2.1; Inner-shell ionization phenomena; Quasimolecular K X-rays

TSONG, Ignatius S.; Materials Research Laboratory; Pennsylvania State University; University Park PA 16802; USA; 814-865-7341; E; 4.2; 5.1; 5.8; Inelastic ion-solid collisions; Nuclear reactions

TSONG, Tien T.; Department of Physics; Pennsylvania State University; 104 Davey Lab; University Park PA 16802; USA; 814-865-2813; 4.4; 5.1

TUAN, Debbie F.; Department of Chemistry; Kent State University; Kent OH 44242; USA; 216-673-6165; T; 1.1; General structure and properties of atoms and molecules

TULLY, John C.; Physical Chemistry Division; Bell Telephone Laboratories; 600 Mountain Avenue; Murray Hill NJ 07974; USA; 201-582-3619; T; 2.1; 2.3; Molecular collisions involving electronic transitions and/or chemical reactions

TUPA, Dale; Department of Physics; University of Wisconsin; 1150 University Avenue; Madison WI 53706; USA; 608-263-3628; E; 3.6; 4.1

TURECHEK, John J.; Technical Operations; Newport Corporation; 8020; 18235 Mt. Baldy Circle; Fountain Valley CA 92728; USA; 714-963-9811; 818-246-9736; E/T; 5.3; 3.10; 3.7; 3.6; 4.1; Emission spectroscopy from plasmas; Discharge light sources; Laser-plasma interactions

TURLEY, R. S.; Optical Physics; Research Laboratory-MS RL65; Hughes Research Lab.; 3011 Malibu Canyon Road; Malibu 9A 90265; USA; 213-317-5622; 805-498-2392; E; 3.7; 3.10; 3.11; 3.9; 3.1; Radiation cooling

TURNER JR., Charles E.; 10033 Glade Avenue; Chatsworth CA 91311; USA

TURNER JR., Charles E.; Advanced Technology; Rocketdyne FA-03; Rockwell International; 6633 Canoga Ave.; Canoga Park CA 91304; USA; 818-700-4806; 818-709-1608; E/T; 5.3; 2.3; 2.2; 2.1; 3.8; Rare gas halide kinetics and laser modeling; Excimer laser research and development; Electron beam pumped laser research and development; Xenon fluoride lasers

TURNER, Mark C.; 520 Clinton Street Apt. #4; Brooklyn NY 11231; USA

TWIST, J. Robert; Department of Physics; Georgia Institute of Technology; Atlanta GA 30332; USA; 404-894-5266; E; 5.3; Ion swarms in rare gases--transport measurements

UEHARA, Yasuo; School of Medicine; Physics Lab.; St. Marianna University; 2095 Sugao; Takatsu-KuKawasaki 213; Japan

UFFORD, C. W.; Department of Physics; University of Pennsylvania; Philadelphia PA 19104; USA; 215-898-7941; 215-860-2685; T; 1.1; Term values in iron

UGBABE, Aako; PO Box 6392; Anglo-Jos; Jos Nigeria

UGLUM, John R.; Austin Research Associates; 1901 Rutland Drive; Austin TX 78758; USA; 512-837-6623; E/T; 5.4; 5.3; 3.2; 3.1

ULLMAN, Robert; Department of Chemistry; Ford Research Laboratory; PO Box 2053; Dearborn MI 48121; USA; 313-337-6328; T/E; 1.3; 5.8; Polymer physical chemistry; Neutron scattering

ULLRICH, Ludger K.; Inst Fur Hoch-Und Hoechst; Ruhr Univ Bochum; Frequenztechnik Gebaude 1C 6; Bochum 4630; West Germany

UMBERGER, David K.; PO Box 317A; Route 1; Espanola New Mexico 87532; USA

URIBE, Roberto M.; Inst De Fisica UNAM; APDO Pos 20-364; Mexico DF 01000; Mexico

UTTERBACK, Nyle G.; Chemical Physics Group; TRW, Inc.; 1 Space Park; Redondo Beach CA 90278; USA; 213-536-1453; E; 4.2; 4.1; 5.1; 5.4; Accelerator used for implanting semiconductors; Laser blow-off mass spectrometry; Plasma isotope separation

UZER, Ahmet T.; Department of Chemistry; University of Colorado; Campus Box 215; Boulder CO 80309; USA

UZER, Turgay; School of Physics; Georgia Tech; Atlanta GA 30332; USA; 404-894-4986; 404-956-8569; T; 3.8; 3.6; 3.5; 3.1; 2.1; Intramolecular energy transfer; Mode-specificity in chemical reactions; Semiclassical collision theory; Multidimensional uniform approximations in scattering theory

UZGIRIS, Egidijus E.; Research and Development Center; General Electric Company; PO Box 8; Schenectady NY 12301; USA

VACCARO, Patrick H.; Department of Chemistry; Room 2-025; Massachusetts Institute of Technology; Cambridge MA 02139; USA

VAHALA, Linda D.; 138 Nina Lane; Williamsburg VA 23185; USA

VALA, Martin; Department of Chemistry; University of Florida; Gainesville FL 32611; USA; 904-392-0529; E; 1.3; 3.2; 3.3; 3.9; Fourier transform IR and matrix-isolation magneto-optic studies of organic molecules--luminescence

VALBERG, Peter A.; Department of Physiology; Harvard School of Public Health; 665 Huntington Avenue; Boston MA 02115; USA

VALERIO JR., Clement V.; 696 Country Wood Court; Cheshire CT 06410; USA

VAN BAAK, David A.; Department of Physics; Calvin College; 3201 Burton Street, S.E.; Grand Rapids MI 49506; USA; 616-957-6275; E; 1.4; 3.3

VAN DRIEL, Henry M.; Department of Physics; University of Toronto; Huron and Russel; Toronto Ontario; Canada

VAN HAERINGEN, Willem; Department of Physics; Theoretical Physics; University of Technology; S13, 5600 MB; Den Dolech 2; Eindhoven 5581; The Netherlands; 040-472748; 04904-15170; T; 1.1; 5.2; 3.6

VAN OOSTRUM, Karel J.; Molenakkers 38; GK Eersel 5521; Holland

VAN SICLEN, Clinton D.; Nuclear Physics; Physical and Biological Sciences; Idaho National Engineering Lab.; PO Box 1625; Idaho Falls ID 83415; USA; 208-526-0617; T; 1.3; 2.3; 3.9; 5.2; 5.4

VAN WINTER, Clasine; Department of Mathematics; University of Kentucky; Lexington KY 40506; USA

VAN WOERKOM, Linn D.; 10332 Parr Avenue; Sunland CA 91040; USA

VAN ZYL, Bert; Department of Physics; Space Science Laboratory; University of Denver; Denver CO 80208; USA; 303-871-2116; E; 2.1; 2.2; 4.3; 5.1; 5.6; Cross sections measurements for ionization and excitation in collisions between low-energy neutral atoms and molecules; Ion-neutral and neutral-surface collisions

VANASSE, George A.; Optical Physics Division; U.S. Air Force Geophysics Lab.; Hanscom AFB MA 01731; USA; 617-861-3614/3018; E/T; 4.1; Spectrometric techniques in high-throughput multiplex spectroscopy

VANBRUNT, Richard J.; Electrosystems Division; Electronics and Electrical Engineering; National Bureau of Standards; Gaithersburg MD 20899; USA; 301-921-3121; 301-921-2346; E; 2.1; 2.3; 5.3; 4.4; Corona induced chemistry in electronegative gases; Ion cluster reactions and ion mobility; Electron avalanche growth and breakdown in gases

VANCE, Dennis W.; 179 Portola Road; Portola Valley CA 94025; USA

VANDER SLUIS, Kenneth L.; Physics; Oak Ridge National Laboratory; PO Box X; Oak Ridge TN 37830; USA; 615-574-4730; E/T; 4.1; 5.4; 1.1; 3.9

VANDYCK JR., Robert S.; Department of Physics; University of Washington; Seattle WA 98195; USA; 206-543-5768; E; 1.4; Precision mass ratio measurements of light ions; G-2 precision anomaly measurements of electrons and positrons

VANE, Charles R.; Materials and Molecular Research Division; Lawrence Berkeley Laboratory; University of California; Building B71-E; 1 Cyclotron Road; Berkeley CA 94720; USA; 415-486-6470; E; 2.1; 2.3; 4.2; 5.2; Accelerator based atomic physics; Highly ionized ions with atoms and solids--collisions; Ion trapping

VARGHESE, S. L.; Department of Physics; University of South Alabama; Mobile AL 36688; USA; 205-460-6224; E; 3.4; 4.2; Accelerator based atomic physics; X-ray fluorescence studies

VARNEY, Robert N.; 4156 Maybell Way; Palo Alto CA 94306; USA

VASHISHTA, Priya D.; Department of Theory; Materials Science and Technology; Physical Research; Argonne National Laboratory; 9700 South Cass Avenue; Argonne IL 60439; USA; 312-972-5494; 312-972-6525; T; 5.2; 1.1; 1.3; Amorphous solids and glasses--computer simulation; Structural phase transitions in solids; Many body theories of classical and quantum systems; Dielectric screening in solids

VASILAKIS, Andrew; Department of Physics; City College; Convent Avenue and 138th Street; New York NY 10031; USA

VAUGHAN, Stephen O.; Department of Chemistry; Remsen Hall; The Johns Hopkins University; 34th and Charles Street; Baltimore MD 21218; USA; 301-435-1191; 301-338-4602; E; 1.1; 2.2; 4.3; 4.4; 5.6; Electron scattering studies of highly reactive atmospheric species; Atomic beam generation using microwave discharge techniques; Metastable molecular beam generation using microwave discharge techniques

VEERARAGHAVAN, Balaji; Department of Chemistry; University of Pittsburgh; PO Box 20; Pittsburgh PA 15260; USA

VELURI, Venkateswar R.; Building 201-2H-16; Argonne National Laboratory; 9700 South Cass Avenue; Argonne IL 60439; USA

VENANZI, Carol A.; Department of Pharmacology; Mount Sinai School of Medicine; 1 Gustave Levy Place; New York NY 10029; USA; 212-650-7250; T; 1.1; 2.2; Small molecules--calculation of interaction energy; Low-energy electron-molecule scattering

VENANZI, Thomas J.; Department of Chemistry; College of New Rochelle; New Rochelle NY 10801; USA; 914-632-5300; T; 1.1; Structure and interactions of bio-molecules

VENKATESAN, T.; Radiation Physics Research Division; Bell Telephone Laboratories; 600 Mountain Avenue; Murray Hill NJ 07974; USA; 201-582-4086; E; 4.2; 4.3; 5.1; Focusing of ion beams of heavy atoms to submicron dimension; Liquid metal ion sources

VENTRICE, C. A.; 183 Paris Street; Cookeville TN 38501; USA

VERDERAME, Frank D.; DAMA-ARZ-D; Headquarters; Department of the Army; Washington DC 20310; USA; 202-697-3558; 202-695-4261; E/T; Atomic and molecular research administrator

VERDEYEN, Joseph T.; Department of Electrical and Computer Engineering; Division of Gaseous Electronics Laboratory; University of Illinois; 1406 West Green Street; Urbana IL 61801; USA; 217-333-2480; E; 3.6; 5.3; 5.1; Gas discharge physics; Laser physics

VERDIECK, James F.; Chemical Physics Department; Research Center; United Technologies Corporation; Silver Lane; East Hartford CT 06108; USA; 203-727-7184; 203-649-0756; E/T; 3.10; 3.6; 3.8; 3.2; 5.5; Laser induced fluorescence diagnostics; CARS diagnostics: Electronic resonance; Two-dimensional imaging for combustion diagnostics; Photochemistry

VERHAAR, Boudewyn J.; Physics Department; Theoretical Physics; Physics Laboratory; Eindhoven University of Technology; P.O. Box 513; Den Dolech 2; Eindhoven 5600MB; The Netherlands; 3149072142; 3140472728; T; 2.1; 2.3; 5.1; 3.9; 3.6; Collisions among atoms at sub-Kelvin energies; traps, Bose-condensation

VERNON, Richard H.; 414 Redwood Drive; Pasadena CA 91105; USA

VESTAL, Marvin L.; Department of Chemistry; University of Houston; 4800 Calhoun; Houston TX 77004; USA;

713-749-2675; E; 2.3; 3.1; 4.4; Laser photodissociation of ions; Ion-molecule reactions; Involution of organic molecules; Mass spectrometry

VICHARELLI, Pablo A.; 2335 Juniper Avenue; Boulder CO 80302; USA

VICTOR, George A.; Atomic and Molecular Physics Division; Harvard-Smithsonian Center for Astrophysics; 60 Garden Street; Cambridge MA 02138; USA; 617-495-7236; FTS-830-7236; T; 1.1; 3.1; 5.6; 5.7; Atomic and molecular processes

VIEHLAND, Larry A.; Department of Science and Mathematics; Parks College/St. Louis University; Cahokia IL 62206; USA; 618-337-7500; T; 5.3; 1.1; 2.1; Ionic motion in neutral gases--kinetic theory; Extraction of microscopic information about ion-neutral collisions from experimental measurements of ionic mobility; diffusion in gases

VIGGIANO, Albert A.; Ionospheric Disturbances; Air Force Geophysics Laboratory; Hascom AFB; Bedford MA 01731; USA; 617-861-4028; E; 2.3; 1.3; 5.6; 1.1; Ion-neutral kinetics; Atmospheric ion chemistry; Electron affinity determinations

VIGIL, Jerome A.; 69 Breck Avenue; Brighton MA 02135; USA

VIGLIANTE, John R.; 306 E. Belcrest Road; Bel Air MD 21014; USA

VINCENT, James S.; Department of Chemistry; University of Maryland at Baltimore County; 5401 Wilkens Avenue; Catonsville MD 21228; USA; 301-455-2531; E; 3.3; 3.10; Electron spin resonance spectroscopy of transition metal complexes; Excited triplet states, and molecular radicals; IR, Raman and resonance spectroscopy of biomembranes

VINCENT, Paul J.; Department of Physics; Building 901A; Brookhaven National Laboratory; Upton NY 11973; USA; 516-282-4106; 516-282-4581; E; 2.1; Atomic and molecular collisions

VISWANATH, A. Kasi; Department of Chemistry; University of Maine; Orono ME 04469; USA; 207-581-7544; E; 3.6; 5.2; Laser spectroscopy; Laser excited luminescence in solid state systems; Nanosecond time resolved spectroscopy and site selective spectroscopy

VISWANATHAN, C. R.; Electrical Science and Engineering Department; University of California; 7731 Boelter Hall; Los Angeles CA 90024; USA

VOGEL, David A.; Georgia Tech; PO Box 31383; Atlanta GA 30332; USA

VOLK, Charles H.; Advanced Sensors; Litton Guidance and Control Systems; MS-19; 5500 Canoga Avenue; Woodland Hills CA 91365; USA; 818-715-3371; E; 1.1; 2.1; 3.6; 4.1; 5.3; Alkali-alkali spin exchange and alkali-noble gas interactions; RF and dc discharge plasma phenomena; Gas lasers-ring lasers; Atomic frequency standards

VOLKIN, H. C.; Rocketdyne Division; Rockwell International; PO Box 5670; Kirkland AFB NM 87185; USA

VON JASKOWSKY, Woldemar F.; Department of Mechanical and Aerospace Engineering; Electric Propulsion Laboratories; Princeton University; Princeton NJ 08544; USA; 609-452-5241; 609-452-5220; E; 5.3; 5.4; 5.1; 4.1; Ionization, recombination and excitation processes in plasmas; Electric propulsion; Erosion of ARC electrodes, insulators

VON ROSENBERG JR., Charles W.; Atomic and Molecular Research Committee; Avco Everett Research Laboratory; 2385 Revere Beach Parkway; Everett MA 02149; USA; 617-389-3000; E; Laser, atmospheric, particle beam, combustion and gasification kinetics

VON TURKOVICH, Branimir F.; Mechanical Engineering Department; College of Engineering and Mathematics; University of Vermont; 113 A Votey; Burlington VT 05405; USA; 802-899-3938; 802-656-3320; E/T; 1.1; 5.2; Defect structure in solids; Energy conversion in solids

VOOK, Richard W.; Department of Physics; Syracuse University; 209 Physics Building; Syracuse NY 13210; USA; 315-423-2564; 315-423-4718; E; 5.1; Surface structure and chemistry--catalysis

VOREADES, Demetrios; 1EG Technology Center; MS 239; Hughes Aircraft Company; 6155 El Camino Real; Carlsbad CA 92008; USA; 619-931-3856; E; 4.3; 3.1; 1.1; 5.2

VOSS, Donald E.; Mission Research Corporation; 1720 Randolph Road, SE; Albuquerque NM 87106; USA

VROOM, David A.; Physical Sciences Division; IRT Corporation; PO Box 80817; San Diego CA 92122; USA; 714-565-7171; E; 2.1; 2.3; 4.2; 5.1; Charge transfer and ion-molecule cross section measurements

VUKASIN, Helen L.; CODEL Inc.; 475 Riverside Drive; New York NY 10016; USA

WACHTER, Joseph R.; Department of Physics; New Mexico State University; PO Box 3D; Las Cruces NM 88003; USA; E; 1.1; 1.3; 2.2

WADEHRA, Jogindra M.; Department of Physics and Astronomy; Wayne State University; Detroit MI 48202; USA; 313-577-2740; T; 2.1; 2.2; 2.3; 3.9; Low energy ion-atom and electron-molecule collisions; Atoms in strong (pulsar-type) magnetic fields; Ion-ion recombination rates of interest in laser physics

WADT, Willard R.; Theoretical Division; Los Alamos National Lab.; PO Box 1663 MSJ569; Los Alamos NM 87545; USA; 505-667-7763; 505-667-2097; T; 1.1; 4.1; 5.3; Theoretical chemistry and electronic structure calculations; Ab initio methods and applications to actinide chemistry; Electronic transition lasers; Organic and inorganic species --excited states

WADZINSKI, Henry T.; Bedford Research Associates; 4 De Angelo Drive; Bedford MA 01730; USA; 617-275-7246; 617-646-1624; T; 2.2; 5.6; Electron collisions with the upper atmosphere; Proton collisions with the upper atmosphere; Plasma effects related to the above

WAGNER, Albert; Chemistry Division; Argonne National Laboratory; 9700 S. Cass Avenue; Argonne IL 60439; USA; 312-972-3597; T; 2.3; 5.1; Atoms and molecules-- chemical dynamics; Sputtering from metal surfaces

WAGNER JR., Lawrence F.; 22 Concord Road; Fishkill NY 12524; USA

WAHL, Arnold C.; Theoretical Chemistry Division; Science Applications, Inc.; 1211 W. 22nd Street; Oak Brook IL 60521; USA; 312-655-5969; T; 2.3; Theoretical and computational chemistry

WAHRHAFTIG, Austin L.; Department of Chemistry; University of Utah; Salt Lake City UT 84112; USA; 801-581-6048; E/T; 1.1; Polyatomic ions-- unimolecular fragmentation

WALCH, Kim P.; Electron Beam Corporation; 9747 Business Park Avenue; San Diego CA 92131; USA

WALKER, James J.; 1300 Camino Amparo, NW; Albuquerque NM 87107; USA

WALKER, Keith G.; Department of Physics; Bethany Nazarene College; 6729 NW 39th Expressway; Bethany OK 73008; USA; 405-789-6400; E; 2.2; Electron-impact cross sections of krypton and xenon atoms, and ions via optical method

WALKER, Robert B.; Group T-12; Los Alamos Scientific Lab.; PO Box 1663 MS569; Los Alamos NM 87545; USA; 505-667-2802; T; 2.1; 2.3; 3.7; Quantum chemical reaction dynamics of atom-molecule systems; Classical dynamics of atom-diatomic (polyatomic) collisions; IR multiple photon absorption dynamics

WALKER, Robert N.; 221 Indiana Apt. A213; Lubbock TX 79415; USA

WALKUP, Robert E.; IBM Watson Research Center, P.O. Box 218; Yorktown Heights NY 10598; USA; 914-945-1512; E/T; 5.1; 3.6; Interactions of photons, electrons, ions, and atoms with surfaces

WALLACE, Bryan G.; 7210 12th Avenue No.; St. Petersburg FL 33710; USA

WALLACE, Richard W.; Chromatix, Inc.; 560 Oakmead Parkway; Sunnyvale CA 94086; USA; 408-736-0300; E; 1.3; 3.1; Light scattering and absorption for various polymer characterizations

WALLACE, Scott; Department of Materials Science and Engineering; Massachusetts Institute of Technology; Room 13-5114; Cambridge MA 02139; USA; 617-253-6918; T; 3.1; 5.1; Photoionization of molecules; Angle-resolved photoemission from surfaces; Energetics and geometry of adsorption

WALLERSTEIN, George; Department of Astronomy; University of Washington; FM-20; Seattle WA 98195; USA; 206-543-2888; E; 5.7; Stellar spectroscopy employing transition probabilities and collision; Cross sections of ions, atoms and molecules

WALLING, Rosemary S.; L-71; Lawrence Livermore National Lab.; PO Box 808; Livermore CA 94550; USA

WALLS, Daniel F.; Department of Physics; University of Waikato; Private Bag; Hamilton; New Zealand; (64) (71) 62889; T; 3.11; 3.10; Quantum optics

WALLS, Fred L.; Time and Frequency Division; National Bureau of Standards; 325 S. Broadway; Boulder CO 80303; USA; 303-497-3207; E; 3.3; 4.4; 4.1; Frequency standards based on infrared, rf, and microwave transitions; Mass spectroscopy utilizing ion traps; Frequency synthesis from microwave to optical

WALLSTROM, Tor E.; 23104 Samuel Street #206; Torrance CA 90505; USA

WALSH, Peter J.; Department of Physics; College of Science and Engineering; Fairleigh Dickinson University; Teaneck NJ 07666; USA; 201-692-2493; T; 5.3; 4.1; 3.2; 3.3; 3.11; Plasma light sources; Optical techniques; Gaseous electronics; Discharges

WALTER, William T.; Department of Electrical Engineering and Computer Science; Polytechnic Institute of New York; Route 110; Farmingdale NY 11735; USA; 516-694-5500 x28; E; 3.3; 3.6; 4.1; Electromagnetic fields with materials--interaction; Metal vapor lasers; Microwave propagation

WALTERS, G. K.; Department of Physics; Rice University; PO Box 1892; Houston TX 77251; USA; 713-527-8101 x3645; E; 5.1; 4.1; 2.1; 2.3; 1.1; Time-resolved spectroscopy on excited transient species in dense gases; Spin-polarized ion, electron and neutral beams for scattering experiments; Penning and chemionization reactions

WALTERS, John P.; Department of Chemistry; Analytical Division; St. Olaf College; Northfield MN 55057; USA; 507-663-3429; 507-663-3104; E; 5.3; Electrical discharges; Spectrochemical analysis; Computers; Chemical and optical instrumentation

WALZ, David G.; 40 Whitman Road #2-4; Waltham MA 02154; USA

WANG, Chao C.; PO Box 54; 12 Chestnut Ridge Road; Holmdel NJ 07733; USA; 201-264-1453; E/T; 3.3; 3.11; 5.1; 5.3; 5.4

WANG, Charles C.; 5225 Provincial Drive; Bloomfield Hills MI 48013; USA

WANG, Charles P.; Department of AMES; University of California, San Diego; La Jolla CA 92093; USA; 213-648-7613; E/T; 3.6; Excimer and metal vapor lasers

WANG, Chen-Show; 58 Mt. Horeb Road; Warren NJ 07060; USA

WANG, Chia P.; Physical Sciences Division; Science and Advanced Technology Lab.; USA Natick R&D Command; Kansas; Natick MA 01760; USA; 617-651-4330; E/T; 1.1; 2.2; 3.6; 3.11; Energy loss, stopping powers of atoms and molecules

WANG, Harry T.; Optical Physics Department; Hughes Research Laboratories; Hughes Aircraft Company; 3011 Malibu Canyon Road; Malibu CA 90265; USA; 213-317-5431; E; 3.3; 4.2; 1.2; High resolution radio frequency and microwave spectroscopy--application

WANG, Kang-Lung; Department of Electrical Science and Engineering; University of California; 405 Hilgard Ave., 7619 Boelter Hall; Los Angeles CA 90024; USA; 213-825-1609; E; 3.8; 5.1; Surface incorporation absorption, desorption and during molecular beam; Epitaxy with UV and electron irradiation; Surface analysis with LEED, AES, EELS

WANG, S. C.; Electro-Optical Systems Division; Xerox Corporation; 300 N. Halstead Street; Pasadena CA 91107; USA; 213-351-2351x2219; E; 3.6; Laser transitions in atomic and ionic species via spectroscopic techniques --investigation of behavior

WANG, Wen-Cheng; Electrical and Computer Engineering; San Diego State University; San Diego CA 92182; USA; 619-265-3700; E; 5.3; 3.1; 3.7; 3.10

WANIEK, Ralph W.; Advanced Kinetics, Inc.; 1231 Victoria Street; Costa Mesa CA 92627; USA; 714-646-7165; E/T; 4.1; IR and far IR lasers

WARD, John F.; Department of Physics; Randall Laboratory; University of Michigan; Ann Arbor MI 48104; USA; 313-763-3179; E; 3.10; 3.6; 4.1; 1.1; 3.7

WARNER, Ray A.; Battelle Northwest; PO Box 999; Richland WA 99352; USA; 509-376-2439; E; 3.1; 3.6; 4.4; Analytical techniques based on atomic spectroscopy

WARREN JR., Walter R.; Pacific Applied Research; PO Box 2157; 6 Crestwind Drive; Rancho Palos Verdes CA 90274; USA; 213-544-0764; 213-377-3698; E; 4.1; 2.3; 3.5; Chemical laser development: vibrational and electronic transitions

WATANABE, Hiroshi; Institute of Theoretical Sciences; University of Oregon; Eugene OR 97403; USA

WATANABE, S. F.; 9671 La Esperanza Avenue; Fountain Valley CA 92708; USA

WATROUS, James A.; 3008 Floral Avenue; Riverside CA 92507; USA

WATSON, Deborah K.; Department of Physics and Astronomy; University of Oklahoma; 440 West Brooks; Norman OK 73019; USA; 405-325-3961; T; 2.2; Electron collisions with atoms, ions and molecules

WATSON JR., James; Department of Physics; Ball State University; Muncie IN 47306; USA

WATSON, R. L.; Cyclotron Institute; Department of Chemistry; Texas A&M University; College Station TX 77843; USA; 713-693-3110; E; 3.4; 4.2; X-ray emission in heavy ion-atom collisions; Multiple inner-shell ionization and decay processes

WATSON, William D.; Department of Physics and Astronomy; Loomis Laboratory of Physics; University of Illinois; 1110 W. Green Street; Urbana IL 61801; USA; 217-333-7240; T; 5.7; Atomic, molecular and optical processes in astrophysical contexts

WATTS, Richard N.; JILA; University of Colorado; Boulder CO 80309; USA; 303-492-6839; E; 4.1; 3.6

WAY, Katherine; Department of Physics; Duke University; Durham NC 27706; USA

WAY, Kermit; Department of Chemistry; Augustana College; Sioux Falls SD 57197; USA; 605-336-4812; E; 2.3; 3.8; Laser induced reaction dynamics

WAYMOUTH, John F.; Research and Development Laboratory; GTE Lighting Products; 100 Endicott Street; Danvers MA 01923; USA; 617-777-1900; E; 5.3; Electrical discharge light sources

WAYNANT, Ronald W.; Laser Physics Branch; Optical Sciences Division; Code 6540; Naval Research Laboratory; 4555 Overlook Avenue, SW; Washington DC 20375; USA; 202-767-2813; 202-767-2512; E; 1.3; 2.3; 4.1; 5.3; 3.3; Novel methods of laser excitation; Rf excitation; Vacuum ultraviolet lasers; UV and VUV solid state lasers; Excimers as pumping sources and optical sources for lithography; Anti-Stokes Raman lasers; X-ray lasers

WEATHERFORD, Charles A.; Department of Physics; Florida A&M University; PO Box 981; Tallahassee FL 32307; USA; 904-599-3825; T; 1.1; 2.2; 2.1; 2.3; Electron-molecule scattering; Molecular bound state calculations; Variational principles

WEAVER, David P.; Lab. for DYP; Air Force Rocket Propulsion; Stop 24; Edwards AFB CA 93523; USA

WEBB, Watt W.; Department of Applied Physics; Cornell University; Clark Hall; Ithaca NY 14853; USA; 607-256-3331; E; 5.2; Molecular dynamics in biological systems-diffusion; Vorticity measurement in fluids; Fluctuations in solids and liquids; Molecular fluorescence

WEBELER, Ray W.; PO Box 40408; Cincinatti OH 45240; USA; 513-671-6503; E; 5.2; Atomic hydrogen in tritium impregnated solid hydrogen below 0.01 K;--storability of large concentrations

WEBER, Ernst; PO Box 1619; Tryon NC 28782; USA

WEBER, Heinz P.; Laser Department; Institute of Applied Physics; University of Berne; Sidlerstrasse 5; Berne CH3012; Switzerland; 031-658931; 031-658913; E; 3.10; 4.1; 5.3

WEBER, Joseph N.; Engineered Nonwovens Structures; E. I. DuPont de Nemours & Co., Inc.; Old Hickory Plant; Old Hickory TN 37138; USA; 615-847-6746; T; 2.3; Atomic and molecular systems, kinetics and related; theoretical processes--modeling

WEBER, Marvin J.; Chemistry and Materials Science Department; Lawrence Livermore National Lab.; PO Box 808; Livermore CA 94550; USA; 415-422-5486; E; 3.2; 3.6; 3.7; 3.10; 4.1

WEBER, Thomas A.; Chemical Physics Division; Bell Telephone Laboratories; 600 Mountain Avenue; Murray Hill NJ 07974; USA; 201-582-2249; T; 1.3; Fluids and polymers--properties

WEBER, Willes H.; Physics Division; Research Staff; Ford Motor Co.; PO Box 2053; Dearborn MI 48121; USA; 313-337-6291; E/T; 3.3; 3.5; High resolution IR molecular spectroscopy of molecules with large; Amplitude or hindered motions using Doppler-limited and sub-Doppler limited techniques

WEBSTER, Harold F.; Research and Development; General Electric Company; PO Box 8; KWC-1619; Schenectady NY 12309; USA; 518-387-5298; 518-399-3463; E; 4.3; 5.1; Beam technology, electron beams; LEED

WEEKS, Dorothy W.; Department of Physics; Wilson College; 28 Dover Road; Wellesley MA 02187; USA; 617-235-1156; E/T; Atomic spectroscopy

WEHNER, Gottfried K.; 6017 Walnut Drive; Minneapolis MN 55436; USA

WEHRING, Bernard W.; Department of Nuclear Engineering; North Carolina State University; Box 7909; Raleigh NC 27695; USA; 518-737-2011; E; 2.1; 5.1; 5.8; Fission-fragmented heavy ions in gases, solids and plasmas --slowing down

WEI, P. S.; Physics Technology Group; Boeing Aerospace Co.; PO Box 3999; MS 2T05; Seattle WA 98124; USA; 206-655-2931; E; 3.4; 4.2; 5.3; 5.4; Plasma diagnostics; Laser effects; Electrical discharges; X-ray absorptions

WEIDMAN, David L.; Cornell University; Clark Hall; Ithaca NY 14853; USA

WEIGOLD, Erich; Physical Sciences; Flinders University of S. Australia; Adelaide South Australia 5042; Australia; 08-275-2317; 08-425460; 1.1; 2.2; 2.3; 3.6

WEINBERG, William H.; Chemical Engineering Department; Chemistry; California Institute of Technology; Pasadena CA 91125; USA; 818-356-4182; E; 5.1; 2.1; 2.2; 2.3; 3.3

WEINER, Brian L.; Chemistry Department; Quantum Theory Project; University of Florida; Williamson Hall; Gainesville FL 32611; USA; 904-392-1597; T; 1.1; 2.3; 3.1

WEINER, Eugene R.; Department of Chemistry; University of Denver; Denver CO 80208; USA; 303-753-2737/2436; E/T; 2.1; 2.3; 4.1; Nozzle-expanded nitrogen beams-- rotational populations; Metastable rare gas atom beams reacting with hydrogen bromine, etc. in a scattering chamber

WEINER, Jerome H.; Department of Engineering; Brown University; Providence RI 02912;

USA; 401-863-2858; T; 2.3; 1.3; Classical and quantum rate theory applied to isomerization; reactions in molecules

WEINER, John; Department of Chemistry; University of Maryland; College Park MD 20742; USA; 301-454-6094; 301-454-6090; E; 2.3; 3.6; 3.8; Laser induced inelastic and reactive collisions

WEINHOLD, Frank A.; Department of Chemistry; University of Wisconsin; Madison WI 53706; USA; 608-262-1511; T; 1.1; Molecular electronic structure; Autoionizing resonances--complex-coordinate studies; Theoretical chemistry

WEINSTEIN, Harel; Department of Physiology and Biophysics; Mount Sinai School of Medicine; 1 Gustave Levy Place; New York NY 10029; USA; 212-650-7018/7014; 212-650-6530; T; 1.1; 1.3; 2.3; Biophysics of macromolecules and proteins; Electronic structure and properties of drugs; Molecular structure and properties of biological molecules

WEISHEIT, Jon C.; L-297; Lawrence Livermore National Laboratory; Livermore CA 94550; USA; 415-423-4254; T; 5.4; 5.7; Atomic collisional and radiative processes in laboratory; Cosmic plasmas

WEISMAN, R. B.; Department of Chemistry; Rice University; PO Box 1892; Houston TX 77251; USA; 713-527-8101x3709; E; 2.3; 3.5; 3.8; 3.6; Time resolved laser spectroscopy applied to study of excited state dynamics in polyatomic molecules

WEISS, Andrew W.; Division 531; National Bureau of Standards; Gaithersburg MD 20899; USA; 301-921-2071; T; 1.1; Atomic structure including correlation and relativistic effects: wavefunctions, energy levels, F-values--calculation; Exploration of basic theoretical methods for such calculation

WEISS, Jeffrey M.; 1572 Peacock Avenue; Sunnyvale CA 94087; USA

WEISS, Phyllis; 315 W. 70th Street Apt. 17E; New York NY 10023; USA

WEISSBLUTH, Mitchel; Department of Applied Physics; Stanford University; Stanford CA 94305; USA; 415-497-0292; 415-857-0440; E/T; 1.1; 1.3; 3.6; 3.1; 3.10; Processes involving the interaction of radiation and matter; Nonlinear interactions of radiation with atoms and molecules

WEISSLER, Gerhard L.; Department of Physics; Natural Sciences and Mathematics; University of Southern California; University Park; Los Angeles CA 90089; USA; 213-743-2792; 818-986-5173; E; 3.1; 3.2; 3.6; 4.4; 5.1; Photoabsorption and ionization cross sections in gases; Photon-induced fluorescence in gases; Photoemission from solids; Line broadening in hot arc plasmas

WEISZMANN, Andrei N.; Department of Physics and Engineering; Staten Island College of CUNY; 130 Stuyvesant Place; Staten Island NY 10301; USA; 718-390-7972; 201-947-9432; T; 3.6; 3.10; 3.11; Raman scattering; Pier 4; Coherence

WEITZ, Eric; Department of Chemistry; Northwestern University; Evanston IL 60201; USA; 312-492-5583; E; 2.3; 3.1; 5.1; Laser chemistry; Energy transfer; Molecule-surface interactions; Photofragment studies

WELLENSTEIN, Hermann F.; Department of Physics; Brandeis University; Waltham MA 02154; USA; 617-647-2850; E; 2.2; 3.4; Electron spectroscopy on atoms and molecules; Electron momentum distributions; K-shell spectroscopy; Electron diffraction

WELLER, Lawrence A.; 8612 Chalet Drive; Wichita KS 67207; USA

WELLS, Gregory J.; Research Division; Varian Instrument Division; 2700 Mitchell Drive; Walnut Creek CA 94598; USA; 415-939-2400; E/T; 2.3; Ion-molecule reactions

WELLS, James W.; Department of Physics; State University College of NY at Buffalo; 1300 Elmwood Avenue; Buffalo NY 14222; USA; 716-878-5230; 716-878-5201; E; 1.1; 5.2; Radical structures created in irradiated organic single crystals; ESR and ENDOR spectroscopy of single crystals

WELLS, Michael B.; Radiation Research Associates, Inc.; 3550 Hulen Street; Fort Worth TX 76107; USA; 817-731-2711; T; 3.2; 5.2; 5.6; Calculation of optical and fluorescent production and transport in atmosphere; Nuclear radiation transport in shields and the atmosphere

WELTNER JR., William; Department of Chemistry; University of Florida; Gainesville FL 32611; USA; 904-392-2155; E; 1.3; 5.2; Matrix-isolated molecules and atoms--ESR, optical spectroscopy; Carbohydrates--NMR

WENDIN, Goran; Institute of Theoretical Physics; Chalmers University of Technology; Goteborg S41296; Sweden

WENTWORTH, Robert H.; Department of Applied Physics; Hansen Lab; Stanford University; Stanford CA 94305; USA

WERNER, Samuel A.; Department of Physics; College of Arts and Science; University of Missouri-Columbia; Columbia MO 65211; USA; 314-882-7664; 314-882-3335; E; 1.4; 3.11; Neutron scattering, interferometry

WESSEL, John E.; Chemistry and Physics Laboratory; The Aerospace Corporation; Box 92957; M2/253; Los Angeles CA 90009; USA; 213-648-6599; E; 1.2; 3.7; Multiphoton molecular spectroscopy and atomic spectroscopy with emphasis on development of advanced detection methods

WEST, Gary A.; Corporate Research and Development; Electronic Materials and Devices; Allied Corporation; PO Box 1021R; Morristown NJ 07960; USA; 201-455-5509; 201-455-3730; E; 3.8; Laser induced chemistry

WEST, John B.; Synchrotron Radiation; Daresbury; Science and Engineering Research Council; Keckwick Lane; Daresbury Cheshire WA44AD; United Kingdom; 0925-65000; E/T; 1.2; 3.1; 3.4; 3.9; 5.6; Photoelectron spectroscopy applied to study resonant phenomena; Photoabsorption and photoionization; Ultraviolet and soft X-ray instrumentation

WEST, Martin L.; Radiological Sciences Division; Battelle Memorial Institute; Pacific Northwest Laboratories; PO Box 999; Area 300, Building 3746; Richland WA 99352; USA; 509-376-3950; E/T; 3.3; 4.2; 5.2; Pulsed radioluminescence studies with charged particle beams; Fast reaction kinetics in liquid systems

WEST, William P.; Experimental Science; Fusion Division; GA Technologies, Inc.; PO Box 85608; San Diego CA 92138; USA; 714-455-2863/3337; E; 5.4; Atoms used for high temperature plasma diagnostics; Neutral beams as Tokamak diagnostic tools --development and application of techniques

WESTBROOK, Edwin P.; 7913 Claudia Drive; Oxon Hill MD 20745; USA

WESTERVELD, Willem B.; Department of Physics; North Carolina State University; PO Box 8202; Cox Hall; Raleigh NC 27695; USA; 919-737-7018; 919-737-7879; E; 2.2; 3.1; 3.2; 3.9; VUV absolute light standard development; Electron-impact photoemission cross sections; Charge transfer collisions

WESTHAUS, Paul A.; Department of Physics; Oklahoma State University; Stillwater OK 74078; USA; 405-624-5815; T; 1.1; Effective Hamiltonians from unitary transformation and perturbation theory; Electronic structure calculations

WESTLING, Lynn A.; B&L Department of Physics and Astronomy; University of Rochester; Rochester NY 14627; USA

WESTON JR., Ralph E.; Chemistry Department; Brookhaven

National Laboratory; Upton NY 11973; USA; 516-282-4373; E; 2.1; 3.1; 3.2; 3.6; 3.8; Electronic and vibrational energy transfer in atoms and molecules; Laser induced photochemistry; Photochemistry and spectroscopy with synchrotron radiation

WEXLER, Sol; Chemistry Division; Argonne National Laboratory; 9700 S. Cass Avenue; Argonne IL 60439; USA; 312-972-3463/3473; E; 1.3; 2.3; Chemi-ionization reactions in accelerated crossed molecular beams; Physics and chemistry of beams of naked metal atom clusters

WHALING, Ward; Department of Physics; California Institute of Technology; MS 106-38; Pasadena CA 91125; USA; 818-356-4276; E; 3.1; 5.7; 5.3; Atomic transition probabilities for astrophysical use

WHARTON, Charles B.; Department of Engineering; Electrical Engineering; Laboratory of Plasma Studies; Cornell University; Phillips Hall; Ithaca NY 14853; USA; 607-256-4307; 607-256-4827; E; 5.1; 5.4; 4.3; Plasma diagnostics development, especially microwave and corpuscular; Microwave sources for interaction with matter; Ion sources and beams; High current, high energy alkali beams

WHARTON, Lennard; Department of Chemistry; University of Chicago; 5735 S. Ellis Avenue; Chicago IL 60637; USA; 312-753-8203; E; 1.1; 2.3; 3.3; 4.3; 5.1; Reaction kinetics; Elastic and inelastic scattering; Molecular structure and microwave spectroscopy; Molecular beams and surface chemistry

WHEALTON, John H.; Thermonuclear Division Y-12; Oak Ridge National Laboratory; Oak Ridge TN 37830; USA; 615-574-1130; T; 4.3; 5.4; 5.3; Ion beam optics; Ion transport in plasmas; Neutralizer physics

WHEELER, Paul C.; L-18; Lawrence Livermore National Lab.; Livermore CA 94550; USA; 415-422-0535; 415-443-5526; E/T; 2.2; 3.1; 5.1; 5.2; 5.3

WHEELER, Thomas M.; 4 Market Street Room 239; Potsdam NY 13676; USA

WHIPPLE JR., Elden C.; Center for Astrophysics and Space Science; University of California, San Diego; La Jolla CA 92093; USA; 619-452-3313; T; 5.6; 5.7; 5.4; Cross-sections such as collision, ionization used for theoretical studies of the upper atmosphere, ionosphere and space

WHITBECK, Michael R.; Environmental Science Division; General Motors Research Laboratories; General Motors Technical Center; 12 Mile and Mound Roads; Warren MI 48090; USA; 313-575-3359; E; 2.3; 5.6; Free radicals pertinent to atmospheric chemistry--spectroscopy; Reaction kinetics

WHITE, Henry W.; Department of Physics; University of Missouri; Columbia MO 65211; USA; 314-882-3335; E; 5.1; 3.6; 4.4; Organic molecules on aluminum oxide--surface reactions

WHITE, Jonathan C.; Room 4C-316; Bell Laboratories; Holmdel NJ 07733; USA

WHITE, Michael G.; Department of Chemistry; Brookhaven National Laboratory; Upton NY 11973; USA; 516-282-4345; E; 3.1; 3.6; 3.5; 1.2; 1.3; Spectroscopy and ionization dynamics of molecular excited states

WHITEHEAD, Walter D.; University of Virginia; 444 Cabell Hall; Charlottesville VA 22903; USA

WHITELEY, Stephen R.; Hypres, Inc.; 175 Clearbrook Road; Elmsford NY 10523; USA; 914-592-1190; E/T; 4.1; 3.3; Superconducting Josephson junction device/circuit development; Superconducting electronic device research; Millimeter wave superconducting component development; Picosecond sampling technique using Josephson junctions

WHITLOCK, Robert R.; Condensed Matter Physics Branch; Condensed Matter and Radiation Science; Naval Research Lab. Washington DC; Box-T; College Park MD 20740; USA; 202-767-2154; E; 3.4; 5.1; 5.4

WHITNEY, Kenneth G.; Plasma Radiation Branch, Code 4720; Plasma Physics Division; Naval Research Laboratory; U.S. Naval Research Laboratory; 4555 Overlook Avenue, SW; Washington DC 20375; USA; 202-767-1686; T; 3.10; 5.4; 5.3; 5.1; 3.1; Analyzing dense plasma nonlinear, non-LTE interactions of electrons, photons, and ions; Modeling emission spectra of optically thick plasmas; Modeling X-ray laser concepts

WHITNEY, Wayne T.; Laser Physics Branch; Code 6540; US Naval Research Laboratory; 4555 Overlook Avenue, SW; Washington DC 20375; USA; 202-767-2507; 202-767-5214; E; 4.1; 3.10; 3.6; 3.3; 5.3; Laser technology and development

WHITTEN, Barbara L.; Department of Physics; L-296; Lawrence Livermore National Lab.; PO Box 808; Livermore CA 94566; USA; 415-423-2785; T; 5.3; 3.2; 2.2

WHITTIER, James S.; Mechanics Research Department; Aerophysics Laboratory; The Aerospace Corporation; PO Box 92957; Los Angeles CA 90009; USA; 213-648-7420; E; 4.1; 4.2; 5.3; Pulsed chemical laser development; Initiation mechanisms and laser kinetics--modeling

WICKE, Brian G.; Physical Chemistry Department; General Motors Research Lab.; General Motors Technical Center; 12 Mile and Mound Roads; Warren MI 48090; USA; 313-575-2358; 313-575-0244; E/T; 2.3; Molecular beam studies of reactions of atoms, small molecules, and free radicals

WICKUN, William G.; Department of Chemistry; State University of NY at Binghamton; Binghamton NY 13901; USA; 607-777-4246; E; 3.6; Small gas phase molecules--laser spectroscopy

WIDING, Kenneth; U.S. Naval Research Laboratory; 4555 Overlook Avenue, SW; Washington DC 20375; USA; 202-767-2507

WIE, Chu R.; Electrical 77and Computer Engineering; Bonner Hall; State University of NY at Buffalo; Amherst NY 14260; USA; 818-356-6811; 818-356-4585; E/T; 5.1; 5.2; 3.4; 3.2; Defect processes in semiconductors; innershell ionization effects; Radiation damage in crystalline solids

WIEDER, Grace M.; Department of Chemistry; Brooklyn College of CUNY; Bedford Avenue and Avenue H; Brooklyn NY 11210; USA; 212-780-5458/5753; E; 3.3; 3.5; 3.10; Condensed phases--IR and Raman spectroscopy

WIEMAN, Carl E.; Department of Physics and JILA; JILA; University of Colorado; PO Box 440; Boulder CO 80309; USA; 303-492-6963/4780; 303-492-6839; E; 1.4; 4.1; 3.9; 3.1; 4.3; Hydrogenic atoms--precision laser spectroscopy; Parity violations in atoms; Charge exchange between proton and excited atomic states

WIESE, Wolfgang L.; Atomic and Plasma Radiation Division; National Bureau of Standards; Gaithersburg MD 20899; USA; 301-921-2071; E; 1.1; 5.4; Plasma line broadening; Stark broadening; Atomic transition probabilities--critical evaluation

WIESENFELD, Jay M.; Transmission and Circuit Research Division; Bell Telephone Laboratories; 4D-515; Crawfords Corner Road; Holmdel NJ 07733; USA; 201-949-6547; E; 2.1; 3.1; 3.5; Picosecond spectroscopy; Molecular photodissociation and vibronic relaxation

WIESENFELD, John; Department of Chemistry; Baker Laboratory; Cornell University; Ithaca NY 14853; USA; 607-256-3869; 607-256-4174; E; 2.1; 3.1; 3.6; 3.8; 5.6; Photodissociation of small molecules; Energy storage and transfer in gas phase collisions of atoms and small molecules

WIESENFELD, Laurent; Lab.

Spectroscopic Hertzienne; Ecole Normale Superieure; 24 Rue Lhomond; Paris 75231; France

WIFF, Donald R.; Research Division; General Corporation; University of Dayton; 2900 Gilchrist Road; Akron OH 44305; USA; E/T; 1.1; 2.3; 3.6; 4.2; 5.2; Polymer morphology/conformational parameters; Single crystal studies; Electronic properties of solids; Theory of transport

WIGGINS, Carl M.; Optical Physics Division; BDM Corporation; 1801 Randolph Road; Albuquerque NM 87106; USA; 505-848-5000; T; Electric discharge laser feasibility studies; HF/DF chemical lasers; High energy laser resonator modeling and analysis; Physical optics

WIGGINS, Thomas A.; Department of Physics; Pennsylvania State University; 104 Davey Laboratory; University Park PA 16802; USA; 814-865-5233; E; 3.6; Spontaneous and stimulated scattering from atoms and molecules

WILCOMB, Bruce E.; Laser Physics Branch; Code 6540; US Naval Research Laboratory; 4555 Overlook Avenue, SW; Washington DC 20375; USA; 202-767-2175; E; 4.1; Mercury bromide photodissociation laser characterization

WILCOX, John B.; 4880 Strickland Drive; Oxnard CA 93030; USA

WILDMAN, David; Accelerator Division; Fermi National Accelerator Laboratory; PO B 500 MS306; Batavia IL 60510; USA; 312-840-4984; FTS-370-4984; E/T; 4.2; 4.3; Accelerator instrumentation

WILETS, Lawrence; Department of Physics, FM-15; University of Washington; Seattle WA 98195; USA; 206-543-2897; T; 1.1; 1.3; 1.4; Atomic structure and collisions; Parity nonconservation; Exotic atoms

WILLARD, John E.; Department of Chemistry; University of Wisconsin; 1101 University Avenue; Madison WI 53706; USA; 608-262-2449; E; 5.2; Trapped electrons, hydrogen atoms, radicals and ions in hydrocarbon and inert gas matrices at cryogenic temperatures--properties

WILLIAMS, Frazer; Department of Electrical Engineering; University of Nebraska; Lincoln NE 68588; USA; 402-472-1910; E; 5.3; Laser-triggered breakdown in gases

WILLIAMS, Graheme J.; Chemistry Division; Brookhaven National Laboratory; Upton NY 11973; USA; 516-345-4383; E/T; 1.1; 3.4; 4.2; 5.2; Molecular structure by X-ray and neutron diffraction--determination of; Intermolecular interactions in the solid state

WILLIAMS, James F.; Department of Physics; University of Western Australia; Perth W.A. 6009; Australia; 380-2737; 386-8881; E; 1.1; 2.1; 3.2; 4.2; 5.6

WILLIAMS, John R.; Department of Physics; Auburn University; Auburn AL 36849; USA; 205-826-4366; E; 1.1; 4.2; 5.3; Basic parameter measurements for gas laser systems

WILLIAMS, Michael D.; High Energy Science; NASA Langley Research Center; MS 160; Hampton VA 23665; USA; 804-865-3781; E; 3.6; Solar pumped laser research

WILLIAMS, Michael R.; 285 Vereda Leyenda; Goleta CA 93117; USA

WILLIAMS, Peter; Department of Chemistry; Arizona State University; Tempe AZ 85287; USA; 602-965-4107; E; 5.1; Phenomena and mechanisms in sputtering and sputtered ion emission from surfaces --investigation

WILLIAMS, R. Stanley; Bell Telephone Laboratories; Room 2B-316; 600 Mountain Avenue; Murray Hill NJ 07974; USA; 201-582-3416; E; 4.2; 5.1; Ion scattering from solids; Adsorbates on solid surfaces

WILLIAMS, Ronald L.; Department of Physics; Plasma Physics Lab.; UCLA; Los Angeles CA 90024; USA

WILLIAMS, W. David; Division 2531; Sandia National Lab.; PO Box 5800; Albuquerque NM 87185; USA; 505-844-7659; E; 3.10; 4.1; 5.2; 5.3; Electrooptic device physics; Waveguide optics and devices; Acoustooptic device physics; Piezoelectric and ferroelectric device physics

WILLIAMS, William L.; Department of Physics; University of Michigan; Ann Arbor MI 48109; USA; 313-763-1150; E; 1.4; Weak neutral current interactions in hydrogen atoms; Parity nonconservation

WILLIAMSON JR., William; Department of Physics and Astronomy; University of Toledo; Toledo OH 43606; USA; 419-537-0188; 419-537-2651; T; 2.1; 2.2; 4.1; 5.1; 5.2; Atomic scattering theory (Glauber and Eikonal); Nonlinear optics; Electron and photon transport using Monte Carlo methods

WILLIS, Charles R.; Department of Physics; Boston University; 111 Cummington Street; Boston MA 02215; USA; 617-353-2600; T; 4.1; Quantum optics; Lasers; Optical bistability

WILLIS, Paul A.; Software Engineering Company; PO Box 2343; 2824 W. George Mason Road; Falls Church VA 22042; USA; 703-534-9181; 703-533-2826; E/T; 3.10; 3.11; 3.3; 1.4; 5.6; Wavelength modulation in optical fibers; Detection thresholds in IR imaging; Information capacity of correlation coding; Atmospheric scattering of IR beams

WILLIS, Richard J.; 4028 Dallas Avenue; San Diego CA 92117; USA

WILLNER, Chris A.; 24608 Star Valley; St. Clair Shores MI 48080; USA

WILSON III, Charles W.; Department of Physics; University of Akron; Akron OH 44325; USA; 216-375-7079; E; 1.3; NMR spectroscopy in polymers

WILSON, Jack; Warrensville Research Laboratory; Standard Oil Company of Ohio; Laboratory for Laser Energetics; University of Rochester; 250 E. River Road; Rochester NY 14623; USA; 716-275-2074; E; 3.6; 5.4; Laser-driven shock waves for equation-of-state measurements--potential; Very high power lasers used for physics, chemistry; and biology experiments

WILSON, John W.; Space Technology Branch; Space Systems Division; Langley Research Center; NASA; MS 160; Mailstop 160; Hampton VA 23665; USA; 804-865-3781; 804-877-1351; T; 1.1; 2.2; 3.8; 5.1; 5.3; Laser chemistry of photodissociation solar pumped lasers; Atomic/molecular collisions with ions; Physical adsorption on surfaces and thin films

WILSON, Syd R.; Semiconductor Group; Semiconductor Research; and Development Laboratory; Motorola, Inc.; 5005 E. McDowell Road; Phoenix AZ 85008; USA; 602-244-4637; E; 4.2; 5.1; Ion implantation; Ion-solid interactions; Laser-solid interactions

WILSON, Walter E.; Radiological Sciences Division; Battelle Memorial Institute; Pacific Northwest Laboratories; PO Box 999; Richland WA 99352; USA; 509-376-5706; E; 4.2; 5.1; 5.2; 2.1; 2.2; Ionizing radiation with biological matter-- interactions; Radiological physics; Radiation dosimetry; Monte Carlo radiation transport

WILSON JR., William L.; Department of Electrical and Computer Engineering; Rice University; PO Box 1892; Houston TX 77252; USA; 713-527-8101x3585; E; 4.1; 4.2; Excimer lasers

WINDER, Dale R.; Department of Physics; Colorado State University; Fort Collins CO 80523; USA

WINDHAM, B.; Department of Physics; Harper College; Algonquin and Roselle Roads; Palatine IL 60067; USA

WINE, Paul H.; Georgia Tech. Research Institute; Physical Sciences Division; Electromagnetics Laboratory; Georgia Institute of Technology;

Atlanta GA 30332; USA; 404-894-3424/3425; 404-894-3425; E; 2.3; 3.6; 3.1; Kinetics of elementary gas phase reactions; Photochemistry and photophysics of small molecules; Aqueous phase elementary reaction kinetics

WINELAND, David J.; Time and Frequency, Division 524.11 (MS); National Measurement Laboratory; National Bureau of Standards; 325 S. Broadway; Boulder CO 80303; USA; 303-497-5286; E/T; 1.1; 3.3; 3.6; 3.9; 4.1; Spectra of electromagnetically confined atomic ions

WING, Byron J.; 250 El Dorado Blvd., #J-173; Webster TX 77598; USA

WING, William H.; Department of Physics; University of Arizona; Tucson AZ 85721; USA; 602-626-4277; E/T; 1.4; 2.1; 3.3; 3.6; Basic atoms and molecules--laser and microwave spectroscopy; Collisional processes; Fundamental constants research

WINICUR, Daniel H.; Department of Chemistry; University of Notre Dame; Notre Dame IN 46556; USA; 219-239-5240; E; 3.2; 3.5; 1.1; 2.1; 2.3; Energy-exchange reactions between electronically excited (metastable) rare gases and small molecules--molecular beam studies

WINKLER, Irwin C.; High Energy Laser Beam Propagation and Control; Optics; Raman Physics; MIT Lincoln Laboratory; PO Box 73; Lexington MA 02173; USA; 863-5500; E; 3.10; 3.7; 2.3; 3.6; 3.9

WINKLER, Linda I.; 1800 Jefferson Park Ave. #136-D; Charlottesville VA 22903; USA

WINKLER, Peter; Department of Physics; University of Nevada at Reno; Reno NV 89557; USA; 702-784-4935; 702-784-6792; T; 1.1; 1.2; 2.1; 3.1; 5.3; Resonances in electron scattering; Recombination, radiative and dielectronic; Many-body methods for atoms and molecules

WINN, John S.; Department of Chemistry; Dartmouth College; Hanover, NH 03755; USA; 603-646-3084; E/T; 2.1; 2.3; 3.5; Weakly bound molecules--molecular spectroscopy; Reaction dynamics and energy transfer

WINTER, Nicholas W.; Theoretical Atomic and Molecular Physics Group; Lawrence Livermore National Lab.; PO Box 808; Livermore CA 94550; USA; 415-422-6215; T; 1.1; 2.3; 3.1; Electronic structure and molecular dynamics; Atom-diatom reactions; Photodissociation; Polyatomic molecules--excited electronic states

WINTER, Thomas G.; Department of Physics; Pennsylvania State University; PO Box PSU; Wilkes-Barre Campus; Lehman PA 18627; USA; 717-675-2171; T; 2.1; 2.3; Electron transfer in ion-atom collisions; Electron excitation in atom-atom collisions; Ionization in ion-atom collisions

WIRTH, Mary M.; Department of Chemistry; Analytical; Lawrence Livermore Laboratory; PO Box 808; Livermore CA 94550; USA; 415-422-4535; 415-423-7839; E; 3.2; 3.6; 3.7; Two-photon spectroscopy; Picosecond spectroscopy; Solatron studies

WISE, John H.; Department of Chemistry; Washington and Lee University; Lexington VA 24450; USA; 703-463-8872; T/E; 1.1; 3.1; High resolution atomic spectroscopy

WITRIOL, Norman M.; Department of Physics; Louisiana Tech. University; Ruston LA 71272; USA; 318-257-4670; 318-255-4568; T; 2.1; 3.1; 3.8; Reactive collisions; Laser induced dissociation/chemistry; Quantum optics; Intramolecular and intermolecular energy transfer

WITTENBERG, Albert M.; Display Technology Department; Electronic and Photonic Division; Advanced Technology Laboratory; AT&T Bell Laboratories; 600 Mountain Avenue; Murray Hill NJ 07974; USA; 201-582-3726; E; 4.1; 4.3; 3.2; Electron beam interaction with solids; Cathodoluminescence of crystal phosphors

WITTIG, Curt; Department of Electrical Engineering; University of Southern California; University Park; Los Angeles CA 90007; USA; 213-741-6389; E; 2.3; 5.3; Gas phase kinetics

WITTKOWER, Andrew; Eaton Corp.; 108 Cherry Hill Drive; Beverly MA 01915; USA; 617-921-0750; E; 4.3; 5.1; Design and characterization of particle accelerators; Interaction of fast ions and atoms with matter

WITTRY, David B.; Department of Math and Science; University of Southern California; Vivian Hall, Room 612; Los Angeles CA 90007; USA

WODARCZYK, Francis J.; Physics and Chemistry Division; Rockwell International Science Center; PO Box 1085; Thousand Oaks CA 91360; USA; 805-498-4545; E; 2.3; 3.6; 3.8; Photochemistry, spectroscopy and kinetics of small molecules and atoms

WODKIEWICZ, Krzysztof; Institute of Theoretical Physics; Warsaw University; HOZA 69; Warsaw 00-681; Poland; T; 3.11; 3.7; 3.10

WOLGA, George J.; Department of Electrical Engineering; College of Engineering; Cornell University; 237 Phillips Hall; Ithaca NY 14853; USA; 607-256-3962; E; 2.1; 2.3; 5.1; 5.5; 3.1; Instrumentation for spectroscopy/acoustoptic filters

WOLICKI, Eligius A.; Wolicki Associates, Inc.; 1310 Gatewood Drive; Alexandria VA 22307; USA; 703-765-4478; E; 4.2; 5.1; Ion beam analysis of metals; Ion implantation

WOLK, Gary L.; Laser Studies Group; Engineering Research Center; AT&T Technologies; PO Box 900; Princeton NJ 08540; USA; 609-639-2478; E; 3.8; 2.3; 4.1; 3.2; 3.6; Atom-molecule reactions; Laser induced chemistry; Chemistry in nonequilibrium vibrationally excited systems; Multiphoton processes

WOLKEN, George; Physical Sciences Division; Battelle Memorial Institute; 505 King Avenue; Columbus OH 43201; USA; 614-424-7274; T; 1.1; Computational molecular physics

WOMACK, Dennis R.; Directed Energy Directorate; U.S. Army Missile Command; DRSMI-RHST; Redstone Arsenal AL 35809; USA; 205-876-8271; E; 4.1; Gamma ray laser research and development

WONG, Joseph K.; Department of Chemistry; Noyes Laboratory of Chemical Physics; California Institute of Technology; Pasadena CA 91106; USA; 213-795-6811x2513; T; 2.1; Quantum mechanical scattering of atom-molecules

WONG, Ngai C.; Joint Institute for Laboratory Astrophysics; University of Colorado; PO Box 440; Boulder CO 80309; USA; 303-492-7855; 303-492-7387; E/T; 3.6; 3.11; 4.1; 3.10; 5.2; Ultra-high resolution, sensitivity laser spectroscopy; Quantum fluctuations, squeezing; Phase modulators, FM sideband techniques; Coherent transient phenomena

WONG, Thomas T.; Department of Electrical Engineering; ITT Center; Illinois Tech.; Chicago IL 60616; USA

WOO, Shien-Biau; Department of Physics; University of Delaware; Newark DE 19711; USA; 302-451-2943; 302-571-3017; E/T; 3.1; 3.6; Ground electronic and vibrational state molecular anions--laser; photodetachment spectrum; Photodetachment cross section calculation (zero-core-contribution model)

WOOD, David R.; Department of Physics; Wright State University; Dayton OH 45435; USA; 513-873-2148; E; 3.2; Atomic emission spectroscopy in the UV

WOOD, John H.; Chemistry-Materials Science Division; Group CMB-5; Los Alamos National Laboratory; PO Box 1663 MS730; Los Alamos NM 87545; USA; 505-667-4414; FTS-843-4414; T; 1.1; Atomic and molecular electronic structure calculations with statistical exchange/correlation potentials

WOOD, Lowell L.; Lawrence Livermore National Lab.; PO Box 808; Livermore CA 94550; USA; 415-422-7281; E/T; 3.4; 3.8; 4.1; 5.7; Basic and applied astrophysics; X-ray lasers; Laser isotope separation; Laser theory

WOOD II, Obert R.; AT&T Bell Laboratories; Crawfords Corner Road; Holmdel NJ 07733; USA; 201-949-6339; E; 1.4; 2.2; 3.6; 4.1; 5.3; XUV lasers; Inner shell photoionization lasers; Plasma recombination lasers

WOOD, Robert M.; Department of Physics and Astronomy; University of Georgia; Athens GA 30602; USA; 404-542-2485; E; 2.1; 4.2; 2.2; Accelerator based molecular dissociation; Accelerator based molecular ionization

WOODRUFF, Pamela R.; Department of Natural Philosophy; University of Aberdeen; Aberdeen AB92UE; Scotland; 0224-40241; E; 3.1; 3.6; 3.7

WOODRUFF, Susan B.; Group Q-8; Los Alamos National Laboratory; PO Box 1663 MS559; Los Alamos NM 87545; USA; 505-667-3852; E/T; 1.1; 2.3; Molecular electronic structure; Classical dynamics of small molecules

WOODRUFF, Truman O.; Department of Physics; Michigan State University; East Lansing MI 48824; USA

WOODS, Charles L.; 31 Eliot Drive; Stow MA 01775; USA

WOODS, R. Claude; Department of Chemistry; University of Wisconsin; 1101 University Avenue; Madison WI 53706; USA; 608-262-2892; E/T; 3.3; 3.5; 3.9; 1.1; 5.7; Microwave spectroscopy of molecular ions and other transients; Interstellar chemistry and radio astronomy; Quantum chemical calculations of molecular structure; Spectroscopic diagnostics of discharge plasmas

WOODWORTH, Joseph R.; Division 4212; Sandia National Laboratories; PO Box 5800; Albuquerque NM 87185; USA; 505-844-8161; E; 4.2; 5.3; UV triggered spark gaps employing resonances between the UV laser and the insulating gas in the spark gap--development

WOODY, Bernard A.; United Technologies Research Center; Chemical Physics Group; United Technologies Corporation; Silver Lane; East Hartford CT 06108; USA; 203-727-7212; E; 2.3; 3.1; 3.2; 3.6; 4.1; Chemical dynamics; Ultra sensitive spectroscopy

WOODYARD, Jack R.; Analytical Research and Support Division; U.S. Bureau of Mines; Reno Research Center; 1605 Evans Avenue; Reno NV 89505; USA; 702-784-5351; T/E; 2.3; Applied emission spectroscopy for chemical analysis

WOOLF, Stanley; Department of Physical Science; Arcon Corporation; 260 Bear Hill Road; Waltham MA 02154; USA; 617-861-8741; T; 2.2; 3.1; 2.1; 5.1; 5.3; Charged particle transport in solids; Neutral particle transport; Radiation effects in electronic device materials

WOOLF, Thomas B.; International House; 1414 East 59th Street; Chicgo IL 60637; USA

WOOTEN, John W.; 106 Crestview Lane; Oak Ridge TN 37830; USA

WORDEN JR., Earl E.; Chemistry and Materials Sciences Department; Chemistry Division; Lawrence Livermore National Lab.; PO B 808 L-467; Livermore CA 94550; USA; 415-422-6203; E; 1.1; 3.8; 3.6; 3.9; 1.2; Actinides and Lanthanides--Laser and conventional spectroscopy; Atomic vapor laser isotope separation--Uranium and plutonium; Alkaline earths--associative ionization studies by laser spectroscopy

WORLOCK, John M.; Physics and Optical Science Research; Solid State Science and Technology; Bell Communications Research, Inc.; Newman Springs Road; Red Bank NJ 07701; USA; E; 3.6; 3.10; 3.9; 3.7; Raman spectroscopy, including surfaces and interfaces; Photoion

WORMHOUDT, Joda C.; Center for Chemical and Environmental Physics; Research Group; Aerodyne Research, Inc.; 45 Manning Road; Billerica MA 01821; USA; 617-663-9500; E/T; 3.2; 3.3; 3.5; 3.6; 2.1; Laser diagnostics in microelectronics and combustion systems; Spectral lineshapes

WRIGHT, John C.; Department of Chemistry; University of Wisconsin; Madison WI 53706; USA; 608-262-0351; E; 3.6; 3.10; Laser spectroscopy and Raman spectroscopy; Trace analysis; Nonradiative relaxation; Quantum electronics

WRIGHT, John J.; Department of Physics; University of New Hampshire; Demeritt Hall; Durham NH 03824; USA; 603-862-2829; E; 3.6; 3.8; 5.2; Van der Waals molecules--laser spectroscopy; Alkali metals in rare gases --matrix isolation spectroscopy; Collision-induced energy transfer

WRIGHT, Jon A.; 8950 Villa La Jolla Dr., Ste. #2145; La Jolla CA 92038; USA

WRIGHT, Lawrence A.; Sandia National Laboratory; Division 5218; P.O. Box 5800; Albuquerque NM 87185; USA; 505-844-6034; T; 2.1; 1.1; 3.6

WRIGHT, Michael D.; Central Res MS/K-103; Varian Associates, Inc.; 611 Hansen Way; Palo Alto CA 94303; USA

WROGE, Michael L.; 7788 W. Bruno Avenue Apt. 2; St. Louis MO 63117; USA

WU, C. Y. R.; Department of Physics; Space Sciences Center; University of Southern California; MC-1341; University Park; Los Angeles CA 90089; USA; 213-743-6460; 213-743-2025; E; 3.1; 3.6; 3.7; 3.10; 4.2; Photoabsorption, photoionization, photodissociation and fluorescence of atoms and molecules in the extreme UV region; Nonlinear optics, wave mixing and harmonic generation through atomic and molecular gases

WU, Frank T.; Code 3918; Naval Weapons Center; China Lake CA 93555; USA; 714-939-1040; E/T; 3.3; 5.3; Gaseous electronics; Microwave interaction with glow discharge

WU, Ming H.; Hamamatsu Corporation; 420 South Avenue; Middlesex NJ 08846; USA

WU, Richard L.; Department of Chemistry; Brehm Laboratory; Wright State University; Dayton OH 45385; USA; 513-873-3194; 513-873-2202; E; 2.3; 4.4; 4.3; 5.6; 5.5; Ion-molecules collisions--kinetics and mechanism; Ionic species--thermodynamic properties

WU, Susan Y.; Department of Aerospace Engineering; University of Tennessee; Space Institute; Tullahoma TN 37388; USA; 615-455-0631x330; E/T; 5.5; MHD energy conservation requiring basic atomic data for combustion; products of coal with potassium

WU, Yao-Hwa; 2120 Roosevelt Avenue Apt. 4; Berkeley CA 94703; USA

WU, Yen C.; W1024 Spofford; Spokane WA 99205; USA

WUILLEUMIER, Francois J.; Villebon Sur Yvette; 23 Rue De La Basse Roche; Palaiseau 91120; France

WUTZKE, S. A.; Gas Laser Research and Development; Westinghouse R&D Center; 1310 Beulah Road; Pittsburgh PA 15235; USA; 412-252-3252; E; 4.1; Gas laser research and development

WYATT, Robert E.; Department of Chemistry; University of Texas; Austin TX 78712; USA; 512-471-3114; T; 2.1; 2.3; 3.6; 3.7; 3.8; Reactive scattering; Multiphoton processes

WYCKOFF, Harold O.; 4108 Montpellier Road; Rockville MD 20853; USA

WYNER, Elliot F.; High Intensity Discharge; Lighting Products Group; Lighting Research Laboratory; GTE-Sylvania; 100 Endicott Street; Danvers MA 01923; USA; 617-777-1900x2361; E; 5.3; 5.4; High pressure sodium lamps; New light sources--development; Optical pumps for lasers; Arc discharges

WYNNE, James J.; Laser Physics and Chemistry; Research; IBM Thomas J. Watson Research Center; PO Box 218; Yorktown Heights NY 10598; USA; 914-945-1575; E; 3.6; 3.10; 5.1; 5.2; 1.2; Atomic spectroscopy using lasers; Energy level and oscillator strength measurements

WYSS, Jerry C.; General Research Company; 5383 Hollister Avenue; Santa Barbara CA 93111; USA; 805-964-7724; E; 4.1; 3.6; 2.3; Development of optical techniques for sensing high frequency; RF fields

XIE, Ming; Department of
Physics; Stanford University;
Stanford CA 94305; USA

XU, Emily Y.; Department of
Physics; Columbia University;
PO Box 31; New York NY 10027;
USA

YAAKOBI, Barukh; Laboratory for Laser Energetics; University of Rochester; 250 E. River Road; Rochester NY 14623; USA; 716-275-5101; E/T; Laser fusion experiments --X-ray spectroscopy

YABLONOVITCH, Eli; Surface-Interface Physics; Physics and Optical Science; Solid State Science and Technology; Bell Communications Research; 600 Mountain Ave.; Murray Hill NJ 07974; USA; 201-582-3737; 201-755-5975; E/T; 3.8; 4.1; 5.4; Laser induced chemistry; Laser plasma interactions; Nonlinear optics

YABLONSKY, Dmeter; 31 Browning Avenue; Yonkers NY 10704; USA

YABUZAKI, Tsutomu; Radio Atmospheric Science Center; Kyoto University; UJI Kyoto 611; Japan; 0774-32-3111; E/T; 1.1; 2.1; 3.6; 3.8; 3.10; Laser spectroscopy; Optical pumping; Laser chemistry

YAGUNOFF, Guy G.; PO Box 31144; San Francisco CA 94131; USA

YAMAMOTO, Tomoko; Department of Physics; Thiel College; Greenville PA 16125; USA; 412-588-7700x272; T; 2.3; Electron tunneling calculations based on the square well model; Bacteriochlorophyll dimer and monomeric bacteriopheophytin molecules

YAMANAKA, Chiyoe; Institute of Laser Engineering; Osaka University; 2-6 Yamada-Oka; Suita Osaka 565; Japan; 06-876-3000; 06-877-5111x6500; E; 1.; 3.; 6.8; 6.10; 4

YAMANAKA, Masanobu; Elec-Mag-Energy Engr. Fac. Engr.; Osaka University; 2-1 Yamada-Oka; Suita Osaka 565; Japan; 06-877-5111; E; 1.1; 3.3; 3.6; 3.9; 4.1; IR gas lasers; Plasma diagnostics applications

YAMANI, Hashim A.; (SANCST); PO Box 6086; Riyadh; Saudi Arabia

YANEY, Perry P.; Department of Physics; University of Dayton; 300 College Park; Dayton OH 45469; USA; 513-229-2221; 513-229-2181; E; 4.1; 5.5; 5.2; 5.4; 5.1; Gas flow and combustion diagnostics using Raman spectroscopy; Solids, structure and surfaces--Raman microprobe; Plasma diagnostics using nonlinear Raman spectroscopy

YANG, Robert Y.; Department of Physics; Polytechnic Institute of NY; Route 110; Farmingdale NY 11735; USA

YANG, Sze-Cheng; Department of Chemistry; University of Rhode Island; Kingston RI 02881; USA; 401-792-2377; 401-792-5081; E; 2.3; 3.1; 3.5; 3.6; 3.8; Diatomic molecules--spectroscopy; Chemical reactions probed via laser-induced fluorescence; Chemical reaction kinetics and dynamics; Photodissociation

YANKWICH, Peter E.; Department of Chemistry; University of Illinois; NL Box 2; 505 S. Mathews Avenue; Urbana IL 61801; USA; 217-333-3518; T; 2.3; Isotope effects on reaction rates of polyatomic molecules--computational modeling

YAO, Shang J.; School of Medicine; Research Laboratory; University of Pittsburgh; 3459 Fifth Avenue; Pittsburgh PA 15213; USA; 412-648-6332; T; 1.1; Molecular tunneling

YARDLEY, James T.; Chemical Sector Technology Lab.; Allied Chemical Corporation; PO Box 1021R; Morristown NJ 07960; USA; 201-455-4676; E; 3.8; 3.10; 4.1; Laser chemistry--fundamental and applied research

YARKONY, David R.; Department of Chemistry; Johns Hopkins University; 34th and Charles Streets; Baltimore MD 21218; USA; 301-338-7462; T; 1.1; Electron structure

YASUOKA, Yoshizumi; Department of Electrical Engineering; National Defense Academy; 1-10-20 Hashirimizu; Yokosuka 239 1218; Japan; 0468-41-3810

YATES, Albert C.; Department of Chemistry; University of Cincinnati; Cincinnati OH 45221; USA; 513-475-4532; T; 1.1; 2.1; 2.2; Potential energy surfaces; Electron, atom, molecule collision theory

YATES, John H.; Chemistry; University of Pennsylvania; 250 S 33rd St.; Philadelphia PA 19104; USA; 215-898-4714; E/T; 1.1

YAU, Hon F.; Physics Department; Chung Cheng Institute of Technology; Tahsi Taoyuen 335; Taiwan; Republic of China

YEH, Shoou-Dyi; Systems and Applied Sciences Corp; 27 Traverse Road #3; Newport News VA 23606; USA; 804-599-5468; 804-827-2719; T; 3.6; 3.8; Laser spectroscopy; Atomic and molecular collisions with radiative interactions

YELLIN, Joseph; 12618 Killion Street; North Hollywood CA 91607; USA

YENCHA, Andrew J.; Chemistry and Physics; State University of NY, Albany; Albany NY 12222; USA; 518-442-4416; 518-442-4394; E; 2.1; Metastable rare gas and molecular systems--energy transfer

YENEN, Orhan; Department of Physics; University of Nebraska, Lincoln; Lincoln NE 68588; USA; 402-472-2786; E; 2.1; 4.2; 5.1

YEUNG, Edward S.; Department of Chemistry; Iowa State University; Gilman Hall; Ames IA 50011; USA; 515-294-8062; E; 2.3; Spectroscopy; Kinetics

YIN, Lo-I; Laboratory for Astronomy and Solar Physics; NASA Goddard Space Flight Center; Code 682; Greenbelt MD 20771; USA; 301-344-5682; E; 3.4; 5.1; X-ray photoelectron spectroscopy; Auger spectra; Ion-bombardment induced surface changes

YIN, Ru-Ying; Department of Physics and Astronomy; University of Pittsburgh; Pittsburgh PA 15260; USA

YIP, Sidney; Department of Nuclear Engineering; Massachusetts Institute of Technology; Cambridge MA 02139; USA; 617-253-3809; T; 5.2; 5.1; 2.1; 2.3; 1.3; Molecular dynamics and Monte Carlo simulations; Structural and transport properties of solids with interfaces; Dynamics of dense fluids and liquid-glass transition; Deflection-induced amorphization of crystalline solids

YODER, Marvel J.; W. J. Schafer Associates; 107 Audubon Road; Wakefield MA 01880; USA; 617-470-1344; 617-246-0450; E/T; 4.1; 5.3; 3.6; 3.10; Laser research and development; Laser radar development; Microwave systems and effects

YORK JR., George W.; Department of Physics; Advanced Laser Group; Los Alamos National Lab.; University of California; PO Box 1663; MS E 543; Los Alamos NM 87545; USA; 505-667-3714; E; 4.1; 5.3; Laser discharges and parameters--characterizations

YOS, Jerrold M.; Thermodynamics Division; Avco Systems Division; 201 Lowell Street; Wilmington MA 01887; USA; 617-657-3156; T; Aerospace applications

YOSHIMINE, Megumu; IBM Research Laboratory K34; 5600 Cottle Road; San Jose CA 95193; USA; 408-256-2235; T; 1.1; 2.3; Quantum mechanical calculations on atoms and molecules--computer program development; Chemical reaction of organic molecules --determination of potential energy surfaces

YOUNG, Bruce K.; L-401; Lawrence Livermore National Lab.; Livermore CA 94550; USA

YOUNG, C. Gilbert; Corporate Technology; Combustion Engineering; PO Box 9308; 900 Long Ridge Road; Stamford CT 06904; USA; 203-328-2358; 203-968-0087; E; 4.1; 5.5; 5.2; Optically-pumped solid-state laser systems; Electro-optical switches, modulators and materials; Physics of solids-electron spin resonance; Infrared sources and transmission, detectors

YOUNG, Charles E.; CHM 200; Argonne National Laboratory; 9700 S. Cass Avenue; Argonne IL 60439; USA

YOUNG, James F.; Electrical Engineering; Edward L. Ginzton Laboratory; Stanford University; Stanford CA

94305; USA; 415-497-1674;
E; 4.1; 3.10; 3.6; 3.1; 3.1;
UV laser sources, development;
Laser-produced plasma source-
reaction; X-ray, soft, sources

YOUNG, Lydia J.; M/S 312;
Perkin-Elmer EBT; 26460
Corporate Avenue; Hayward CA
94545; USA

YOUNG, R. Gaines; 38 Alexander
Avenue; Nutley NJ 07110; USA

YOUNG, Victor J.; 34 Birch
Street; Port Washington NY
11050; USA

YOUNGER, Stephen M.; Defense
Sciences; A-Division; Lawrence
Livermore National Lab.;
University of California;
Livermore CA 94550; USA;
415-423-3938; E/T; 5.4;
2.2; 3.1; 5.3; 5.1; X-ray
lasers; Thermonuclear processes
in high density plasmas

YOUNGQUIST, Sarah E.; 108
Stewart Avenue; Ithaca NY
14850; USA

YU, Ki-Su; Department of
Physics; Won Kwang University;
344-2 Shinyong-Dong 510;
IRI Chun-Buk; Republic of Korea

YU, Simon; M-Division; Lawrence
Livermore National Lab.; PO
Box 808; Livermore CA 94550;
USA; 415-422-7876; T; 2.2;
4.2; 5.6; Air chemistry in
presence of an electron beam

YUAN, Jian-Min; Department
of Physics and Atmospheric
Science; Drexel University;
32nd and Chestnut Streets;
Philadelphia PA 19104; USA;
215-895-2722; T; 1.1; 2.3;
3.7; 3.8; 3.10; Nonlinear
dynamics--bistability and
chaos; Photo-desorption of
surface adsorbed molecules

YUN, Kwang-Sik; Department
of Chemistry; University of
Mississippi; University MS
38677; USA; 601-232-7301;
E/T; 2.3; 5.1; 5.3; Gas
phase reactions--shock tube
studies; Liquid plane solu-
tions--optical and thermodynamic properties

ZAHNISER, Mark S.; Center for Chemical and Environmental Physics; Applied Sciences Division; Aerodyne Research, Inc.; 45 Manning Road; Billerica MA 02174; USA; 617-663-9500; E; 1.1; 2.3; 3.1; 5.2; 4.1; Chemical kinetics of atoms and radicals; Quantitative infrared spectroscopy; Atmospheric chemistry; Combustion chemistry

ZAIDI, Haider R.; Department of Physics; University of New Brunswick; Fredericton NB E3B5A3; Canada; T; 3.11; 3.10; 3.7; 3.6; 2.1; Quantum optics; Role of collisions

ZAJONC, Arthur G.; Department of Physics; Amherst College; Amherst MA 01002; USA; 413-542-2199; E; 1.4; 2.1; 3.6; Excitation transfer in atoms; Parity nonconservation in atomic physics; Laser physics

ZAK, Bernard D.; Environmental Research Division; Sandia National Laboratories; PO Box 5800; Albuquerque NM 87185; USA; 505-844-7328; E/T; 5.6; Environment-related research

ZAKHEIM, David S.; Department of Chemistry; Howard University; Washington DC 20059; USA; 202-636-5631; E; 3.1; 3.8; UV laser spectroscopy and photodissociation processes in small molecules

ZALUBAS, Romuald; Atomic Energy Levels Data Center; National Bureau of Standards; Physics Building, Room A155; Gaithersburg MD 20899; USA; 301-921-2011; E; 1.1; 3.2; Atomic spectra: wavelengths, line classifications, energy levels, ionization potentials --observation and analysis; Atomic energy levels--critical compilation

ZANDER, Arlen R.; Department of Physics; East Texas State University; Commerce TX 75428; USA; 214-886-5480; E; 2.1; 4.2; Heavy ion induced ionization processes

ZANELLI, Claudio I.; Endosonics; 180-C Blue Ravine Rd.; Folsom CA 95630; USA; 916-351-0220; E; 2.2; 3.4; 5.8; Energy loss by K-capture-I.C. electrons; Point-source measurements and calculations

ZAPATA, Luis E.; High Energy Science Branch; Space Systems Division; Miami University, NASA-Langley Research Center; 160; Hampton VA 23665; USA; 804-865-3781; 804-865-3127; E/T; 4.1; 5.3; 5.4; 3.1; 1.1; Solar pumped lasers; Nuclear pumped lasers; Time dependent perturbed angular correlations

ZARE, Richard N.; Department of Chemistry; Stanford University; Stanford CA 94305; USA; 415-497-3062; E; 1.1; 2.3; 3.6; 3.7; 5.1; Molecular energy level structure; Heavy particle collisions; Atomic photoionization

ZDASIUK, George A.; MS K-127; Varian Associates; 611 Hansen Way; Palo Alto CA 94303; USA

ZEMAITIS, Vincent; PO Box 1133; Madison Square Station; New York NY 10159; USA

ZEMKE, Warren T.; Department of Chemistry; Wartburg College; Waverly IA 50677; USA; 319-352-1200; T; 1.1; Diatomic molecules/ions--calculations; Potential energy curves for excited and ground states; Transition and dipole moments; Transition probabilities and lifetimes

ZEWAIL, Ahmed H.; Chemistry; Chemistry and Chemical Engineering; Noyes Lab. of Chemical Physics; California Institute of Technology; Pasadena CA 91125; USA; 213-795-6811x2536; E; 3.6; 3.8; 3.7; 4.1; Laser chemistry; Laser spectrocopy; Laser techniques

ZHANG, Jianhua; PO Box 1014; College Station TX 77841; USA

ZHAO, Ping; Department of Physics; Yale University; New Haven CT 06520; USA

ZHOU, Ying-Hua; 321 Gordon McKay Lab; Harvard University; Cambridge MA 02138; USA

ZIDE, Arnold; 45-10 Kissena Boulevard Apt. 2B; Flushing NY 11355; USA

ZIEGLER, Daniel L.; Opto-electronics Division; Hewlett-Packard; 640 Page Mill Road; Palo Alto CA 94304; USA

ZIEGLER, James F.; IBM Thomas J. Watson Research Center; PO Box 218; Yorktown Heights NY 10598; USA; 914-945-2165; E/T; 4.2; 5.2; Atomic collisions in solids

ZIELESCH, Marguerite; 19347 Woodland Street; Harper Woods MI 48225; USA

ZIMM, Bruno H.; Department of Chemistry B-017; University of California at San Diego; La Jolla CA 92093; USA; 619-452-4416; E/T; 1.3; 5.2; Macromolecules in solution--physical chemistry

ZIMMER, Robert S.; Institute of Educational Technology; The Open University; Milton Keynes MK76AJ; England

ZIMMERMAN, I. H.; Department of Physics; Clarkson College; Potsdam NY 13676; USA; 315-268-2341; T; 2.1; 2.3; 3.8; Reactive atom-diatom collisions; Atomic and molecular collisions in intense laser fields; Ion-atom scattering; Computational simulation of collisions

ZIMMERMAN, Myron L.; Department of Physics; Massachusetts Institute of Technology; Rm 26-228; 77 Massachusetts Avenue; Cambridge MA 02139; USA; 617-253-6810; T/E; 3.6; 3.9; Highly excited atoms in strong electric magnetic fields--laser spectroscopy

ZINK, J. W.; Star Route Railroad Flat Road; Mountain Ranch CA 95246; USA; 209-754-4520; T; 1.1; 3.1

ZIPF, Edward; Department of Physics and Astronomy; University of Pittsburgh; Pittsburgh PA 15230; USA; 412-624-4361; E; 5.6; Atomic and atmospheric physics

ZITTEL, Paul F.; Chemistry and Physics Laboratory; The Aerospace Corporation; PO Box 92957; Los Angeles CA 90009; USA; 213-648-6642; E; 2.1; 3.1; 3.8; Vibrationally excited molecules --photodissociation and quenching; Laser isotope separation

ZITTER, Robert N.; Department of Physics; Southern Illinois University; Carbondale IL 62901; USA; 618-453-5110; E/T; 3.8; IR laser induced chemical reactions

ZITZEWITZ, Paul W.; Department of Natural Sciences; University of Michigan at Dearborn; 4901 Evergreen Road; Dearborn MI 48128; USA; 313-593-5277; 313-763-3464; E; 1.3; 1.4; 2.2; 4.3; 5.1; Slow positron beams to study positronium; Interactions of positrons and positronium with surfaces

ZIV, Alan R.; Department of Chemistry; University of Illinois of Chicago Circle; PO Box 4348; Chicago IL 60680; USA; 312-996-6949; T; 3.2; Non-Hilbert space; Quantum theory for theoretical description of spectroscopic processes--development

ZOLLARS, Byron G.; Department of Physics; Rice University; 1892; 6100 S. Main Street; Houston TX 77005; USA; 713-527-8101x3624; 713-527-4932; E; 1.2; 2.1; 3.9; 2.3; 4.4

ZOLLWEG, Robert J.; Plasma and Nuclear Science; Applied Sciences; Westinghouse R&D Center; 1310 Beulah Road; Pittsburgh PA 15235; USA; 412-256-1641; E/T; 5.4; 5.3; 3.2; 3.3; 1.3; High pressure arc discharges and plasmas; Modeling of radiative devices; Radiation imprisonment; Complex molecule properties

ZORN, Jens C.; Department of Physics; Randall Laboratory; University of Michigan; Ann Arbor MI 48109; USA; 313-764-4450; 313-662-0683; E; 1.1; 3.3; 3.6; 3.8; 3.9; Atomic and molecular beams

ZUMBULYADIS, Nicholas; Special Projects Laboratory; Eastman Kodak Research Laboratories; 1999 Lake Avenue; Rochester NY 14650; USA; 716-722-1409; E; 1.3; 5.2; NMR studies of amorphous semiconductors and other electronic materials

ZUPUTLITZ, Gisbert; Physics Institute; Philosophenweg 12; University of Heidelberg; Philosophenweg 12; Heidelberg 1 D-6900; West Germany; 49-6221-569211; E; 1.1; 1.3; 1.4; 3.3; 3.6; Atomic spectroscopy; Atomic clusters; Elementary particles, Exotic atoms; Atom-nucleus interactions

Part *II*

Listing of
Atomic, Molecular, and Optical
Scientists by Research Specialty

Atomic, Molecular, and Optical Science Specialty Codes

1. Structure and properties of atoms and molecules.
   1.1 General structure and properties of atoms and molecules; e.g. energy levels, wave functions, polarizability, bonds, interaction potentials, hyperfine interactions.
   1.2 Properties and interactions of Rydberg states.
   1.3 Properties of other special atoms and molecules; e.g. exotic atoms, macromolecules, polymers, clusters.
   1.4 Fundamental properties of atoms and molecules; e.g. QED, parity, Lamb shifts.

2. Atomic and molecular collisions.
   2.1 atomic and molecular collisions excluding electron collisions.
   2.2 Electron and positron collisions with atoms and molecules; e.g. energy loss spectra.
   2.3 Chemical physics excluding photochemistry; e.g. reactive collisions, kinetics, charge transfer.

3. Interactions of radiation and dc fields with atoms and molecules.
   3.1 Conventional photon-atom and photon-molecule effects; e.g. transition probabilities, photoabsorption, photoionization, photodetachment, photodissociation, photoelectron spectra, Compton effect.
   3.2 Optical and uv spectra including fluorescence and beam foil spectra.
   3.3 Infrared, rf, and microwave spectra.
   3.4 Inner shell transitions including X-ray absorption and emission.
   3.5 Specifically molecular spectra.
   3.6 Interactions of laser radiation with atoms and molecules; e.g. laser spectroscopy, laser-induced fluorescence.
   3.7 Specifically intense-field effects; e.g. multiphoton effects, power broadening.
   3.8 Laser chemistry; photochemistry; e.g. collisions or reactions influenced by a radiation field, laser isotope separation.
   3.9 Interactions of dc fields with atoms and molecules; e.g. Stark effect, Zeeman effect.
   3.10 Nonlinear interactions of radiation with atoms and molecules; e.g. nonlinear optics, wave mixing, Raman effect.
   3.11 Quantum optics; e.g. correlation and statistical properties of radiation and their effects on interactions, superradiance.

4. Development and use of special techniques and instrumentation in A&M science.
   4.1 Optical techniques; e.g. optical pumping, laser development, electrooptics, photoacoustic spectra.
   4.2 Accelerator-based A&M physics.

    4.3 Beam technology.
    4.4 Mass spectrometry.

5. **Interface between A&M and other areas of science and technology.**
   5.1 Interaction of particles and radiation with surfaces; e.g. sputtered particles, LEED, laser fusion, inertial fusion.
   5.2 A&M physics in solids and liquids; e.g. NMR, transport in solids.
   5.3 Gaseous electronics; e.g. discharges, transport phenomena, laser models.
   5.4 A&M physics in plasmas; e.g. diagnostics, magnetic fusion, arcs, isotope separation, collisional line broadening.
   5.5 Combustion and other energy-related processes excluding fusion plasmas.
   5.6 Atmospheric and environmental applications of A&M physics; e.g. planetary atmospheres.
   5.7 Astrophysical applications of A&M physics.
   5.8 A&M physics in nuclear physics; e.g. K-capture.

Abrahamson, Adolf A.
Acquista, Nicolo
Adams, William H.
Adelman, Saul J.
Akins, Daniel L.
Alder, Berni J.
Ali, Mahamed A.
Allen Jr., Harry C.
Altick, Philip L.
Amano, Takayoshi
Andersen, Nils O.
Anderson, Alfred B.
Anderson, James B.
Andresen, Bjarne B.
Arents, John S.
Armentrout, Peter B.
Armstrong Jr., Lloyd
Armstrong, J. A.
Arnold, James O.
Atlas, Susan R.
Averill, Frank W.
Babcock, Lucia M.
Bae, Young K.
Baglin, Frank G.
Bagus, P. S.
Balling, Ludwig C.
Band, Yehuda B.
Barreto, Ernesto
Bartell, Lawrence S.
Bartiromo, Rosario
Bartlett, Rodney J.
Bartolotti, Libero J.
Basbas, George J.
Bashkin, Stanley
Bates Jr., Richard D.
Bay, Zoltan L.
Beaudet, Robert A.
Beckel, Charles L.
Becker, Karl H.
Bederson, Benjamin
Bennett, Robert B.
Bentley, John
Bergmann, Otto
Berk, Alexander
Bernstein, Elliot R.
Berry, R. Steven
Best, Philip E.
Beyerinck, Herman C.
Bhadra, Kalidas
Bhalla, Chander P.
Bhatia, Anand K.
Bichsel, Hans
Biedenharn, Lawrence C.
Blaha, Milan
Blatt, Rainer
Boggs, James E.
Bohn, Robert K.
Boring, Michael
Borsella, Elisabetta
Bottrell, Gerald J.
Bouman, Thomas D.
Boyer, Timothy H.
Branscomb, Lewis M.
Brault, James W.
Brenn, Rudiger
Brennan, James G.
Brewer, Leo
Brink, Gilbert O.
Broadhurst, Martin G.

Brooker, Murray H.
Browne, James C.
Bruch, Ludwig W.
Burke, Philip G.
Burr, Alex F.
Butler, Scott E.
Cade, Paul E.
Callan, Edwin J.
Cardon, Bartley L.
Carlton, Terry S.
Carmichael, Ian
Carr, Herman Y.
Cattolica, Robert J.
Caves, Thomas C.
Ceperley, David M.
Certain, Phillip R.
Chackerian Jr., Charles
Chancey, Charles C.
Chander, Jagdish
Chang, Edward S.
Chang, Tu-nan
Chappell, Richard F.
Chen, Augustine C.
Chen, Chin-Lin
Cheng, Kwok-Tsang
Chiang, Joseph F.
Childs, William J.
Ching, Wai-Yim
Cho, Chung Won
Cho, Hyuck
Chu, Shih-I
Chung, Sunggi
Clark, Charles W.
Clough, Shepard A.
Coffman, Robert E.
Cohen, James S.
Coldwell, Robert L.
Conway, John G.
Cooper, C. Dewey
Copeland, Gary E.
Coplan, Michael A.
Corliss, Charles H.
Corongiu, Giorgina
Cosby, Philip C.
Cowan, Robert D.
Cowley, Charles R.
Craig, Norman C.
Crampton, Stuart B.
Crawford Jr., Bryce
Cromer, Christopher L.
Csavinszky, Peter
Cummings, Frank E.
Cunningham, Augustine J.
Curtis, Lorenzo J.
Das, Tara P.
Davidson, Ernest R.
Davis, David S.
Davis, Robert W.
De La Vega, Jose R.
Deakyne, Carol A.
Decius, John C.
DeFotis, Gary C.
Dehmer, Joseph L.
Del Bene, Janet E.
Delgado-Barrio, Gerardo
DeYoung, Russell J.
DiMauro, Louis F.
Dixon, David A.
Donahue, Joey B.

Donnally, Bailey
Doolen, Gary D.
Dow, John D.
Dows, David A.
Drake, Gordon W.
Dreizler, Reiner M.
Dressler, Kurt
Drobny, Vladimir F.
Drullinger, Robert E.
Druzbick, John
Duke, Charles B.
Dunlap, Brett I.
Dunning Jr., Thomas H.
Eberhardt, William H.
Eck, Thomas G.
Eckhardt, Craig J.
Ederer, David L.
Elander, Nils O.
Ellis, David G.
Ellison, Frank O.
Emery, Guy T.
Engelman Jr., Rolf
England, Walter B.
Erickson, Glen W.
Ermler, Walter C.
Estreicher, Stefan K.
Exton, Reginald J.
Eyler, Edward E.
Fales, Norman J.
Fano, Ugo
Farley, John W.
Fayer, Michael D.
Fehsenfeld, Frederick
Feld, Michael S.
Field, Robert W.
Findley, Gary L.
Fink, William H.
Finn, Edward J.
Fischer, C. R.
Fischer, Charlotte F.
Flannery, Martin R.
Flygare, Willis H.
Foley, Charles K.
Foner, Samuel N.
Fontana, Peter R.
Ford, A. L.
Fowler, Bruce W.
Fox, Kenneth
Franck, Carl P.
Francke, Ricardo E.
Freed, Charles
Freed, Karl F.
Freeman, Robert D.
Freund, Hans J.
Friedrich, Harald S.
Fry, Edward S.
Frye, Daniel D.
Fuhr, Jeffrey R.
Gallup, Gordon A.
Ganas, Perry S.
Garrett, William R.
Garscadden, Alan
Garstang, Roy H.
Gatland, Ian R.
Geller, M.
German, Kenneth R.
Gianturco, Franco A.
Gibbons, Patrick C.
Giese, John P.

Gilbert, Sarah L.
Gilles, Paul W.
Gimarc, Benjamin M.
Ginell, Robert
Ginter, Marshall L.
Goddard III, William A.
Golden, Sidney
Goldflam, Rudolf
Goldwire Jr., Henry C.
Gole, James L.
Good Jr., Roland H.
Goodman, Leonard S.
Goorvitch, David
Gordon, Mark S.
Gordon, Roy G.
Graves, John L.
Graybeal, Jack D.
Green, Sheldon
Greene, Chris H.
Greenlees, G. W.
Greytak, Thomas J.
Griffin, Donald C.
Groeneveld, Karl-Ontjese E.
Guberman, Steven L.
Gutzwiller, Martin C.
Gwinn, William D.
Hagstrom, Stanley
Hameka, Hendrik F.
Hamermesh, Morton
Hamilton, Peter A.
Hanna, Stanley S.
Hanson, Harold P.
Hardcastle, Donald L.
Hardis, Jonathan E.
Harmony, Marlin D.
Harrell II, Evans M.
Harriman, John E.
Harris, Frank E.
Harris, Stephen E.
Hartmann, Sven R.
Hastie, John W.
Hauser, Ulrich A.
Hay, Philip J.
Hayes, Edward F.
Hayhurst, Thomas L.
Hazi, Andrew U.
Heaton, Marie M.
Heckmann, Paul H.
Hedberg, Kenneth W.
Hefferlin, Ray A.
Heil, Timothy G.
Heller, Eric J.
Henderson, George A.
Henley, Ernest M.
Henneberger, Walter C.
Herbst, Jan F.
Herman, Frank
Herman, Robert
Herman, Roger M.
Herrick, David R.
Herring, Conyers
Herschbach, Dudley R.
Hess Jr., Doren W.
Hilborn, Robert C.
Hill, Robert N.
Hinds, Edward A.
Hipps, Kerry W.
Hirschfelder, Joseph O.
Hobbs, Robert

1.1 General Structure and Properties of Atoms and Molecules

Holmes, John R.
Holoien, Erling
Hopper, Darrel G.
Hsu, Donald K.
Huang, Keh-Ning
Hudson, David F.
Huebner, Russell H.
Huebner, Walter F.
Huestis, David L.
Huffaker, James N.
Hughes, William M.
Huhnermann, Harry
Hurst, Robert P.
Hutchinson Jr., Clyde A.
Innes, Frederick R.
Innes, K. K.
Intemann, Robert L.
Ishida, Takanobu
Islam, Muhammad A.
Itano, Wayne M.
Ivey, Robert C.
Iwinski, Zbigniew R.
Jacob, Elizabeth J.
Jaduszliwer, Bernardo
Jaffe, Richard L.
Jen, Chih K.
Jeon, Yoon H.
Jester Jr., William A.
Jette, A. N.
Johnson, Carol
Johnson, Lee K.
Johnson, Philip M.
Johnson, Roy R.
Johnson, Walter R.
Johnston Jr., Milton D.
Jones, Douglas W.
Jones, M. T.
Jones, Steven E.
Jordan, Kenneth D.
Joshi, Bhairav D.
Judge, Darrell L.
Junker, Bobby R.
Kahn, Luis
Kaldor, Uzi
Kanter, Elliot P.
Karo, Arnold M.
Karplus, Martin
Kaufman, Joyce J.
Kellman, Michael E.
Kelly, Hugh P.
Kelsey, Edward J.
Kemple, Marvin D.
Kenney, John W.
Kern, C. W.
Kestner, Neil R.
Kim, Longhuan
King, William T.
Kinsey, James L.
Kirby, Kate P
Kirtman, Bernard
Klose, Jules Z.
Knochenmuss, Richard D.
Kobayashi, Hisao
Koel, Bruce E.
Kohl, John L.
Komornicki, Andrew
Konowalow, Daniel D.
Kopelman, Raoul
Kosman, Warren M.

Koszykowski, Michael L.
Kramer, Peter B.
Krause, Jeffrey L.
Krause, Lucjan
Krause, Manfred O.
Krauss, Morris
Krishna, N. R.
Krohn, Burton J.
Kromhout, Robert A.
Kubis, Joseph J.
Kurucz, Robert L.
Kushick, Joseph N.
Kvale, Thomas J.
Kwiram, Alvin L.
Kwok, Thomas Y.
Laane, Jaan
Lamb Jr., Willis E.
Lambert, David L.
Lang Jr., John C.
Lang, Neil
Langhoff, Stephen R.
Langhoff, Peter W.
Lapatovich, Walter P.
Larson, Daniel J.
Larter, Raima
Laskowski, Bernard C.
Lassettre, Edwin N.
Lawler, James E.
Layzer, David
Learner, Richard C.
Lee, Chi-Hsiang
Lee, Edward K.
Lee, Ja H.
Leroi, George E.
Lesk, Arthur M.
Lester Jr., William A.
Letamendia, Louis
Leventhal, Jacob J.
Levin, Frank S.
Lewis, David A.
Lide Jr., David R.
Lieb, Elliot H.
Liebman, Joel F.
Light, John C.
Lin, Chii-Dong
Linder, Bruno
Lipscomb, William N.
Lipsky, Lester
Lisy, James M.
Livingston, A.E. Gene
Lodhi, Sattar K.
Lohr Jr., Lawrence L.
Lovas, Frank J.
Lovoi, Paul
Lowe, John P.
Lowitz, David A.
Lubell, Michael S.
Lucatorto, Thomas B.
Ludena, Eduardo V.
Luken, William L
Lynds, Lahmer
Macek, Joseph H.
Maggiora, Gerald M.
Maier II, William B.
Mallow, Jeffry V.
Mansikka, Kauko A.
Manson, Joseph R.
Manson, Steven T.
Manzanares, Elizabeth R.

Maricq, M. Matti
Martin, William C.
Mason, Arthur A.
Mason, Edward A.
Matcha, Robert L.
Mathews, C. Weldon
Matsen, F. Albert
Matthias, Eckart
Mattson, Timothy G.
Mazur, Jacob
Mazur, Ursula
McClure, Donald S.
McColm, Douglas W.
McCullough Jr., E. A.
McDermott, Mark N.
McDowell, Harding K.
McGlynn, Sean P.
McGregor Jr., Wheeler K.
McGuire, James H.
McGuire, Michael D.
Mead, C. Alden
Mead, Roy D.
Melveger, Alvin J.
Menocal, Serafin G.
Merrifield, Richard E.
Messmer, Richard P.
Meyer, Carl B.
Micha, David A.
Michels, H. Harvey
Miles, Richard B.
Miller, Kenneth J.
Miller, Steven M.
Mills Jr., Allen P.
Mires, Raymond W.
Mitchner, M.
Mizushima, Masataka
Mohr, Peter J.
Monce, Michael N.
Monkhorst, Hendrik J.
Moore, C. Fred
Morrison, Michael A.
Moscatelli, Frank A.
Moscowitz, Albert J.
Mowat, J. R.
Mower, Lyman
Murad, Edmond
Murday, James S.
Nazaroff, George V.
Nesbitt, David J.
Neumann, David B.
Newman, James H.
Newton, Marshall D.
Nicholls, Ralph W.
Nitz, David E.
O, Chun-Sing
O'Neil, Stephen V.
Ohrn, Yngve
Okabe, Hideo
Oleksik, John J.
Olsgaard, David A.
Orient, Otto J.
Oza, Dipak H.
Pack, Russell T
Pan, Y.
Pantinakis, Apostolos
Park, Changhwan
Parker, Paul M.
Parr, Christopher A.
Parr, Robert G.

Parsons, Michael L.
Partridge III, Harry
Patty, Richard R.
Payne, Philip W.
Paysen, Robert A.
Pechukas, Philip
Peng, Jin-Sheng
Pepper, David M.
Percus, Jerome K.
Perkowitz, Sidney
Petersson, George A.
Peticolas, Warner L.
Pichanick, Francis M.
Piebold, Gerald J.
Pitzer, Kenneth S.
Pitzer, Russell M.
Politzer, Peter
Pople, John A.
Porter, Richard N.
Poshusta, Ronald D.
Pradhan, Anil K.
Pratt, David W.
Pratt, Richard H.
Present, Richard D.
Prior, Michael H.
Pritchard, David E.
Pritchard, Richard H.
Pyykko, V. P.
Ragle, John L.
Rajnak, Katheryn
Rapp, Donald
Rast Jr., Howard E.
Ratner, Mark A.
Reader, Joseph
Redding, Rogers W.
Reddy, Satti P.
Redi, Martha H.
Redi, Olav
Ree, Francis H.
Reed, Kennedy
Reid, Charles E.
Reid Jr., Roderick V.
Reidy, James J.
Rescigno, Thomas N.
Rhim, Won-Kyu
Richardson, James W.
Ring, James W.
Rinker Jr., George A.
Risley, John S.
Robin, Melvin B.
Rogers, Max T.
Rollefson, R.
Rosen, Gerald
Rosenberg, Richard A.
Rosner, S. David
Rotenberg, Manuel
Rothenberg, Joshua E.
Rothman, Laurence S.
Rothstein, Stuart M.
Roy, Denis
Ruedenberg, Klaus
Rumble, John R.
Ruskai, Mary B.
Russek, Arnold
Russell, Geoffrey A.
Ryan, Dave
Sabelli, Nora H
Sabin, John R.
Sage, Martin L.

142

Saha, Haripada
Salk, Sung-Ho S.
Sampson, Douglas H.
Sanders Jr., Frank C.
Sandlin, Glenn D.
Sandner, Wolfgang
Sando, Kenneth M.
Sansonetti, Craig J.
Sarachman, Theodore N.
Saxon, Roberta P.
Schaefer III, Henry F.
Schappert, Gottfried T.
Schectman, Richard M.
Schmalz, Thomas G.
Schmiedekamp, Ann M.
Schneider, Barry I.
Schnepp, Otto
Schowengerdt, Franklin C.
Schuessler, Hans A.
Scofield, James H.
Secrest, Don
Segal, Gerald A.
Seybold, Paul G.
Shalimoff, George V.
Sharpton, Francis A.
Shavitt, Isaiah
Shimkaveg, Gregory M.
Shirts, Randall B.
Shugart, Howard A.
Shyn, Tong-Wha
Sickafus, Ed N.
Silfvast, William T.
Silver, David M.
Silvers, Stuart J.
Silverstone, Harris J.
Simon, Barry
Simpson, W. T.
Sims, James S.
Sinanoglu, Oktay
Siska, Peter E.
Skutlartz, Alexander E.
Smalley, R. E.
Smio, Tani
Smith, Joseph A.
Smith, Peter L.
Snyder, Ralph B.
St. John, Robert M.
St. John III, Willard M.
Stallcop, James R.
Stanton, Richard E.
Stauffer, Allan D.
Stechel, Ellen B.
Steel, Duncan G.
Steele, William A.
Sternheimer, Rudolph M.
Stevens, Walter J.
Stockdale, John A.
Stogryn, Daniel E.
Stoicheff, Boris P.
Stone, Philip M.
Stoner Jr., John O.
Story, Troy L.
Stratt, Richard M.
Strickler, Stewart J.
Stumpf, Bernhard J.
Stwalley, William C.
Suck, Sung Ho
Sugar, Jack
Synek, M.

Szalewicz, Krzysztof
Talman, James D.
Tam, Andrew C.
Tang, Fu-Ching
Tang, K. T.
Tarter, C. Bruce
Tatewaki, Hiroshi
Tayal, Swaraj S.
Taylor, Lyle H.
Temkin, Aaron
Thomas, Bruce R.
Thompson Jr., H. Bradford
Thomsen, John S.
Tipping, Richard H.
Topp, Michael R.
Trabert, Elmar C.
Treacy, Peter B.
True, Nancy S.
Truhlar, Donald G.
Tseng, Hsiang-Kuang
Tseng, Tien-Jiunn
Tuan, Debbie F.
Ufford, C. W.
Van Haeringen, Willem
Vander Sluis, Kenneth L.
Vashishta, Priya D.
Vaughan, Stephen O.
Venanzi, Carol A.
Venanzi, Thomas J.
Victor, George A.
Viehland, Larry A.
Viggiano, Albert A.
Volk, Charles H.
von Turkovich, Branimir F.
Voreades, Demetrios
Wachter, Joseph R.
Wadt, Willard R.
Wahrhaftig, Austin L.
Walters, G. K.
Wang, Chia P.
Ward, John F.
Weatherford, Charles A.
Weigold, Erich
Weiner, Brian L.
Weinhold, Frank A.
Weinstein, Harel
Weiss, Andrew W.
Weissbluth, Mitchel
Wells, James W.
Westhaus, Paul A.
Wharton, Lennard
Wiese, Wolfgang L.
Wiff, Donald R.
Wilets, Lawrence
Williams, Graheme J.
Williams, James F.
Williams, John R.
Wilson, John W.
Wineland, David J.
Winicur, Daniel H.
Winkler, Peter
Winter, Nicholas W.
Wise, John H.
Wolken, George
Wood, John H.
Woodruff, Susan B.
Woods, R. Claude
Worden Jr., Earl E.
Wright, Lawrence A.

Yabuzaki, Tsutomu
Yamanaka, Masanobu
Yao, Shang J.
Yarkony, David R.
Yates, Albert C.
Yates, John H.
Yoshimine, Megumu
Yuan, Jian-Min
Zahniser, Mark S.
Zalubas, Romuald
Zapata, Luis E.
Zare, Richard N.
Zemke, Warren T.
Zink, J. W.
Zorn, Jens C.
zuPutlitz, Gisbert

1.2 Interactions of Rydberg States

Baer, Tomas
Bayfield, James E.
Becker, Richard L.
Beiting III, Edward J.
Berry, Scott D.
Bhattacharya, Samir K.
Brown, Charles M.
Burgdoerfer, Joachim E
Chang, Edward S.
Chen, C. H.
Chung, Sunggi
Clark, Charles W.
Cooke, William E.
Crane, John K.
Dagata, John A.
Davidson, Ernest R.
DeBeer, David P.
Dehmer, Joseph L.
Delos, John B.
DiMauro, Louis F.
Dixit, Shamasundar N.
Drachman, Richard J.
Drake, Charles W.
Dressler, Kurt
Ducas, Theodore W.
Dunning, F. Barry
Elander, Nils O.
Ellis, David G.
Eyler, Edward E.
Farley, John W.
Findley, Gary L.
Flannery, Martin R.
Franzen, Wolfgang
Freund, Robert S.
Friedrich, Harald S.
Gallagher, Thomas F.
Garscadden, Alan
Gelbwachs, Jerry A.
Gentile, Thomas R.
Ginter, Marshall L.
Goss, Larry P.
Greene, Chris H.
Hahn, Yukap
Herrick, David R.
Hill, Robert M.
Hinds, Edward A.
Holoien, Erling
Hudson, David F.
Huwel, Lutz
Jaffe, Hans H.
Jeys, Thomas H.
Johnson, Philip M.
Kelsey, Edward J.
Kern, C. W.
Kleppner, Daniel
Klots, Cornelius E.
Kocher, Carl A.
Konowalow, Daniel D.
Korevaar, Eric J.
Kosman, Warren M.
Lahiri, Jayanti
Leach, Sydney
Lee, Edward T.
Lindsay, Mark D.
Lipsky, Lester
MacAdam, Keith B.
Mansky II, Edmund J.
Matsuzawa, Michio

Matthias, Eckart
McGlynn, Sean P.
McLaughlin, Daniel J.
McMillian, Gary B.
Merts, A. L.
Meyer, Fred W.
Miller, John C.
Miller, Kenneth J.
Morgan, Thomas J.
Omidvar, Kazem
Patton, Robert J.
Pipkin, Francis M.
Rau, A. R.
Robin, Melvin B.
Rossi, Angelo R.
Ruff, George A.
Sandner, Wolfgang
Sansonetti, Craig J.
Schiavone, James A.
Schuessler, Hans A.
Silverman, Mark P.
Smith, David B.
Stebbings, Ronald F.
Stoicheff, Boris P.
Stoneman, Robert C.
Stroud Jr., Carlos R.
Tatewaki, Hiroshi
Temkin, Aaron
Thirumalai, Devarajan V.
Treacy, Peter B.
Wang, Harry T.
Wessel, John E.
West, John B.
White, Michael G.
Winkler, Peter
Worden Jr., Earl E.
Wynne, James J.
Zollars, Byron G.

Abrahamson, Adolf A.
Ahn, Myong-Ku
Amme, Robert C.
Andres, Ronald P.
Armentrout, Peter B.
Averill, Frank W.
Beckel, Charles L.
Bendler, John T.
Bergeman, Thomas H.
Berney, Charles V.
Berry, R. Steven
Berry, Scott D.
Bigio, Irving J.
Black, Graham
Black, Truman D.
Bobin, Jean Louis
Bohn, Robert K.
Bose, Subir K.
Bottger, Gary L.
Bowers, Michael T.
Broadhurst, Martin G.
Brooks, Charles L.
Brown, Nancy J.
Budick, Burton
Cade, Paul E.
Callaway, Joseph
Canter, Karl
Castleman Jr., A. W.
Chakraborti, Parimal K.
Chen, C. H.
Chu, Ben
Chu, Steven
Cohen, James S.
Companion, Audrey L.
Curnutte Jr., Basil
De Rijk, Waldemar G.
Deakyne, Carol A.
Demir, Oktay
DeSerio, Robert
Deutch, John M.
Deutsch, Peter W.
Diana, Leonard M.
DiMauro, Louis F.
Dow, John D.
Dows, David A.
Drachman, Richard J.
Drake, Gordon W.
Dunlap, Brett I.
Dyke, Thomas R.
Eckhardt, Craig J.
Eisenstadt, Maurice
Emery, Guy T.
Endres, Paul F.
Estreicher, Stefan K.
Ewing, George E.
Fairbank Jr., William M.
Fox, Kenneth
Frankel, Robert D.
Freed, Karl F.
Fried, Zoltan
Friedman, Joel M.
Friedrich, Harald S.
Froelich, David V.
Gidley, David W.
Gislason, Eric A.
Gladisch, Michael W.
Gouterman, Martin
Gray, Tom J.
Grover, James R.

Ham, Mooyoung
Hefferlin, Ray A.
Henchman, Michael J.
Herschbach, Dudley R.
Holoien, Erling
Holton, Gerald J.
Hsu, Shaw L.
Huang, Keh-Ning
Hughes, Vernon W.
Jaffe, Richard L.
Jen, Chih K.
Johnson, Lawrence W.
Jones, Steven E.
Kaldor, Andrew P.
Kanter, Elliot P.
Kaufman, Joyce J.
Kenney, John W.
Kern, C. W.
Kestner, Neil R.
Kirchhoff, William H.
Klots, Cornelius E.
Knauss, Donald C.
Koel, Bruce E.
Kohler, Bryan E.
Konowalow, Daniel D.
Kopelman, Raoul
Krenos, John R.
Kushick, Joseph N.
Kwiram, Alvin L.
Ladanyi, Branka M.
Lang Jr., John C.
Lee, Paul L.
Lee, Sanboh
Legg, James C.
Leon, Melvin
Lepage, G. P.
Lim, Teck-Kah
Long Jr., Edward R.
Longworth, James W.
Lovas, Frank J.
Ludena, Eduardo V.
Maggiora, Gerald M.
Mallow, Jeffry V.
Marron, Michael T.
Martin, Richard L.
McCammon, James A.
Melveger, Alvin J.
Miller, David R.
Miller, Kenneth J.
Miller, Roger E.
Mills Jr., Allen P.
Mohr, Peter J.
Monkhorst, Hendrik J.
Murray, Paul T.
Naumann, Robert A.
Nayfeh, Munir H.
Northrup, Scott H.
O'Reilly, James M.
Oleksik, John J.
Parks, Eric K.
Patton, Robert J.
Paul, Derek AL
Payne, Philip W.
Pecora, Robert
Pellin, Michael J.
Pendleton, Hugh N.
Petersson, George A.
Pfeiffer, Gary V.
Phillips, Donald H.

Pierce, Thomas M.
Plummer, Patricia LM
Ramler, Warren J.
Ratner, Mark A.
Redi, Martha H.
Reeves, Richard L.
Reidy, James J.
Reiss, Howard
Repko, Wayne W.
Rich, Arthur
Rinker Jr., George A.
Russell, Geoffrey A.
Russell, James E.
Rutherford, John A.
Sage, Martin L.
Schrader, David M.
Seki, Ryoichi
Seliger, Howard H.
Silverman, Mark P.
Skowronek, Maurice
Soltis, Paul J.
Souder, Paul A.
Stevens, Walter J.
Szalewicz, Krzysztof
Tatewaki, Hiroshi
Thirumalai, Devarajan V.
Topp, Michael R.
Ullman, Robert
Vala, Martin
Van Siclen, Clinton D.
Vashishta, Priya D.
Viggiano, Albert A.
Wachter, Joseph R.
Wallace, Richard W.
Waynant, Ronald W.
Weber, Thomas A.
Weiner, Jerome H.
Weinstein, Harel
Weissbluth, Mitchel
Weltner Jr., William
Wexler, Sol
White, Michael G.
Wilets, Lawrence
Wilson III, Charles W.
Yip, Sidney
Zimm, Bruno H.
Zitzewitz, Paul W.
Zollweg, Robert J.
Zumbulyadis, Nicholas
zuPutlitz, Gisbert

1.4 Fundamental Properties of Atoms and Molecules

Armstrong Jr., Lloyd
Au, C. K.
Bennett, Donald L.
Bernheim, Robert A
Bollinger, John J.
Brown, Lowell S.
Bucksbaum, Philip H.
Carmichael, Howard I
Chamberlin, Edwin P.
Chen, C. H.
Cohen, E. R.
Commins, E. D.
Conti, Ralph S.
Couillaud, Bernard J.
Coveney, Peter V.
Curnutte Jr., Basil
DeSerio, Robert
Deslattes, Richard D.
DeVoe, Ralph G.
Donahue, Joey B.
Drake, Gordon W.
Egan, Patrick O.
Ellis, David G.
Erickson, Glen W.
Eyler, Edward E.
Fairbank Jr., William M.
Finch, Eric C.
Fornari, Luigi S.
Fortson, E. Norval
Gabrielse, Gerald
Gaily, T. D.
Gidley, David W.
Gilbert, Sarah L.
Gladisch, Michael W.
Goldwire Jr., Henry C.
Gould, Harvey A.
Greenberg, Jack S.
Grotch, Howard
Hall, John L.
Hameka, Hendrik F.
Hansteen, Johannes M.
Harvey, Kenneth C.
Heckel, Blayne R.
Heckmann, Paul H.
Henley, Ernest M.
Hinds, Edward A.
Holoien, Erling
Holt, Richard A.
Huang, Keh-Ning
Hughes, Vernon W.
Itano, Wayne M.
Jacobs, Verne L.
Johnson, Walter R.
Kelsey, Edward J.
Kessler, Karl G.
Kimble, Harry J.
Kleppner, Daniel
Klose, Jules Z.
Knize, Randall J.
Koch, H. W.
Lamb Jr., Willis E.
Layer, Howard P.
Lee, Siu-Au
Leventhal, Marvin
Lewis, Lindon L.
Lewis, Robert R.
Lieber, Michael
Livingston, A.E. Gene
Lubell, Michael S.

MacArthur, Duncan W.
Majumder, Protik K.
Mansikka, Kauko A.
Marrmar, Earl S.
Mellen, Walter R.
Mills Jr., Allen P.
Mohr, Peter J.
Mower, Lyman
Murnick, Daniel E.
Myers, Edmund G.
Nachman, Paul
Oleksik, John J.
Patel, C. Kumar K.
Paul, Derek AL
Pendleton, Hugh N.
Phillips, William D.
Pichanick, Francis M.
Pipkin, Francis M.
Pratt, David W.
Raab, Frederick J.
Ramsey, Norman F.
Reidy, James J.
Repko, Wayne W.
Rich, Arthur
Rinker Jr., George A.
Robinson, Hugh G.
Robiscoe, R. T.
Rosner, S. David
Ruskai, Mary B.
Sanders Jr., Frank C.
Schuessler, Hans A.
Seely, John F.
Sellin, Ivan A.
Shyn, Tong-Wha
Silver, Joshua D.
Stroud Jr., Carlos R.
Sucher, Joseph
Taylor, Barry N.
Teague, Michael R.
Trabert, Elmar C.
Van Baak, David A.
VanDyck Jr., Robert S.
Werner, Samuel A.
Wieman, Carl E.
Wilets, Lawrence
Williams, William L.
Willis, Paul A.
Wing, William H.
Wood II, Obert R.
Zajonc, Arthur G.
Zitzewitz, Paul W.
zuPutlitz, Gisbert

Atomic and Molecular Collisions 2.1

Abi-Ghanem, Georges V.
Abraham, George
Aldridge III, Jack P.
Alexander, Millard H.
Allan, Michael
Alper, Joseph S.
Alton, Gerald D.
Amme, Robert C.
Andersen, Nils O.
Anderson, Louis W.
Anderson, Roger W.
Anderson, Thomas G.
Andres, Ronald P.
Andresen, Bjarne B.
Annis, Brian K.
Atkinson, J. B.
Aubrey, Bertrand B.
Auerbach, Daniel J.
Babcock, Lucia M.
Bae, Young K.
Baer, Thomas M.
Baglin, Frank G.
Baird, James C.
Band, Yehuda B.
Bardsley, J. N.
Barnett, Charles F.
Barnett, Clarence F.
Basbas, George J.
Bates Jr., Richard D.
Baughcum, Steven L.
Bay, Zoltan L.
Bayfield, James E.
Bearman, Gregory
Becker, Richard L.
Behringer, Robert E.
Benesch, William M.
Bentley, John
Berkner, Klaus H.
Berne, Bruce J.
Bernius, Mark T.
Bernstein, E. M.
Bernstein, Richard B.
Berry, H. G.
Berry, Scott D.
Beyerinck, Herman C.
Bhalla, Chander P.
Bhasavanich, Daun
Bhattacharya, Ashok K.
Bichsel, Hans
Bieniek, Ronald J.
Biondi, Manfred A.
Blais, Normand C.
Bobin, Jean Louis
Boring, John W.
Borst, Walter L.
Bottrell, Gerald J.
Bowman, Joel M.
Brandenberger, John R.
Breckenridge, W. H.
Brenn, Rudiger
Brennan, James G.
Brooks, Philip R.
Brower, Michael C.
Browne, James C.
Brunner, Timothy A.
Burgdoerfer, Joachim E
Burnett, Keith
Butler, Scott E.
Cacak, Robert K.

Cattolica, Robert J.
Chackerian Jr., Charles
Chakraborti, Parimal K.
Chamberlin, Edwin P.
Champion, R. L.
Chan, F. T.
Chander, Jagdish
Chang, Cheng-hui
Chapman, Sally
Chaturvedi, Ram P.
Chen, Che-Jen
Chen, Hao-lin
Chen, Joseph C.
Chiu, Lue-Yung C.
Chu, Shih-I
Chutjian, Ara
Cipolla, Sam J.
Clendenin, James E.
Cocke, C. L.
Coggiola, Michael J.
Cohen, James S.
Collins, Lee A.
Cooney, Patrick J.
Cordaro, Richard B.
Coveney, Peter V.
Crampton, Stuart B.
Crandall, David H.
Crane, John K.
Crawford, Oakley H.
Crim, F. F.
Cross, Jon B.
Cross Jr., R. J.
Cunningham, David L.
Current, David H.
Curry, Bill P.
Curtiss, Charles F.
Dagdigian, Paul J.
Dahler, John S.
Dalgarno, Alexander
Darewych, J. W.
Dasch, Cameron J.
Davis, Christopher C.
Dawson, Horace R.
DeBeer, David P.
Decius, John C.
Delgado-Barrio, Gerardo
Delos, John B.
DePristo, Andrew E.
DeSerio, Robert
DeVries, Paul L.
DeYoung, Russell J.
Diestler, D. J.
Donnally, Bailey
Doughty, Ben M.
Doverspike, Lynn D.
Dreizler, Reiner M.
Druger, Stephen D.
Druzbick, John
Dube, Louis J.
Duff, James W.
Duncan, M. M.
Dunn, Gordon H.
Dunning, F. Barry
Ederer, David L.
Edwards, Alan K.
Elander, Nils O.
Elston, Stuart B.
Eno, Larry
Ernie, Douglas W.

Fano, Ugo
Fehsenfeld, Frederick
Fenn, John B.
Finzi, Jack
Firestone, Richard F.
Fisher, H. Leonard
Fisk, George A.
Fite, Wade L.
Flannery, Martin R.
Fleischmann, Hans H.
Flygare, Willis H.
Ford, A. L.
Fornari, Luigi S.
Fou, Cheng-Ming
Francke, Ricardo E.
Franco, Victor
Franz, Frank A.
Freed, Charles
Friedrich, Harald S.
Fritsch, Wolfgang
Fulton, Robert L.
Furst, Mitchell L.
Gaily, T. D.
Gaiser, James E.
Gallagher, Alan C.
Gallagher, Jean W.
Gallagher, Thomas F.
Gallup, Gordon A.
Garcia Jr., Jose D.
Gay, Timothy J.
Gelfand, Jack J.
Geltman, Sydney
Gerjuoy, Edward
Gianturco, Franco A.
Gien, Tran T.
Giese, Clayton F.
Giese, John P.
Gilbody, Henry B.
Gillen, Keith T.
Gillespie, George H.
Gislason, Eric A.
Glauber, Roy J.
Goldberger, Arthur L.
Golde, Michael F.
Goldflam, Rudolf
Goldstein, John C.
Golub, John E.
Gordon, Robert J.
Gordon, Roy G.
Gould, Harvey A.
Graham, William G.
Gray, Tom J.
Green, Sheldon
Greenberg, Jack S.
Greene, Edward F.
Greytak, Thomas J.
Groeneveld, Karl-Ontjese E.
Gundersen, Roy
Hagmann, Siegbert J.
Hahn, Yukap
Hahne, Gerhard E.
Hall, James M.
Hall, Richard I.
Halpern, Alvin M.
Hammond, Gordon L.
Hance, Robert L.
Hanna, Stanley S.
Hansteen, Johannes M.
Harris, Harold H.

Harris, Stephen E.
Hartmann, Sven R.
Harvey, Nancy M.
Hase, William L.
Hauser, Ulrich A.
Havener, Charles C.
Havey, Mark D.
Hayes, Edward F.
Head, Charles E.
Heckmann, Paul H.
Heil, Timothy G.
Heller, Eric J.
Herbst, Eric
Herzenberg, Arvid
Hickman, Albert P.
Hill, Robert M.
Hinchen, John J.
Hird, Brian
Hobbs, Robert
Holoien, Erling
Holton, Gerald J.
Hopper, Darrel G.
Houston, Paul L.
Husmann, Otto K.
Huwel, Lutz
Hyman, Howard A.
Ioup, George E.
Ioup, Juliette W.
Jaecks, Duane H.
Jakas, Mario M.
Jamison, Keith A.
Jester Jr., William A.
Johnsen, Rainer
Johnson, Brant M.
Johnson, Carol
Johnson, Edward A.
Johnson, Lee K.
Jones, E. G.
Jones, Patrick L.
Jones, Steven E.
Joyce, J. M.
Julienne, Paul S.
Junker, Bobby R.
Kamaratos, E.
Kang, Ik-Ju
Kanter, Elliot P.
Katayama, Daniel H.
Katayama, Ichiro
Katsanos, Anastasios A.
Katsonis, Konstantinos
Kelley, J. Daniel
Kelley, Michael H.
Kenefick, Robert A.
Kessel, Quentin C.
Kielkopf, John F.
Kim, H. J.
Kinsey, James L.
Kleiber, Paul D.
Knize, Randall J.
Kobayashi, Hisao
Kocher, Carl A.
Konowalow, Daniel D.
Kopelman, Raoul
Kostroun, Vaclav O.
Koszykowski, Michael L.
Kouri, Donald J.
Kraus, Joseph S.
Krause, Lucjan
Krotkov, Robert V.

2.1 Atomic and Molecular Collisions

Kvale, Thomas J.
Kwan, Ching-Kwan CK
Land, David J.
Lane, Neal F.
Langsam, Yedidyah
Larter, Raima
Latta, Bryan M.
Lee, Edward T.
Legg, James C.
Leone, Stephen R.
Lester Jr., William A.
Lieber, Michael
Light, John C.
Lim, Teck-Kah
Lin, Chun C.
Lineberger, William C.
Lockwood, Grant J.
Lodhi, Sattar K.
Lynds, Lahmer
MacAdam, Keith B.
Macek, Joseph H.
Madison, Don H.
Magnuson, Gustav D.
Mahadevan, P.
Maier II, William B.
Mangelson, Nolan F.
Manning, Irwin
Manson, Steven T.
Maricq, M. Matti
Marrus, Richard
Martin, Richard L.
Martin, Richard M.
Mason, Edward A.
Mathur, Bhagwan P.
Matsuzawa, Michio
McAfee Jr., Kenneth B.
McArthur, David A.
McCann, Kevin J.
McCurdy, C. W.
McDaniel, Floyd D.
McDonald, J. D.
McGuire, James H.
McIntyre, Adelbert
McLaughlin, Daniel J.
McMillian, Gary B.
Menendez, Manuel G.
Merzbacher, Eugen
Meyer, Fred W.
Meyerhof, Walter E.
Micha, David A.
Michaels, Gordon E.
Michels, H. Harvey
Mickish, Roger A.
Miers, Richard E.
Mies, Frederick H.
Miles, Richard B.
Miller, Glenn H.
Miller, James A.
Miller, Phillip D.
Miller, Roger E.
Miller, Steven M.
Miller, Walter B.
Miller, William H.
Min, Kwang S.
Mitchner, M.
Molitoris, John D.
Monce, Michael N.
Moore, C. Fred
Moore-Head, M. E.

Moran, Thomas F.
Morgan, Thomas J.
Morgan, William L.
Morrison, Michael A.
Mossberg, Thomas W.
Mowat, J. R.
Mueller, Charles R.
Muschlitz Jr., Earle E.
Nelson, Robert N.
Neumann, Herschel
Newman, James H.
Neynaber, Roy H.
Nitz, David E.
Odom, Robert W.
Ohrn, Yngve
Olson, Ronald E.
Oppenheimer, Michael
Otter, Fred A.
Pack, Russell T
Park, Changhwan
Park, John T.
Parker, Gregory A.
Parks, Eric K.
Parmenter, Charles S.
Parson, John M.
Payne, Marvin G.
Peacher, Jerry L.
Pechukas, Philip
Peplinski, Daniel R.
Pepmiller, Philip L.
Pepper, George H.
Perel, Julius
Peterson, James R.
Peterson, Lennart R.
Peterson, Randolph S.
Phaneuf, Ronald A.
Poe, Robert T.
Pollack, Edward
Pomilla, Frank R.
Porter, Leonard E.
Prentice, John K.
Preses, Jack M.
Price, Jack L.
Pritchard, David E.
Pritt Jr., Alfred T.
Pyle, Robert V.
Quinn, Jarus W.
Rabinovitch, B. Seymour
Rabitz, Herschel
Raith, Burkhard
Rapp, Donald
Ratkowski, Anthony J.
Raynor, Susanne
Reading, John F.
Reddy, Satti P.
Redmon, Michael J.
Reed, Kennedy
Reeves, Tricia M.
Rhodes, Charles K.
Rich, Joseph W.
Richard, Patrick
Risley, John S.
Robinson, Edward J.
Roesler, Fred L.
Ronn, Avigdor M.
Rothe, Erhard W.
Rudd, M. E.
Ruff, George A.
Rush Jr., John E.

Russek, Arnold
Ryan, Stewart R.
Rybka, Theodore W.
Sabin, John R.
Sachs, Judith G.
Salk, Sung-Ho S.
Sando, Kenneth M.
Saxon, Roberta P.
Schatz, George C.
Schearer, Laird D.
Schectman, Richard M.
Schlachter, Alfred S.
Schlessinger, Leonard
Schmalz, Thomas G.
Schmieder, Robert W.
Schuessler, Hans A.
Secrest, Don
Seely, David G.
Sellin, Ivan A.
Semon, Mark D.
Senba, Masayoshi
Shakeshaft, Robin
Sheldon, John W.
Sheridan, John R.
Shin, Hyung K.
Shipsey, Edward J.
Shoemaker, Richard L.
Shorer, Philip
Silver, David M.
Silver, Joel A.
Simony, Paul R.
Skutlartz, Alexander E.
Slanger, Tom
Smith, David B.
Smith, Felix T.
Smith, Leslie N.
Smith, Winthrop W.
Snyder, Ralph B.
Spicer, Leonard D.
Spooner, David W.
Spruch, Larry
Stalder, Kenneth R.
Stanton, Alan C.
Starace, Anthony F.
Stearns, John W.
Stebbings, Ronald F.
Steel, Duncan G.
Steinberg, Martin
Stettler, John D.
Stockdale, John A.
Stockli, Martin P.
Stoneman, Robert C.
Stwalley, William C.
Suck, Sung Ho
Sutton, David G.
Tang, K. T.
Tang, Sheng Y.
Tanis, John A.
Tarter, C. Bruce
Tayal, Swaraj S.
Taylor, Howard S.
Taylor, John R.
Taylor, Ronald D.
Thomas, Brian K.
Thomas, Edward W.
Thomson, George M.
Thorson, Walter R.
Toburen, Larry H.
Toennies, Jan Peter

Trabert, Elmar C.
Trainor, Daniel W.
Treacy, Peter B.
Tserruya, Itshak
Tully, John C.
Turner, Jr., Charles E.
Uzer, Turgay
Van Zyl, Bert
VanBrunt, Richard J.
Vane, Charles R.
Verhaar, Boudewyn J.
Viehland, Larry A.
Vincent, Paul J.
Volk, Charles H.
Wadehra, Jogindra M.
Walker, Robert B.
Walters, G. K.
Weatherford, Charles A.
Weinberg, William H.
Weiner, Eugene R.
Weston Jr., Ralph E.
Wiesenfeld, Jay M.
Wiesenfeld, John
Williams, James F.
Williamson Jr., William
Wilson, Walter E.
Wing, William H.
Winicur, Daniel H.
Winkler, Peter
Winn, John S.
Winter, Thomas G.
Witriol, Norman M.
Wolga, George J.
Wong, Joseph K.
Wood, Robert M.
Woolf, Stanley
Wormhoudt, Joda C.
Wright, Lawrence A.
Wyatt, Robert E.
Yabuzaki, Tsutomu
Yates, Albert C.
Yencha, Andrew J.
Yenen, Orhan
Yip, Sidney
Zaidi, Haider R.
Zajonc, Arthur G.
Zander, Arlen R.
Zimmerman, I. H.
Zittel, Paul F.
Zollars, Byron G.

Ali, Mahamed A.
Allis, William P.
Altick, Philip L.
Altman III, Joseph C.
Andersen, Nils O.
Anderson, Richard J.
Aubrey, Bertrand B.
Bailey, Thomas L.
Bardsley, J. N.
Barreto, Ernesto
Bartell, Lawrence S.
Bartiromo, Rosario
Battleson, Kirk W.
Baum, Guenter G.
Becker, Dr. Kurt H.
Bederson, Benjamin
Beers, Brian L.
Bennett, Robert B.
Berk, Alexander
Berry, R. Steven
Best, Philip E.
Bhadra, Kalidas
Bhalla, Chander P.
Bhatia, Anand K.
Bichsel, Hans
Bien, Fritz
Biondi, Manfred A.
Blaha, Milan
Bonham, Russell A.
Borst, Walter L.
Brandt, Werner
Branscomb, Lewis M.
Brodie, Ivor
Brown Jr., Howard H.
Burch, David S.
Burke, Philip G.
Burns, Donal J.
Burns III, Jay
Burrow, Paul D.
Byron Jr., Frederick W.
Callaway, Joseph
Campbell, David H.
Campbell, John L.
Canter, Karl
Carnahan, Byron
Carragher, Beverly A.
Cartwright, David C.
Celotta, Robert J.
Champagne, Louis F.
Chang, Edward S.
Chang, Tu-nan
Chantry, Peter J.
Chen, C. L.
Chen, Hao-lin
Cho, Hyuck
Christophorou, Loucas G.
Chung, Sunggi
Chupka, William A.
Chutjian, Ara
Clark, Charles W.
Clark, Robert E.
Clegg, Thomas B.
Coleman, Paul G.
Collins, Lee A.
Cooper, William S.
Cosby, Philip C.
Coulter, Philip W.
Crandall, David H.
Crawford, Oakley H.

Cue, Nelson
Cunningham, Augustine J.
Dahler, John S.
Dang, Richard K.
Darewych, J. W.
Dassen, H. W.
Datla, Raju U.
Davis, H. Ted
Davis, Jack
Diana, Leonard M.
Dick, Charles E.
Dill, Dan
Dittner, Peter F.
Doering, J. P.
Donnally, Bailey
Doolen, Gary D.
Dowell, Jerry T.
Drachman, Richard J.
Dresser, Miles J.
Dube, Louis J.
Dunn, Gordon H.
Elander, Nils O.
Emken, Walter C.
Fadley, Charles S.
Fano, Ugo
Fessenden, Richard W.
Fink, Manfred
Firestone, Richard F.
Fischbeck, H. J.
Fite, Wade L.
Foner, Samuel N.
Fornari, Luigi S.
Foster, Gershom C.
Fou, Cheng-Ming
Fowler, Richard G.
Fox, Robert A.
Fradkin, David M.
Franco, Victor
Franzen, Wolfgang
Freeman, Gordon R.
Freund, Robert S.
Fry, Edward S.
Gallagher, Alan C.
Gallagher, Jean W.
Gallup, Gordon A.
Ganas, Perry S.
Gardner, Larry D.
Garrett, William R.
Garscadden, Alan
Gay, Timothy J.
Geltman, Sydney
Gerjuoy, Edward
Gianturco, Franco A.
Gibbons, Patrick C.
Gidley, David W.
Gien, Tran T.
Golden, David E.
Golden, Lawrence B.
Green, Alex E.
Greene, Arthur E.
Greene, Chris H.
Gregory, Donald C.
Griffin, Donald C.
Guberman, Steven L.
Hahn, Yukap
Hall, Richard I.
Halpern, Alvin M.
Haskell, Hugh B.
Hayashi, Shigeo

Hazi, Andrew U.
Hellwig, Helmut W.
Henry, Ronald J.
Hipps, Kerry W.
Holt, Helen K.
Hudson, George E.
Huebner, Russell H.
Hughes, Vernon W.
Hughes, William M.
Hummer, David G.
Humpherys, Kent C.
Hunter, Scott R.
Huo, Winifred M.
Husmann, Otto K.
Hyman, Howard A.
Intemann, Robert L.
Isozumi, Yasuhito
Ivey, Robert C.
Iwinski, Zbigniew R.
Jacob, Elizabeth J.
Jacobsen, E. H.
Jaduszliwer, Bernardo
Johnsen, Rainer
Johnson, Brant M.
Johnson, Edward A.
Johnston, David B.
Jona, F. P.
Jones, Dale R.
Jordan, Kenneth D.
Junker, Bobby R.
Kaldor, Uzi
Kang, Ik-Ju
Katsonis, Konstantinos
Kauppila, Walter E.
Kay, Richard B.
Kelley, Michael H.
Kelly, Roger
Kelsey, Edward J.
Kestner, Neil R.
Kim, Longhuan
Kirtman, Bernard
Kline, Laurence E.
Knochenmuss, Richard D.
Koel, Bruce E.
Kohl, John L.
Kupperman, Aron
Kuyatt, Chris E.
Kwan, Ching-Kwan
LaGattuta, Kenneth J.
Lake, Max L.
Lassettre, Edwin N.
Lieber, Michael
Lipsky, Lester
Lipsky, Sanford
Long Jr., Edward R.
Lubell, Michael S.
Lucchese, Robert R.
Macek, Joseph H.
Madey, J. M.
Madison, Don H.
Magee Jr., Norman H.
Mahadevan, P.
Maier II, William B.
Malik, F. Bary
Mann, Joseph B.
Mansky II, Edmund J.
Manson, Joseph R.
Manson, Steven T.
Mathur, Bhagwan P.

Mazur, Ursula
McClellan, Gene E.
McClelland, Jabez J.
McCurdy, C. W.
McGowan, J. W.
McGuire, James H.
McGuire, Stephen C.
McKoy, Vincent
McLaughlin, Daniel J.
Menegozzi, Lionel H.
Menendez, Manuel G.
Merts, A. L.
Michels, H. Harvey
Milde, Helmut I.
Miller, Thomas M.
Miller Jr., William R.
Mitchell, Robert R.
Mitchner, M.
Moeny, William M.
Monkhorst, Hendrik J.
Moore, John H.
Morrison, Michael A.
Mozumder, A.
Murday, James S.
Murphy, Randall E.
Murray, Frank
Muschlitz Jr., Earle E.
Nahar, Sultana N.
Nazaroff, George V.
Nesbet, Robert K.
Norcross, David W.
Nygaard, Kaare J.
O'Malley, Thomas F.
Oleksik, John J.
Omidvar, Kazem
Orient, Otto J.
Ott, William R.
Oza, Dipak H.
Pai, Robert Y.
Paikeday, Joseph M.
Paske, William C.
Pate, Bradford S.
Peng, Jin-Sheng
Peterson, Lennart R.
Phaneuf, Ronald A.
Phelps, Arthur V.
Pichanick, Francis M.
Pierce, Daniel T.
Pindzola, Michael S.
Poe, Robert T.
Pomilla, Frank R.
Porter, Leonard E.
Powell, Cedric J.
Pradhan, Anil K.
Pratt, Richard H.
Prentice, John K.
Quarles, C. A.
Raith, Wilhelm
Ratkowski, Anthony J.
Rau, A. R.
Reed, Kennedy
Register, David F.
Rescigno, Thomas N.
Rich, Joseph W.
Riley, Merle E.
Risley, John S.
Roberts, James R.
Robinson, Edward J.
Rogers, Wade T.

2.2 Electron and Positron Collisions

Rosenberg, Leonard
Rotenberg, Manuel
Roy, Denis
Rozsnyai, Balazs F.
Rubin, Kenneth
Rudd, M. E.
Rumble, John R.
Ryan, Stewart R.
Saha, Haripada
Sampson, Douglas H.
Sandner, Wolfgang
Scheps, Richard
Schiavone, James A.
Schlessinger, Leonard
Schneider, Barry I.
Schowengerdt, Franklin D.
Schrader, David M.
Scofield, James H.
Seltzer, Stephen M.
Shafroth, Stephen M.
Shelton, W. Neil
Shimizu, Sakae
Shively, John E.
Shyn, Tong-Wha
Simony, Paul R.
Simpson, Charles G.
Smio, Tani
Smith, Felix T.
Sparrow, Julian H.
Spence, David
Srivastava, Santosh K.
St. John, Robert M.
Starace, Anthony F.
Stauffer, Allan D.
Stein, Talbert S.
Stockdale, John A.
Stone, Philip M.
Stoner Jr., John O.
Stricklett, Kenneth L.
Stumpf, Bernhard J.
Sun, James C.
Szalewicz, Krzysztof
Tang, Fu-Ching
Taylor, Howard S.
Temkin, Aaron
Teubner, Peter J.
Thomas, Brian K.
Thomas, Bruce R.
Tiernan, Thomas O.
Tio, T. K.
Trajmar, Sandor
Truhlar, Donald G.
Tseng, Hsiang-Kuang
Turner, Jr., Charles E.
Van Zyl, Bert
Vaughan, Stephen O.
Venanzi, Carol A.
Wachter, Joseph R.
Wadehra, Jogindra M.
Wadzinski, Henry T.
Walker, Keith G.
Wang, Chia P.
Watson, Deborah K.
Weatherford, Charles A.
Weigold, Erich
Weinberg, William H.
Wellenstein, Hermann F.
Westerveld, Willem B.
Wheeler, Paul C.
Whitten, Barbara L.
Williamson Jr., William
Wilson, John W.
Wilson, Walter E.
Wood, Robert M.
Wood II, Obert R.
Woolf, Stanley
Yates, Albert C.
Younger, Stephen M.
Yu, Simon
Zanelli, Claudio I.
Zitzewitz, Paul W.

Chemical Physics Excluding Photochemistry

Albritton, Daniel L.
Ali, Abdul W.
Alper, Joseph S.
Anderson, Alfred B.
Anderson, James B.
Anderson, Larry G.
Anderson, Richard J.
Anderson, Roger W.
Andresen, Bjarne B.
Armentrout, Peter B.
Au, C. K.
Auerbach, Roy A.
Babcock, Lucia M.
Baer, Tomas
Bailey, Thomas L.
Bair, Edward J.
Barker, John R.
Barrett, John L.
Bartlett, Rodney J.
Bauer, Simon H.
Baughcum, Steven L.
Bauman, Robert P.
Bayes, Kyle D.
Beauchamp, Jesse L.
Bechtel, James H.
Becker, Karl H.
Becker, Kurt H.
Behringer, Robert E.
Bel Bruno, Joseph J.
Belford, R. L.
Bendler, John T.
Benson, Sidney W.
Berne, Bruce J.
Bernstein, Richard B.
Berry, R. Steven
Bersohn, Richard
Betts, Jeanette A.
Bhattacharya, Ashok K.
Bierbaum, Veronica M.
Binns, Walter R.
Biondi, Manfred A.
Birely, John H.
Bissinger, George A.
Black, Graham
Blais, Normand C.
Bogan, Denis J.
Bose, Subir K.
Bowers, Michael T.
Bowman, Joel M.
Breinig, Marianne
Brenner, Douglas M.
Brink, Gilbert O.
Brooks, Charles L.
Brooks, Philip R.
Brown Jr., Howard H.
Burnett, Keith
Burnham, Ralph
Bushman, Gary
Butler, James E.
Butler, Scott E.
Cade, Paul E.
Cardon, Bartley L.
Carmichael, Ian
Carr Jr., Robert W.
Castleman Jr., A. W.
Caves, Thomas C.
Center, R. E.
Chan, F. T.
Chandler, David

Chapman, Sally
Charatis, George
Chilukuri, Santaram
Chiu, Lue-Yung C.
Chopra, Dev R.
Christophorou, Loucas G.
Chupka, William A.
Church, David A.
Clark, Kenneth C.
Cohen, Ronald B.
Collins, C. B.
Coplan, Michael A.
Corderman, Reed R.
Cosby, Philip C.
Creighton, John R.
Crim, F. F.
Crosley, David R.
Cross, Jon B.
Cross Jr., R. J.
Cunningham, Augustine J.
Curtis, Paul M.
Dagdigian, Paul J.
Dahler, John S.
Dalgarno, Alexander
Datz, Sheldon
Davenport, John E.
Davidovits, Paul
Davis, H. Ted
De La Vega, Jose R.
DeKoven, Benjamin M.
DePristo, Andrew E.
DeVries, Paul L.
Dharamsi, Amin N.
Diestler, D. J.
Dobbs, Gregory M.
Dorfman, Leon M.
Drake, J. M.
Drobny, Vladimir F.
Drummond, David L.
Dunlap, Brett I.
Dunn, Gordon H.
Dunning Jr., Thomas H.
Eckstrom, Donald J.
Eisele, Fred L.
Eliasson, Baldur
Eliezer, Isaac
Ellison, Frank O.
Ellison, G. B.
Endres, Paul F.
Engelken, Robert D.
English, Thomas C.
Eno, Larry
Ermler, Walter C.
Ernie, Douglas W.
Fahey, David W.
Fano, Ugo
Farrar, James M.
Fayer, Michael D.
Fehsenfeld, Frederick
Feld, Michael S.
Fineman, Morton A.
Fink, Richard D.
Finzi, Jack
Firestone, Richard F.
Fisher, Galen B.
Fite, Wade L.
Flannery, Martin R.
Foner, Samuel N.
Forbrich Jr., Carl A.

Ford, A. L.
Forrester, A. T.
Freed, Karl F.
Freedman, Andrew
Futrell, J. H.
Gallagher, Jean W.
Garcia Jr., Jose D.
Gardiner Jr., W. C.
Garetz, Bruce A.
Garscadden, Alan
Gatland, Ian R.
Gay, Timothy J.
Gelfand, Jack J.
Genack, Azriel
Gentry, W. R.
Gersh, Michael E.
Gianturco, Franco A.
Giese, Clayton F.
Giese, John P.
Gilbody, Henry B.
Gillen, Keith T.
Gillispie, Gregory D.
Gilmore, Forrest R.
Gimarc, Benjamin M.
Girardeau Jr., Marvin D.
Gislason, Eric A.
Goddard III, William A.
Goldberg, Ira B.
Goldberg, Lawrence S.
Golde, Michael F.
Golden, David M.
Golden, Sidney
Gole, James L.
Goncalves, Antonio M.
Gordon, Mark S.
Gordon, Robert J.
Graff, Margaret M.
Grant, Edward R.
Graves, John L.
Gray, Eoin W.
Gray, Tom J.
Green, Thomas A.
Greene, Edward F.
Grover, James R.
Guillory, William A.
Gunton, Robert C.
Gusinow, Michael A.
Hall, Richard I.
Hamilton, Peter A.
Hampson Jr., Robert F.
Hanrahan, Robert J.
Harris, Charles B.
Harris, Harold H.
Harvey, Nancy M.
Hase, William L.
Hastie, John W.
Haugsjaa, Paul O.
Hay, Philip J.
Hays, Gerald N.
Heicklen, Julian P.
Heidner III, R. F.
Henchman, Michael J.
Herbst, Eric
Herm, Ronald R.
Herschbach, Dudley R.
Hessler, Jan P.
Hierl, Peter M.
Hill, Robert M.
Hinchen, John J.

Hirsch, Robert G.
Hiskes, John R.
Hochstrasser, Robin M.
Hodges, Ronald V.
Hoffmann, Roald
Holt, Helen K.
Howard, Carleton J.
Howard, Robert E.
Huestis, David L.
Hughes, William M.
Hunter, Scott R.
Hurst, George S.
Huwel, Lutz
Innes, Frederick R.
Inokuti, Mitio
Ioup, George E.
Jaffe, Richard L.
Jeffries, Jay B.
Johnson Jr., Charles S.
Jones, Claude R.
Jones, Patrick L.
Jones, Steven E.
Jordan, Kenneth D.
Joyce, J. M.
Kadlecek, John A.
Kamaratos, E.
Karplus, Martin
Kasper, Jerome V.
Kaufman, Myron J.
Kay, Kenneth G.
Kelley, J. Daniel
Keto, John W.
Kinsey, James L.
Kleiber, Paul D.
Klemm, R. Bruce
Kline, Laurence E.
Klosterman, Elliot L.
Knize, Randall J.
Knuth, Eldon L.
Koenig, Thomas W.
Koeppl, Gerald W.
Koffend, Brooke
Koffend, John B.
Kolb Jr., Charles E.
Komornicki, Andrew
Koski, Walter S.
Kostroun, Vaclav O.
Koszykowski, Michael L.
Kouri, Donald J.
Kramer, Steven D.
Krause, Herbert F.
Krenos, John R.
Krishnan, Mahadevan
Krohn, Kenneth A.
Ku, Peh Sun
Kupperman, Aron
Kurylo III, Michael J.
Kwei, George H.
Kwok, Munson A.
Kwong, Victor HS
Ladanyi, Branka M.
Laguna, Glenn
Lambert, David L.
Lampe, Frederick W.
Lang, Neil
Lapatovich, Walter P.
Larter, Raima
Laskowski, Bernard C.
Laudenslager, James B.

2.3 Chemical Physics Excluding Photochemistry

Laufer, Allan H.
Lawton, Stan
Layne, Clyde B.
Lee, Edward K.
Lee, Francis W.
Legg, James C.
Leone, Stephen R.
Lester Jr., William A.
Lieber, Michael
Light, Glenn C.
Light, John C.
Lim, Teck-Kah
Lin, Chii-Dong
Linder, Bruno
Livingston, Ralph
Lockwood, Grant J.
Loder, Rurik K.
Lohr Jr., Lawrence L.
Long Jr., William H.
Lorents, Donald C.
Loyd Jr., David H.
Lu, Jia-Jih
Luntz, Alan C.
MacPherson, Alistair K.
Magee, John L.
Maggiora, Gerald M.
Maier II, William B.
Mandl, Alexander E.
Mann, David M.
Mansky II, Edmund J.
Manzanares, Elizabeth R.
Marcus, Rudolph A.
Mariella Jr., Raymond P.
Martin, L. R.
Mason, Edward A.
Mattson, Timothy G.
Mazur, Jacob
McAfee Jr., Kenneth B.
McClelland, Gary M.
McCullough Jr., E. A.
McCurdy, C. W.
McDowell, Harding K.
McGregor Jr., Wheeler K.
McGuirk, Michael
McLennan, James A.
Meisels, Gerhard G.
Merkelo, Henri
Metiu, Horia I.
Micha, David A.
Michels, H. Harvey
Miller, David R.
Miller, Donald J.
Miller, Glenn H.
Miller, James A.
Miller, Terry A.
Miller, Thomas M.
Miller, William H.
Mitchner, M.
Mohnen, Volker A.
Molina, Mario J.
Montgomery, Donald J.
Moore, C. Bradley
Moran, Thomas F.
Morgan, William L.
Mortensen, Earl M.
Moseley, John T.
Mozumder, A.
Murad, Edmond
Murphy, Randall E.

Mutch, George W.
Nagy, Paul J.
Nesbitt, David J.
Netzel, Thomas L.
Neumann, Herschel
Nicovich, John M.
Nighan, William L.
Noid, Donald W.
Northrup, Scott H.
Noyes, Richard M.
O, Chun-Sing
O'Malley, Thomas F.
Ohrn, Yngve
Oldenborg, Richard C.
Ollison, William M.
Osgood, Richard M.
Ostlund, Neil S.
Palmer, Howard B.
Pan, Y.
Park, John T.
Parmenter, Charles S.
Parr, Christopher A.
Parson, John M.
Partlow, William D.
Partridge III, Harry
Paske, William C.
Paulson, John F.
Pechukas, Philip
Pengra, James G.
Perel, Julius
Perry, Robert A.
Peterson, James R.
Petersson, George A.
Peticolas, Warner L.
Phaneuf, Ronald A.
Phillips, Donald H.
Piebold, Gerald J.
Pindzola, Michael S.
Pipkin, Francis M.
Pitts Jr., James N.
Plummer, Patricia LM
Pollack, Edward
Priestley, Eldon B.
Prior, Michael H.
Pritchard, Richard H.
Pritt Jr., Alfred T.
Pruett, James G.
Pugh Jr., Henry L.
Rabinovitch, B. Seymour
Rabitz, Herschel
Ramler, Warren J.
Ratkowski, Anthony J.
Ravishankara, Akkihebbal R.
Ray, Gary W.
Reck, Gene P.
Redding, Rogers W.
Reddy, K. V.
Redi, Martha H.
Reeves, Richard L.
Reilly, James P.
Reiner, Robert H.
Reinhardt, William P.
Rentzepis, P. M.
Retzloff, David G.
Reuther, James J.
Rich, Joseph W.
Richard, Patrick
Riley, Merle E.
Robinson, Hugh G.

Roesler, Fred L.
Rol, Pieter K.
Root, John W.
Rose, Timothy L.
Rosen, Gerald
Rosenberg, Richard A.
Ross, John
Ruedenberg, Klaus
Russell, James E.
Rutherford, John A.
Safron, Sanford A.
Saha, Haripada
Sanzone, George
Saporoschenko, Mykola
Schatz, George C.
Scheps, Richard
Schlachter, Alfred S.
Schmalz, Thomas G.
Schofield, Keith
Schuler, Robert H.
Seliger, Howard H.
Sellin, Ivan A.
Senba, Masayoshi
Setser, D. W.
Sharpton, Francis A.
Shay, Thomas M.
Sheldon, John W.
Sheridan, John R.
Shipsey, Edward J.
Shirts, Randall B.
Siegel, Melvin W.
Silver, David M.
Silver, Joel A.
Sinanoglu, Oktay
Siska, Peter E.
Skinner, Charles H.
Skinner, Gordon B.
Slanger, Tom
Smyth, Kermit C.
Sparks, Randal K.
Spicer, Leonard D.
Spruch, Larry
St. John III, Willard M.
Staley, Ralph H.
Stebbings, Ronald F.
Stechel, Ellen B.
Steinfeld, Jeffrey I.
Stern, Richard C.
Stone, Philip M.
Stricklett, Kenneth L.
Suck, Sung Ho
Sun, James C.
Suto, Masako
Sutton, David G.
Tam, Andrew C.
Tang, K. T.
Tang, Kenneth Y.
Taylor, Howard S.
Taylor, Raymond L.
Taylor, Ronald D.
Terhune, Robert W.
Thirumalai, Devarajan V.
Thompson, Donald L.
Tiernan, Thomas O.
Tio, T. K.
Toennies, Jan Peter
Torr, Douglas G.
Trainor, Daniel W.
True, Nancy S.

Truhlar, Donald G.
Tully, John C.
Turner, Jr., Charles E.
Van Lint, Victor
Van Siclen, Clinton D.
VanBrunt, Richard J.
Vane, Charles R.
Verhaar, Boudewyn J.
Vestal, Marvin L.
Viggiano, Albert A.
Wadehra, Jogindra M.
Wagner, Albert
Walker, Robert B.
Walters, G. K.
Warren Jr., Walter R.
Way, Kermit
Waynant, Ronald W.
Weatherford, Charles A.
Weber, Joseph N.
Weigold, Erich
Weinberg, William H.
Weiner, Brian L.
Weiner, Eugene R.
Weiner, Jerome H.
Weiner, John
Weinstein, Harel
Weisman, R. B.
Weitz, Eric
Wells, Gregory J.
Wexler, Sol
Wharton, Lennard
Whitbeck, Michael R.
Wicke, Brian G.
Wiff, Donald R.
Wine, Paul H.
Winicur, Daniel H.
Winkler, Irwin C.
Winn, John S.
Winter, Nicholas W.
Winter, Thomas G.
Wittig, Curt
Wodarczyk, Francis J.
Wolga, George J.
Wolk, Gary L.
Woodruff, Susan B.
Woody, Bernard A.
Woodyard, Jack R.
Wu, Richard L.
Wyatt, Robert E.
Wyss, Jerry C.
Yamamoto, Tomoko
Yang, Sze-Cheng
Yankwich, Peter E.
Yeung, Edward S.
Yip, Sidney
Yoshimine, Megumu
Yuan, Jian-Min
Yun, Kwang-Sik
Zahniser, Mark S.
Zare, Richard N.
Zimmerman, I. H.
Zollars, Byron G.

Abraham, George
Ackerhalt, Jay R.
Affatato, Joseph F.
Aldridge III, Jack P.
Ali, Mahamed A.
Alton, Gerald D.
Anderson, William R.
Antcliff, Richard R.
Armentrout, Peter B.
Armstrong Jr., Lloyd
Au, C. K.
Aubrey, Bertrand B.
Averill, Frank W.
Babbitt, William R.
Bae, Young K.
Baer, Tomas
Bair, Edward J.
Baker, Howard C.
Bakshi, Pradip M.
Baliga, Shankar B.
Band, Hans E.
Band, Yehuda B.
Basbas, George J.
Bass, Arnold M.
Baughcum, Steven L.
Bayes, Kyle D.
Bayfield, James E.
Beauchamp, Jesse L.
Beausoleil Jr., R. G.
Beiting III, Edward J.
Bender, Peter L.
Bengston, Roger D
Berg, Jacqueline O.
Bergeman, Thomas H.
Bergquist, James C.
Berkowitz, Joseph
Berman, Paul R.
Bernhardt, Anthony F.
Bernstein, Elliot R.
Bersohn, Richard
Beyerinck, Herman C.
Bhattacharya, Samir K.
Bieniek, Ronald J.
Bischel, William K.
Bjorkholm, John E.
Bjorklund, Gary C.
Blass, William E.
Blatt, Rainer
Blumberg, Leroy N.
Bobin, Jean Louis
Bokor, Jeffrey
Borsella, Elisabetta
Bouman, Thomas D.
Branscomb, Lewis M.
Brault, James W.
Brenn, Rudiger
Brewer, Richard G.
Bridges, William B.
Britt, Edward I.
Brooker, Murray H.
Broudy, Robert M.
Brown, Ellen R.
Brown, Lorin W.
Bryant, Howard C.
Bucksbaum, Philip H.
Bukow, Hans
Burke, Philip G.
Burns III, Jay

Burris Jr., John
Cacak, Robert K.
Cardon, Bartley L.
Carlson, Thomas A.
Carlsten, John L.
Carmichael, Howard I
Carroll, Clark E.
Cartwright, David C.
Cathey, LeConte
Chakraborti, Parimal K.
Champagne, Louis F.
Chang, Edward S.
Chang, Tu-nan
Charatis, George
Chen, Hao-lin
Cheng, Kwok-Tsang
Cho, Chung Won
Christophorou, Loucas G.
Chu, Ben
Chung, Sunggi
Chupka, William A.
Church, David A.
Chutjian, Ara
Clark, Charles W.
Clark, Kenneth C.
Cobb, Donald D.
Code, R. F.
Cody, Regina J.
Cohen, Stanley
Cohn, Gerald E.
Coldwell, Robert L.
Coleman, Paul D.
Collins, C. B.
Collins, Lee A.
Comaskey, Brian
Compaan, Alvin
Compton, Robert N.
Cooper, C. Dewey
Cooper, John
Cooper, John W.
Cosby, Philip C.
Coulter, Claude A.
Coulter, Philip W.
Cox Jr., Hollace L.
Crasemann, Bernd
Crawford Jr., Bryce
Cromer, Christopher L.
Cue, Nelson
Cunningham, David L.
Dagata, John A.
Davenport, John E.
Davidson, Steven A.
Davis, Lawrence W.
DeBeer, David P.
Dehmer, Joseph L.
Dehmer, Patricia M.
Delgado-Barrio, Gerardo
Derr, Vernon E.
Deslattes, Richard D.
DeTemple, Thomas A.
DeYoung, Russell J.
Dill, Dan
Dixit, Shamasundar N.
Donahue, Joey E.
Doolen, Gary D.
Dow, John D.
Drake, Gordon W.
Dressler, Kurt

Druger, Stephen D.
Ducas, Theodore W.
Duke, Charles B.
Dunbar, Robert C.
Duzy, Carolyn
Dyke, Thomas R.
Eberly, Joseph H.
Ebert, Paul J.
Eck, Thomas G.
Eckstrom, Donald J.
Ehlotzky, Fritz
Ellis, David G.
Ellis, Walton P.
Elmquist, Randolph E.
Emery, Guy T.
Engelken, Robert D.
Engelking, Paul C.
Engelman Jr., Rolf
Ewing, James J.
Exton, Reginald J.
Eyler, Edward E.
Fadley, Charles S.
Fairchild, Clifford E.
Falk, Joel
Farrar, James M.
Fayer, Michael D.
Ferrett, Tricia A.
Figueira, Joseph F.
Findley, Gary L.
Firestone, Richard F.
Fischer, Charlotte F.
Fisher, H. Leonard
Fisher, Robert A.
Fletcher, Gary D.
Fleury, Paul A.
Foltz, Greg W.
Foltz, Nevin D.
Fowler, Bruce W.
Fowler, Richard G.
Fox, Robert A.
Franck, Carl P.
Freed, Karl F.
Freedman, Andrew
Freund, Hans J.
Friedrich, Donald M.
Frueholz, Robert P.
Frye, Daniel D.
Fuhr, Jeffrey R.
Furumoto, Horace
Gallagher, Jean W.
Ganas, Perry S.
Gardner, Larry D.
Garing, John S.
Garrett, William R.
Garstang, Roy H.
Geballe, Ronald
Gentile, Thomas R.
Gilmore, Forrest R.
Ginter, Marshall L.
Giordmaine, Joseph A.
Gold, L. Peter
Goldwire Jr., Henry C.
Golub, John E.
Good Jr., Roland H.
Good Jr., William E.
Goss, Larry P.
Gould, Phillip L.
Graff, Margaret M.

Gram, Peter AM
Greene, Chris H.
Greve, Peter
Grover, James R.
Gruen, Dieter M.
Guberman, Steven L.
Gutcheck, Robert A.
Haan, Stanley L.
Hall, Richard B.
Hall, Robert J.
Hameka, Hendrik F.
Hamermesh, Morton
Hamm, Robert W.
Hammer, J. M.
Hammond, Gordon L.
Hardis, Jonathan E.
Harris, Bernard
Hartman, Paul L.
Hartmann, Sven R.
Harvey, James F.
Hauser, Ulrich A.
Hayhurst, Thomas L.
Hazi, Andrew U.
Heer, Clifford V.
Heinzen, Daniel J.
Heller, Donald F.
Heller, Eric J.
Herm, Ronald R.
Hess Jr., Doren W.
Hessler, Jan P.
Hinrichs, C. K.
Holroyd, Richard
Holt, Helen K.
Horsley, John A.
Horwitz, Alexander B.
Houston, Paul L.
Hsu, Donald K.
Hsu, Hsiung
Hubbell, John H.
Huebner, Russell H.
Huebner, Walter F.
Humpherys, Kent C.
Husmann, Otto K.
Hyman, Howard A.
Inokuti, Mitio
Ivey, Henry F.
Jackson, William M.
Jacobs, Verne L.
Jacox, Marilyn E.
Jen, Chih K.
Jennings, Donald A.
Jensen, Barbara L.
Jeon, Yoon H.
Jimenez-Mier, Jose
Johnson, Carol
Johnson, Walter R.
Johnston, Harold S.
Johnston Jr., Thomas F.
Jones, Patrick L.
Judge, Darrell L.
Kachru, Ravinder
Kamaratos, E.
Kang, Ik-Ju
Katayama, Mikio
Keeffe, William M.
Kelley, Paul L.
Kellman, Michael E.
Kelly, Raymond L.

3.1 Conventional Photon-Atom and Photon-Molecule Effects

Kenefick, Robert A.
Keto, John W.
Kim, Jin J.
Kimble, Harry J.
Kinsey, James L.
Kirby, Kate P
Klein, Lewis
Klose, Jules Z.
Klots, Cornelius E.
Knochenmuss, Richard D.
Koenig, Thomas W.
Kohl, John L.
Kolb Jr., Charles E.
Korevaar, Eric J.
Kosman, Warren M.
Kostiuk, Theodor
Kramer, Steven D.
Krasinski, Jerzy S.
Krause, Manfred O.
Krohn, Burton J.
Kumar, Prem
Kung, Robert T.
Kupperman, Aron
Kurnit, Norman A.
Kurucz, Robert L.
Kwan, Ching-Kwan CK
Laane, Jaan
Lambert, David L.
Langhoff, Peter W.
LaVilla, Robert E.
Lawler, James E.
Leach, Sydney
Learner, Richard C.
Lee, Chi-Hsiang
Lee, Ching-Tsung
Lee, Ja H.
Lee, Jonathan K.
Lee, Long C.
Lee, Sang S.
Lee, Siu-Au
Lee, Yuan T.
Leone, Stephen R.
Lerman, Juan-Carlos
Leroi, George E.
Lester Jr., William A.
Leuchs, Gerhard
Levenson, Marc D.
Levy, Laurent P.
Liberman, Irving
Lieber, Michael
Lightman, Allan J.
Lin, Chii-Dong
Lin, John
Lindle, Dennis W.
Lineberger, William C.
Lipsky, Lester
Lipsky, Sanford
Liu, Yung S.
Loree, Thomas R.
Lorents, Donald C.
Lovoi, Paul
Lubell, Michael S.
Lucchese, Robert R.
Lunney, James G.
Madden, Robert P.
Magnuson, Dale W.
Mahon, Rita
Mainardi, Raul T.

Malley, Michael M.
Mansikka, Kauko A.
Manson, Steven T.
Maricq, M. Matti
Marron, Michael T.
Martin, Richard L.
Martin, Richard M.
Massey, Gail A.
Mathews, C. Weldon
Matulic, Ljubomir
Mavroyannis, Constantine
McCurdy, C. W.
McGregor Jr., Wheeler K.
McGuire, James H.
McIlrath, Thomas J.
McIntyre Jr., L. C.
McKenzie, Robert L.
McKoy, Vincent
Mead, Roy D.
Meisels, Gerhard G.
Mellen, Walter R.
Mendelsohn, Lawrence B.
Mendlowitz, Harold
Menyuk, Norman
Merts, A. L.
Mickish, Roger A.
Mies, Frederick H.
Miller, John C.
Miller, John H.
Miller, Roger E.
Miller, Thomas M.
Mitchner, M.
Moeny, William M.
Mohr, Peter J.
Molina, Mario J.
Moore, C. Fred
Moroi, David S.
Moseley, John T.
Mossberg, Thomas W.
Mumma, Michael J.
Murphy, John C.
Murray, Frank
Murray, John R.
Nanes, Roger
Narducci, Lorenzo M.
Nayfeh, A. H.
Nelson, William H.
Netzel, Thomas L.
Ni, Wei-Tou
Nicholls, Ralph W.
Nygaard, Kaare J.
O, Chun-Sing
O'Connell, Robert F.
Ohrn, Yngve
Okabe, Hideo
Oleksik, John J.
Olsen, Thomas
Olsgaard, David A.
Osgood, Richard M.
Ott, William R.
Oza, Dipak H.
Pack, Russell T
Pantinakis, Apostolos
Park, Changhwan
Parkinson, W. H.
Parks, James E.
Parmenter, Charles S.
Parr, Albert C.

Parsons, Michael L.
Pate, Bradford B.
Paulson, John F.
Payne, Marvin G.
Peatman, William B.
Pendleton, Hugh N.
Peng, Jin-Sheng
Perkowitz, Sidney
Peterson, James R.
Peterson, Lennart R.
Peuse, Bruce W.
Picard, Richard H.
Pierce, Thomas M.
Piltch, Martin S.
Pindzola, Michael S.
Pipkin, Francis M.
Poultney, Sherman K.
Powell, Cedric J.
Pratt, Richard H.
Pratt, Stephen T.
Pugh Jr., Henry L.
Ramaker, David E.
Rast Jr., Howard E.
Ratkowski, Anthony J.
Rau, A. R.
Rayborn, Grayson H.
Raymer, Michael G.
Reilly, James P.
Rescigno, Thomas N.
Rhodes, Charles K.
Rice, James K.
Rich, Nathan H.
Robin, Melvin B.
Robinson, Edward J.
Roesler, Fred L.
Rosenberg, Richard A.
Rothe, Erhard W.
Rothenberg, Joshua E.
Roy, Denis
Ruff, George A.
Rumble, John R.
Rustgi, Om P.
Rutherford, John A.
Ryan, Frederick M.
Rybka, Theodore W.
Saha, Haripada
Saloman, Edward B.
Sampson, Douglas H.
Samson, James A.
Sanders Jr., Frank C.
Sarka, Benjamin NM
Saxon, Roberta P.
Scarl, Donald B.
Schappert, Gottfried T.
Schatz, Paul N.
Schenck, Peter
Schneider, Richard T.
Schuler, Robert H.
Schultz, John W.
Schulz, Peter A.
Schumer, Douglas B.
Scofield, James H.
Seki, Hajimes
Sell, Jeffrey A.
Seltzer, Stephen M.
She, Chiao Y.
Shen, Y. R.
Sher, Mark H.

Sheridan, John R.
Shirley, Jon H.
Shorer, Philip
Silfvast, William T.
Silver, Joel A.
Silverman, Mark P.
Simpson, Charles G.
Singh, Surendra P.
Siomos, Konstadinos
Skinner, Charles H.
Skowronek, Maurice
Slanger, Tom
Slater, Richard
Smith, Winthrop W.
Smith, Stephen J.
Smith, Peter L.
Snyder, James J.
Sparks, Randal K.
Spezeski, Joseph J.
Starace, Anthony F.
Starr, Walter L.
Steel, Duncan G.
Steele, William A.
Stehle, Philip
Stehman, Robert M.
Stevens, Walter J.
Stinespring, Charter D.
Stockbauer, Roger
Stoicheff, Boris P.
Stotlar, Suzanne C.
Stroud Jr., Carlos R.
Suto, Masako
Swanson, Nils
Switzer, Gary L.
Szalewicz, Krzysztof
Sze, Robert C.
Talman, James D.
Tam, Andrew C.
Tambe, Balkrishna R.
Tang, Kenneth Y.
Tarter, C. Bruce
Tayal, Swaraj S.
Taylor, David J.
Taylor, James W.
Taylor, Lyle H.
Tehrani, Mohammad M.
Teich, Malvin C.
Thomas, T. D.
Tomkins, Frank S.
Tomlinson, W. J.
Topp, Michael R.
Trainor, Daniel W.
Trump, Darryl D.
Tsai, Bilin P.
Tseng, Hsiang-Kuang
Turechek, John J.
Turley, R. S.
Uglum, John R.
Uzer, Turgay
Verdieck, James F.
Vestal, Marvin L.
Victor, George A.
Vincent, James S.
Voreades, Demetrios
Wallace, Richard W.
Walls, Daniel F.
Walsh, Peter J.
Wang, Chao C.

Wang, Chia P.
Wang, Wen-Cheng
Ward, John F.
Warner, Ray A.
Weber, Heinz P.
Weber, Marvin J.
Weiner, Brian L.
Weissbluth, Mitchel
Weissler, Gerhard L.
Weiszmann, Andrei N.
Weitz, Eric
Werner, Samuel A.
West, John B.
Westerveld, Willem B.
Weston Jr., Ralph E.
Whaling, Ward
Wheeler, Paul C.
White, Michael G.
Whitney, Kenneth G.
Whitney, Wayne T.
Wieder, Grace M.
Wieman, Carl E.
Wiesenfeld, Jay M.
Wiesenfeld, John
Williams, W. David
Willis, Paul A.
Wine, Paul H.
Winkler, Irwin C.
Winkler, Peter
Winter, Nicholas W.
Wise, John H.
Witriol, Norman M.
Wodkiewicz, Krzysztof
Wolga, George J.
Wong, Ngai C.
Woo, Shien-Biau
Woodruff, Pamela R.
Woody, Bernard A.
Woolf, Stanley
Worlock, John M.
Wright, John C.
Wu, C. Y. R.
Wynne, James J.
Yabuzaki, Tsutomu
Yang, Sze-Cheng
Yardley, James T.
Yoder, Marvel J.
Young, James F.
Younger, Stephen M.
Yuan, Jian-Min
Zahniser, Mark S.
Zaidi, Haider R.
Zakheim, David S.
Zapata, Luis E.
Zink, J. W.
Zittel, Paul F.

3.2 Optical and UV Spectra

Aldridge III, Jack P.
Ali, Mahamed A.
Allen Jr., Harry C.
Aller, Lawrence H.
Anderson, James B.
Anderson, Larry G.
Anderson, William R.
Andrew, Kenneth L.
Band, Hans E.
Battleson, Kirk W.
Baughcum, Steven L.
Baur, James F.
Becher, Jacob
Behring, William E.
Benesch, William M.
Berg, Jacqueline O.
Bernheim, Robert A
Bernstein, Elliot R.
Berry, H. G.
Betts, Jeanette A.
Bhaskar, Natarajan D.
Bickel, William S.
Bouman, Thomas D.
Bowman, Richard L.
Brault, James W.
Brenn, Rudiger
Brewer, Leo
Brink, Gilbert O.
Brown, Charles M.
Bryant, Howard C.
Bukow, Hans
Burns III, Jay
Caird, John A.
Chen, Kuo-In
Chilukuri, Santaram
Chimenti, Robert J.
Chupka, William A.
Cobb, Donald D.
Code, R. F.
Cohn, Gerald E.
Conway, John G.
Cooke, William E.
Corderman, Reed R.
Crosswhite, Henry M.
Curnutte Jr., Basil
Curtis, Lorenzo J.
Davenport, John E.
Davis, David S.
Davis, Sumner P.
Degenkolb, Eugene
Dehmelt, Hans G.
DeSerio, Robert
Dow, John D.
Drake, Gordon W.
Dressler, Kurt
Drullinger, Robert E.
Duke, Charles B.
Duncan, M. M.
Eckhardt, Craig J.
Ederer, David L.
Egan, Patrick O.
Eliezer, Isaac
Ellis, David G.
Engelken, Robert D.
Engelman Jr., Rolf
Exton, Reginald J.
Field, Robert W.
Freeman, Daryl E.
Furst, Mitchell L.

Gelbwachs, Jerry A.
Ginter, Marshall L.
Goble, Alfred T.
Gold, L. Peter
Golden, Sidney
Golightly, Donald W.
Gouterman, Martin
Grosjean, Dennis F.
Guberman, Steven L.
Guillory, William A.
Haglund Jr., Richard F.
Hamilton, Peter A.
Hardis, Jonathan E.
Harris, Bernard
Harris, Harold H.
Hartman, Paul L.
Harvey, Kenneth C.
Hayhurst, Thomas L.
Heckmann, Paul H.
Heroux, Leon J.
Hess, Roger
Hinchen, John J.
Hinds, Edward A.
Hipps, Kerry W.
Howard (Retired), John N.
Huestis, David L.
Huffman, Robert E.
Hughes, Raymond H.
Husmann, Otto K.
Hutchison, Sheldon B.
Itano, Wayne M.
Ivey, Henry F.
Jacobson, Harry C.
James, David R.
Jimenez-Mier, Jose
Johnson, Brant M.
Johnson, Peter D.
Johnston, David B.
Jona, F. P.
Judd, Brian R.
Katayama, Daniel H.
Keane, Christopher J.
Keefer, Dennis R.
Keeffe, William M.
Kelly, Raymond L.
Kenney, John W.
Kielkopf, John F.
Klose, Jules Z.
Knochenmuss, Richard D.
Kobayashi, Hisao
Koffend, Brooke
Korevaar, Eric J.
Kosman, Warren M.
Kruse, Theodore H.
Kunze, Hans-Joach D.
Lawler, James E.
Layzer, David
Leach, Sydney
Learner, Richard C.
Leavitt, John A.
Lee, Long C.
Leroi, George E.
Lipsky, Sanford
Livingston, A.E. Gene
Loree, Thomas R.
Luken, William L
Madden, Robert P.
Maggiora, Gerald M.
Maki, Arthur G.

Marrus, Richard
Martin, L. R.
Martin, William C.
Mathews, C. Weldon
Mavroyannis, Constantine
Mazur, Ursula
McCubbin, T. K.
McDiarmid, Ruth S.
McDonald, J. D.
McGlynn, Sean P.
McGuire, Eugene J.
McIlrath, Thomas J.
McPherson Jr., Leroy A.
Menzies, Robert T.
Molina, Mario J.
Moore, C. Bradley
Moore Jr., Frank L.
Moseley, John T.
Myers, Stephen A.
Nanes, Roger
Neil, George R.
Nicholls, Ralph W.
Ohrn, Yngve
Palmer, Byron A.
Parkinson, W. H.
Parsons, Michael L.
Partlow, William D.
Passenheim, Burr C.
Poirier, John A.
Porter, Richard N.
Pradhan, Anil K.
Quick Jr., Charles R.
Rajnak, Katheryn
Rast Jr., Howard E.
Ratkowski, Anthony J.
Raynor, Susanne
Reader, Joseph
Reddy, Satti P.
Redi, Olav
Reuther, James J.
Richardson, James W.
Robin, Melvin B.
Ryan, Frederick M.
Saloman, Edward B.
Sansonetti, Craig J.
Schatz, Paul N.
Scherer, James R.
Shalimoff, George V.
Sheridan, John R.
Shull, Michael
Silver, Joshua D.
Sitterly, Charlotte M.
Skinner, Gordon B.
Slanger, Tom
Smith, Arthur A.
Smith, Peter L.
Soltis, Paul J.
Spence, David
Sposito, Garrison
Starr, Walter L.
Steinhaus, David W.
Sugar, Jack
Sundaram, S.
Switzer, Gary L.
Tellinghuisen, Joel B.
Thomas, H. Ronald
Tomkins, Frank S.
Trabert, Elmar C.
Tracy, David H.

Trajmar, Sandor
Trump, Darryl D.
Tsai, Bilin P.
Uglum, John R.
Vala, Martin
Verdieck, James F.
Walsh, Peter J.
Weber, Marvin J.
Weissler, Gerhard L.
Wells, Michael B.
Westerveld, Willem B.
Weston Jr., Ralph E.
Whitten, Barbara L.
Wie, Chu K.
Williams, James F.
Winicur, Daniel H.
Wirth, Mary M.
Wittenberg, Albert M.
Wolk, Gary L.
Wood, David R.
Woody, Bernard A.
Wormhoudt, Joda C.
Zalubas, Romuald
Ziv, Alan R.
Zollweg, Robert J.

Affatato, Joseph F.
Alexeff, Igor
Allen Jr., Harry C.
Altgilbers, Larry L.
Amano, Takayoshi
Anderson, Larry G.
Anderson, Robert W.
Anderson, Thomas G.
Bae, Young K.
Baliga, Shankar B.
Ballard, Stanley S.
Band, Hans E.
Beaudet, Robert A.
Benson, Richard C.
Bernheim, Robert A
Black, Truman D.
Blumberg, William A.
Bohn, Robert K.
Bollinger, John J.
Bottger, Gary L.
Bouman, Thomas D.
Brault, James W.
Brooker, Murray H.
Broudy, Robert M.
Burns III, Jay
Chackerian Jr., Charles
Chan, I. Y.
Chang, C. K.
Chiang, Joseph F.
Cho, Chung Won
Coleman, Paul D.
Copeland, Gary E.
Cornwell, C. D.
Craig, Norman C.
Crawford Jr., Bryce
Cummins, Sally E.
Curtis, Earl C.
Danielewicz Jr., Edward J.
Davis, David S.
Davis, Robert W.
De Lucia, Frank C.
Dehmelt, Hans G.
Dely, Alex
deZafra, Robert L.
Dixon, David A.
Dowe Jr., R. Michael
Dowell, Jerry T.
Dowling, Jerome M.
Dyke, Thomas R.
Eckhardt, Craig J.
Edwards, Thomas H.
Emken, Walter C.
English, Thomas C.
Farley, John W.
Fletcher, Gary D.
Flygare, Willis H.
Fortson, E. Norval
Fowler, Bruce W.
Fowler, Richard G.
Fox, Kenneth
Freed, Charles
Frueholz, Robert P.
Fuson, Nelson
Geller, M.
George, Simon
Gingerich, Karl
Gladisch, Michael W.
Gold, L. Peter
Goorvitch, David

Graybeal, Jack D.
Guillory, William A.
Gwinn, William D.
Hall, Robert J.
Harmony, Marlin D.
Hart, Raymond K.
Herbst, Eric
Herm, Ronald R.
Hess Jr., Doren W.
Hill, John C.
Hillman, John J.
Hinds, Edward A.
Hipps, Kerry W.
Ho, William W.
Holt, Richard A.
Howard (Retired), John N.
Hrubesh, Lawrence W.
Hsu, Shaw L.
Itano, Wayne M.
Jacob, Elizabeth J.
Jen, Chih K.
Jennings, Donald A.
Jensen, Barbara L.
Johnson, Charles E.
Johnston, Lawrence H.
Jones, Claude R.
Kasper, Jerome V.
Kelley, Paul L.
Kelly, Raymond L.
Kershenstein, John C.
Kleppner, Daniel
Koepf, Gerhard
Kostiuk, Theodor
Krohn, Burton J.
Kuo, Chien-Yu
Laane, Jaan
Lacey, Richard F.
Lambert, David L.
Larson, Daniel J.
Lauer, James L.
Lide Jr., David R.
Livingston, Ralph
Long Jr., Edward R.
Lovas, Frank J.
Magnuson, Dale W.
Mariani, David R.
Maricq, M. Matti
Mason, Arthur A.
Mazur, Ursula
McDermott, Mark N.
McDowell, Robin S.
Mehlhorn, Herbert A.
Miller, Roger E.
Moore, C. Bradley
Mumma, Michael J.
Murphy, John C.
Murphy, Randall E.
Murray, Frank
Myers, Stephen A.
Nelson, Albert C.
Nelson, Leonard Y.
Nesbitt, David J.
Nielsen, Alvin H.
O'Reilly, James M.
Parker, Paul M.
Pearl, John C.
Pearson, Edwin F.
Perkowitz, Sidney
Peyton, Bernard J.

Pickett, Herbert M.
Pine, Alan S.
Poultney, Sherman K.
Pradhan, Anil K.
Pratt, David W.
Price, Stephan D.
Quick Jr., Charles R.
Radford, Harrison E.
Ragle, John L.
Ramsey, Norman F.
Rast Jr., Howard E.
Redding, Rogers W.
Reddy, Satti P.
Rich, Nathan H.
Rollefson, R.
Rosner, S. David
Rothman, Laurence S.
Royce, Barrie S.
Ryan, Frederick M.
Schneider, Richard T.
Schowengerdt, Franklin D.
Schultz, John W.
Schwendeman, Richard H.
Seki, Hajimes
Senitzky, Benjamin
Shelton, W. Neil
Shirts, Randall B.
Shugart, Howard A.
Slimey Jr., James G.
Sposito, Garrison
Stein, Samuel R.
Stettler, John D.
Stoneman, Robert C.
Stotlar, Suzanne C.
Sundaram, S.
Taylor, Robert C.
Temmer, Georges M.
Thaddeus, Patrick
Thomas, H. Ronald
True, Nancy S.
Vala, Martin
Van Baak, David A.
Vincent, James S.
Walls, Fred L.
Walsh, Peter J.
Wang, Chao C.
Wang, Harry T.
Waynant, Ronald W.
Weber, Willes H.
Weinberg, William H.
West, Martin L.
Wharton, Lennard
Whiteley, Stephen R.
Whitney, Wayne T.
Wieder, Grace M.
Willis, Paul A.
Wineland, David J.
Wing, William H.
Woods, R. Claude
Wormhoudt, Joda C.
Wu, Frank T.
Yamanaka, Masanobu
Zollweg, Robert J.
Zorn, Jens C.
zuPutlitz, Gisbert

3.4 Inner Shell Transitions

Bartiromo, Rosario
Basbas, George J.
Beers, Brian L.
Bernard, Davy L.
Berreman, Dwight W.
Best, Philip E.
Beyer, Louis M.
Bhalla, Chander P.
Bhattacharya, Samir K.
Bichsel, Hans
Bissinger, George A.
Bloch, Jeffrey J.
Brenn, Rudiger
Brown, Matt D.
Budick, Burton
Burr, Alex F.
Campbell, John L.
Catz, Leonard A.
Charatis, George
Chen, Mau-Hsiung
Childs, Wendell A.
Chopra, Dev R.
Cipolla, Sam J.
Clark, Arnold F.
Crasemann, Bernd
Cromer, Christopher L.
Curnutte Jr., Basil
Delvaille, John P.
Deslattes, Richard D.
Dick, Charles E.
Dow, John D.
Ebert, Paul J.
Ederer, David L.
Egan, Patrick O.
Ellis, Walton P.
Emery, Guy T.
Fadley, Charles S.
Ferguson, Stephen M.
Fetzer, Homer D.
Fink, Richard W.
Flygare, Willis H.
Forsyth, James M.
Fortner, Richard J.
Franck, Carl P.
Friedman, Herbert
Furst, Mitchell L.
Gabbard, Fletcher
George, T. V.
Gibbons, Patrick C.
Hanson, David M.
Hanson, Harold P.
Hansteen, Johannes M.
Harris, Bernard
Hauer, Allan A.
Hinrichs, C. K.
Houston, John M.
Hubbell, John H.
Isozumi, Yasuhito
Jacobs, Verne L.
Jamison, Keith A.
Johnson, Brant M.
Joyce, J. M.
Kallne, Elisabeth
Katsanos, Anastasios A.
Kaufman, Victor
Kenefick, Robert A.
Kessel, Quentin C.
Kessler Jr., Ernest G.
Kessler, Karl G.

Khandelwal, Govind S.
Kim, Longhuan
Kramer, Stephen L.
Kraner, H. W.
Krause, Manfred O.
Kuckuck, Robert W.
Kunz, Christof
Kusko, Bruce H.
Land, David J.
Langhoff, Peter W.
LaVilla, Robert E.
Lee, Paul L.
Liefeld, Robert J.
Livingston, A.E. Gene
Lodhi, Sattar K.
Lunney, James G.
MacDowell, Alastair A.
Mainardi, Raul T.
Mangelson, Nolan F.
Manson, Steven T.
Martin, Fred W.
McCorkle, R. A.
McCormick, Larry D.
McIlrath, Thomas J.
McLaughlin, Daniel J.
McPherson Jr., Leroy A.
Meriwether, J. R.
Milchberg, Howard M.
Miller, Phillip D.
Molitoris, John D.
Morgan, Thomas J.
Morris, Roberta J.
Morton III, John R.
Nagel, David J.
Oosterhuis, William T.
Pantinakis, Apostolos
Pate, Bradford B.
Pengra, James G.
Pepper, George H.
Perez III, Joseph D.
Peterson, Randolph S.
Petrasso, Richard
Pings, Cornelius J.
Plecious, Robert C.
Powell, Cedric J.
Prentice, John K.
Quarles, C. A.
Ramaker, David E.
Rayburn, Louis A.
Reinovsky, Robert E.
Rhodes, Charles K.
Robinson, Edward L.
Rosenberg, Richard A.
Rudd, M. E.
Rustgi, Om P.
Saloman, Edward B.
Samson, James A.
Sandner, Wolfgang
Sartwell, Bruce D.
Sauder, William C.
Schappert, Gottfried T.
Schlachter, Alfred S.
Schnopper, Herbert W.
Scofield, James H.
Seely, John F.
Seka, Wolf
Seltzer, Stephen M.
Shafroth, Stephen M.
Sham, Tsun K.

Sher, Mark H.
Shimizu, Sakae
Shirley, David A.
Shull, Michael
Silfvast, William T.
Simons, Donald G.
Skinner, Charles H.
Skupsky, Stanley
Smith, Winthrop W.
Sparrow, Julian H.
Stearns, John W.
Strand, Oliver T.
Talman, James D.
Tape, James W.
Thomas, Richard G.
Thomsen, John S.
Thomson, George M.
Tseng, Hsiang-Kuang
Varghese, S. L.
Watson, R. L.
Wei, P. S.
Wellenstein, Hermann F.
West, John B.
Whitlock, Robert R.
Wie, Chu R.
Williams, Grahame J.
Wolfe, Gordon W.
Wood, Lowell L.
Yin, Lo-I
Zanelli, Claudio I.

Albrecht, A. C.
Albritton, Daniel L.
Aldridge III, Jack P.
Allen Jr., Harry C.
Amano, Takayoshi
Ames, Donald P.
Anderson, Alfred B.
Atkinson, George H.
Baglin, Frank G.
Benesch, William M.
Bernheim, Robert A
Bernstein, Elliot R.
Birnbaum, George
Blass, William E.
Bohn, Robert K.
Bottger, Gary L.
Bowman, Richard L.
Bray, Robert G.
Brewer, Leo
Brooker, Murray H.
Cantrell, Cyrus D.
Carlson, Nils W.
Chackerian Jr., Charles
Chang, C. K.
Chilukuri, Santaram
Cornwell, C. D.
Crawford Jr., Bryce
Cummins, Sally E.
Danielewicz Jr., Edward J.
Daniels, Robert L.
Davenport, John E.
Davis, Robert W.
Davis, Sumner P.
Dehmer, Patricia M.
Dows, David A.
Drake, J. M.
Dressler, Kurt
Eckhardt, Craig J.
Edwards, Thomas H.
Eliezer, Isaac
Engelking, Paul C.
England, Walter B.
Ermler, Walter C.
Eyler, Edward E.
Feichtner, John D.
Field, Robert W.
Finn, Edward J.
Fleury, Paul A.
Fox, Kenneth
Freeman, Daryl E.
Friedman, Joel M.
Garetz, Bruce A.
Garing, John S.
Gelbart, William M.
Gelbwachs, Jerry A.
Gelfand, Jack J.
Geller, M.
Gingerich, Karl
Gordon, Roy G.
Greenberg, Jack S.
Harmony, Marlin D.
Harris, David O.
Harter, William G.
Helm, Hanspeter
Herbst, Eric
Hinchen, John J.
Hipps, Kerry W.
Hochstrasser, Robin M.
Hougen, Jon T.

Howard (Retired), John N.
Hsu, Shaw L.
Hudson, Robert D.
Innes, K. K.
Jacox, Marilyn E.
Jaffe, Richard L.
Jennings, Donald A.
Jimenez-Mier, Jose
Johnson, Lawrence W.
Johnson, Philip M.
Johnston Jr., Milton D.
Jones, M. T.
Julienne, Paul S.
Kachru, Ravinder
Karplus, Martin
Kemple, Marvin D.
Khadjavi, Abbas
King, William T.
Kinsey, James L.
Koffend, Brooke
Kohin, Roger P.
Kohler, Bryan E.
Komornicki, Andrew
Krohn, Burton J.
Kwiram, Alvin L.
Laane, Jaan
Laguna, Glenn
Langhoff, Stephen R.
Laudenslager, James B.
Levy, Donald H.
Lightman, Allan J.
Lindsay, Mark D.
Litvak, Marvin M.
Lovas, Frank J.
Lucchese, Robert R.
MacDowell, Alastair A.
Mason, Arthur A.
Mazur, Ursula
McClure, Donald S.
McDowell, Robin S.
Miller, Roger E.
Nanes, Roger
Nicholls, Ralph W.
Nixon, Eugene R.
Olsgaard, David A.
Overend, John
Palmer, Byron A.
Panock, Richard
Parker, Paul M.
Parkinson, W. H.
Pavlopoulos, Theodore G.
Perkowitz, Sidney
Peticolas, Warner L.
Phillips, Donald H.
Phillips, John G.
Pine, Alan S.
Polo, Santiago R.
Porter, Richard N.
Pritchard, David E.
Pruett, James G.
Quinn, Jarus W.
Ramsey, Norman F.
Reilly, James P.
Rich, Nathan H.
Robin, Melvin B.
Schatz, Paul N.
Schawlow, Arthur L.
Scheps, Richard
Schiavone, James A.

Schnepp, Otto
Schreiber, Paul W.
Schultz, John W.
Seliger, Howard H.
Seybold, Paul G.
Sheridan, John R.
Shirk, James S.
Simpson, W. T.
Smith, Gregory P.
Snyder, Lewis E.
Spence, David
St. Peters, Richard L.
Steinfeld, Jeffrey I.
Stidham, Howard D.
Strickler, Stewart J.
Switzer, Gary L.
Taylor, Robert C.
Teets, Richard E.
Tellinghuisen, Joel B.
Terhune, Robert W.
Timsit, Roland S.
Tipping, Richard H.
True, Nancy S.
Uzer, Turgay
Warren Jr., Walter R.
Weber, Willes H.
Weisman, R. B.
White, Michael G.
Wieder, Grace M.
Wiesenfeld, Jay M.
Winicur, Daniel H.
Winn, John S.
Woods, R. Claude
Wormhoudt, Joda C.
Yang, Sze-Cheng

3.6 Interactions of Laser Radiation with Atoms and Molecules

Abella, Isaac D.
Akerman, M. A.
Albrecht, Georg F.
Amano, Takayoshi
Ames, Donald P.
Anderson, Larry G.
Anderson, Richard J.
Anderson, William R.
Atkinson, George H.
Atkinson, J. B.
Atlas, Susan R.
Babbitt, William R.
Bae, Young K.
Baer, Thomas M.
Bair, Edward J.
Balling, Ludwig C.
Band, Hans E.
Barker, John R.
Barry, J. D.
Bates Jr., Richard D.
Baughcum, Steven L.
Baur, James F.
Bay, Zoltan L.
Bayfield, James E.
Beauchamp, Jesse L.
Beaudet, Robert A.
Beausoleil Jr., R. G.
Bechis, Kenneth P.
Bechtel, James H.
Becker, Karl H.
Bederson, Benjamin
Beiting III, Edward J.
Bel Bruno, Joseph J.
Bender, Peter L.
Bennett Jr., William R.
Benson, Richard C.
Berg, Jacqueline O.
Bergquist, James C.
Berman, Paul R.
Bernhardt, Anthony F.
Bernheim, Robert A
Bernstein, Elliot R.
Bernstein, Richard B.
Berry, H. G.
Bersohn, Richard
Betz, Albert L.
Beverly III, Robert E.
Bhattacharya, Ashok K.
Bhaumik, Mani L.
Bickel, William S.
Bicknell, William E.
Bieniek, Ronald J.
Bierbaum, Veronica M.
Bigio, Irving J.
Birely, John H.
Bischel, William K.
Bjorkholm, John E.
Bjorklund, Gary C.
Blais, Normand C.
Blass, William E.
Bloembergen, Nicolaas
Bloom, Arnold L.
Bobin, Jean Louis
Bohn, Robert K.
Bollinger, John J.
Bonczyk, Paul A.
Boness, M. J.
Borsella, Elisabetta
Bottger, Gary L.

Bowden, Charles M.
Bowers, Michael T.
Bowman, Joel M.
Boyer, Keith
Brandenberger, John R.
Brau, Charles D.
Brewer, Richard G.
Bridges, William B.
Brink, Gilbert O.
Bristow, Thomas C.
Brooks, Charles L.
Brown, Charles M.
Brown, Lorin W.
Burnett, Keith
Burnham, Ralph
Butler, James E.
Campbell, David H.
Campion, Alan
Carlson, Nils W.
Carmichael, Howard I
Cason, Charles M.
Castro, George
Cathey, LeConte
Cattolica, Robert J.
Celotta, Robert J.
Center, R. E.
Chakraborti, Parimal K.
Chan, I. Y.
Chang, Cheng-hui
Chang, Edward S.
Charatis, George
Chen, C. H.
Chen, Chin-Lin
Chiang, Joseph F.
Childs, William J.
Chimenti, Robert J.
Cho, Chung Won
Church, David A.
Clark, Charles W.
Code, R. F.
Cody, Regina J.
Cohen, Ronald B.
Coleman, Paul D.
Collins, Robert J.
Comaskey, Brian
Cook III, Thomas B.
Cooney, John
Cooper, John
Cooper, John W.
Cosby, Philip C.
Couillaud, Bernard J.
Coulter, Claude A.
Cox Jr., Hollace L.
Crane, John K.
Cromer, Christopher L.
Crosley, David R.
Crosswhite, Henry M.
Cunningham, David L.
Curl, Robert F.
Curry, Bill P.
Dagdigian, Paul J.
Dagenais, Mario
Dasch, Cameron J.
Davis, Christopher C.
Davis, Lawrence W.
Davis, Richard W.
DeBeer, David P.
Dehmer, Joseph L.
Dehmer, Patricia M.

DeKoven, Benjamin M.
Dely, Alex
Dendramis, Achille L.
DeTemple, Thomas A.
Deutch, John M.
DeVoe, Ralph G.
DeVries, Paul L.
Dharamsi, Amin N.
Diels, Jean-Claude
Diestler, D. J.
Dietrich, Daniel D.
DiMauro, Louis F.
Dixit, Shamasundar N.
Dodd, Jack G.
Dougal, Arwin A.
Duncanson Jr., John A.
Eberly, Joseph H.
Eckbreth, Alan C.
Eckstrom, Donald J.
Ederer, David L.
Ehler, Arthur W.
Ehlotzky, Fritz
Ehrlich, Daniel J.
Elmquist, Randolph E.
Elton, Dr. Raymond C.
Engelking, Paul C.
Estler, Ron C.
Evenson, Kenneth M.
Exton, Reginald J.
Ezekiel, Shaoul
Fairbank Jr., William M.
Fairchild, Clifford E.
Falcone, Roger W.
Fang, Ta-Ming
Farley, John W.
Fayer, Michael D.
Feichtner, John D.
Feldman, Donald W.
Field, Robert W.
Figueira, Joseph F.
Findley, Gary L.
Firestone, Richard F.
Fisher, H. Leonard
Fisher, Robert A.
Fletcher, Gary D.
Fletcher, William H.
Fleury, Paul A.
Flynn, G. W.
Foltz, Nevin D.
Fonck, Raymond J.
Forbrich Jr., Carl A.
Forsley, Lawrence P.
Fortson, E. Norval
Fox, Kenneth
Fox, Robert A.
Francke, Ricardo E.
Franz, Frank A.
Freedman, Andrew
Freeman, Richard R.
Freund, Robert S.
Gaily, T. D.
Galbraith, Harold W.
Gallagher, Thomas F.
Gardner, Larry D.
Garrett, William R.
Gerardo, James B.
Gibbs, Hyatt M.
Gilbert, Sarah L.
Gillen, Keith T.

Gillispie, Gregory D.
Gladisch, Michael W.
Gold, L. Peter
Goldberg, Lawrence S.
Golde, Michael F.
Golden, David E.
Goldman, Leonard M.
Goldsmith, John EM
Goldstein, John C.
Gole, James L.
Golub, John E.
Good Jr., William E.
Goodman, Leonard S.
Gordon, James P.
Gould, Phillip L.
Grant, Edward R.
Greenlees, G. W.
Greve, Peter
Grischkowsky, Daniel R.
Grosjean, Dennis F.
Gruen, Dieter M.
Gudmundsen, Richard A.
Gundersen, Martin A.
Gupta, Rajendra
Gutman, William M.
Hackel, Lloyd A.
Haglund Jr., Richard F.
Hall, John L.
Hall, Richard B.
Hall, Robert J.
Hamilton, Peter A.
Hansch, Theodor W.
Hanson, David M.
Happer, William
Hardis, Jonathan E.
Hargis Jr., Philip J.
Harmony, Marlin D.
Harris, Charles B.
Harris, David O.
Harris, Harold H.
Hartmann, Sven R.
Hauser, Ulrich A.
Havey, Mark D.
Head, Charles E.
Heaps, William S.
Heer, Clifford V.
Heller, Donald F.
Hellwarth, Robert W.
Herman, Roger M.
Hess Jr., Doren W.
Hessler, Jan P.
Hilborn, Robert C.
Hill, John C.
Hinchen, John J.
Hirschfelder, Joseph O.
Hochstrasser, Robin M.
Holt, Richard A.
Hoppe, John C.
Howard, Carleton J.
Howgate, David W.
Hrubesh, Lawrence W.
Hsu, Donald K.
Hudson, George E.
Hudson, Robert D.
Huestis, David L.
Huhnermann, Harry
Hurst, George S.
Hutchison, Sheldon B.
Huwel, Lutz

Itano, Wayne M.
Ivey, Henry F.
Jacobs, Ralph R.
Jaduszliwer, Bernardo
Javan, Ali
Jeffries, Jay B.
Jensen, Craig C.
Jeys, Thomas H.
Jimenez-Mier, Jose
Johnson, Carol
Johnson Jr., Charles S.
Johnson, Lee K.
Johnston, Lawrence H.
Jones, Patrick L.
Judd, O'Dean P.
Kachru, Ravinder
Kaldor, Andrew P.
Kasper, Jerome V.
Katayama, Daniel H.
Katayama, Mikio
Kaufman, Stanley L.
Kay, Richard B.
Keefer, Dennis R.
Keller, Richard A.
Kelley, Paul L.
Kelly, Roger
Kelsey, Edward J.
Kenney, John W.
Kepros, John G.
Kershenstein, John C.
Keto, John W.
Khadjavi, Abbas
Killinger, Dennis K.
Kimble, Harry J.
Kleiber, Paul D.
Klein, Lewis
Kleppner, Daniel
Knochenmuss, Richard D.
Kobe, Donald H.
Koch, Peter M.
Koepf, Gerhard
Koffend, John B.
Kolb Jr., Charles E.
Korevaar, Eric J.
Kourlas, James
Kramer, Peter B.
Kramer, Steven D.
Krasinski, Jerzy S.
Krause, Lucjan
Krohn, Burton J.
Ku, Robert T.
Kumar, Prem
Kung, Robert T.
Kunze, Hans-Joach D.
Kupperman, Aron
Kurnit, Norman A.
Kwok, Munson A.
Kwong, Victor HS
Laane, Jaan
Lamb Jr., Willis E.
Lambropoulos, P. P.
Langsam, Yedidyah
Lapatovich, Walter P.
Larson, Daniel J.
Lau, Albert M.
Lawler, James E.
Lawton, Stan
Lebow, Paul
Lee, Chi-Hsiang

Lee, Jonathan K.
Lee, Long C.
Lee, Sang S.
Lee, Siu-Au
Leiby Jr., Clare C.
Leroi, George E.
Letamendia, Louis
Levenson, Marc D.
Li, Ming C.
Liao, Paul F.
Lichten, William L.
Liebenberg, Donald H.
Liedholz, Gerhard A.
Lightman, Allan J.
Lin, John
Lindsay, Mark D.
Lineberger, William C.
Lippmann, Bernard A.
Littman, Michael G.
Liu, Jenn-Ying
Liu, Yung S.
Loree, Thomas R.
Lorents, Donald C.
Lubell, Michael S.
Lucatorto, Thomas B.
Lynds, Lahmer
MacArthur, Duncan W.
Mahon, Rita
Malley, Michael M.
Mallow, Jeffry V.
Malone, Dennis P.
Manzanares, Elizabeth R.
Mason, Arthur A.
Mathews, C. Weldon
Matthias, Eckart
Mavroyannis, Constantine
Mazur, Eric
McClelland, Jabez J.
McDowell, Robin S.
McGee, Thomas J.
McGuire, Stephen C.
McKenzie, Robert L.
McMillian, Gary B.
McPherson Jr., Leroy A.
Mead, Roy D.
Menocal, Serafin G.
Menyuk, Norman
Metcalf, Harold J.
Miers, Richard E.
Mies, Frederick H.
Miles, Richard B.
Miller, John C.
Miller, Steven M.
Miller, Terry A.
Mittleman, Marvin H.
Moody, Mitchell L.
Moore, C. Bradley
Morgan, Thomas J.
Moscatelli, Frank A.
Moskowitz, Paul A.
Moskowitz, Philip E.
Mossberg, Thomas W.
Muenter, John S.
Murnick, Daniel E.
Myers, Edmund G.
Myers, Stephen A.
Nanes, Roger
Narducci, Lorenzo M.
Nayfeh, A. H.

Nayfeh, Munir H.
Nelson, Albert C.
Nesbitt, David J.
Netzel, Thomas L.
Nitz, David E.
Nygaard, Kaare J.
O, Chun-Sing
O'Connell, Robert F.
Oldenborg, Richard C.
Olsgaard, David A.
Olson, Robert A.
Omidvar, Kazem
Paisner, Jeffrey A.
Pantinakis, Apostolos
Papayoanou, Aristotle
Parks, James E.
Parmenter, Charles S.
Pau, Louis F.
Payne, Marvin G.
Pecora, Robert
Pegg, David J.
Petersen, Alan B.
Petuchowski, Samuel J.
Peuse, Bruce W.
Peyton, Bernard J.
Pfender, Emil
Phillips, William D.
Picard, Richard H.
Piltch, Martin S.
Pine, Alan S.
Pipkin, Francis M.
Powers, E. L.
Pratt, David W.
Pratt, Stephen T.
Prelas, Mark A.
Priestley, Eldon B.
Prior, Michael H.
Pritt Jr., Alfred T.
Ramsey, John M.
Rank, David H.
Rast Jr., Howard E.
Ravishankara, Akkihebbal R.
Raymer, Michael G.
Reck, Gene P.
Reddy, K. V.
Reddy, Satti P.
Reilly, James P.
Reiner, Robert H.
Rentzepis, P. M.
Richardson, Jeffery H.
Richardson, Ralph J.
Rigrod, William W.
Rink, John P.
Rizzo, Joseph
Roberts, Thomas G.
Robinson, Edward J.
Rockwood, Stephen D.
Rosner, S. David
Rothe, Erhard W.
Rubin, Kenneth
Russell, Gary R.
Saha, Haripada
Saloman, Edward B.
Sandner, Wolfgang
Sando, Kenneth M.
Sansonetti, Craig J.
Sargent III, Murray
Sarka, Benjamin NM
Scarl, Donald B.

Schawlow, Arthur L.
Schearer, Laird D.
Scherer, James R.
Schlachter, Alfred S.
Schlessinger, Leonard
Schofield, Keith
Schueler, Bruno W.
Schuessler, Hans A.
Schultz, John W.
Schwendeman, Richard H.
Scott Jr., John E.
Scully, Marlan O.
Sell, Jeffrey A.
Setser, D. W.
Shand, Michael L.
Shank, Charles V.
Sharpton, Francis A.
Shay, Thomas M.
Shen, Y. R.
Sher, Mark H.
Shimkaveg, Gregory M.
Shirk, James S.
Shirts, Randall B.
Shoemaker, Richard L.
Shore, Bruce W.
Shugart, Howard A.
Siebert, Donald R.
Siegman, Anthony E.
Silfvast, William T.
Silver, Joel A.
Silver, Joshua D.
Silverman, Mark P.
Siomos, Konstadinos
Skinner, Charles H.
Skinner, Gordon B.
Skowronek, Maurice
Slanger, Tom
Smalley, R. E.
Smith, Allan L.
Smith, Arthur A.
Smith, David C.
Smith, DeForest F.
Smith, Gregory P.
Smith, Leslie N.
Smith, Peter L.
Smith, Winthrop W.
Smyth, Kermit C.
Snyder, James J.
Solarz, Richard W.
Sorem, Michael S.
Spicer, Leonard D.
Sposito, Garrison
Sprouse, Gene D.
Srivastava, B. N.
St. Peters, Richard L.
Stalder, Kenneth R.
Stallcop, James R.
Stanton, Alan C.
Stearns, John W.
Steel, Duncan G.
Stehle, Philip
Stein, Samuel R.
Stidham, Howard D.
Stockdale, John A.
Stoicheff, Boris P.
Stroke, H. Henry
Stroud Jr., Carlos R.
Stumpf, Bernhard J.
Stwalley, William C.

3.6 Interactions of Laser Radiation with Atoms and Molecules

Suckewer, S.
Swamy, Nyayapathi V.
Switzer, Gary L.
Sze, Robert C.
Szoke, Abraham
Taylor, Raymond L.
Taylor, Ronald D.
Teets, Richard E.
Thomas, Dan M.
Thomson, George M.
Toennies, Jan Peter
Topp, Michael R.
Trump, Darryl D.
Tupa, Dale
Turechek, John J.
Uzer, Turgay
Van Haeringen, Willem
Verdeyen, Joseph T.
Verdieck, James F.
Verhaar, Boudewyn J.
Viswanath, A. Kasi
Volk, Charles H.
Walkup, Robert E.
Wang, Charles P.
Wang, Chia P.
Wang, S. C.
Ward, John F.
Warner, Ray A.
Watts, Richard N.
Weber, Marvin J.
Weigold, Erich
Weiner, John
Weisman, R. B.
Weissbluth, Mitchel
Weissler, Gerhard L.
Weiszmann, Andrei N.
Weston Jr., Ralph E.
White, Henry W.
White, Michael G.
Whitney, Wayne T.
Wickun, William G.
Wiesenfeld, John
Wiff, Donald R.
Wiggins, Thomas A.
Williams, Michael D.
Wilson, Jack
Wine, Paul H.
Wineland, David J.
Wing, William H.
Winkler, Irwin C.
Wirth, Mary M.
Wodarczyk, Francis J.
Wolk, Gary L.
Wong, Ngai C.
Woo, Shien-Biau
Wood II, Obert R.
Woodruff, Pamela R.
Woody, Bernard A.
Worden Jr., Earl E.
Worlock, John M.
Wormhoudt, Joda C.
Wright, John C.
Wright, John J.
Wright, Lawrence A.
Wu, C. Y. R.
Wyatt, Robert E.
Wynne, James J.
Wyss, Jerry C.
Yabuzaki, Tsutomu
Yamanaka, Masanobu
Yang, Sze-Cheng
Yeh, Shoou-Dyi
Yoder, Marvel J.
Young, James F.
Zaidi, Haider R.
Zajonc, Arthur G.
Zare, Richard N.
Zewail, Ahmed H.
Zimmerman, Myron L.
Zorn, Jens C.
zuPutlitz, Gisbert

Ackerhalt, Jay R.
Affatato, Joseph F.
Anderson, Thomas G.
Armstrong Jr., Lloyd
Avouris, Phaedon
Baer, Tomas
Bakshi, Pradip M.
Bayfield, James E.
Becker, Michael F.
Bel Bruno, Joseph J.
Berg, Jacqueline O.
Bergeman, Thomas H.
Bernhardt, Anthony F.
Berry, R. Steven
Bischel, William K.
Bjorkholm, John E.
Bjorklund, Gary C.
Bokor, Jeffrey
Borsella, Elisabetta
Brewer, Richard G.
Brown, Lowell S.
Bucksbaum, Philip H.
Cantrell, Cyrus D.
Carlsten, John L.
Carroll, Clark E.
Cho, Chung Won
Chu, Shih-I
Cody, Regina J.
Comaskey, Brian
Compton, Robert N.
Cooper, C. Dewey
Cooper, John
Coulter, Philip W.
Czuchlewski, Stephen J.
DeBeer, David P.
Dehmer, Patricia M.
DeTemple, Thomas A.
DeVries, Paul L.
Diels, Jean-Claude
DiMauro, Louis F.
Dixit, Shamasundar N.
Drake, Charles W.
Drummond, David L.
Eberly, Joseph H.
Ehlotzky, Fritz
Engelking, Paul C.
Findley, Gary L.
Fisanick, Georgia J.
Fletcher, Gary D.
Fowler, Bruce W.
Fradkin, David M.
Fried, Zoltan
Friedrich, Donald M.
Galbraith, Harold W.
Garrett, William R.
Geltman, Sydney
Gold, L. Peter
Goldsmith, John EM
Goldwire Jr., Henry C.
Golub, John E.
Gould, Phillip L.
Grant, Edward R.
Gruen, Dieter M.
Haan, Stanley L.
Hauer, Allan A.
Heller, Donald F.
Hessler, Jan P.
Hilborn, Robert C.
Horwitz, Alexander B.

Hudgens, Jeffrey W.
Hudson, Robert D.
Huo, Winifred M.
Jacobs, Verne L.
Jensen, Barbara L.
Johnson, Philip M.
Judd, O'Dean P.
Katayama, Mikio
Kay, Kenneth G.
Kay, Richard B.
Kelley, Paul L.
Keto, John W.
Kimble, Harry J.
Kleiber, Paul D.
Klein, Lewis
Koch, Peter M.
Krasinski, Jerzy S.
Kumar, Prem
Kung, Robert T.
Lambropoulos, P. P.
Lampe, Frederick W.
Lau, Albert M.
Lee, Francis W.
Loy, Michael T.
Lucatorto, Thomas B.
Majumder, Protik K.
Mavroyannis, Constantine
Mazur, Eric
McGee, Thomas J.
McGuire, Eugene J.
McGuirk, Michael
McIlrath, Thomas J.
McKoy, Vincent
McPherson Jr., Leroy A.
Mies, Frederick H.
Milchberg, Howard M.
Miller, John C.
Mossberg, Thomas W.
Nayfeh, A. H.
Olsen, Thomas
Pan, Liwen
Payne, Marvin G.
Peuse, Bruce W.
Phillips, William D.
Pitts Jr., James N.
Pratt, Stephen T.
Raymer, Michael G.
Reading, Melissa M.
Reinhardt, William P.
Reiss, Howard
Rhodes, Charles K.
Riley, Merle E.
Robinson, Edward J.
Rockwood, Stephen D.
Ronn, Avigdor M.
Schlessinger, Leonard
Shakeshaft, Robin
Shay, Thomas M.
Shirley, Jon H.
Shirts, Randall B.
Shore, Bruce W.
Silverman, Mark P.
Skinner, Charles H.
Smith, Stephen J.
Starace, Anthony F.
Stehle, Philip
Stockdale, John A.
Strickler, Stewart J.
Szoke, Abraham

Taylor, Howard S.
Teague, Michael R.
Teich, Malvin C.
Tellinghuisen, Joel B.
Turechek, John J.
Turley, R. S.
Walker, Robert B.
Wang, Wen-Cheng
Ward, John F.
Weber, Marvin J.
Wessel, John E.
Winkler, Irwin C.
Wirth, Mary M.
Wodkiewicz, Krzysztof
Woodruff, Pamela R.
Worlock, John M.
Wu, C. Y. R.
Wyatt, Robert E.
Yuan, Jian-Min
Zaidi, Haider R.
Zare, Richard N.
Zewail, Ahmed H.

3.8 Laser Chemistry; Photochemistry

Adrian, Frank J.
Andersen, Nils O.
Anderson, James B.
Anderson, Roger W.
Anderson, Thomas G.
Atkinson, George H.
Auerbach, Roy A.
Avouris, Phaedon
Barker, John R.
Bartell, Lawrence S.
Bates Jr., Richard D.
Bauer, Simon H.
Bel Bruno, Joseph J.
Benson, Sidney W.
Berman, Paul R.
Bernhardt, Anthony F.
Bernstein, Richard B.
Bhaskar, Natarajan D.
Bieniek, Ronald J.
Birely, John H.
Borsella, Elisabetta
Bowden, Charles M.
Bowers, Michael T.
Bray, Robert G.
Breckenridge, W. H.
Brenner, Douglas M.
Brooks, Philip R.
Burnett, Keith
Butler, James E.
Caird, John A.
Cantrell, Cyrus D.
Carr Jr., Robert W.
Castleman Jr., A. W.
Ch'en, Shang-Yi
Chakraborti, Parimal K.
Chang, Tai Y.
Chen, Hao-lin
Chen, Joseph C.
Christiansen, Walter H.
Christophorou, Loucas G.
Cohen, E. R.
Cohen, Leslie
Coleman, Paul D.
Collins, George J.
Cooper, John
Cooper, Walter
Crane, John K.
Cross, Jon B.
Curl, Robert F.
Dagata, John A.
Davis, James I.
Dehmer, Joseph L.
Dehmer, Patricia M.
DeKoven, Benjamin M.
Demir, Oktay
DeVries, Paul L.
Diestler, D. J.
Dorfman, Leon M.
Dows, David A.
Drummond, David L.
Duncanson Jr., John A.
Ehlotzky, Fritz
Ehrlich, Daniel J.
Engelken, Robert D.
Eno, Larry
Estler, Ron C.
Falcone, Roger W.
Fisher, H. Leonard

Flynn, G. W.
Foltz, Greg W.
Frankel Jr., Donald S.
Freed, Karl F.
Frommhold, Lothar W.
Gallagher, Thomas F.
Gamo, Hideya
Garbuny, Max
Gelbart, William M.
Geltman, Sydney
Genack, Azriel
George, Thomas F.
Glauber, Roy J.
Goldberg, Ira B.
Golden, David M.
Gordon, Robert J.
Gouterman, Martin
Grant, Edward R.
Greve, Peter
Guillory, William A.
Gutcheck, Robert A.
Hahne, Gerhard E.
Hall, Richard B.
Hancock, Kent J.
Hanrahan, Robert J.
Hargis Jr., Philip J.
Harter, William G.
Haus, Hermann A.
Havey, Mark D.
Hayden, Howard C.
Hayhurst, Thomas L.
Heicklen, Julian P.
Heidner III, R. F.
Heller, Donald R.
Hobbs, Robert
Holroyd, Richard
Horsley, John A.
Howard, Carleton J.
Howard, Robert E.
Hrubesh, Lawrence W.
Huestis, David L.
Humpherys, Kent C.
Huntley, Wright H.
Hutchison, Sheldon B.
Huwel, Lutz
Ivey, Henry F.
Jackson, William M.
Jacobs, Ralph R.
Jeffries, Jay B.
Jimenez-Mier, Jose
Jones, Roger C.
Judd, O'Dean P.
Julienne, Paul S.
Kaldor, Andrew P.
Katayama, Mikio
Katz, Joseph L.
Keefer, Dennis R.
Keller, Richard A.
Kelley, J. Daniel
Kepros, John G.
Keto, John W.
Kielkopf, John F.
Kleiber, Paul D.
Klots, Cornelius E.
Koch, Peter M.
Koffend, John B.
Krause, Herbert F.
Kung, Robert T.

Kupperman, Aron
Kurnit, Norman A.
Lam, Juan T.
Lampe, Frederick W.
Langhoff, Stephen R.
Langsam, Yedidyah
Lau, Albert M.
Lee, Edward K.
Lee, Yuan T.
Leventhal, Jacob J.
Liu, Jenn-Ying
Liu, Yung S.
Loree, Thomas R.
Luntz, Alan C.
Lynds, Lahmer
Malley, Michael M.
Marcus, Rudolph A.
Mariella Jr., Raymond P.
Martin, Richard M.
Mathur, Bhagwan P.
Mattson, Timothy G.
Mazur, Eric
McDowell, Robin S.
McGlynn, Sean P.
Melamed, Nathan T.
Melton, Lynn A.
Miller, John C.
Mittleman, Marvin H.
Mizushima, Masataka
Moore, C. Bradley
Mozumder, A.
Murphy, John C.
Nayfeh, Munir H.
Noid, Donald W.
Okabe, Hideo
Oldenborg, Richard C.
Ollison, William M.
Olson, Gregory L.
Parmenter, Charles S.
Piebold, Gerald J.
Piltch, Martin S.
Pimentel, George C.
Polak-Dingels, Penelope M.
Powers, E. L.
Pratt, David W.
Pratt, Stephen T.
Preses, Jack M.
Priestley, Eldon B.
Pritchard, David E.
Pritt Jr., Alfred T.
Pugh Jr., Henry L.
Quick Jr., Charles R.
Quigley, Gerard P.
Radziemski, Leon
Ravishankara, Akkihebbal R.
Ray, Gary W.
Reck, Gene P.
Register, David F.
Reilly, James P.
Reiner, Robert H.
Reinhardt, William P.
Reiss, Howard
Rentzepis, P. M.
Rice, James K.
Rich, Joseph W.
Rice, Stuart A.
Richardson, Jeffery H.
Rink, John P.

Robinson, G. W.
Rockwood, Stephen D.
Rogers, Wade T.
Ronn, Avigdor M.
Rose, Timothy L.
Rosenberg, Leonard
Rothe, Erhard W.
Schuler, Robert H.
Schwirzke, Fred R.
Sheats, James R.
Shirk, James S.
Siomos, Konstadinos
Slocum, Robert E.
Solarz, Richard W.
Spicer, Leonard D.
Spruch, Larry
Staley, Ralph H.
Steinberg, Martin
Steinfeld, Jeffrey I.
Stern, Richard C.
Strickler, Stewart J.
Stricklett, Kenneth L.
Stroke, H. Henry
Stumpf, Bernhard J.
Stwalley, William C.
Tang, Sheng Y.
Taylor, Lyle H.
Taylor, Raymond L.
Taylor, Ronald D.
Thomas, H. Ronald
Topp, Michael R.
Trainor, Daniel W.
Turner, Jr., Charles E.
Uzer, Turgay
Verdieck, James F.
Wang, Kang-Lung
Way, Kermit
Weiner, John
Weisman, R. B.
West, Gary A.
Weston Jr., Ralph E.
Wiesenfeld, John
Wilson, John W.
Witriol, Norman M.
Wodarczyk, Francis J.
Wolk, Gary L.
Wood, Lowell L.
Worden Jr., Earl E.
Wright, John J.
Wyatt, Robert E.
Yablonovitch, Eli
Yabuzaki, Tsutomu
Yang, Sze-Cheng
Yardley, James T.
Yeh, Shoou-Dyi
Yuan, Jian-Min
Zakheim, David S.
Zewail, Ahmed H.
Zimmerman, I. H.
Zittel, Paul F.
Zitter, Robert N.
Zorn, Jens C.

Bakshi, Pradip M.
Band, Hans E.
Bederson, Benjamin
Bergeman, Thomas H.
Bergquist, James C.
Berk, Alexander
Bhattacharya, Samir K.
Bollinger, John J.
Brandenberger, John R.
Bryant, Howard C.
Chen, Augustine C.
Chiu, Lue-Yung C.
Cohen, Stanley
Comaskey, Brian
Cooper, John W.
Crane, John K.
Davidson, Ernest R.
Davis, Robert W.
Delos, John B.
DeSerio, Robert
DeTemple, Thomas A.
Donahue, Joey B.
Dunning, F. Barry
Dyke, Thomas R.
Eck, Thomas G.
Erber, Thomas
Field, Robert W.
Fletcher, Gary D.
Fried, Zoltan
Friedrich, Harald S.
Garscadden, Alan
Garstang, Roy H.
Geller, M.
Geltman, Sydney
Gilbert, Sarah L.
Goldberg, Leon P.
Good Jr., Roland H.
Goss, Larry P.
Greene, Chris H.
Ham, Mooyoung
Hameka, Hendrik F.
Hammer, J. M.
Hanson, David M.
Harmony, Marlin D.
Heckel, Blayne R.
Hellwig, Helmut W.
Hess Jr., Doren W.
Hochstrasser, Robin M.
Holt, Helen K.
Hrubesh, Lawrence W.
Hughes, Vernon W.
Jacobs, Verne L.
Jaduszliwer, Bernardo
Jason, Andrew J.
Jen, Chih K.
Jensen, Craig C.
Jeys, Thomas H.
Johnston, Lawrence H.
Jones, Larry A.
Kelleher, Daniel E.
Kern, C. W.
Klein, Lewis
Kleppner, Daniel
Knight, Randall D.
Kocher, Carl A.
Korevaar, Eric J.
Lax, Benjamin
Levy, Donald H.
Levy, Laurent P.

Lewis, Robert R.
Littman, Michael G.
Liu, Jenn-Ying
MacAdam, Keith B.
Marchetti, Alfred P.
McDermott, Mark N.
McGlynn, Sean P.
McLaughlin, Daniel J.
McMillian, Gary B.
Metcalf, Harold J.
Moscatelli, Frank A.
Myers, Stephen A.
Nanes, Roger
Nesbitt, David J.
O'Connell, Robert F.
Olson, Robert A.
Pepper, David M.
Phillips, William D.
Raab, Frederick J.
Rau, A. R.
Retzloff, David G.
Rich, Joseph W.
Ries, Richard R.
Roman, Ward C.
Saloman, Edward B.
Schatz, Paul N.
Schearer, Laird D.
Schlessinger, Leonard
Schnepp, Otto
Schrader, David M.
Shugart, Howard A.
Simon, Barry
Starace, Anthony F.
Stoneman, Robert C.
Stumpf, Bernhard J.
Swamy, Nyayapathi V.
Taylor, Howard S.
Thomas, Dan M.
Turley, R. S.
Vala, Martin
Van Siclen, Clinton D.
Vander Sluis, Kenneth L.
Verhaar, Boudewyn J.
Wadehra, Jogindra M.
West, John B.
Westerveld, Willem B.
Wieman, Carl E.
Wineland, David J.
Winkler, Irwin C.
Woods, R. Claude
Worden Jr., Earl E.
Worlock, John M.
Yamanaka, Masanobu
Zimmerman, Myron L.
Zollars, Byron G.
Zorn, Jens C.

3.10 Nonlinear Interactions of Radiation with Atoms and Molecules

Abraham, George
Ackerhalt, Jay R.
Antcliff, Richard R.
Armstrong Jr., Lloyd
Babbitt, William R.
Baliga, Shankar B.
Bayfield, James E.
Beausoleil Jr., R. G.
Beiting III, Edward J.
Berg, Jacqueline O.
Berman, Paul R.
Bernhardt, Anthony F.
Bischel, William K.
Bjorkholm, John E.
Bjorklund, Gary C.
Blatt, Rainer
Bobin, Jean Louis
Bokor, Jeffrey
Brewer, Richard G.
Brooker, Murray H.
Brown, Lorin W.
Bucksbaum, Philip H.
Burris Jr., John
Carlsten, John L.
Carmichael, Howard I
Carroll, Clark E.
Cho, Chung Won
Chu, Ben
Code, R. F.
Coleman, Paul D.
Compaan, Alvin
Cooper, C. Dewey
Cooper, John
Coulter, Claude A.
Cox Jr., Hollace L.
Cromer, Christopher L.
Davis, Lawrence W.
DeBeer, David P.
DeTemple, Thomas A.
Dixit, Shamasundar N.
Doolen, Gary D.
Dyke, Thomas R.
Ehlotzky, Fritz
Ewing, James J.
Exton, Reginald J.
Falk, Joel
Fayer, Michael D.
Figueira, Joseph F.
Fisher, Robert A.
Fleury, Paul A.
Foltz, Nevin D.
Franck, Carl P.
Garrett, William R.
Giordmaine, Joseph A.
Golub, John E.
Good Jr., William E.
Gruen, Dieter M.
Haan, Stanley L.
Hall, Richard B.
Hall, Robert J.
Hameka, Hendrik F.
Hamermesh, Morton
Hammer, J. M.
Hartmann, Sven R.
Heer, Clifford V.
Heller, Donald F.
Herm, Ronald R.
Hessler, Jan P.
Hsu, Hsiung
Hyman, Howard A.
Jennings, Donald A.
Jensen, Barbara L.
Johnston Jr., Thomas F.
Kachru, Ravinder
Katayama, Mikio
Kelley, Paul L.
Keto, John W.
Kim, Jin J.
Klein, Lewis
Kostiuk, Theodor
Kramer, Steven D.
Krasinski, Jerzy S.
Kumar, Prem
Kung, Robert T.
Kurnit, Norman A.
Laane, Jaan
Langhoff, Peter W.
Lee, Chi-Hsiang
Lee, Ching-Tsung
Lee, Sang S.
Lee, Siu-Au
Lerman, Juan-Carlos
Leroi, George E.
Levenson, Marc D.
Levy, Laurent P.
Liberman, Irving
Lightman, Allan J.
Liu, Yung S.
Loree, Thomas R.
Lunney, James G.
Mahon, Rita
Malley, Michael M.
Mansikka, Kauko A.
Marron, Michael T.
Massey, Gail A.
Mavroyannis, Constantine
McIlrath, Thomas J.
McKenzie, Robert L.
Mead, Roy D.
Menyuk, Norman
Miller, John C.
Murphy, John C.
Murray, John R.
Narducci, Lorenzo M.
Nayfeh, A. H.
Olsen, Thomas
Olsgaard, David A.
Pantinakis, Apostolos
Perkowitz, Sidney
Peuse, Bruce W.
Pierce, Thomas M.
Piltch, Martin S.
Rast Jr., Howard E.
Raymer, Michael G.
Rhodes, Charles K.
Rich, Nathan H.
Robinson, Edward J.
Rothenberg, Joshua E.
Rybka, Theodore W.
Sarka, Benjamin NM
Schuler, Robert H.
Schultz, John W.
Schumer, Douglas B.
Seki, Hajimes
She, Chiao Y.
Shen, Y. R.
Shirley, Jon H.
Silverman, Mark P.
Singh, Surendra P.
Skinner, Charles H.
Skowronek, Maurice
Slater, Richard
Smith, Stephen J.
Snyder, James J.
Steel, Duncan G.
Steele, William A.
Stoicheff, Boris P.
Stotlar, Suzanne C.
Switzer, Gary L.
Sze, Robert C.
Tang, Kenneth Y.
Taylor, David J.
Tehrani, Mohammad M.
Tomlinson, W. J.
Trainor, Daniel W.
Trump, Darryl D.
Turechek, John J.
Turley, R. S.
Verdieck, James F.
Vincent, James S.
Walls, Daniel F.
Wang, Wen-Cheng
Ward, John F.
Weber, Heinz P.
Weber, Marvin J.
Weissbluth, Mitchel
Weiszmann, Andrei N.
Whitney, Kenneth G.
Whitney, Wayne T.
Wieder, Grace M.
Williams, W. David
Willis, Paul A.
Winkler, Irwin C.
Wodkiewicz, Krzysztof
Wong, Ngai C.
Worlock, John M.
Wright, John C.
Wu, C. Y. R.
Wynne, James J.
Yabuzaki, Tsutomu
Yardley, James T.
Yoder, Marvel J.
Young, James F.
Yuan, Jian-Min
Zaidi, Haider R.

Ackerhalt, Jay R.
Affatato, Joseph F.
Beausoleil Jr., R. G.
Bender, Peter L.
Bergeman, Thomas H.
Beyerinck, Herman C.
Carlsten, John L.
Carmichael, Howard I
Cohn, Gerald E.
Coulter, Claude A.
Cox Jr., Hollace L.
Dixit, Shamasundar N.
Eberly, Joseph H.
Fowler, Bruce W.
Gentile, Thomas R.
Giordmaine, Joseph A.
Golub, John E.
Gould, Phillip L.
Hall, Robert J.
Hammer, J. M.
Heinzen, Daniel J.
Hess Jr., Doren W.
Jen, Chih K.
Jensen, Barbara L.
Jeon, Yoon H.
Kim, Jin J.
Kimble, Harry J.
Kumar, Prem
Kurnit, Norman A.
Lee, Ching-Tsung
Lee, Sang S.
Levenson, Marc D.
Mavroyannis, Constantine
Mossberg, Thomas W.
Murray, Frank
Narducci, Lorenzo M.
Ni, Wei-Tou
O'Connell, Robert F.
Olsen, Thomas
Pendleton, Hugh N.
Peng, Jin-Sheng
Picard, Richard H.
Pierce, Thomas M.
Raymer, Michael G.
Rothenberg, Joshua E.
Scarl, Donald B.
Shirley, Jon H.
Singh, Surendra P.
Skowronek, Maurice
Smith, Stephen J.
Steel, Duncan G.
Stehle, Philip
Stroud Jr., Carlos R.
Tehrani, Mohammad M.
Teich, Malvin C.
Turley, R. S.
Walls, Daniel F.
Walsh, Peter J.
Wang, Chao C.
Wang, Chia P.
Weiszmann, Andrei N.
Werner, Samuel A.
Willis, Paul A.
Wodkiewicz, Krzysztof
Wong, Ngai C.
Zaidi, Haider R.

4.1 Optical Techniques

Abraham, George
Abraham, Neal B.
Abrams, Richard L.
Akerman, M. A.
Alcaraz, Ernest C.
Ali, Abdul W.
Altgilbers, Larry L.
Ames, Donald P.
Antcliff, Richard R.
Ashkin, Arthur
Asmus, John F.
Atkinson, George H.
Atkinson, J. B.
Aubrey, Bertrand B.
Babbitt, William R.
Baldwin, George C.
Band, Yehuda B.
Barr Jr., Thomas A.
Bartiromo, Rosario
Baum, Guenter G.
Bay, Zoltan L.
Beausoleil Jr., R. G.
Becker, Michael F.
Beiting III, Edward J.
Bekefi, G.
Berg, Jacqueline O.
Bermudez, Victor M.
Bernard, Davy L.
Bernhardt, Anthony F.
Betts, Jeanette A.
Beverly III, Robert E.
Beyerinck, Herman C.
Bhaskar, Natarajan D.
Bhaumik, Mani L.
Bien, Fritz
Bigio, Irving J.
Binns, Walter R.
Birnbaum, Milton
Bischel, William K.
Bjorkholm, John E.
Bjorklund, Gary C.
Black, Truman D.
Blass, William E.
Blatt, Rainer
Bloembergen, Nicolaas
Bloom, Arnold L.
Boness, M. J.
Bowden, Charles M.
Bradford Jr., Robert S.
Brandenberger, John R.
Brau, Charles D.
Brewer, Richard G.
Bricks, Bernard G.
Bridges, William B.
Bristow, Thomas C.
Bucksbaum, Philip H.
Burnham, Ralph
Burris Jr., John
Butler, J. K.
Byer, Robert L.
Cantrell, Cyrus D.
Carbone, Robert J.
Cardon, Bartley L.
Cason, Charles M.
Cattolica, Robert J.
Celotta, Robert J.
Champagne, Louis F.
Chang, William SC
Chen, Che-Jen

Chen, Chin-Lin
Chien, K. R.
Chimenti, Robert J.
Christiansen, Walter H.
Chu, Ben
Code, R. F.
Cohn, Gerald E.
Coleman, Paul D.
Collins, George J.
Collins, Robert J.
Comaskey, Brian
Compaan, Alvin
Cone, Rufus L.
Cooke, William E.
Cooper, Gary W.
Couillaud, Bernard J.
Cox Jr., Hollace L.
Crawford Jr., Bryce
Culver, William H.
Curtis, Earl C.
Czuchlewski, Stephen J.
Dasch, Cameron J.
Davidson, Steven A.
Davis, Christopher C.
Davis, Lawrence W.
De Lucia, Frank C.
Demir, Oktay
Dendramis, Achille L.
Denes, Louis J.
DeTemple, Thomas A.
DeVoe, Ralph G.
DeYoung, Russell J.
Dobbs, Gregory M.
Dodd, Jack G.
Dorain, Paul B.
Dougal, Arwin A.
Dowe Jr., R. Michael
Dowling, Jerome M.
Drake, J. M.
Dreyfus, Russell W.
Druger, Stephen D.
Drullinger, Robert E.
Drummond, Peter D.
Duncanson Jr., John A.
Dunning, F. Barry
Eckbreth, Alan C.
Eckstrom, Donald J.
Edighoffer, John A.
Eerkens, Jeff W.
Ehrlich, Daniel J.
Elton, Dr. Raymond C.
Emmons, Donald A.
English, Thomas C.
Ernie, Douglas W.
Esherick, Peter
Ewing, James J.
Ezekiel, Shaoul
Falk, Joel
Fayer, Michael D.
Feld, Michael S.
Feldman, Donald W.
Figueira, Joseph F.
Fisher, Edward R.
Fisher, Robert A.
Fletcher, William H.
Fohl, Timothy
Foltz, Nevin D.
Forsyth, James M.
Fortson, E. Norval

Francke, Ricardo E.
Freed, Charles
Freeman, Richard R.
Friedman, Joel M.
Furumoto, Horace
Galbraith, Harold W.
Gallagher, Thomas F.
Gamo, Hideya
Garetz, Bruce A.
Gelinas, Robert J.
Genack, Azriel
German, Kenneth R.
Gibbs, Hyatt M.
Gilbert, Sarah L.
Goldstein, John C.
Gordon, James P.
Goss, Larry P.
Greenlees, G. W.
Greve, Peter
Griffiths, James E.
Grosjean, Dennis F.
Gruen, Dieter M.
Guch Jr., Steve
Gundersen, Martin A.
Gupta, Rajendra
Gustafson, Ture K.
Haglund Jr., Richard F.
Hammer, J. M.
Hancock, Kent J.
Harney, Robert C.
Harris, Stephen E.
Hassan, H. A.
Hawk, James F.
Heckel, Blayne R.
Heer, Clifford V.
Heller, Donald F.
Hellwarth, Robert W.
Hellwig, Helmut W.
Hernquist, Karl G.
Hill Jr., Ralph H.
Hinrichs, C. K.
Hirshfield, Jay L.
Hochuli, Urs E.
Hodges Jr., Dean T.
Holt, Richard A.
Hoppe, John C.
Houston, John M.
Howgate, David W.
Hsu, Hsiung
Hudson, George E.
Huebner, Russell H.
Hummer, David G.
Huntley, Wright H.
Hutchison, Sheldon B.
Hwang, Dah-Min D.
Hyman, Howard A.
Islam, Muhammad A.
Jackson, William M.
Jacobs, Ralph R.
Jacobs, Stephen F.
Jaffe, Hans H.
Jennings, Donald A.
Jeys, Thomas H.
Johnson, Lee K.
Johnson, Roy R.
Johnston, Lawrence H.
Johnston Jr., Thomas F.
Jones, Claude R.
Jones, Patrick L.

Karras, Thomas W.
Kasner, William H.
Kebabian, Paul
Kelley, Paul L.
Kenan, Richard P.
Killinger, Dennis K.
Kim, Jin J.
Kimble, Harry J.
Kline, Laurence E.
Knize, Randall J.
Kolb, Alan C.
Kondilis Jr., Francis N.
Konopnicki, Marek J.
Kostiuk, Theodor
Kowalski, Frank V.
Krasinski, Jerzy S.
Kroll, Norman M.
Ku, Robert T.
Kumar, Prem
Kung, Robert T.
Kuo, Chien-Yu
Kurnit, Norman A.
Kushner, Mark J.
Kwok, Munson A.
Lacina, William B.
Lam, Juan T.
Lam, Leo K.
Lambropoulos, P. P.
Lang Jr., John C.
Lapatovich, Walter P.
Lapp, Marshall
Larson, Daniel J.
Laudenslager, James B.
Lee, Chi-Hsiang
Lee, Edward K.
Lee, Ja H.
Lee, Long C.
Lee, Sang S.
Lepage, G. P.
Leslie, Scott G.
Leuchs, Gerhard
Levenson, Marc D.
Levy, Donald H.
Lewis, Lindon L.
Lewis, Robert R.
Liao, Paul F.
Liberman, Irving
Lightman, Allan J.
Liu, Jenn-Ying
Liu, Yung S.
Loder, Rurik K.
Loree, Thomas R.
Louisell, William H.
Lowry, Jerald F.
Loy, Michael T.
Lunney, James G.
Lyons, Peter B.
MacKnight, Allen K.
Madey, J. M.
Mahr, H.
Mandel, Leonard
Mandelberg, Hirsch I.
Manor, Robert E.
Martin, L. R.
Massey, Gail A.
Mattson, Timothy G.
McArthur, David A.
McClelland, Jabez J.
McDermott, Mark N.

McFarlane, Ross A.
McIlrath, Thomas J.
McPherson Jr., Leroy A.
Mead, Roy D.
Menocal, Serafin G.
Menyuk, Norman
Mickish, Roger A.
Miles, R. O.
Miles, Richard B.
Miley, George H.
Miller, Steven M.
Miller, Thomas G.
Moeny, William M.
Mollow, Benjamin R.
Moody, Stephen E.
Moody, Elizabeth A.
Morton, Richard G.
Mossberg, Thomas W.
Mumma, Michael J.
Murphy, John C.
Murray, John R.
Myers, Edmund G.
Myers, Stephen A.
Neil, George R.
Nelson, Leonard Y.
Netzel, Thomas L.
Newman, Leon A.
Nighan, William L.
North, Dwight O.
Oldenborg, Richard C.
Olson, Robert A.
Owens, James C.
Paisner, Jeffrey A.
Panock, Richard
Papayoanou, Aristotle
Parsons, Donald F.
Passenheim, Burr C.
Patel, C. Kumar K.
Patterson, Edward L.
Paulson, Ronald F.
Peatman, William B.
Perkowitz, Sidney
Petersen, Alan B.
Petuchowski, Samuel J.
Peuse, Bruce W.
Phadke, Laxman G.
Phillips, William D.
Piebold, Gerald J.
Pilloff, H. S.
Piltch, Martin S.
Pistoresi, Denis J.
Poultney, Sherman K.
Powell, Edward
Powell, Howard T.
Powers, E. L.
Prelas, Mark A.
Raab, Frederick J.
Rank, David H.
Rayman, Marc D.
Raymer, Michael G.
Raymonda, John W.
Rentzepis, P. M.
Richardson, Ralph J.
Richardson, Jeffery H.
Rigrod, William W.
Rink, John P.
Robertson, M. M.
Rockwood, Stephen D.
Ronn, Avigdor M.
Rothenberg, Joshua E.
Royce, Barrie S.
Ruff, George A.

Russell, Gary R.
Ryan, Frederick M.
Rybka, Theodore W.
Sarka, Benjamin NM
Sarraf, Sanwal P.
Scarl, Donald B.
Schadt, Randall J.
Schappert, Gottfried T.
Schearer, Laird D.
Schieve, W. C.
Schlie, LaVerne A.
Schneider, Richard T.
Schumer, Douglas B.
Schweitzer Jr., Walter G.
Scully, Marlan O.
Sell, Jeffrey A.
Shay, Thomas M.
Shimkaveg, Gregory M.
Shore, Bruce W.
Sierra, Rafael A.
Silfvast, William T.
Silvers, Stuart J.
Singh, Jag J.
Skifstad, James G.
Slater, Richard
Smith, Arthur A.
Smith, George F.
Smith, Joseph A.
Snavely, Benjamin B.
Snyder, James J.
Solomon, Jerry E.
Sorem, Michael S.
Stein, Samuel R.
Stoicheff, Boris P.
Stotlar, Suzanne C.
Stroke, H. Henry
Stroud Jr., Carlos R.
Stwalley, William C.
Suhre, Dennis R.
Switzer, Gary L.
Synek, M.
Sze, Robert C.
Szoke, Abraham
Tang, Kenneth Y.
Taylor, David J.
Taylor, Lyle H.
Teague, Michael R.
Teegarden, Kenneth J.
Tehrani, Mohammad M.
Teich, Malvin C.
Terhune, Robert W.
Teubner, Peter J.
Thaler, William J.
Thomas, Dan M.
Tisone, Gary C.
Trainor, Daniel W.
Tregay, George W.
Trump, Darryl D.
Tsang, Won-Tien
Tupa, Dale
Turechek, John J.
Utterback, Nyle G.
Vanasse, George A.
Vander Sluis, Kenneth L.
Volk, Charles H.
Von Jaskowsky, Woldemar F.
Wadt, Willard R.
Walls, Fred L.
Walsh, Peter J.
Walters, G. K.
Waniek, Ralph W.
Ward, John F.

Warren Jr., Walter R.
Watts, Richard N.
Waynant, Ronald W.
Weber, Heinz P.
Weber, Marvin J.
Weiner, Eugene R.
Whiteley, Stephen R.
Whitney, Wayne T.
Whittier, James S.
Wieman, Carl E.
Wilcomb, Bruce E.
Williams, W. David
Williamson Jr., William
Willis, Charles R.
Wilson Jr., William L.
Wineland, David J.
Wittenberg, Albert M.
Wolk, Gary L.
Womack, Dennis R.
Wong, Ngai C.
Wood, Lowell L.
Wood II, Obert R.
Woody, Bernard A.
Wutzke, S. A.
Wyss, Jerry C.
Yablonovitch, Eli
Yamanaka, Masanobu
Yaney, Perry P.
Yardley, James T.
Yoder, Marvel J.
York Jr., George W.
Young, C. Gilbert
Young, James F.
Zahniser, Mark S.
Zapata, Luis E.
Zewail, Ahmed H.

4.2 Accelerator-Based A&M Physics

Aldridge III, Jack P.
Altgilbers, Larry L.
Alton, Gerald D.
Andersen, Nils O.
Appleton, B. R.
Ausloos, Pierre
Bainum, David E.
Barnett, Charles F.
Bashkin, Stanley
Battleson, Kirk W.
Beauchamp, Jesse L.
Becher, Jacob
Bekefi, G.
Bergquist, James C.
Berkner, Klaus H.
Berman, B. L.
Bernard, Davy L.
Berney, Charles V.
Bernius, Mark T.
Bernstein, E. M.
Berry, H. G.
Bersohn, Richard
Beyer, Louis M.
Bickel, William S.
Bissinger, George A.
Blattner, Richard J.
Blewett, John P.
Bliven, Steven M
Blumberg, Leroy N.
Bowman, Charles D.
Bradford Jr., Robert S.
Brandt, Werner
Brau, Charles D.
Breinig, Marianne
Brennan, James G.
Bromley, D. A.
Brown, Charles M.
Brown, Matt D.
Bryant, Howard C.
Brynjolfsson, Ari
Budick, Burton
Bukow, Hans
Burkhalter, Philip G.
Burns, Donal J.
Campbell, John L.
Carlson, Thomas A.
Cecil, Joseph N.
Center, R. E.
Chamberlin, Edwin P.
Childs, Wendell A.
Chu, Wei-Kan
Church, David A.
Clark, Arnold F.
Clegg, Thomas B.
Cocke, C. L.
Cohen, Leslie
Cohen, Stanley
Comas, James
Compton, Robert N.
Cook, Donald L.
Cooney, Patrick J.
Cordaro, Richard B.
Corderman, Reed R.
Cowgill, Donald F.
Crandall, David H.
Crasemann, Bernd
Crawford, Oakley H.
Cue, Nelson
Curnutte Jr., Basil
Curtis, Lorenzo J.
Datz, Sheldon

Davis, Jay C.
Dely, Alex
Deslattes, Richard D.
Dick, Charles E.
Dietrich, Daniel D.
Dixon, Dwight R.
Donahue, D. J.
Donahue, Joey B.
Dorfman, Leon M.
Doughty, Ben M.
Dowe Jr., R. Michael
Duggan, Jerome L.
Duncan, M. M.
Ebert, Paul J.
Edge, Ronald D.
Edighoffer, John A.
Edwards, Alan K.
Egan, Patrick O.
Ehler, Arthur W.
Ehlers, Kenneth W.
Elander, Nils O.
Elston, Stuart B.
Emery, Guy T.
Erber, Thomas
Feld, Michael S.
Ferguson, Stephen M.
Fessenden, Richard W.
Fischbeck, H. J.
Fischer, Traugott E.
Fisher, Edward R.
Fornari, Luigi S.
Fortner, Richard J.
Fou, Cheng-Ming
Franck, Carl P.
Freeman, Gordon R.
Freund, Hans J.
Fritsch, Wolfgang
Furst, Mitchell L.
Gabbard, Fletcher
Gaiser, James E.
Gay, Timothy J.
Gemmell, Donald S.
Giese, John P.
Gilbody, Henry B.
Ginter, Marshall L.
Gladisch, Michael W.
Golden, David E.
Golovchenko, J. A.
Gould, Harvey A.
Graham, William G.
Gram, Peter AM
Gray, Tom J.
Greenberg, Jack S.
Greenlees, G. W.
Gregory, Donald C.
Groeneveld, Karl-Ontjese E.
Grover, James R.
Guiragossian, Zaven G.
Gutcheck, Robert A.
Hagmann, Siegbert J.
Hall, James M.
Hamm, Robert W.
Hanna, Stanley S.
Hanrahan, Robert J.
Hansteen, Johannes M.
Hardis, Jonathan E.
Hartman, Paul L.
Haugsjaa, Paul O.
Hauser, Ulrich A.
Hayden, Howard C.

Head, Charles E.
Heckmann, Paul H.
Hird, Brian
Hirsh, Merle N.
Holland, Orin W.
Holroyd, Richard
Holt, Richard A.
Hudson, G. M.
Hughes, William M.
Huhnermann, Harry
Huneke, John C.
Jakas, Mario M.
Jason, Andrew J.
Johnsen, Russell H.
Johnson, Brant M.
Johnson, Edward A.
Jones, Keith W.
Joyce, J. M.
Kallne, Elisabeth
Kaminsky, Manfred S.
Kanter, Elliot P.
Katayama, Ichiro
Katsanos, Anastasios A.
Kaufman, Stanley L.
Kauppila, Walter E.
Kelly, John C.
Kenefick, Robert A.
Kessel, Quentin C.
Kessler Jr., Ernest G.
Kessler, Karl G.
Kim, H. J.
Kliwer, James K.
Kobayashi, Hisao
Koch, Peter M.
Kolb, Alan C.
Kramer, Stephen L.
Kraner, H. W.
Kraus, Joseph S.
Krause, Herbert F.
Krause, Manfred O.
Krohn, Kenneth A.
Kruse, Theodore H.
Kuckuck, Robert W.
Kusko, Bruce H.
Kvale, Thomas J.
Kwan, Ching-Kwan CK
Land, David J.
LaVilla, Robert E.
Leach, Sydney
Leavitt, John A.
Lee, Francis W.
Lee, Paul L.
Lenhard, Joseph A.
Leon, Melvin
Leventhal, Marvin
Lewis, David A.
Li, Tien K.
Livingston, A.E. Gene
Lockwood, Grant J.
Lodhi, Sattar K.
Long Jr., Edward R.
Lovoi, Paul
Lu, Jia-Jih
MacAdam, Keith B.
MacArthur, Duncan W.
Madden, Robert P.
Maddox, William E.
Madey, J. M.
Maeda, Kaichi
Magnuson, Gustav D.
Mangelson, Nolan F.

Markisz, John A.
Marrus, Richard
Martin, Fred W.
McCall, David W.
McClure, Gordon W.
McCorkle, R. A.
McCormick, Larry D.
McDaniel, Floyd D.
McGowan, J. W.
McGuire, James H.
McIntyre Jr., L. C.
McLaughlin, Ralph
Menegozzi, Lionel H.
Menendez, Manuel G.
Menne, Thomas J.
Meriwether, J. R.
Meyer, Fred W.
Meyerhof, Walter E.
Miers, Richard E.
Miller, Glenn H.
Miller, John H.
Miller, Phillip D.
Moak, C. D.
Molitoris, John D.
Monahan, Kevin M.
Monce, Michael N.
Moore, C. Fred
Morawitz, Hans
Morgan, Thomas J.
Mowat, J. R.
Murnick, Daniel E.
Myers, Edmund G.
Naumann, Robert A.
Neil, George R.
O, Chun-Sing
Oosterhuis, William T.
Otter, Fred A.
Overbury, Stephen H.
Park, John T.
Parr, Albert C.
Patel, C. Kumar K.
Patterson, Edward L.
Peatman, William B.
Pegg, David J.
Pepmiller, Philip L.
Pepper, George H.
Perez III, Joseph D.
Peterson, Randolph S.
Phaneuf, Ronald A.
Pindzola, Michael S.
Plecious, Robert C.
Poirier, John A.
Pollack, Edward
Powell, Edward
Powers, E. L.
Preses, Jack M.
Price, Jack L.
Prior, Michael H.
Pronko, Peter P.
Pyle, Robert V.
Quarles, C. A.
Radicati, Filippo
Raith, Wilhelm
Ramler, Warren J.
Ramsey, Norman F.
Ray, J. A.
Rayburn, Louis A.
Redi, Olav
Retzloff, David G.
Rice, James K.
Richard, Patrick

Roberts, Thomas G.
Robinson, Edward L.
Root, John W.
Rosenberg, Richard A.
Ryan, Stewart R.
Rybka, Theodore W.
Sartwell, Bruce D.
Saylor III, Tillman K.
Schectman, Richard M.
Schlachter, Alfred S.
Schlie, LaVerne A.
Schmieder, Robert W.
Schuler, Robert H.
Scully, Marlan O.
Seely, David G.
Sellin, Ivan A.
Senba, Masayoshi
Shafroth, Stephen M.
Sham, Tsun K.
Silver, Joshua D.
Simons, Donald G.
Singh, Jag J.
Skutlartz, Alexander E.
Smith, Winthrop W.
Souder, Paul A.
Spooner, David W.
Sprouse, Gene D.
St. John, Robert M.
St. Peters, Richard L.
Stalder, Kenneth R.
Stearns, John W.
Stein, Talbert S.
Stelson, Paul H.
Stockbauer, Roger
Stockli, Martin P.
Stone, Edward J.
Stoner Jr., John O.
Stroke, H. Henry
Styris, D. L.
Swann, Charles P.
Swanson, Nils
Tanis, John A.
Teague, Michael R.
Temmer, Georges M.
Thoe, Robert S.
Thomas, Dan M.
Thomas, H. Ronald
Thomson, George M.
Tisone, Gary C.
Toburen, Larry H.
Tomkins, Frank S.
Trabert, Elmar C.
Tsong, Ignatius S.
Utterback, Nyle G.
Van Lint, Victor
Vane, Charles R.
Varghese, S. L.
Venkatesan, T.
Wang, Harry T.
Watson, R. L.
Wei, P. S.
West, Martin L.
Whittier, James S.
Wiff, Donald R.
Wildman, David
Williams, R. Stanley
Williams, Graheme J.
Williams, John R.
Williams, James F.

Wilson, Walter E.
Albrecht, Georg F.
Wilson Jr., William L.
Wilson, Syd R.
Wolicki, Eligius A.
Wolfe, Gordon W.
Wood, Robert M.
Woodworth, Joseph R.
Wu, C. Y. R.
Yenen, Orhan
Yu, Simon
Zander, Arlen R.
Ziegler, James F.

4.3 Beam Technology

Alcaraz, Ernest C.
Allison, Paul
Altgilbers, Larry L.
Amme, Robert C.
Anderson, James B.
Anderson, Roger W.
Asmus, John F.
Auerbach, Daniel J.
Bacon, Frank M.
Battleson, Kirk W.
Bearman, Gregory
Bechis, Kenneth P.
Bederson, Benjamin
Bernius, Mark T.
Bernstein, Richard B.
Berry, H. G.
Beyerinck, Herman C.
Blewett, John P.
Bliven, Steven M
Brodie, Ivor
Brooks, Philip R.
Cardino, Mark J.
Cecil, Joseph N.
Chamberlin, Edwin P.
Chander, Jagdish
Chen, C. L.
Choyke, Wolfgang J.
Chu, Wei-Kan
Clark Jr., William M.
Clendenin, James E.
Cobble, James A.
Cook, Donald L.
Coope, Dan
Cooper, William S.
Crampton, Stuart B.
Cross, Jon B.
Cue, Nelson
Daniele, Joseph J.
Dely, Alex
Dendramis, Achille L.
Diana, Leonard M.
Edge, Ronald D.
Edighoffer, John A.
Ehlers, Kenneth W.
Emmons, Donald A.
English, Thomas C.
Farrar, James M.
Feinberg, Benedict
Fink, Richard D.
Fischbeck, H. J.
Fornari, Luigi S.
Forrester, A. T.
Forsley, Lawrence P.
Galantowicz, Thomas A.
Gavin, Basil
Gilbody, Henry B.
Gillispie, Gregory D.
Gilmore, John
Gray, Tom J.
Guiragossian, Zaven G.
Hadeishi, Tetsuo
Ham, Mooyoung
Hart, Raymond K.
Harvey, Kenneth C.
Haugsjaa, Paul O.
Havener, Charles C.
Hellwig, Helmut W.
Hoerlin, Herman W.
Hudgens, Jeffrey W.
Hudson, G. M.
Hughes, Raymond H.
Hutson, Richard L.
Jacobs, Stephen F.
Jacobsen, E. H.
Jaduszliwer, Bernardo
Johnson, Norman J.
Katsanos, Anastasios A.
Kebabian, Paul
Kelly, John C.
Kenefick, Robert A.
Khadjavi, Abbas
King, John G.
Knuth, Eldon L.
Kowalski, Frank V.
Krenos, John R.
Kwan, Ching-Kwan CK
Lacey, Richard F.
Ladish, Joseph S.
Leiby Jr., Clare C.
Lowry, Jerald F.
Lundquist, Theodore R.
Magnuson, Gustav D.
Manor, Robert E.
Martin, Fred W.
Martin, L. R.
McClure, Gordon W.
Meyer, Fred W.
Miers, Richard E.
Miller, David R.
Moore, John H.
Nelson, Robert N.
Oblas, Daniel W.
Orient, Otto J.
Orloff, Jon H.
Owens, James C.
Peplinski, Daniel R.
Perel, Julius
Raith, Wilhelm
Ray, J. A.
Ries, Richard R.
Robertson, M. M.
Rohatgi, Vijay K.
Roy, Denis
Ryan, Dave
Sartwell, Bruce D.
Schwirzke, Fred R.
Scott Jr., John E.
Sharpton, Francis A.
Skofronick, James G.
Slocum, Robert E.
Smith, David A.
Stotlar, Suzanne C.
Tang, Fu-Ching
Tang, Sheng Y.
Van Zyl, Bert
Vaughan, Stephen O.
Venkatesan, T.
Voreades, Demetrios
Webster, Harold F.
Wharton, Charles B.
Wharton, Lennard
Whealton, John H.
Wieman, Carl E.
Wildman, David
Wittenberg, Albert M.
Wittkower, Andrew
Wu, Richard L.
Zitzewitz, Paul W.

Altgilbers, Larry L.
Armentrout, Peter B.
Aubrey, Bertrand B.
Babcock, Lucia M.
Baer, Tomas
Barnes, Ramon M.
Barton Jr., George W.
Becker, Karl H.
Bel Bruno, Joseph J.
Benson, Richard C.
Bernius, Mark T.
Bhasavanich, Daun
Bierbaum, Veronica M.
Blais, Normand C.
Bliven, Steven M
Bowers, Michael T.
Brodie, Ivor
Buttrill, Jr., Sidney E.
Cecil, Joseph N.
Cho, Hyuck
Chupka, William A.
Collins, George J.
Cooper, Charles B.
Coplan, Michael A.
Degenkolb, Eugene
Dely, Alex
Dixon, David A.
Donahue, D. J.
Drobny, Vladimir F.
Dunbar, Robert C.
Dunn, Gordon H.
Eisele, Fred L.
Ernie, Douglas W.
Fales, Norman J.
Farrar, James M.
Fetzer, Homer D.
Fite, Wade L.
Fleming, Ronald H.
Foner, Samuel N.
Futrell, J. H.
Gabrielse, Gerald
Giese, Clayton F.
Gillen, Keith T.
Gilles, Paul W.
Gingerich, Karl
Goldstein, Raymond
Greenlees, G. W.
Grover, James R.
Hanrahan, Robert J.
Henchman, Michael J.
Hubert, Jay M.
Hudgens, Jeffrey W.
Huhnermann, Harry
Huneke, John C.
Hunter, Scott R.
Ioup, George E.
Ioup, Juliette W.
Johnsen, Russell H.
Kaufman, Myron J.
Knuth, Eldon L.
Kramer, Steven D.
Lacey, Richard F.
Laudenslager, James B.
Leach, Sydney
Leppelmeier, Gilbert W.
Lerman, Juan-Carlos
Litvak, Herbert E.
Lundquist, Theodore R.
Madson, James M.

Meisels, Gerhard G.
Michael, Irving
Mohnen, Volker A.
Moran, Thomas F.
Murad, Edmond
Murray, Paul T.
Nier, Alfred O.
Oblas, Daniel W.
Odom, Robert W.
Ollison, William M.
Olson, John R.
Otter, Fred A.
Partlow, William D.
Perel, Julius
Radicati, Filippo
Ray, Gary W.
Rayborn, Grayson H.
Rhodin, Thor N.
Ries, Richard R.
Roman, Ward C.
Rutherford, John A.
Ryan, Dave
Samson, James A.
Sanzone, George
Saporoschenko, Mykola
Schroeer, Juergen M.
Schueler, Bruno W.
Shaw, David T.
Siegel, Melvin W.
Singh, Jag J.
Sloane, Thompson M.
Snyder, James J.
Stockbauer, Roger
Styris, D. L.
Taylor, James W.
Tiernan, Thomas O.
Tsong, Tien T.
VanBrunt, Richard J.
Vaughan, Stephen O.
Vestal, Marvin L.
Walls, Fred L.
Warner, Ray A.
Weissler, Gerhard L.
White, Henry W.
Wu, Richard L.
Zollars, Byron G.

5.1 Interaction of Particles and Radiation with Surfaces

Abi-Ghanem, Georges V.
Abraham, George
Adrian, Frank J.
Akins, Daniel L.
Amme, Robert C.
Anderson, Alfred B.
Anderson, Wayne
Andresen, Bjarne B.
Andrews, Hugh R.
Appleton, B. R.
Asmus, John F.
Auerbach, Daniel J.
Bacon, Frank M.
Bainum, David E.
Barnett, Clarence F.
Barton Jr., George W.
Bashkin, Stanley
Beauchamp, Jesse L.
Becker, Gordon E.
Bedell, Louis R.
Bengston, Roger D
Benson, Richard C.
Beri, Avinash C.
Berkner, Klaus H.
Berman, B. L.
Bermudez, Victor M.
Bernius, Mark T.
Best, Philip E.
Bickel, William S.
Blattner, Richard J.
Bloise, Anthony
Bloom, Arnold L.
Bokor, Jeffrey
Boring, John W.
Brandt, Werner
Brennan, James G.
Brice, David K.
Bristow, Thomas C.
Britt, Edward I.
Brodie, Ivor
Bromley, D. A.
Bruch, Ludwig W.
Brynjolfsson, Ari
Bucksbaum, Philip H.
Bukow, Hans
Burns, Donal J.
Burns III, Jay
Bushman, Gary
Campion, Alan
Canter, Karl
Cardino, Mark J.
Celotta, Robert J.
Charatis, George
Chen, C. L.
Cherrington, Blake E.
Chopra, Dev R.
Choyke, Wolfgang J.
Chu, Wei-Kan
Church, David A.
Cipolla, Sam J.
Coburn, John W.
Cohn, Gerald E.
Coldwell, Robert L.
Coleman, Paul G.
Collins, George J.
Comas, James
Companion, Audrey L.
Cooper, Charles B.

Cooper, Walter
Cooper, William S.
Cowgill, Donald F.
Cox Jr., Hollace L.
Crampton, Stuart B.
Crawford, Edward A.
Crawford, Oakley H.
Cross, Jon B.
Cue, Nelson
Curry, Bill P.
Dagenais, Mario
Davis, Jay C.
Degenkolb, Eugene
Delfino, Michelangelo
Demir, Oktay
Dendramis, Achille L.
DePristo, Andrew E.
Desplat, Jean-Louis
Dexter, Richard N.
Dixon, Dwight R.
Dozier, Charles M.
Dresser, Miles J.
Drobny, Vladimir F.
Duncan, M. M.
Dunning, F. Barry
Dylla, H. F.
Ebert, Paul J.
Eck, Thomas G.
Edge, Ronald D.
Ehlers, Kenneth W.
Ellis, Walton P.
Elton, Dr. Raymond C.
Engelhardt, A. G.
Fadley, Charles S.
Feldman, Leonard C.
Fenn, John B.
Ferguson, Stephen M.
Figueira, Joseph F.
Fineman, Morton A.
Fink, William H.
Fischer, C. R.
Fischer, Traugott E.
Fisher, Galen B.
Flamm, Daniel L.
Fohl, Timothy
Foner, Samuel N.
Freund, Hans J.
Fry, Edward S.
Gay, Timothy J.
Gelbart, William M.
George, Patricia M.
George, Thomas F.
Gershon, Nahum D.
Gersten, Joel I.
Gethner, Jon S.
Gidley, David W.
Gillen, Keith T.
Giordmaine, Joseph A.
Goddard III, William A.
Golden, David M.
Goldman, Leonard M.
Golovchenko, J. A.
Good Jr., Roland H.
Graham, William G.
Gray, Eoin W.
Greene, Edward F.
Groeneveld, Karl-Ontjese E.
Gruen, Dieter M.

Hagen, Gunter
Haglund Jr., Richard F.
Hagstrum, Homer D.
Hall, Richard B.
Hance, Robert L.
Hargis Jr., Philip J.
Harrison Jr., Don E.
Hatfield, Lynn L.
Haugsjaa, Paul O.
Helbig, Herbert F.
Heritage, Jonathan P.
Herman, Frank
Hernquist, Karl G.
Herschbach, Dudley R.
Herzenberg, Arvid
Hexter, Robert M.
Hinrichs, C. K.
Hoffman, Nelson M.
Holland, Orin W.
Honig, Richard E.
Houston, Paul L.
Hubert, Jay M.
Hughes, Raymond H.
Husmann, Otto K.
Jackson, William M.
Jakas, Mario M.
James, David R.
Janow, Richard H.
Jarnagin, Richard C.
Jensen, Barbara L.
Jette, A. N.
Johnson, Edward A.
Judge, Darrell L.
Kaldor, Andrew P.
Kaminsky, Manfred S.
Katsonis, Konstantinos
Kauffman, Robert L.
Kazmerski, Lawrence L.
Keeffe, William M.
Kelly, John C.
Kelly, Roger
Kikuchi, Chihiro
Kirkpatrick, Ronald C.
Kliwer, James K.
Koel, Bruce E.
Kouri, Donald J.
Kramer, Stephen L.
Krauss, Alan R.
Kromhout, Robert A.
Kunze, Hans-Joach D.
Kurtz, Richard L.
Kwok, Thomas Y.
Kwong, Victor HS
Lambropoulos, Hector D.
Latta, Bryan M.
Leavitt, John A.
Lee, Chi-Hsiang
Lee, Long C.
Lee, Sanboh
Lengel, Russell K.
Leppelmeier, Gilbert W.
Liberman, Irving
Litke, John D.
Lodhi, Sattar K.
Logothetis, Eleftherio M.
Lowry, Jerald F.
Loxton, Chris M.
Lucchese, Robert R.

Ludena, Eduardo V.
Lund, Clarence
Lundquist, Theodore R.
Lunney, James G.
Luntz, Alan C.
Lyons, Peter B.
MacPherson, Alistair K.
MacRae, A. U.
Madden, Robert P.
Madey, Theodore E.
Magnuson, Gustav D.
Mahadevan, P.
Manson, Joseph R.
Mariella Jr., Raymond P.
Martin, Fred W.
Martin, Richard L.
Martin, Richard M.
Massey, Gail A.
McCall, Samuel L.
McClelland, Gary M.
McCormick, Larry D.
McGowan, J. W.
McIntyre Jr., L. C.
McLean, Edgar A.
McRae, E. G.
McRae, Thomas
McVey, John B.
Mendelsohn, Lawrence B.
Milchberg, Howard M.
Miller, David R.
Monahan, Kevin M.
Morawitz, Hans
Mueller, George P.
Murday, James S.
Murphy, John C.
Murray, Paul T.
Murray, R. B.
Neumann, Herschel
Odom, Robert W.
Olson, Gregory L.
Orloff, Jon H.
Osgood, Richard M.
Ott, William R.
Overbury, Stephen H.
Overend, John
Pallmer, Paul
Palmer, Robert L.
Passenheim, Burr C.
Pate, Bradford B.
Patterson, Gary D.
Pau, Louis F.
Pellin, Michael J.
Peng, Jin-Sheng
Peplinski, Daniel R.
Perel, Julius
Pfeiffer, Gary V.
Philpott, Michael R.
Piebold, Gerald J.
Pierce, Daniel T.
Politzer, Peter
Post Jr., Douglass E.
Post, Richard S.
Poulsen, Peter
Powell, Cedric J.
Powell, Edward
Prentice, John K.
Ramaker, David E.
Rasor, Ned S.

Reed, David A.
Rhodin, Thor N.
Ritchie, R. H.
Rizzo, Joseph
Rohatgi, Vijay K.
Rol, Pieter K.
Rose, Timothy L.
Rothenberg, Joshua E.
Roy, Denis
Rule, Donald W.
Rustgi, Om P.
Rybka, Theodore W.
Sabin, John R.
Saltsburg, Howard M.
Sarraf, Sanwal P.
Sartwell, Bruce D.
Schectman, Richard M.
Schneider, Richard T.
Schroeer, Juergen M.
Schwirzke, Fred R.
Segredo, Eugenio
Senba, Masayoshi
Shafroth, Stephen M.
Sham, Tsun K.
Shay, Thomas M.
Shen, Y. R.
Sickafus, Ed N.
Silcox, John
Simons, Donald G.
Skibowski, Michael
Skofronick, James G.
Smither, Robert K.
Somorjai, Gabor A.
Souza, Steven P.
Spicer, William M.
Stalder, Kenneth R.
Steele, William A.
Stinespring, Charter D.
Stockbauer, Roger
Stotlar, Suzanne C.
Styris, D. L.
Sutcliffe Jr., Victor C.
Suto, Masako
Swann, Charles P.
Swanson, Nils
Synek, M.
Tang, Kenneth Y.
Taylor, Lyle H.
Taylor, James W.
Thirumalai, Devarajan V.
Thompson, Eric
Toennies, Jan Peter
Truhlar, Donald G.
Tsong, Ignatius S.
Tsong, Tien T.
Utterback, Nyle G.
Van Zyl, Bert
Venkatesan, T.
Verdeyen, Joseph T.
Verhaar, Boudewyn J.
Von Jaskowsky, Woldemar F.
Vook, Richard W.
Wagner, Albert
Walkup, Robert E.
Walters, G. K.
Wang, Chao C.
Wang, Kang-Lung
Webster, Harold F.
Weinberg, William H.
Weissler, Gerhard L.
Weitz, Eric
Wharton, Charles B.
Wharton, Lennard
Wheeler, Paul C.
White, Henry W.
Whitlock, Robert R.
Whitney, Kenneth G.
Wie, Chu R.
Williams, Peter
Williams, R. Stanley
Williamson Jr., William
Wilson, John W.
Wilson, Syd R.
Wilson, Walter E.
Wittkower, Andrew
Wolga, George J.
Wolicki, Eligius A.
Woolf, Stanley
Wynne, James J.
Yaney, Perry P.
Yenen, Orhan
Yin, Lo-I
Yip, Sidney
Younger, Stephen M.
Yun, Kwang-Sik
Zare, Richard N.
Zitzewitz, Paul W.

5.2 A&M Physics in Solids and Liquids

Abi-Ghanem, Georges V.
Ahn, Myong-Ku
Albrecht, A. C.
Andrews, Lester S.
Ashley, James C.
Avouris, Phaedon
Baglin, Frank G.
Baliga, Shankar B.
Basbas, George J.
Bates Jr., Richard D.
Bergmann, Otto
Berman, B. L.
Berney, Charles V.
Bjorklund, Gary C.
Black, Truman D.
Blades, John D.
Brennan, James G.
Brewer, Richard G.
Brice, David K.
Broadhurst, Martin G.
Brooks, Charles L.
Brynjolfsson, Ari
Burgdoerfer, Joachim E
Caves, Thomas C.
Chan, I. Y.
Chancey, Charles C.
Choi, Sang-Il
Chu, Wei-Kan
Code, R. F.
Comas, James
Compaan, Alvin
Corongiu, Giorgina
Cowgill, Donald F.
Crawford, Oakley H.
Crosswhite, Henry M.
Datz, Sheldon
De La Vega, Jose R.
Delfino, Michelangelo
Devor, Donald P.
Dexter, Richard N.
Diana, Leonard M.
Driscoll Jr., Timothy J.
Drobny, Vladimir F.
Ehler, Arthur W.
Eisenstadt, Maurice
Fadley, Charles S.
Feldman, Leonard C.
Foley, Charles K.
Freed, Jack H.
Freeman, Gordon R.
Freund, Hans J.
Froelich, David V.
Fuson, Nelson
Gelinas, Robert J.
Gerardi, Gary J.
Gethner, Jon S.
Gibbons, Patrick C.
Gillispie, Gregory D.
Gingerich, Karl
Goldberg, Lawrence S.
Groeneveld, Karl-Ontjese E.
Hanna, Stanley S.
Hanson, David M.
Harriman, John E.
Harrison Jr., Don E.
Hayden, Howard C.
Hedin, Lars T.
Henrichs, P. M.
Herbst, Jan F.

Herring, Conyers
Hodgson, Rodney T.
Holroyd, Richard
Holstein, T. D.
Hubbard, Paul S.
Hutson, Richard L.
James, David R.
Johnsen, Russell H.
Jones, M. T.
Judd, Brian R.
Kachru, Ravinder
Kaldor, Andrew P.
Kallne, Elisabeth
Kaminsky, Manfred S.
Kanter, Elliot P.
Kemple, Marvin D.
Kestner, Neil R.
Kivelson, Daniel
Knauss, Donald C.
Kobayashi, Hisao
Kramer, Peter B.
Krause, Herbert F.
Kromhout, Robert A.
Kwiram, Alvin L.
Ladanyi, Branka M.
Laskar, Amulya L.
Latta, Bryan M.
Law, H. David
Leavitt, John A.
Lee, Sanboh
Levy, Laurent P.
Linder, E. G.
Maier II, William B.
Manning, Irwin
Marchetti, Alfred P.
Maricq, M. Matti
Matthias, Eckart
Mazur, Eric
McCammon, James A.
McClure, Benjamin T.
McClure, Donald S.
McDaniel, Floyd D.
McGuirk, Michael
McNeil, Laurie E.
Miller, Phillip D.
Mires, Raymond W.
Moak, C. D.
Monkhorst, Hendrik J.
Montgomery, Donald J.
Moore Jr., Frank L.
Morawitz, Hans
Morgan, William L.
Mueller, George P.
Ni, Wei-Tou
O'Connell, Robert F.
O'Malley, Thomas F.
Otter, Fred A.
Overbury, Stephen H.
Panock, Richard
Parsons, Donald F.
Patton, Robert J.
Phillips, Donald H.
Porter, Leonard E.
Powers, E. L.
Prelas, Mark A.
Pronko, Peter P.
Ragle, John L.
Rahman, Aneesur
Rand, Stephen C.

Reeves, Richard L.
Retzloff, David G.
Rhim, Won-Kyu
Rice, Stuart A.
Richardson, James W.
Rigrod, William W.
Rimbey, Peter R.
Rinker Jr., George A.
Robinson, G. W.
Rogers, Max T.
Rosenberger, Franz E.
Schaeffer, Norman M.
Schrader, David M.
Sellin, Ivan A.
Seltzer, Stephen M.
Senba, Masayoshi
Sham, Tsun K.
Shand, Michael L.
Shirk, James S.
Shively, John E.
Silbey, Robert J.
Simpson, Charles G.
Siomos, Konstadinos
Skibowski, Michael
Smith, George W.
Sposito, Garrison
Stearns, Mary B.
Stratt, Richard M.
Surko, Clifford M.
Tape, James W.
Teegarden, Kenneth J.
Temmer, Georges M.
Thaler, William J.
Thirumalai, Devarajan V.
Thomas, Edward W.
Van Haeringen, Willem
Van Siclen, Clinton D.
Vane, Charles R.
Vashishta, Priya D.
Viswanath, A. Kasi
von Turkovich, Branimir F.
Voreades, Demetrios
Webb, Watt W.
Webeler, Ray W.
Wells, James W.
Wells, Michael B.
Weltner Jr., William
West, Martin L.
Wheeler, Paul C.
Wie, Chu R.
Wiff, Donald R.
Willard, John E.
Williams, Graheme J.
Williams, W. David
Williamson Jr., William
Wilson, Walter E.
Wong, Ngai C.
Wright, John J.
Wynne, James J.
Yaney, Perry P.
Yip, Sidney
Young, C. Gilbert
Zahniser, Mark S.
Ziegler, James F.
Zimm, Bruno H.
Zumbulyadis, Nicholas

Alexeff, Igor
Ali, Abdul W.
Allis, William P.
Alvarez Jr., Raymond A.
Anderson, John M.
Anderson, Louis W.
Babcock, Lucia M.
Band, Yehuda B.
Bardsley, J. N.
Barnes, Ramon M.
Bauman, Robert P.
Belford, R. L.
Benesch, William M.
Bengston, Roger D
Betts, Jeanette A.
Beverly III, Robert E.
Bhasavanich, Daun
Bhattacharya, Ashok K.
Bhaumik, Mani L.
Bierbaum, Veronica M.
Bigio, Irving J.
Blades, John D.
Bliven, Steven M
Bloch, Jeffrey J.
Boness, M. J.
Bricks, Bernard G.
Bridges, William B.
Burch, David S.
Burke, Philip G.
Burrow, Paul D.
Cartwright, David C.
Cecil, Joseph N.
Champagne, Louis F.
Chanin, Lorne M.
Chantry, Peter J.
Chen, Che-Jen
Cherrington, Blake E.
Chien, K. R.
Chimenti, Robert J.
Christophorou, Loucas G.
Coburn, John W.
Collins, George J.
Conway, John G.
Cooke, Chatham M.
Cooper, Gary W.
Corderman, Reed R.
Crawford, Edward A.
Cunningham, Augustine J.
Cunningham, David L.
Curry, Bill P.
Davis, H. Ted
Degenkolb, Eugene
Demir, Oktay
Denes, Louis J.
DeYoung, Russell J.
Dobbs, Gregory M.
Douglas-Hamilton, D. H.
Duncanson Jr., John A.
Eckert, Hans U.
Ehlers, Kenneth W.
Eisele, Fred L.
Eliasson, Baldur
Ellis, Harry W.
English, Thomas C.
Ennis Jr., Robert M.
Ernie, Douglas W.
Ewing, James J.
Figueira, Joseph F.
Fisher, Edward R.

Flamm, Daniel L.
Fohl, Timothy
Foreman, Larry R.
Fowler, Richard G.
Freed, Charles
Freeman, Gordon R.
Gallagher, Alan C.
Gallagher, Jean W.
Gallo Jr., Charles F.
Gatland, Ian R.
Gerardo, James B.
Goldstein, Raymond
Gray, Eoin W.
Green, Thomas A.
Greene, Arthur E.
Grosjean, Dennis F.
Guberman, Steven L.
Haglund Jr., Richard F.
Harrison Jr., Don E.
Hatfield, Lynn L.
Haugsjaa, Paul O.
Hays, Gerald N.
Heberlein, Joachim VR
Henson, Bob L.
Hernquist, Karl G.
Hirsh, Merle N.
Hirshfield, Jay L.
Hodges Jr., Dean T.
Hodges, Ronald V.
Hong, Siu-Ping
Houston, John M.
Hudson, David F.
Huerta, Manuel A.
Hunter, Scott R.
Hutchison, Sheldon B.
Ingold, John H.
Islam, Muhammad A.
James, David R.
Johnson, Peter D.
Johnston Jr., Thomas F.
Jones, Claude R.
Jones, Roger C.
Kassal, Thomas T.
Katz, Ira
Keane, Christopher J.
Keefer, Dennis R.
Keeffe, William M.
Kinsinger, Richard E.
Kleban, Peter H.
Kline, Laurence E.
Klosterman, Elliot L.
Knize, Randall J.
Krause, Lucjan
Kwok, Munson A.
Lacina, William B.
Lamont Jr., Lawrence T.
Lawler, James E.
Lee, Ja H.
Leep, David A.
Leland, Wallace T.
Leslie, Scott G.
Levatter, Jeffrey I.
Linder, E. G.
Litvak, Herbert E.
Long Jr., William H.
Lowry, Jerald F.
Madson, James M.
Magnuson, Gustav D.
Mansky II, Edmund J.

Mason, Edward A.
McArthur, David A.
McClure, Benjamin T.
McCoy, Benjamin J.
McKnight, Ronald H.
McVey, John B.
Menegozzi, Lionel H.
Merts, A. L.
Mickish, Roger A.
Milchberg, Howard M.
Milde, Helmut I.
Miles, Richard B.
Miller, Hillard C.
Misakian, Martin
Mitchell, Robert R.
Moeny, William M.
Morgan, William L.
Moskowitz, Philip E.
Murray, Frank
Newman, Leon A.
Nighan, William L.
Nygaard, Kaare J.
O'Malley, Thomas F.
Olson, Robert A.
Oskam, Hendrik J.
Ott, William R.
Otter, Fred A.
Pai, Robert Y.
Palumbo, Louis J.
Papayoanou, Aristotle
Partlow, William D.
Peek, James M.
Phelps, Arthur V.
Pitchford, Leanne C.
Pomraning, Gerald C.
Powell, Howard T.
Prelas, Mark A.
Rasor, Ned S.
Raymonda, John W.
Reichardt, John W.
Robertson, H. S.
Rockwood, Stephen D.
Rutherford, John A.
Sahni, Omesh
Sanzone, George
Saporoschenko, Mykola
Sarka, Benjamin NM
Sauers, Isidor
Schadt, Randall J.
Schearer, Laird D.
Schwirzke, Fred R.
Setser, D. W.
Sharpton, Francis A.
Slater, Richard
Smith, George W.
Sobel, Alan
Springer, Robert H.
Srivastava, B. N.
Stalder, Kenneth R.
Stauffer, Allan D.
Stone, Sam
Suhre, Dennis R.
Taylor, Raymond L.
Teets, Richard E.
Tregay, George W.
Turechek, John J.
Turner, Jr., Charles E.
Twist, J. Robert
Uglum, John R.

VanBrunt, Richard J.
Verdeyen, Joseph T.
Viehland, Larry A.
Volk, Charles H.
Von Jaskowsky, Woldemar F.
Wadt, Willard R.
Walsh, Peter J.
Walters, John P.
Wang, Chao C.
Wang, Wen-Cheng
Waymouth, John F.
Waynant, Ronald W.
Weber, Heinz P.
Wei, P. S.
Whaling, Ward
Whealton, John H.
Wheeler, Paul C.
Whitney, Kenneth G.
Whitney, Wayne T.
Whitten, Barbara L.
Whittier, James S.
Williams, Frazer
Williams, John R.
Williams, W. David
Wilson, John W.
Winkler, Peter
Wittig, Curt
Wood II, Obert R.
Woodworth, Joseph R.
Woolf, Stanley
Wu, Frank T.
Wyner, Elliot F.
Yoder, Marvel J.
York Jr., George W.
Younger, Stephen M.
Yun, Kwang-Sik
Zapata, Luis E.
Zollweg, Robert J.

5.4 A&M Physics in Plasmas

Abella, Isaac D.
Ali, Abdul W.
Ali, Mahamed A.
Alvarez Jr., Raymond A.
Anderson, John M.
Ashley, James C.
Asmus, John F.
Bacon, Frank M.
Bakshi, Pradip M.
Barnes, Ramon M.
Barnes, Ronald
Barnett, Clarence F.
Bartiromo, Rosario
Bauman, Leslie E.
Baur, James F.
Bekefi, G.
Bengston, Roger D
Berkner, Klaus H.
Beverly III, Robert E.
Bhasavanich, Daun
Bhattacharya, Ashok K.
Bitter, Manfred L.
Blades, John D.
Blaha, Milan
Blewett, John P.
Bottrell, Gerald J.
Boyer, Keith
Brink, Gilbert O.
Britt, Edward I.
Brooks, Neil H.
Brown, Charles M.
Brown Jr., Howard H.
Brown, Robert T.
Burke, Philip G.
Burkhalter, Philip G.
Cade, Paul E.
Ch'en, Shang-Yi
Champagne, Louis F.
Chanin, Lorne M.
Chen, Kuo-In
Cherrington, Blake E.
Cobble, James A.
Coensgen, F. H.
Collins, Lee A.
Condit, W.
Cooper, John
Cooper, John W.
Cooper, William S.
Crandall, David H.
Crawford, Edward A.
Cremers, C. J.
Crume Jr., E. C.
Dalgarno, Alexander
Datla, Raju U.
Davis, Jay C.
Davis, William A.
Degenkolb, Eugene
Deslattes, Richard D.
Devoto, R. Stephen
Dexter, Richard N.
Dietrich, Daniel D.
Dixon, Robert H.
Dollinger, Richard E.
Doolen, Gary D.
Doschek, George A.
Dreizler, Reiner M.
Dunn, Gordon H.
Dylla, H. F.
Ebert, Paul J.
Eckert, Hans U.

Eckstrom, Donald J.
Eddy, Thomas L.
Egan, Patrick O.
Ehler, Arthur W.
Ehlers, Kenneth W.
Ehlotzky, Fritz
Elton, Dr. Raymond C.
Ensberg, Earl S.
Eubank, Harold P.
Fader, Walter J.
Fineman, Morton A.
Fisher, Edward R.
Fisher, H. Leonard
Fleischmann, Hans H.
Fohl, Timothy
Fonck, Raymond J.
Forsyth, James M.
Fortner, Richard J.
Fowler, Richard G.
Frind, Gerhard
Fritsch, Wolfgang
Geballe, Ronald
Gilbody, Henry B.
Glickstein, Stanley S.
Goldberg, Leon P.
Goldman, Leonard M.
Goldstein, Raymond
Golightly, Donald W.
Graboske Jr., Harold C.
Graham, William G.
Gray, Eoin W.
Green, Joseph M.
Gregory, Donald C.
Griem, Hans R.
Grosjean, Dennis F.
Hahn, Yukap
Ham, Mooyoung
Hammond, Gordon L.
Hansen, Lorin K.
Hatfield, Lynn L.
Hauer, Allan A.
Hays, Gerald N.
Heberlein, Joachim VR
Hernquist, Karl G.
Hess, Roger
Hinnov, Einar
Hinrichs, C. K.
Hirsch, Robert G.
Hirsh, Merle N.
Hirshfield, Jay L.
Hoffman, Nelson M.
Hooper Jr., E. B.
Huerta, Manuel A.
Hunt, Angus L.
Ingold, John H.
Innes, Frederick R.
Isler, Ralph C.
Jahoda, Franz C.
Jeffries, Jay B.
Jennings, William C.
Johnson, Norman J.
Johnson, Peter D.
Johnson, Stephen G.
Jones, Claude R.
Jones, Larry A.
Jong, R. A.
Kallne, Elisabeth
Kammash, Terry
Katsonis, Konstantinos
Katz, Ira

Kauffman, Robert L.
Kaufman, Victor
Keane, Christopher J.
Keeffe, William M.
Keliher, Peter N.
Kelleher, Daniel E.
Khalid, Joseph M.
Klose, Jules Z.
Kolb, Alan C.
Krauss, Alan R.
Krishnan, Mahadevan
Kunze, Hans-Joach D.
Kwong, Victor HS
Ladish, Joseph S.
Lapatovich, Walter P.
Latta, Bryan M.
Lazar, Norman H.
Lee, Ja H.
Leppelmeier, Gilbert W.
Liebenberg, Donald H.
Litvak, Marvin M.
Long Jr., William H.
Lyons, Peter B.
Madson, James M.
Magee Jr., Norman H.
Mahon, Rita
Malone, Dennis P.
Mansky II, Edmund J.
Marmar, Earl S.
McCorkle, R. A.
McCullen, John D.
McGee, James F.
McGuire, Stephen C.
McLean, Edgar A.
Menne, Thomas J.
Merts, A. L.
Meyer, Fred W.
Michaels, Gordon E.
Milchberg, Howard M.
Miley, George H.
Miller, Hillard C.
Monahan, Kevin M.
Moos, H. Warren
Morgan, William L.
Moscatelli, Frank A.
Moskowitz, Philip E.
Mowat, J. R.
Murray, Frank
Nagel, David J.
Nehring, Frederick W.
Olson, Robert A.
Oskam, Hendrik J.
Ott, William R.
Oza, Dipak H.
Parsons, Michael L.
Partlow, William D.
Perez III, Joseph D.
Petrasso, Richard
Peyton, Bernard J.
Pfender, Emil
Phadke, Laxman G.
Pittman, Timothy L.
Pomraning, Gerald C.
Post Jr., Douglass E.
Poulsen, Peter
Powell, Edward
Powell, Howard T.
Pratt, Richard H.
Prelas, Mark A.
Price, Jack L.

Pyle, Robert V.
Rajagopal, A. K.
Rau, A. R.
Ray, J. A.
Reed, Kennedy
Reiner, Robert H.
Reinovsky, Robert E.
Rinker Jr., George A.
Roberts, James R.
Roberts, Thomas G.
Robertson, H. S.
Rohatgi, Vijay K.
Roman, Ward C.
Romesser, Thomas
Roszman, Larry
Rowan, William L.
Rozsnyai, Balazs F.
Rule, Donald W.
Sandlin, Glenn D.
Sarraf, Sanwal P.
Schima, Francis J.
Schreiber, Paul W.
Schwirzke, Fred R.
Scofield, James H.
Seely, John F.
Seka, Wolf
Shorer, Philip
Siegel, Melvin W.
Skinner, Gordon B.
Skupsky, Stanley
Smith, Earl W.
Smither, Robert K.
Snavely, Benjamin B.
Sorem, Michael S.
Springer, Robert H.
Stauffer, Allan D.
Stone, Philip M.
Stone, Sam
Strand, Oliver T.
Stroke, H. Henry
Suchannek, Rudolf G.
Suckewer, S.
Surko, Clifford M.
Takeo, Makoto
Tarter, C. Bruce
Taylor, Ronald D.
Teague, Michael R.
Thomas, Dan M.
Thomson, David B.
Tseng, Hsiang-Kuang
Uglum, John R.
Utterback, Nyle G.
Van Siclen, Clinton D.
Vander Sluis, Kenneth L.
Von Jaskowsky, Woldemar F.
Wang, Chao C.
Wei, P. S.
West, William P.
Wharton, Charles B.
Whealton, John H.
Whipple Jr., Elden C.
Whitlock, Robert R.
Whitney, Kenneth G.
Wiese, Wolfgang L.
Wilson, Jack
Wyner, Elliot F.
Yablonovitch, Eli
Yaney, Perry P.
Younger, Stephen M.
Zapata, Luis E.

Anderson, William R.
Antcliff, Richard R.
Barker, John R.
Barnett, Clarence F.
Barreto, Ernesto
Bauer, Simon H.
Bauman, Leslie E.
Bayes, Kyle D.
Bechtel, James H.
Beiting III, Edward J.
Bhasavanich, Daun
Birely, John H.
Blais, Normand C.
Blint, Richard J.
Bonczyk, Paul A.
Britt, Edward I.
Brodie, Ivor
Cattolica, Robert J.
Chou, Mau-Song
Cott, Donald W.
Creighton, John R.
Crosley, David R.
Curtis, Earl C.
Curtis, Paul M.
Dasch, Cameron J.
deZafra, Robert L.
Duncanson Jr., John A.
Eckbreth, Alan C.
Farrar, James M.
Frankel Jr., Donald S.
Freedman, Andrew
Gilmore, Forrest R.
Goldsmith, John EM
Goss, Larry P.
Grant, Edward R.
Gusinow, Michael A.
Hall, Robert J.
Hampson Jr., Robert F.
Harris, Stephen J.
Hill, John C.
Howard, Carleton J.
Jacobs, Stephen F.
Jaffe, Richard L.
Jeffries, Jay B.
Jones, Walter W.
Kaufman, Myron J.
Keefer, Dennis R.
Klemm, R. Bruce
Knipe, Richard H.
Knuth, Eldon L.
Kolb Jr., Charles E.
Kurylo III, Michael J.
Lapp, Marshall
Lawton, Stan
Layne, Clyde B.
Lee, Edward K.
Lightman, Allan J.
Madson, James M.
Mallard, W. Gary
Massey, Gail A.
Mathews, C. Weldon
McGregor Jr., Wheeler K.
McRae, Thomas
Melton, Lynn A.
Mickish, Roger A.
Neumann, David B.
Okabe, Hideo
Oldenborg, Richard C.
Ollison, William M.

Palmer, Howard B.
Parsons, Michael L.
Perry, Robert A.
Phadke, Laxman G.
Pomraning, Gerald C.
Poultney, Sherman K.
Pruett, James G.
Ravishankara, Akkihebbal R.
Reiner, Robert H.
Reuther, James J.
Rohatgi, Vijay K.
Ronn, Avigdor M.
Rosen, Gerald
Ryan, Frederick M.
Sarka, Benjamin NM
Schenck, Peter
Schofield, Keith
Schreiber, Paul W.
Sell, Jeffrey A.
Shimada, Katsunori
Silver, Joel A.
Sloane, Thompson M.
Smith, Arthur A.
Smith, George W.
Smith, Gregory P.
Smither, Robert K.
Smyth, Kermit C.
Steinberg, Martin
Taylor, David J.
Teets, Richard E.
Trump, Darryl D.
Verdieck, James F.
Wolga, George J.
Wu, Richard L.
Wu, Susan Y.
Yaney, Perry P.
Young, C. Gilbert

5.6 Atmospheric and Environmental Applications of A&M Physics

Abi-Ghanem, Georges V.
Albritton, Daniel L.
Ali, Abdul W.
Amme, Robert C.
Anderson, Larry G.
Apt III, Jerome
Barker, John R.
Barreto, Ernesto
Bass, Arnold M.
Bayes, Kyle D.
Becker, Karl H.
Benesch, William M.
Bicknell, William E.
Bien, Fritz
Bierbaum, Veronica M.
Biondi, Manfred A.
Birely, John H.
Blass, William E.
Breig, Edward L.
Britt, Edward I.
Brown, Lorin W.
Burris Jr., John
Cardon, Bartley L.
Carleton, Nathaniel P.
Cartwright, David C.
Chackerian Jr., Charles
Champion, Kenneth S. W.
Chang, Tai Y.
Chung, Sunggi
Clark, Kenneth C.
Cody, Regina J.
Cook III, Thomas B.
Dalgarno, Alexander
Davenport, John E.
Davidovits, Paul
Davis, David S.
Derr, Vernon E.
deZafra, Robert L.
Dowe Jr., R. Michael
Dowling, Jerome M.
Drullinger, Robert E.
Eisele, Fred L.
Eliasson, Baldur
Fahey, David W.
Fang, Ta-Ming
Feldman, Paul D.
Fite, Wade L.
Foltz, Nevin D.
Frankel Jr., Donald S.
Freedman, Andrew
Fukui, Katsura
Garing, John S.
Gelinas, Robert J.
Geller, M.
Gerjuoy, Edward
Gilmore, Forrest R.
Goodman, Leonard S.
Goorvitch, David
Green, Alex E.
Gunton, Robert C.
Gutman, William M.
Hampson Jr., Robert F.
Heaps, William S.
Henchman, Michael J.
Herm, Ronald R.
Herman, Roger M.
Heroux, Leon J.
Hill, John C.
Hoerlin, Herman W.
Howard, Carleton J.
Howard, Robert E.
Howard (Retired), John N.
Hudson, G. M.
Hudson, Robert D.
Huebner, Walter F.
Huebner, Russell H.
Huffman, Robert E.
Hurst, George S.
Inokuti, Mitio
Johnson, Norman J.
Kadlecek, John A.
Kaufman, Myron J.
Kebabian, Paul
Keliher, Peter N.
Khadjavi, Abbas
Kolb Jr., Charles E.
Komornicki, Andrew
Kostiuk, Theodor
Ku, Peh Sun
Kurylo III, Michael J.
Kusko, Bruce H.
Kwok, Munson A.
Langhoff, Peter W.
Lebow, Paul
Lee, Edward T.
Lee, Yuan T.
Lerman, Juan-Carlos
Litvak, Marvin M.
Luckey, George W.
Maeda, Kaichi
Marquet, Louis C.
Martin, L. R.
Mason, Arthur A.
McCullough Jr., E. A.
McIntyre, Adelbert
Menne, Thomas J.
Menyuk, Norman
Menzies, Robert T.
Meyer, Carl B.
Miller, Steven M.
Misakian, Martin
Mohnen, Volker A.
Molina, Mario J.
Monce, Michael N.
Moody, Elizabeth A.
Mumma, Michael J.
Murad, Edmond
Nelson, Albert C.
Newman, James H.
Nicholls, Ralph W.
Nicovich, John M.
Nier, Alfred O.
O'Brien, Thomas J.
Okabe, Hideo
Ollison, William M.
Olson, John R.
Omidvar, Kazem
Patel, C. Kumar K.
Patty, Richard R.
Paulson, John F.
Pearl, John C.
Pearson, Edwin F.
Phillips, Donald H.
Picard, Richard H.
Pitts Jr., James N.
Plummer, Patricia LM
Pomraning, Gerald C.
Poultney, Sherman K.
Price, Stephan D.
Raith, Burkhard
Ravishankara, Akkihebbal R.
Ray, Gary W.
Rees, M. H.
Rogers, James W.
Samson, James A.
Schadt, Randall J.
Schaeffer, Norman M.
Silver, David M.
Singh, Jag J.
Slater, Richard
Sloane, Christine S.
Smith, David C.
Smith, George W.
Sposito, Garrison
Starr, Walter L.
Stebbings, Ronald F.
Stettler, John D.
Suto, Masako
Swider, William
Tang, Fu-Ching
Tarter, C. Bruce
Taylor, Raymond L.
Temkin, Aaron
Thomson, George M.
Torr, Douglas G.
Van Zyl, Bert
Vaughan, Stephen O.
Victor, George A.
Viggiano, Albert A.
Wadzinski, Henry T.
Wells, Michael B.
West, John B.
Whipple Jr., Elden C.
Whitbeck, Michael R.
Wiesenfeld, John
Williams, James F.
Willis, Paul A.
Wolfe, Gordon W.
Wu, Richard L.
Yu, Simon
Zak, Bernard D.
Zipf, Edward

Adelman, Saul J.
Aller, Lawrence H.
Amano, Takayoshi
Avrett, Eugene H.
Bay, Zoltan L.
Behring, William E.
Bender, Peter L.
Berman, B. L.
Betz, Albert L.
Bhatia, Anand K.
Bhattacharya, Samir K.
Bidelman, William P.
Bieniek, Ronald J.
Biondi, Manfred A.
Bloch, Jeffrey J.
Bottrell, Gerald J.
Brault, James W.
Bukow, Hans
Butler, Scott E.
Carleton, Nathaniel P.
Cartwright, David C.
Chamberlain, Joseph W.
Chu, Shih-I
Cobb, Donald D.
Collins, Lee A.
Coplan, Michael A.
Cowley, Charles R.
Cox, Donald P.
Cummins, Sally E.
Dalgarno, Alexander
Davis, David S.
Davis, Sumner P.
Doschek, George A.
Drachman, Richard J.
Erber, Thomas
Fairchild, Clifford E.
Farley, John W.
Feldman, Paul D.
Fetzer, Homer D.
Friedman, Herbert
Fuhr, Jeffrey R.
Fukui, Katsura
Garstang, Roy H.
Gianturco, Franco A.
Glassgold, A. E.
Goldberg, Leo
Goldberg, Leon P.
Goldstein, Raymond
Goldwire Jr., Henry C.
Goorvitch, David
Graboske Jr., Harold C.
Graff, Margaret M.
Green, Sheldon
Greve, Peter
Hahn, Yukap
Hammond, Gordon L.
Heil, Timothy G.
Herbst, Eric
Herman, Irving P.
Hessel, Merrill M.
Huebner, Walter F.
Hummer, David G.
Jackson, William M.
Johnson, Carol
Kastner, Sidney O.
Kirkpatrick, Ronald C.
Klein, Lewis
Knight, Randall D.
Kostiuk, Theodor

Krotkov, Robert V.
Kurucz, Robert L.
Kwong, Victor HS
Lamberg, D. L.
Lambert, David L.
Lin, John
Mohler, Orren C.
Monce, Michael N.
Moody, Elizabeth A.
Moore, Edwin N.
Moos, H. Warren
Mumma, Michael J.
Oppenheimer, Michael
Patel, C. Kumar K.
Paysen, Robert A.
Pepper, George H.
Peterson, Deane M.
Peterson, William K.
Petuchowski, Samuel J.
Pomraning, Gerald C.
Price, Stephan D.
Robinson, Edward L.
Rozsnyai, Balazs F.
Sabin, John R.
Sandlin, Glenn D.
Shorer, Philip
Shull, Michael
Sitterly, Charlotte M.
Smith, Peter L.
Smither, Robert K.
Snyder, Lewis E.
Sollner, Gerhard
Soltis, Paul J.
Stebbings, Ronald F.
Stone, Sam
Suto, Masako
Synek, M.
Temkin, Aaron
Thaddeus, Patrick
Victor, George A.
Wallerstein, George
Watson, William D.
Whaling, Ward
Whipple Jr., Elden C.
Wood, Lowell L.
Woods, R. Claude

5.8 A&M Physics in Nuclear Physics

Abi-Ghanem, Georges V.
Andrews, Hugh R.
Bainum, David E.
Baldwin, George C.
Berman, B. L.
Bichsel, Hans
Bowman, Charles D.
Bromley, D. A.
Budick, Burton
Campbell, John L.
Childs, Wendell A.
Dick, Charles E.
Fink, Richard W.
Hamermesh, Morton
Hansteen, Johannes M.
Heckel, Blayne R.
Henchman, Michael J.
Hughes, Vernon W.
Hurst, George S.
Intemann, Robert L.
Isozumi, Yasuhito
Jones, Steven E.
Katayama, Ichiro
Lee, Jonathan K.
MacArthur, Duncan W.
Merzbacher, Eugen
Meyerhof, Walter E.
Molitoris, John D.
Monkhorst, Hendrik J.
Moore, C. Fred
Moscatelli, Frank A.
Myers, Edmund G.
O'Reilly, James M.
Ramsey, Norman F.
Reidy, James J.
Schima, Francis J.
Shimizu, Sakae
Shimkaveg, Gregory M.
Singh, Jag J.
Spooner, David W.
Szalewicz, Krzysztof
Thoe, Robert S.
Tsong, Ignatius S.
Ullman, Robert
Zanelli, Claudio I.

Tear Sheet

Please correct my entry in the Directory of AM&O Scientists to read as follows:

Name: _____

Address: _____

_____

_____

_____

Phone: _____

Experimental and/or Theoretical _____

Research Specialty: _____

_____

_____

_____

Indicate broad category research specialties in the current directory format.

_____

_____

Suggested names and addresses to be added to the Directory.

_____

_____

_____

Mail to: National Research Council
Board on Physics and Astronomy
2101 Constitution Avenue
Washington, DC 20418

ISBN 0-309-03696-8

 NATIONAL ACADEMY PRESS

The National Academy Press was created by the National Academy of Sciences to publish the reports issued by the Academy and by the National Academy of Engineering, the Institute of Medicine, and the National Research Council, all operating under the charter granted to the National Academy of Sciences by the Congress of the United States.